T0310251

NANO AND CELL MECHANICS

Microsystem and Nanotechnology Series

Series Editors:
Ron Pethig and Horacio Dante Espinosa

Nano and Cell Mechanics: Fundamentals and Frontiers,
Espinosa and Bao, January 2012

Digital Holography for MEMS and Microsystem Metrology,
Asundi, July 2011

Multiscale Analysis of Deformation and Failure of Materials,
Fan, December 2010

Fluid Properties at Nano/Meso Scale,
Dyson et al., September 2008

Introduction to Microsystem Technology,
Gerlach, March 2008

AC Electrokinetics: Colloids and Nanoparticles,
Morgan and Green, January 2003

Microfluidic Technology and Applications,
Koch et al., November 2000

NANO AND CELL MECHANICS
FUNDAMENTALS AND FRONTIERS

Edited by

Horacio D. Espinosa
Northwestern University, USA

Gang Bao
Georgia Institute of Technology, USA

A John Wiley & Sons, Ltd., Publication

Library of Congress Cataloging-in-Publication Data

Nano and cell mechanics : fundamentals and frontiers / edited by Horacio D. Espinosa, Gang Bao.
    p. cm. – Microsystem and nanotechnology series
  Includes bibliographical references and index.
  ISBN 978-1-118-46039-9 (hardback : alk. paper)
  1. Biotechnology. 2. Mechanics. 3. Nanotechnology. I. Espinosa, H. D. (Horacio D.), 1957-
II. Bao, Gang, 1952-
  TP248.2.N353 2013
  660.6–dc23

                           2012027597

A catalogue record for this book is available from the British Library.

ISBN: 9781118460399

Typeset in 10/12pt Times by Laserwords Private Limited, Chennai, India
Printed and bound in Singapore by Markono Print Media Pte Ltd

Contents

About the Editors   xiii

List of Contributors   xv

Foreword   xix

Series Preface   xxi

Preface   xxiii

Part One   BIOLOGICAL PHENOMENA

1   **Cell–Receptor Interactions**   3
     David Lepzelter and Muhammad Zaman
     1.1   Introduction   3
     1.2   Mechanics of Integrins   4
     1.3   Two-Dimensional Adhesion   7
     1.4   Two-Dimensional Motility   9
     1.5   Three-Dimensional Adhesion   11
     1.6   Three-Dimensional Motility   12
     1.7   Apoptosis and Survival Signaling   13
     1.8   Cell Differentiation Signaling   13
     1.9   Conclusions   14
            References   15

2   **Regulatory Mechanisms of Kinesin and Myosin Motor Proteins:**
     Inspiration for Improved Control of Nanomachines   19
     Sarah Rice
     2.1   Introduction   19
     2.2   Generalized Mechanism of Cytoskeletal Motors   19
     2.3   Switch I: A Controller of Motor Protein and G Protein Activation   21
     2.4   Calcium-Binding Regulators of Myosins and Kinesins   23
     2.5   Phospho-Regulation of Kinesin and Myosin Motors   26

2.6   Cooperative Action of Kinesin and Myosin Motors as a "Regulator"   28
2.7   Conclusion   29
        References   30

3   Neuromechanics: The Role of Tension in Neuronal Growth and Memory   35
*Wylie W. Ahmed, Jagannathan Rajagopalan, Alireza Tofangchi,
and Taher A. Saif*
3.1   Introduction   35
        3.1.1   What is a Neuron?   36
        3.1.2   How Does a Neuron Function?   38
        3.1.3   How Does a Neuron Grow?   40
3.2   Tension in Neuronal Growth   41
        3.2.1   In Vitro Measurements of Tension in Neurons   41
        3.2.2   In Vivo Measurements of Tension in Neurons   43
        3.2.3   Role of Tension in Structural Development   45
3.3   Tension in Neuron Function   48
        3.3.1   Tension Increases Neurotransmission   48
        3.3.2   Tension Affects Vesicle Dynamics   48
3.4   Modeling the Mechanical Behavior of Axons   52
3.5   Outlook   58
        References   58

Part Two   NANOSCALE PHENOMENA

4   Fundamentals of Roughness-Induced Superhydrophobicity   65
Neelesh A. Patankar
4.1   Background and Motivation   65
4.2   Thermodynamic Analysis: Classical Problem (Hydrophobic to
        Superhydrophobic)   67
        4.2.1   Problem Formulation   68
        4.2.2   The Cassie–Baxter State   71
        4.2.3   Predicting Transition from Cassie–Baxter to Wenzel State 73
        4.2.4   The Apparent Contact Angle of the Drop   77
        4.2.5   Modeling Hysteresis   79
4.3   Thermodynamic Analysis: Classical Problem (Hydrophilic to
        Superhydrophobic)   84
4.4   Thermodynamic Analysis: Vapor Stabilization   86
4.5   Applications and Future Challenges   90
        Acknowledgments   91
        References   91

**5   Multiscale Experimental Mechanics of Hierarchical Carbon-Based
        Materials   95**
Horacio D. Espinosa, Tobin Filleter, and Mohammad Naraghi
5.1   Introduction   95

	5.2	Multiscale Experimental Tools	97
	5.2.1	*Revealing Atomic-Level Mechanics: In-Situ TEM Methods*	98
	5.2.2	*Measuring Ultralow Forces: AFM Methods*	101
	5.2.3	*Investigating Shear Interactions: In-Situ SEM/AFM Methods*	102
	5.2.4	*Collective and Local Behavior: Micromechanical Testing Methods*	103
	5.3	Hierarchical Carbon-Based Materials	106
	5.3.1	*Weak Shear Interactions between Adjacent Graphitic Layers*	106
	5.3.2	*Cross-linking Adjacent Graphitic Layers*	110
	5.3.3	*Local Mechanical Properties of CNT/Graphene Composites*	113
	5.3.4	*High Volume Fraction CNT Fibers and Composites*	115
	5.4	Concluding Remarks	120
		References	123

6	**Mechanics of Nanotwinned Hierarchical Metals**		**129**
	Xiaoyan Li and Huajian Gao		
	6.1	Introduction and Overview	129
	6.1.1	*Nanotwinned Materials*	130
	6.1.2	*Numerical Modeling of Nanotwinned Metals*	132
	6.2	Microstructural Characterization and Mechanical Properties of Nanotwinned Materials	134
	6.2.1	*Structure of Coherent Twin Boundary*	134
	6.2.2	*Microstructures of Nanotwinned Materials*	135
	6.2.3	*Mechanical and Physical Properties of Nanotwinned Metals*	137
	6.3	Deformation Mechanisms in Nanotwinned Metals	145
	6.3.1	*Interaction between Dislocations and Twin Boundaries*	146
	6.3.2	*Strengthening and Softening Mechanisms in Nanotwinned Metals*	147
	6.3.3	*Fracture of Nanotwinned Copper*	155
	6.4	Concluding Remarks	156
		References	157

7	**Size-Dependent Strength in Single-Crystalline Metallic Nanostructures**		**163**
	Julia R. Greer		
	7.1	Introduction	163
	7.2	Background	164
	7.2.1	*Experimental Foundation*	164
	7.2.2	*Models*	167
	7.3	Sample Fabrication	170
	7.3.1	*FIB Approach*	170
	7.3.2	*Directional Solidification and Etching*	172

	7.3.3	Templated Electroplating	173
	7.3.4	Nanoimprinting	173
	7.3.5	Vapor–Liquid–Solid Growth	174
	7.3.6	Nanowire Growth	175
7.4	Uniaxial Deformation Experiments		175
	7.4.1	Nanoindenter-Based Systems (Ex Situ)	176
	7.4.2	In-Situ Systems	176
7.5	Discussion and Outlook on Size-Dependent Strength in Single-Crystalline Metals		178
	7.5.1	Cubic Crystals	178
	7.5.2	Non-Cubic Single Crystals	183
7.6	Conclusions and Outlook		184
	References		185

Part Three   EXPERIMENTATION

8   *In-Situ* TEM Electromechanical Testing of Nanowires and Nanotubes   193
Horacio D. Espinosa, Rodrigo A. Bernal, and Tobin Filleter

8.1	Introduction		193
	8.1.1	Relevance of Mechanical and Electromechanical Testing for One-Dimensional Nanostructures	194
	8.1.2	Mechanical and Electromechanical Characterization of Nanostructures: The Need for In-Situ TEM	196
8.2	In-Situ TEM Experimental Methods		197
	8.2.1	Overview of TEM Specimen Holders	199
	8.2.2	Methods for Mechanical and Electromechanical Testing of Nanowires and Nanotubes	200
	8.2.3	Sample Preparation for TEM of One-Dimensional Nanostructures	208
8.3	Capabilities of In-Situ TEM Applied to One-Dimensional Nanostructures		212
	8.3.1	HRTEM	212
	8.3.2	Diffraction	216
	8.3.3	Analytical Techniques	217
	8.3.4	In-Situ Specimen Modification	218
8.4	Summary and Outlook		220
	Acknowledgments		221
	References		221

9   Engineering Nano-Probes for Live-Cell Imaging of Gene Expression   227
Gang Bao, Brian Wile, and Andrew Tsourkas

9.1	Introduction		227
9.2	Molecular Probes for RNA Detection		229
	9.2.1	Fluorescent Linear Probes	229
	9.2.2	Linear FRET Probes	232

	9.2.3	Quenched Auto-ligation Probes	233
	9.2.4	Molecular Beacons	234
	9.2.5	Dual-FRET Molecular Beacons	236
	9.2.6	Fluorescent Protein-Based Probes	237
9.3	Probe Design, Imaging, and Biological Issues		239
	9.3.1	Specificity of Molecular Beacons	239
	9.3.2	Fluorophores, Quenchers, and Signal-to-Background	241
	9.3.3	Target Accessibility	242
9.4	Delivery of Molecular Beacons		244
	9.4.1	Microinjection	245
	9.4.2	Cationic Transfection Agents	245
	9.4.3	Electroporation	245
	9.4.4	Chemical Permeabilization	246
	9.4.5	Cell-Penetrating Peptide	246
9.5	Engineering Challenges and Future Directions		248
	Acknowledgments		249
	References		249

10 Towards High-Throughput Cell Mechanics Assays for Research and Clinical Applications — **255**

David R. Myers, Daniel A. Fletcher, and Wilbur A. Lam

10.1	Cell Mechanics Overview		255
	10.1.1	Cell Cytoskeleton and Cell-Sensing Overview	256
	10.1.2	Forces Applied by Cells	259
	10.1.3	Cell Responses to Force and Environment	260
	10.1.4	General Principles of Combined Mechanical and Biological Measurements	261
10.2	Bulk Assays		262
	10.2.1	Microfiltration	262
	10.2.2	Rheometry	264
	10.2.3	Ektacytometry	266
	10.2.4	Parallel-Plate Flow Chambers	267
10.3	Single-Cell Techniques		268
	10.3.1	Micropipette Aspiration	268
	10.3.2	Atomic Force Microscopy	270
	10.3.3	Microplate Stretcher	272
	10.3.4	Optical Tweezers	273
10.4	Existing High-Throughput Cell Mechanical-Based Assays		274
	10.4.1	Optical Stretchers	274
	10.4.2	Traction Force Microscopy via Bead-Embedded Gels	275
	10.4.3	Traction Force Microscopy via Micropost Arrays	275
	10.4.4	Substrate Stretching Assays	277
	10.4.5	Magnetic Twisting Cytometry	277
	10.4.6	Microfluidic Pore and Deformation Assays	278
10.5	Cell Mechanical Properties and Diseases		280
	References		284

11   Microfabricated Technologies for Cell Mechanics Studies   293
        Sri Ram K. Vedula, Man C. Leong, and Chwee T. Lim

        11.1   Introduction   293
        11.2   Microfabrication Techniques   294
                11.2.1   Photolithography and Soft Lithography   294
                11.2.2   Microphotopatterning (μPP)   297
        11.3   Applications to Cell Mechanics   298
                11.3.1   Micropatterned Substrates   298
                11.3.2   Micropillared Substrates   301
                11.3.3   Microfluidic Devices   304
        11.4   Conclusions   307
                References   307

Part Four   MODELING

**12   Atomistic Reaction Pathway Sampling: The Nudged Elastic Band
        Method and Nanomechanics Applications   313**
        Ting Zhu, Ju Li, and Sidney Yip

        12.1   Introduction   313
                12.1.1   Reaction Pathway Sampling in Nanomechanics   314
                12.1.2   Extending the Time Scale in Atomistic Simulation   314
                12.1.3   Transition-State Theory   315
        12.2   The NEB Method for Stress-Driven Problems   315
                12.2.1   The NEB method   315
                12.2.2   The Free-End NEB Method   317
                *12.2.3   Stress-Dependent Activation Energy and
                           Activation Volume*   320
                *12.2.4   Activation Entropy and Meyer–Neldel
                           Compensation Rule*   322
        12.3   Nanomechanics Case Studies   324
                12.3.1   Crack Tip Dislocation Emission   324
                12.3.2   Stress-Mediated Chemical Reactions   326
                12.3.3   Bridging Modeling with Experiment   327
                *12.3.4   Temperature and Strain-Rate Dependence
                           of Dislocation Nucleation*   329
                12.3.5   Size and Loading Effects on Fracture   330
        12.4   A Perspective on Microstructure Evolution at Long Times   332
                12.4.1   Sampling TSP Trajectories   333
                12.4.2   Nanomechanics in Problems of Materials Ageing   334
                References   336

13   Mechanics of Curvilinear Electronics   339
        *Shuodao Wang, Jianliang Xiao, Jizhou Song, Yonggang Huang, and John A.
        Rogers*

        13.1   Introduction   339

13.2   Deformation of Elastomeric Transfer Elements during Wrapping
        Processes   342
        13.2.1   Strain Distribution in Stretched Elastomeric Transfer
                  Elements   342
        13.2.2   Deformed Shape of Elastomeric Transfer Elements   344
13.3   Buckling of Interconnect Bridges   347
13.4   Maximum Strain in the Circuit Mesh   351
13.5   Concluding Remarks   355
        Acknowledgments   355
        References   355

14   Single-Molecule Pulling: Phenomenology and Interpretation   359
        Ignacio Franco, Mark A. Ratner, and George C. Schatz

14.1   Introduction   359
14.2   Force–Extension Behavior of Single Molecules   360
14.3   Single-Molecule Thermodynamics   364
        14.3.1   Free Energy Profile of the Molecule Plus Cantilever   365
        14.3.2   Extracting the Molecular Potential of Mean Force $\phi(\xi)$ 366
        14.3.3   Estimating Force–Extension Behavior from $\phi(\xi)$ 369
14.4   Modeling Single-Molecule Pulling Using Molecular Dynamics   370
        14.4.1   Basic Computational Setup   370
        14.4.2   Modeling Strategies   371
        14.4.3   Examples   373
14.5   Interpretation of Pulling Phenomenology   376
        14.5.1   Basic Structure of the Molecular Potential
                  of Mean Force   377
        14.5.2   Mechanical Instability   378
        14.5.3   Dynamical Bistability   381
14.6   Summary   384
        Acknowledgments   385
        References   385

15   Modeling and Simulation of Hierarchical Protein Materials   389
        Tristan Giesa, Graham Bratzel, and Markus J. Buehler

15.1   Introduction   389
15.2   Computational and Theoretical Tools   391
        15.2.1   Molecular Simulation from Chemistry Upwards   391
        15.2.2   Mesoscale Methods for Modeling Larger Length
                  and Time Scales   392
        15.2.3   Mathematical Approaches to Biomateriomics   394
15.3   Case Studies   400
        15.3.1   Atomistic and Mesoscale Protein Folding
                  and Deformation in Spider Silk   400
        15.3.2   Coarse-Grained Modeling of Actin Filaments   402
        15.3.3   Category Theoretical Abstraction of a Protein
                  Material and Analogy to an Office Network   403

15.4   Discussion and Conclusion   406
        Acknowledgments   406
        References   406

16   Geometric Models of Protein Secondary-Structure Formation   411
        Hendrik Hansen-Goos and Seth Lichter

16.1   Introduction   411
16.2   Hydrophobic Effect   412
        16.2.1   Variable Hydrogen-Bond Strength   415
16.3   Prior Numerical and Coarse-Grained Models   415
16.4   Geometry-Based Modeling: The Tube Model   416
        16.4.1   Motivation   416
        16.4.2   Impenetrable Tube Models   417
        16.4.3   Including Finite-Sized Particles Surrounding the Protein   419
        16.4.4   Models Using Real Protein Structure   421
16.5   Morphometric Approach to Solvation Effects   422
        16.5.1   Hadwiger's Theorem   422
        16.5.2   Applications   424
16.6   Discussion, Conclusions, Future Work   429
        16.6.1   Results   429
        16.6.2   Discussion and Speculations   430
        Acknowledgments   433
        References   433

17   Multiscale Modeling for the Vascular Transport of Nanoparticles   437
        Shaolie S. Hossain, Adrian M. Kopacz, Yongjie Zhang, Sei-Young Lee,
        Tae-Rin Lee, Mauro Ferrari, Thomas J.R. Hughes, Wing Kam Liu,
        and Paolo Decuzzi

17.1   Introduction   437
17.2   Modeling the Dynamics of NPs in the Macrocirculation   438
        17.2.1   The 3D Reconstruction of the Patient-Specific
                   Vasculature   439
        17.2.2   Modeling the Vascular Flow and Wall Adhesion
                   of NPs   440
        17.2.3   Modeling NP Transport across the Arterial Wall
                   and Drug Release   440
17.3   Modeling the NP Dynamics in the Microcirculation   448
        17.3.1   Semi-analytical Models for the NP Transport   449
        17.3.2   An IFEM for NP and Cell Transport   452
17.4   Conclusions   456
        Acknowledgments   456
        References   457

Index   461

About the Editors

 **Horacio Dante Espinosa** is the James and Nancy Farley Professor of Mechanical Engineering and the Director of the Theoretical and Applied Mechanics Program at the McCormick School of Engineering, Northwestern University. He received his PhD in Applied Mechanics from Brown University in 1992. He has made contributions in the areas of dynamic failure of advanced materials, computational modeling of fracture, and multiscale experiments and simulations of micro- and nanosystems. He has published over 200 technical papers in these fields. His work has received broad attention in the media, including United Press International, NSF Discoveries, Frost and Sullivan, Science Daily, EurekAlert, Small Times, PhysOrg, Nanotechwire, Bio-Medicine, NanoVIP, Nanowerk, Genetic Engineering and Biotechnology News, Medical News Today, Materials Today, Next Big Future, Beyond Breast Cancer, AZoNano, Spektrumdirekt, and MEMSNet.

Professor Espinosa has received numerous awards and honors recognizing his research and teaching. He was elected foreign member of the Russian Academy of Engineering in 2011 and of the European Academy of Sciences and Arts in 2010, Fellow of the American Academy of Mechanics in 2001, the American Society of Mechanical Engineers in 2004, and the Society for Experimental Mechanics in 2008. He was the Timoshenko Visiting Professor at Stanford University in 2011. He received two Young Investigator Awards: the NSF-Career in 1996 and the Office of Naval Research-Young Investigator Award in 1997. He also received the American Academy of Mechanics (AAM) 2002-Junior Award, the Society for Experimental Mechanics (SEM) 2005 HETENYI Award, the Society of Engineering Science (SES) 2007 Junior Medal, and the 2008 LAZAN award from the Society for Experimental Mechanics. He currently serves as Founding Principal Editor of *MRS Communications* and co-editor of the *Wiley Book Series in Microsystems and Nanotechnologies*. He also served as Editor-in-Chief of the *Journal of Experimental Mechanics* and Associate Editor of the *Journal of Applied Mechanics*. He is the 2012 President of the Society of Engineering Science and a member of the US National Committee for Theoretical and Applied Mechanics.

Professor Espinosa is also the founder and chief scientific officer of iNfinitesimal LLC, a nanotechnology company developing robust next-generation nanoscale devices, scalable nanomanufacturing tools, and microdevices for single-cell transfection and analysis.

 **Dr Gang Bao** is Robert A. Milton Chair of Biomedical Engineering and a College of Engineering Distinguished Professor in the Department of Biomedical Engineering, Georgia Institute of Technology and Emory University. He is Director of the Center for Translational Cardiovascular Nanomedicine, an NIH/NHLBI Program of Excellence in Nanotechnology (PEN) at Georgia Tech and Emory University, and Director of Nanomedicine Center for Nucleoprotein Machines, an NIH Nanomedicine Development Center (NDC) at Georgia Tech, and Director of the Center for Pediatric Nanomedicine at Children's Healthcare of Atlanta and Georgia Tech. Dr Bao received his undergraduate and Master's degrees from Shandong University in China and his PhD from Lehigh University in the USA. Dr Bao is a Fellow of the American Association of Advancement in Science (AAAS), a Fellow of the American Society of Mechanical Engineers (ASME), a Fellow of the American Physical Society (APS), and a Fellow of the American Institute for Medical and Biological Engineering (AIMBE).

Dr Bao's current research is focused on the development of nanotechnology and biomolecular engineering tools for biological and disease studies, including molecular beacons, magnetic nanoparticle probes, quantum dot bioconjugates, protein tagging/targeting methods, and engineered nucleases. These approaches have been applied to the diagnosis and treatment of cancer and cardiovascular disease, viral infection detection, and the development of gene correction approaches for treating single-gene disorders.

List of Contributors

Wylie W. Ahmed
Department of Mechanical Science & Engineering, University of Illinois, USA

Rodrigo A. Bernal
Department of Mechanical Engineering, Northwestern University, USA

Graham Bratzel
Laboratory for Atomistic and Molecular Mechanics, Department of Civil and Environmental Engineering, USA
Department of Mechanical Engineering, Massachusetts Institute of Technology, USA

Markus J. Buehler
Laboratory for Atomistic and Molecular Mechanics, Department of Civil and Environmental Engineering, Massachusetts Institute of Technology, USA

Paolo Decuzzi
Department of Translational Imaging and Department of Nanomedicine, The Methodist Hospital Research Institute, USA

Mauro Ferrari
Department of Translational Imaging and Department of Nanomedicine, The Methodist Hospital Research Institute, USA

Tobin Filleter
Department of Mechanical and Industrial Engineering, University of Toronto, Canada

Daniel A. Fletcher
Department of Bioengineering, University of California at Berkeley, USA
Graduate Group in Biophysics, University of California, Berkeley, California, USA

Ignacio Franco
Department of Chemistry, Northwestern University, Evanston, USA

Huajian Gao
School of Engineering, Brown University, USA

Tristan Giesa
Laboratory for Atomistic and Molecular Mechanics, Department of Civil and Environmental Engineering, Massachusetts Institute of Technology, USA
Department of Mechanical Engineering, RWTH Aachen University, Germany

Julia R. Greer
California Institute of Technology, USA

Hendrik Hansen-Goos
Department of Geology and Geophysics, Yale University, USA

Shaolie S. Hossain
Department of Translational Imaging and Department of Nanomedicine, The Methodist Hospital Research Institute, USA

Yonggang Huang
Department of Mechanical Engineering, USA
Department of Civil and Environmental Engineering, Northwestern University, USA

Thomas J.R. Hughes
Institute for Computational Engineering and Sciences, The University of Texas at Austin, USA

Adrian M. Kopacz
Department of Mechanical Engineering, Northwestern University, USA

Wilbur A. Lam
Department of Pediatrics, Aflac Cancer Center and Blood Disorders Service of Children's Healthcare of Atlanta and Emory University School of Medicine
Wallace H. Coulter Department of Biomedical Engineering, Georgia Institute of Technology and Emory University, USA

Sei-Young Lee
Samsung – Global Production Technology Center, Korea

Tae-Rin Lee
Department of Mechanical Engineering, Northwestern University, USA

Xiaoyan Li
School of Engineering, Brown University, USA

Man C. Leong
NUS Graduate School for Integrative Sciences and Engineering, Singapore

David Lepzelter
Department of Biomedical Engineering, Boston University, USA

Ju Li
Department of Nuclear Science and Engineering
Department of Materials Science and Engineering, Massachusetts Institute of Technology, USA

Seth Lichter
Department of Mechanical Engineering, Northwestern University, USA

Chwee T. Lim
Mechanobiology Institute, National University of Singapore, Singapore
Department of Bioengineering, Singapore

Wing Kam Liu
Department of Mechanical Engineering, Northwestern University, USA

David R. Myers
Department of Pediatrics, Aflac Cancer Center and Blood Disorders Service of Children's Healthcare of Atlanta and Emory University School of Medicine, USA
Wallace H. Coulter Department of Biomedical Engineering, Georgia Institute of Technology and Emory University, USA

Mohammad Naraghi
Texas A&M University, USA

Neelesh A. Patankar
Department of Mechanical Engineering, Northwestern University, USA

Jagannathan Rajagopalan
Arizona State University, USA

Mark A. Ratner
Department of Chemistry, Northwestern University, USA

Sarah Rice
Department of Cell and Molecular Biology, Northwestern University Medical School, USA

John A. Rogers
Department of Materials Science and Engineering, Department of Mechanical Science and Engineering, Department of Electrical and Computer Engineering, Department of Chemistry, and Frederick Seitz Materials Research Laboratory, University of Illinois, USA

Taher A. Saif
Department of Mechanical Science and Engineering, University of Illinois, USA

George C. Schatz
Department of Chemistry, Northwestern University, USA

Jizhou Song
Department of Mechanical and Aerospace Engineering, University of Miami, USA

Alireza Tofangchi
Department of Mechanical Science and Engineering, University of Illinois, USA

Andrew Tsourkas
Department of Bioengineering, University of Pennsylvania, USA

Sri Ram K. Vedula
Mechanobiology Institute, National University of Singapore, Singapore

Shuodao Wang
Department of Mechanical Engineering, Northwestern University, USA

Brian Wile
Department of Biomedical Engineering, Georgia Institute of Technology and Emory, USA

Jianliang Xiao
Department of Mechanical Engineering, University of Colorado, USA

Sidney Yip
Department of Nuclear Science and Engineering
Department of Materials Science and Engineering, Massachusetts Institute of Technology, USA

Muhammad Zaman
Department of Biomedical Engineering, Boston University, USA

Yongjie Zhang
Department of Mechanical Engineering, Carnegie Mellon University, USA

Ting Zhu
Woodruff School of Mechanical Engineering, Georgia Institute of Technology, USA

Foreword

The articles included in this volume, compiled by Dr. Gang Bao and Dr. Horacio Espinosa, can be viewed as a benchmark along the evolutionary path of what our scholarly community views as its "field of research." Consequently, the preparation of this foreword provides an opportunity to lend perspective to the book as a whole.

Among senior researchers, it is common for one to feel that their own field reached its pinnacle when they were young, but that it has been suffering a steady decline ever since. The contents of this book will surely dissuade anyone involved in research on the mechanics of hard or soft materials from harboring such a view! When I was first exposed to the field of mechanics nearly a half century ago, the focus was on the mathematical solution of boundary-value problems, either exactly or approximately, and this point of view pervaded graduate courses. There was a small but active experimental community involved in measurement of deformation with strain gages, photoelastic strategies, and, occasionally, optical interference techniques. At about this time, a few visionaries in the community demonstrated to the Department of Defense that there was enormous potential for advances in engineering and technology through more effective exploitation of materials; as a result, a number of national materials research laboratories were established at universities around the country, with mechanics taking a central role in a number of these centers.

As a result, the organizational barriers between mechanics and materials were gradually overcome, and many of us opted to align our research efforts with the new area of mechanics of materials. Computational methods joined analytical and experimental methods as core activities, and advances in both fabrication and characterization of experimental samples introduced us to the exciting and complex world of deformation of materials at small size scales. Graduate education became much more diverse with the integration of mechanics and materials science, and the introduction of computation provided a powerful means for describing the behavior of materials quantitatively and for addressing issues of competition among potential mechanisms of deformation or failure.

The important role of mechanical stress on thin film materials and other small structures, which are largely intended for nonmechanical functions, arose as a focus of research efforts. Connections between macroscopic fracture strength of materials and the details of material microstructure were established, the inuence of plastic strain localization due to material instability in deformation was quantified, and the role of dislocation formation or crystallographic twinning or other phase transformations on the nonmechanical functional characteristics of materials was established, for example. The topics became important focal issues for graduate courses in the field, and a greatly increased range and sophistication in sample preparation and diagnostic tools became available in laboratories. This was the prevailing atmosphere in the field at about the time the editors of this volume emerged as active members of the community.

The field of mechanics of materials had changed fundamentally, having formed a focus on the physical and chemical aspects of phenomena as much as the mathematical l aspects, computation became an indispensable methodology, and methods of laboratory sample fabrication, observational techniques, and data analysis created opportunities that seemed unattainable only a short time earlier. Researchers could now apply this suite of tools to examine and quantitatively understand the behavior of materials at an even smaller scale. It was soon found that the mechanisms of biological functions, which before had been identified through the long and painstaking process of assaying, could be observed directly.

As we gaze forward in time from the current frontiers of mechanics research, we see that the "landscape" to be traversed is less well defined than it had been in the past. Mechanical phenomena at the smaller size scales are often inseparable from chemical, biological, and/or quantum mechanical inuences, and the rules governing behavior are no longer certain. Consequently, instead of reliance on precise analytical and/or numerical analyses, we must examine data obtained through brief glimpses of the nano-world in the laboratory or seek guidance from the global principles of thermodynamics. Although these aspects loom as impediments to progress, the task of overcoming them lends a sense of timeliness and excitement to the mechanics research field overall.

As illustrations of the broad consequences of these developments, this volume includes a report on the inuence of mechanical tension on neuronal growth and memory in the human brain. Another article reports on direct measurement of the mechanical properties of cylindrical test samples with diameters as small as 50 nm. There is a discussion of the connections between mechanical properties of cells and human disease, as well as a quantitative description of what can be learned by separating a pair of chemically bound molecules under controlled conditions, plus many others. We can learn a great deal from study of the articles individually, of course. It is equally important to consider their collective significance as an indicator of the excitement, broad relevance, and future promise of our field.

L. B. Freund
Champaign, IL, USA

Series Preface

Books in this series are intended, through scholarly works of the highest quality, to serve researchers and scientists wishing to keep abreast of advances in the expanding field of nano- and micro-technology. These books are also intended to be a rich interdisciplinary resource for teachers and students of specialized undergraduate and postgraduate courses.

A recent example includes the university textbook *Introduction to Microsystem Technology* by Gerlach and Dötzel, covering the design, production and application of miniaturised technical systems from the viewpoint that for engineers to be able to solve problems in this field they need to have interdisciplinary knowledge over several areas as well as the capability of thinking at the system level. In their book *Fluid Properties at Nano/Meso Scale*, Dyson *et al* take us step by step through the fluidic world bridging the nanoscale, where molecular physics is required as our guide, and the microscale where macro continuum laws operate. Jinghong Fan in *Multiscale Analysis of Deformation and Failure of Materials* provides a comprehensive coverage of a wide range of multiscale modeling methods and simulations of the solid state at the atomistic/nano/submicron scales and up through those covering the micro/meso/macroscopic scale. Most recently *Digital Holography for MEMS and Microsystem Metrology*, edited by Anand Asundi, offers timely contributions from experts at the forefront of the development and applications of this important technology.

In this book Professors Espinosa and Bao have assembled, through their own inputs and those of 47 other experts of their chosen fields of endeavour, 17 timely and exciting chapters that must surely represent the most comprehensive coverage yet presented of all aspects of the mechanics of cells and biomolecules. The editors have ensured, through the careful choice of the contents and their order of presentation in four main sections, that we have a coherent presentation of the extraordinary wide range of interdisciplinary components that make up this exciting frontier in applied mechanics. Apart from their clarity of presentations, all the authors have adopted a pedagogical style of writing, making much of this book's content suitable for inclusion in undergraduate and postgraduate courses. A foreword, both historic and insightful, has also been composed by Professor L Ben Freund - whose own contributions to various aspects of the mechanics of biological materials will have influenced the thinking of many of the contributors to this excellent book.

Ronald Pethig
Professor of Bioelectronics
University of Edinburgh, UK

Preface

In the past decade, nano- and bio-technologies have received unprecedented attention from the government and private sectors as well as the general public owing to their potential in impacting our lives through fundamental discoveries, innovation, and translational research efforts. Engineers and applied scientists have played a major role in developing these technologies and made essential contributions to applying them to a wide range of industries, including manufacturing, healthcare, agriculture, energy, and defense.

Consistent with this, research in nano-mechanics and the mechanics of living cells and biomolecules has become a frontier in applied mechanics. Studies in this exciting research area are interdisciplinary in nature and draw engineers and scientists from a diverse range of fields, including nanoscale science and engineering, biology, statistical and continuum mechanics, and multiscale-multiphysics modeling and experimentation. As a result, original contributions to the development of nano- and cell-mechanics are published in a large number of specialized journals, which prompted us in editing this book. The book documents, for the first time, many recent developments in nano- and cell-mechanics and showcases emergent new research areas and techniques in engineering that are at the boundaries of mechanics, materials science, chemistry, biology, and medicine. As such, this book allows those entering the field a quick overview of experimental, analytical, and computational tools used to investigate biological and nanoscale phenomena. This book may also serve as a textbook for a graduate course in theoretical and applied mechanics, mechanical engineering, materials science, and applied physics.

The 17 chapters in this book are organized in four sections: (1) Biological phenomena, (2) Nanoscale phenomena, (3) Experimentation, and (4) Modeling. The biological phenomena section covers cell–receptors interactions, regulatory molecular motors, and the role of tension in neuronal growth and memory. The nanoscale phenomena section examines superhydrophobicity, multiscale mechanics of hierarchical carbon-based materials, mechanics of twinning in hierarchical metals, and size-dependent strength in single-crystalline metallic nanostructures. The experimentation section discusses *in-situ* electron characterization of nanomaterials, the engineering of nano-probes for live-cell imaging of gene expression, high-throughput cell mechanic assays for research and clinical applications, and microfabrication technologies for cell mechanics studies. The last section, modeling, spans a number of methods and applications: atomistic reaction pathway sampling, mechanics of curvilinear electronics, single molecular pulling, modeling of hierarchical protein materials, geometric models of protein secondary structure formation, and multiscale modeling for the vascular transport of nanoparticles.

We would like to thank Ben Freund for providing a historic perspective and an inspiring Foreword. Likewise, we are particularly thankful to all authors for providing authoritative and comprehensive reviews of recent advances in their field of expertise. We would also like to thank the Wiley staff, Anne Hunt and Tom Carter in particular, for guidance and assistance over the preparation of this book. Their experience and professionalism was essential to this project. A special thanks is also due to our assistants, Andrea DeNunzio and Amy Tang, who communicate regularly with the authors to collect all the needed materials.

Horacio D. Espinosa and Gang Bao

Part One

Biological Phenomena

1

Cell–Receptor Interactions

David Lepzelter and Muhammad Zaman
Boston University, USA

1.1 Introduction

One of the most basic functions of the cells of a multicellular organism is to stay attached to the rest of the organism. This is called cell adhesion. It is not a trivial task, especially since different kinds of cells require different levels of attachment. Further, in complex organisms some cells must be able to change their levels of attachment based on their local environments: platelets must transform from almost completely nonadhesive to adhesive when they reach the site of an injury, and white blood cells must recognize cell types and respond by attaching to (and attacking) foreign materials and infected cells. Further still, sensing the levels and types of attachment can be vital to extremely important decision-making pathways in cells, such as those related to apoptosis (cell death) and cell differentiation. The precise mechanisms by which these processes occur are varied, and while they are still under study, much about them is known.

In general, cells attach to extracellular objects or other cells using proteins that are imbedded in their membranes. Transmembrane proteins have at least one hydrophobic portion, or membrane domain, which anchors them in the membrane. They also have hydrophilic portions, which interact with the cytosol (cytoplasmic domains) or cell environment (extracellular domains). These proteins often serve as sensors, binding to molecules with their extracellular regions and passing along information about the environment to the inside of the cell via configuration changes; this is called outside-in signaling. Some reverse the process, providing an extracellular signal regarding conditions inside the cell, and some perform both functions. A few bind strongly enough to extracellular materials or other cells' extracellular protein domains that they remain attached even with significant forces pulling them away. These form the basis of cell adhesion.

Among these adhesion proteins are immunoglobulins, selectins, cadherins, and integrins. This chapter will briefly discuss all of them, but will focus on the primary agent in cell adhesion to the extracellular matrix: integrins.

Nano and Cell Mechanics: Fundamentals and Frontiers, First Edition. Edited by Horacio D. Espinosa and Gang Bao.
© 2013 John Wiley & Sons, Ltd. Published 2013 by John Wiley & Sons, Ltd.

Immunoglobulins, or antibodies, are used by the immune systems of vertebrates to identify and eliminate infection and cell malfunction. They do this through selective and customized binding; an immune-related cell will find a marker (often a protein) on a foreign object and customize an antibody in such a way that it binds preferentially to that marker based largely on its shape [1]. This will allow identification of proper targets for immune response, and begin the immune response appropriate for the object in question.

Selectins are adhesion molecules specific to leukocytes, platelets, and endothelial cells. They exhibit very fast binding, allowing them to act similar to a braking system on moving cells when they have reached a site of bleeding or infection [2].

Cadherins are direct cell–cell adhesion proteins. They generally operate by binding to each other; cadherins on one cell will bind to cadherins on another. There are a number of different varieties (e.g., E-cadherin), many of which preferentially bind to cadherins of the same type [3]. Members of the cadherin superfamily are involved in cell signaling, adhesion, recognition, communication, morphogenesis, and angiogenesis [4].

Integrins, the main focus of this chapter, are found in animals and bind primarily to various proteins which are part of a ubiquitous biological scaffolding in animals called the extracellular matrix. The form of integrins commonly found *in vivo* is actually a heterodimer: an α protein and a β protein each with an extracellular domain, a membrane domain, and a cytoplasmic domain. The two interact with each other to make the overall form of the integrin functional. There are currently 18 known varieties of the α subunit and eight known varieties of the β, making 24 distinct viable combinations [5].

This does not include alternatively spliced integrin subunits; β_{1D} integrin subunits are genetically identical to β_{1A} subunits, but, before the mRNA is translated into amino acids, different amino acids are removed from the mRNA sequence coding for the protein. The two are no longer functionally identical and associate with different α subunits [6]. However, in all cases the central mechanism of the integrin involves binding to the extracellular matrix.

The matrix to which integrins bind is composed of a number of proteins and glycosaminoglycans, with the exact composition depending on the precise function of the matrix and the animal's species. Generally, these proteins may include collagens, fibronectins, laminins, elastins, and others [7]. Each has different roles; elastins, for example, increase matrix elasticity. Collectively, they can provide a mechanical scaffolding for cell movement or anchoring, or a film separating different types of cells. Additionally, they can provide the biochemical and biomechanical signals necessary for various kinds of cell behavior. Such signals, both biochemical and biomechanical, are often transmitted to cells via integrins. The combined effects of signaling and physical scaffolding form the basis for adhesion via clusters of integrins, and consequently the various related signaling pathways inside the cell.

1.2 Mechanics of Integrins

Complete integrins, with both an α subunit and a β, have two basic configurations: activated and deactivated (see Figure 1.1). Deactivated integrins have low affinity for matrix molecules; only activated integrins have significant substrate-adhesion properties. Activated integrins are believed to have an extended extracellular domain and an open ligand-binding head, though the necessity of the extended domain is a matter of

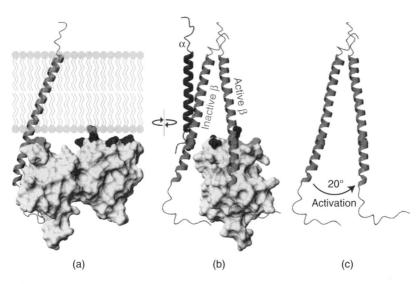

Figure 1.1 Activation of the $\alpha_5\beta_{1D}$ integrin (α in blue, β inactive conformation in purple, β active conformation in red, talin in yellow). Adapted by permission from [17], © 2009 Nature Publishing Group

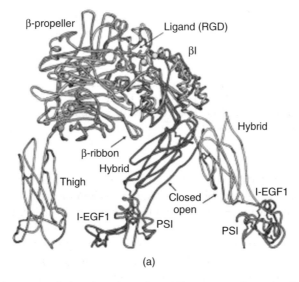

Figure 1.2 Integrin open and closed configurations. The open configuration is in yellow, and the closed in purple. Reproduced with permission from Elsevier from [23]

debate; deactivated integrins have a closed head and a bent extracellular domain [8–10] (see Figure 1.2). In addition, there is some evidence for a third, intermediate state with an extended extracellular domain and a closed head [10,11]. Activation is begun with the pairing of the α and β subunits, and often continues with talin, a cytoplasmic protein which binds to two parts of the β cytoplasmic domain [12–14]. The second binding

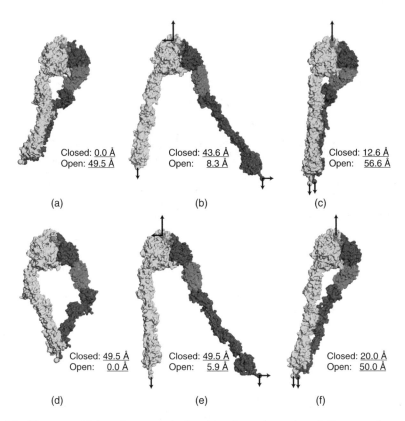

Figure 1.3   Force-sensitivity of open and closed configurations. A and D are resting open and closed configurations, respectively; B and C, and E and F, are simulation results from forces applied to the head. Numbers are total distances from open and closed configurations. Reproduced with permission from Elsevier from [23]

causes the cytoplasmic tails of the subunits to separate slightly and reorients the structure as a whole in a way that encourages an activated configuration [15–17].

Once the integrin is activated and bound to an appropriate ligand, the integrin–ligand bond may typically last up to seconds or even more if unstressed. A force of some tens of piconewtons applied to a single integrin–ligand bond typically ruptures the bond on a much shorter time scale, as modeled by the equation $t_{\mathrm{off}}(f) \approx t_{\mathrm{off}}^{*}\mathrm{e}^{-f/f_{\beta}}$, where t_{off} is the amount of time the bond takes to dissociate, f is the applied force, t_{off}^{*} is the unstressed lifetime of the bond, and f_{β} is a force scale appropriate to the specific integrin and ligand [18].

Though the mechanisms are still under study, it is certain that integrins can produce force-sensitive signals independent of dissociation [19–24] (Figure 1.3). Activation itself is force sensitive, which has been suggested to be because of the extended nature of the activated conformation [19]. Generally, this force is easiest to understand when exerted by the cytoskeleton inside the cell, though the stiffness of the substrate is also force related and is known to have significant effects on integrin activity. However, it should be noted that the simplest explanation for stiffness-related action, a stress sensor (responding directly to force), is definitely not correct; integrins, instead, seem to have a kind of strain-sensing mechanism (responding to displacement) [19–21]. It should also be noted that adhesion studies are often performed using fibroblasts, which may be more sensitive to forces and matrix mechanics than other cell types [25].

Additionally, the cytoplasmic proteins associated with integrins are believed to have mechanosensitive properties [26]. Suggested possibilities for the mechanism are component loss or particle contraction from applied force; both of these are known to occur in proteins in general. Further discussion of the various mechanosensitive parts of the complex of integrins and cytoplasmic proteins requires a description of those parts and their activity, which in turn is most easily described in terms of the historical means by which they were discovered.

1.3 Two-Dimensional Adhesion

The study of cell adhesion began with cell monolayers in petri dishes. This was, of course, the easiest and most obvious place to study adhesion in the laboratory. As we will discuss later, it was not ideal for comparison with *in vivo* conditions, but it is still useful to understand because there are strong connections between two-dimensional adhesion behavior and its three-dimensional cousin. Indeed, it has been suggested that two-dimensional adhesion is an exaggerated version of three-dimensional adhesion [27]. Further, because studies of three-dimensional behavior are generally more complex and have only been done recently, the results of two-dimensional studies are generally easier to interpret.

Two-dimensional cell adhesion experiments can be done on a thin layer of material (often collagen) which has been attached to the bottom of a petri dish or similar cell container. The precise kind of material makes a difference to the behavior, as does its physical and chemical patterning in space. As mentioned previously, the rigidity of the material is among the most important aspects of dependence on material type. In addition, materials with sparse integrin binding sites alter adhesion and related motility functions [28]. Materials physically arranged in protruding shapes, altering the basic two-dimensional assumption without truly approximating a realistic three-dimensional environment, can significantly change adhesion and motility behavior [29–31]. More recent experiments showed that cells cultured between two substrates, one above and one below, can also demonstrate altered adhesion behavior [32].

Regardless of the details of the surfaces used, though, there are some universal responses within the cell to an adhesion event. In total, over 100 cytoplasmic proteins are known to be closely involved in various ways with cell–matrix adhesion [33–35], and this number is likely to be a significant underestimate [36]. A small number of these proteins directly bind to integrins, providing some amount of inside-out or outside-in signaling (these are necessary for inside-out signaling because integrin cytoplasmic domains have no enzyme-related activity). A few are additional transmembrane proteins believed to synergize with integrins. At least 57 directly involve the actin cytoskeleton, encouraging growth or forming a physical bridge between integrin and actin. Significant numbers are enzymes and enzyme regulators, generally used in signaling pathways for cell survival, differentiation, and various other cell functions.

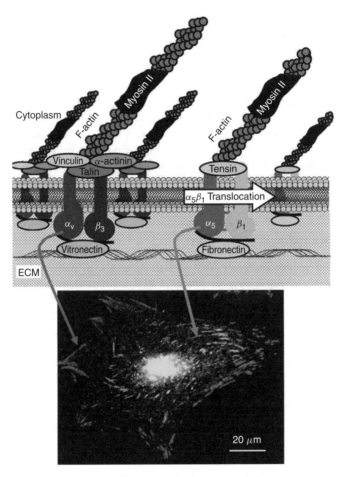

Figure 1.4 Various adhesions in two dimensions, with a graphical description of focal adhesions and fibrillar adhesions. Reprinted with permission from [45]. Refer to Plate 1 for the colored version

Interestingly, in addition to adhesion-related signaling, the outside-in activation of integrins can in some cases directly cause inside-out adhesion signals. Some integrin activations lead to suppression of other kinds of integrins [37,38]. The cases seen in the study involved the protein kinase A and protein kinase C-α pathways as intermediaries.

Not all proteins exist in the same concentrations in every adhesion site. Even in the realm of two-dimensional cell adhesion, there are distinct varieties of adhesion sites (Figure 1.4). The smallest of these (focal complexes) are the result of initial integrin–matrix adhesion events. They are approximately $1\,\mu m^2$ or less in area and relatively diverse in chemical composition, though some common features include talin, vinculin, paxillin, and $\alpha_v\beta_3$ integrins [39,40].

Focal adhesions are large, up to multiple micrometers in diameter. At the center of the adhesion, and to a lesser extent up until the peripheral approximately 200 nm, the nearby actin cytoskeleton is aligned in one direction [26]. Inside the focal adhesion, some local variability with respect to molecular composition is believed to be present, and the

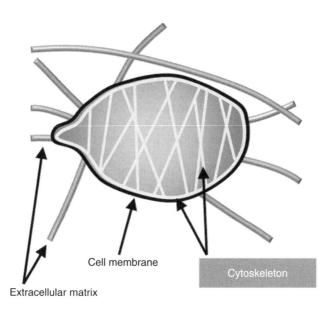

Cell membrane

Cytoskeleton

Extracellular matrix

Figure 1.5 Illustration of motility in a three-dimensional matrix. Protrusions with a cortical actin cytoskeleton but few stress fibers attach to the matrix via integrin binding. The exact signaling involved is still under study

various particles involved are clustered differently in the core of the adhesion than in the periphery. Common components include paxillin, vinculin, focal adhesion kinase (FAK) and phosphorylated FAK, tensin, zyxin, phosphotyrosine, and $\alpha_v\beta_3$ integrins [27,41].

Fibrillar adhesions are in some ways more mature versions of focal adhesions. They form in response to the extracellular matrix protein fibronectin and contain high levels of $\alpha_5\beta_1$ integrins and tensin, with relatively low levels of paxillin and several other proteins [42–45]. These $\alpha_5\beta_1$ integrins bind almost exclusively to fibronectin fibers and are believed to move those fibers away from the center of the adhesion and open them to further binding [36,46].

These various collections of proteins are sensitive to forces. Forces generally encourage integrins to form focal complexes when applied from either inside or outside the cell [47]. They are necessary for the transition from focal complex to focal adhesion and affect the size of sufficiently mature adhesions and the biochemical signals that are passed along to the inside of the cell [30,48]. Adhesion complexes in general, combined with the related cytoskeletal molecules, are also responsible for cells exerting forces on their surroundings. In combination, adhesion, force sensitivity, and force application form the basis for the energy-using directed movement of cells, known as motility.

1.4 Two-Dimensional Motility

As mentioned previously, there are a number of cytoplasmic proteins which form a bridge between integrins and the actin cytoskeleton. These generally both anchor integrins to the cytoskeleton and encourage cytoskeletal growth near an integrin cluster, though they are also involved in the disassembly of unwanted focal adhesions. For the most part, cytoskeletal growth and focal adhesion disassembly are encouraged indirectly by many of the other proteins associated with integrin clustering.

This collection of functions combines with the basic integrin function of adhering to an extracellular substance, and the ability of the molecular motor myosin to move actin filaments against each other, to provide an important capability to a cell: motility. Animal cells are, in general, capable of moving like an amoeba instead when their environment is insufficiently favorable to integrin binding [49,50]. However, the kind of motility most interesting to those working with integrins is the step-by-step protrusion of the cell in the forward direction, formation of adhesion sites in the same direction, and dissolution of anterior sites. This is important in morphogenesis, wound healing, and cancer behavior, including metastasis [50–52].

The precise mechanism for this process has been studied extensively. To begin, the cell produces a lamellipodium, a protrusion of the cell with a high actin concentration and no major organelles, and a lamellum, a distinct and slightly lower actin region of the cell that connects the lamellipodium to the cell body [53,54].

The use of actin to create protrusions such as the lamellipodium has been of interest to scientists for decades. It is reviewed in depth in, for instance, Li *et al*. [55]; an overview is presented here. A variety of chemical signals, generally highly overlapping with integrin signaling pathways, cause actin polymerization at the barbed end of an actin chain, while depolymerization occurs at the pointed end [56]. With thermal fluctuations and actin elasticity, this allows the actin to serve as an elastic Brownian ratchet, steadily pushing the membrane forward with an estimated maximum force of 5–7 pN per actin filament [57,58]. Such a statement is, of course, a gross oversimplification of an astoundingly complex process. The action of profilin, for instance, is of significant importance in stopping actin polymerization when the signals appropriate for stopping are present, and the protein myosin is used to produce force on the actin from the inside. For that matter, the actin cytoskeleton is not always involved; blebbing, the formation of small protrusions that lack significant cytoskeletal structure, can also be involved, though this is a distinct and poorly studied alternative [59].

The rest of the picture, in which focal complexes then form where the protrusions meet the substrate and then mature as the cell drags itself forwards, is also overly simplistic. It implies a steady maturation and progression from leading-edge adhesion complex to posterior focal adhesion or fibrillar adhesion to dissolution; however, it is important to note that new adhesion clusters frequently fail to mature [60]. Focal complex maturation requires specific tensile forces. As a major supplier of those forces, local myosin activity is one of the main intermediaries for maturation into focal adhesions. It is controlled by a number of adhesion-related proteins, including FAK, which is itself a downstream signal from integrins [61].

The factors that affect cell adhesion in two dimensions can also affect and cause motility in two dimensions [55]. A gradient of matrix proteins to which integrins can bind can cause movement; this is called hapotaxis [28,62]. Differences in substrate stiffness also give rise to movement towards the stiffer portions of substrate, in what is called durotaxis [63]. Further, mechanical forces (e.g., gravity or adjacent liquid flow) can cause motility [64], as can chemical attractants [65]. Physically protruding substrates, mentioned previously, also affect motility [29]. The resulting forces from the cell motility machinery vary wildly based on these factors, and cell type. Fibroblasts tend to move slowly, exerting high forces in all directions (on order of tens to thousands of $pN\,\mu m^{-2}$) which when totaled directionally come to approximately $10\text{--}1000\,pN$ per cell [58,66]. Fish epithelial keratocytes usually move more quickly, exerting forces of roughly $\sim\!200\,pN\,\mu m^{-2}$ mostly in one direction in one study and totaling up to $85\,000\,pN$ in another study. If one includes contracting as well as truly motile cells, the total summed nondirectional forces involved can reach $\sim\!10\,\mu N$ [67].

Of the factors mentioned, three directly involve the extracellular matrix: matrix protein concentration, matrix stiffness, and physical shape of the matrix. It is clear, therefore, that the various details of the extracellular matrix are central to integrin-based cell motility. This, however, is only the beginning of the interrelationship between the two; the story is far more complex, in ways that can only properly be described in terms of three-dimensional adhesion and motility.

1.5 Three-Dimensional Adhesion

Beginning in 2001, researchers performed studies showing that cell behavior in a realistic three-dimensional matrix is fundamentally different from two-dimensional adhesion behavior in specific ways. Among these is a lack of traditional fully formed focal adhesions. Though there is some debate about the subject, it is clear that three-dimensional adhesion experiments see integrins clustering faster into much smaller collections, their size highly dependent on matrix stiffness. These clusters also associate with a different (though similar) collection of cytoplasmic proteins.

Indeed, in some ways the three-dimensional adhesion complexes seen in early studies are closer to fibrillar adhesions than to focal adhesions. Their predominant integrin type is the same, $\alpha_5\beta_1$ instead of the $\alpha_v\beta_3$ integrins in focal adhesions [68]. Phosphorylation of FAK is closer to that in a fibrillar adhesion as well. However, it might be more accurate to say that three-dimensional adhesions are something of a mixture of the two two-dimensional adhesion cluster varieties, as $\alpha_v\beta_3$ integrins are more prominent in three-dimensional clusters than in fibrillar adhesions, and paxillin is also closer to the level found in focal adhesions.

These studies of three-dimensional adhesions were performed on cells partly imbedded in a realistic matrix; because of equipment limitations, the apical surface of the cell was instead attached to a two-dimensional surface. A more recent study used a different methodology, in which the cell was not in contact with any purely two-dimensional surface. It found that three-dimensional adhesions are smaller than two-dimensional adhesion structures (under 300 nm in diameter) and more transient (less than 1 s in lifetime, as opposed to typically up to 15 min and occasionally more for a two-dimensional focal adhesion) in a wide variety of matrix stiffnesses [69]. It should be noted that this last study questioned whether adhesion structures below those scales exist at all in a proper three-dimensional environment, but did not rule out the possibility; another, later, study discussed the possibility that background noise had interfered with seeing those adhesions, and showed adhesions in a similar though not identical environment; these also appeared somewhat short-lived [70].

1.6   Three-Dimensional Motility

The study of motility in three-dimensional matrices is still at its beginning stages. It is certain that those things involved in three-dimensional adhesion and two-dimensional motility are important to motility in three dimensions, but it is also clear that three-dimensional motility is very much distinct from its two-dimensional cousin in a large number of ways.

Among the most important differences is the mechanism. As mentioned previously, it is not completely clear that anything like a focal adhesion exists at all in realistic three-dimensional motility [69]. Even those studies that have found them have seen adhesions that usually lack stress fibers, the well-developed parts of the cytoskeleton that reach into the interior of the cell and provide a convenient elastic Brownian ratchet pressure [71,72]. They may be associated with the actin cortical cytoskeleton, the portion of the cytoskeleton immediately beneath the membrane that helps to provide the membrane with shape and structure [73]. The cortical cytoskeleton and its related myosin are known to be involved in amoeboid motility, which might seem suggestive, but amoeboid motility is integrin independent [49,74].

Nonetheless, it has been proposed that the predominant role for adhesion-related proteins in the three-dimensional environment could be in pseudopodia activity and matrix deformation instead of the more traditional adhesion-and-pulling described above [69]. This is a possibility because of a lack of stress fibers in three-dimensional adhesions, and because pseudopodia are dependent on myosin action [75], which in turn is a downstream signal from integrins. Therefore, it is not yet certain precisely what mechanisms integrins use to influence motility in a three-dimensional matrix.

However, a number of observations regarding three-dimensional motility in general are quite clear. Matrix stiffness and composition (including types and densities of integrin ligands) can still significantly change a cell's locomotion behavior, though the effects are not always the same as on two-dimensional substrates [76]. Stiffness can even change a cell's basic type of motility [77]. Three-dimensional matrices also offer the possibility of steric hindrance, in which the matrix physically gets in the way of a moving cell [78,79]. Cells can deal with steric hindrance by degrading the matrix, most notably with collagenase, or by physically pushing the matrix, but these processes do not eliminate its effects entirely. Further related concepts are pore size, the general size of open areas within the heterogeneous matrix, and matrix density; both are important influences on motility behavior.

Artificial extracellular matrices have been used to explore the effects of pore size on motility. For very small pores (smaller than cell size), matrix stiffness is of even greater importance than usual, and interferes with motility rather than helping [76]. In this case, the primary consideration is the cell's ability to degrade the matrix. At larger pore size, considerably larger than the cells in question, larger pores cause a decrease in the motile fraction (fraction of cells which are motile) and speed of motile cells [80].

Another potentially important factor is the ability of cells to migrate as clusters. This occurs both on two-dimensional substrates and in realistic three-dimensional matrices [81,82]. In general, cadherins or other cell–cell adhesion molecules cause cells to essentially drag each other along during normal motility functions. The relationship between collective motility and individual motility, and the possibility that either behaviors could be encouraged by cells' environments (including the extracellular matrix) is still uncertain, though recent work suggests a stiffer substrate may interfere with collective action [83,84]. Further complicating matters is that cells can change the effective local matrix stiffness observed by other cells, in essence communicating via the matrix [85].

1.7 Apoptosis and Survival Signaling

Cell adhesion and motility, while extremely important, are not the only results of adhesion protein interactions. As mentioned previously, a large number of adhesion cluster proteins are involved in gene expression or are enzymes involved in various pathways. A number of the most important associated proteins are involved in apoptosis or survival pathways; without adhesion, cells die even more quickly than they do without nutrients [86].

FAK and Src family proteins are the major known survival-pathway signals. Through interconnected chains of reactions, these activate the mitogen-activated protein kinase (MAPK) and phosphoinositide 3-kinase (PI3K) pathways, which in turn are vital for regulation of cyclin-dependent kinases (CDKs) and related proteins [86,87]. CDKs are involved in transcription and mRNA processing, and often in cell cycle regulation. In this way, integrin–substrate binding is intimately involved in the basic necessities of life for eukaryotic cells.

In addition, some integrin-related pathways are involved in preventing internal apoptosis signaling due to environmental stresses and external cell-death signals [86]. These include a wide variety of cell insults, including starvation, cytotoxic drugs, and radiation.

Cadherins may also be involved in survival signaling, though the relative extent of their influence compared with that of integrins is unknown [88].

1.8 Cell Differentiation Signaling

Integrins are also extremely important in differentiation of cell lines. Stem cells (undifferentiated cells) of various kinds respond to integrin-mediated signals, and functional integrins are necessary for proper differentiation in general [89–92].

Indeed, even without the appropriate extracellular chemical signals, cell differentiation can occur to a point based solely on extracellular matrix composition and stiffness; this clearly implicates mechanical sensitivity of the process via integrin action. Specifically, a soft collagen matrix makes a stem cell exhibit neuron-like differentiation, while a stiff one causes bone-like structure, and one of an intermediate stiffness gives rise to a striated muscle cell appearance [93]. These three conditions, perhaps unsurprisingly, match the stiffness of the appropriate cells and their environments, which generally match each other [94]. Further, the study of matrix stiffness and differentiation also examined the effects of myosin inhibition and found that it removed the observed differentiation behavior. More recent research into the details of bone-cell stiffness have clearly indicated integrins (α_2 subunits specifically), FAK, and Rho kinase (ROCK) are all involved, and FAK and ROCK are both downstream signals from integrins [95].

It is certain, however, that stiffness is not by any means the only important contribution the extracellular matrix makes to differentiation, though at least some of the others are integrin related as well. A certain mixture of ligands intended to replicate basement membrane concentrations has been shown to prolong the nondifferentiated life of mesenchymal stromal cell lines [96]. In addition, the physical placement patterns of integrin ligands or

Table 1.1 Ideal matrix stiffness for differentiation

Intended cell differentiation	Matrix stiffness (kPa)	
	Two-dimensional [93]	Three-dimensional [98]
Neuronal	0.1–1	7
Myogenic	8–17	25
Osteogenic	25–40	75

presence of nearby cells may be important. On a two-dimensional substrate, large grids (pattern size on the order of cell size) of neural differentiation-inducing ligands caused more neural differentiation than strips or blocks [97].

Making the cell environment three-dimensional again changes these results. In the case of differentiation there seem to be two differences: a relatively small multiplicative factor for the relevant stiffness and a significant increase in the protein concentrations of the appropriate cell type [98]. The ideal stiffnesses for two- and three-dimensional environments and the three cell lineages are shown in Table 1.1.

Cadherins are known to play some role in differentiation as well, though again their relative role is unknown and may likely be smaller [99].

1.9 Conclusions

Cell adhesion interactions are an extremely important and widespread part of biological function. They are involved in signals that keep cells alive and tell stem cells how to differentiate. Perhaps even more importantly, they keep the cells of a multicellular organism together. Various kinds of adhesion interactions exist (Table 1.2), but among the most interesting is that between integrins and the extracellular matrix. This is a complex interaction, involving over a hundred proteins, and the interaction itself is highly sensitive to piconewton-level mechanical forces. Changing of those mechanical forces, generally done in a biologically relevant way by altering the properties of the matrix to which the integrins are attached, is known to affect adhesion, motility, cell survival, and differentiation. This makes understanding those forces and how they interact with the cell an issue of great interest and importance to the scientific and medical communities.

Table 1.2 Summary of binding protein functions

Protein	Function			
	Adhesion to	Motility	Survival	Differentiation
Integrin	Matrix	Yes	Yes	Yes
Cadherin	Other cells	Yes	Yes	Yes
Selectin	Blood vessel	No	No	No
Immunoglobulin	Foreign bodies	No	No	No

References

[1] Berg, J.M., Tymoczko, J.L., and Stryer, L. (2002) *Biochemistry*, 5th edn., W.H. Freeman.

[2] Somers, W.S., Tang, J., Shaw, G.D., and Camphausen, R.T. (2000) Insights into the molecular basis of leukocyte tethering and rolling revealed by structures of P- and E-selectin bound to SLeX and PSGL-1. *Cell*, **103**, 467–479.

[3] Duguay, D., Foty, R.A., and Steinberg, M.S. (2003) Cadherin-mediated cell adhesion and tissue segregation: qualitative and quantitative determinants. *Developmental Biology*, **253**, 309–323.

[4] Angst, B.D., Marcozzi, C., and Magee, A.I. (2001) The cadherin superfamily: diversity in form and function. *Journal of Cell Science*, **114**, 629–641.

[5] Hynes, R.O. (2002) Integrins: bidirectional, allosteric signaling machines. *Cell*, **110**, 673–687.

[6] Belkin, A.M., Retta, S.F., Pletjushkina, O.Y. *et al.* (1997) Muscle β1D integrin reinforces the cytoskeleton–matrix link: modulation of integrin adhesive function by alternative splicing. *The Journal of Cell Biology*, **139**, 1583–1595.

[7] Hay, E.D. (1981) Extracellular matrix. *The Journal of Cell Biology*, **91**, 205s–223s.

[8] Xiong, J.-P. (2001) Crystal structure of the extracellular segment of integrin $\alpha_V\beta_3$. *Science*, **294**, 339–345.

[9] Xiong, J.-P., Mahalingham, B., Alonso, J.L. *et al.* (2009) Crystal structure of the complete integrin $\alpha_V\beta_3$ ectodomain plus an α/β transmembrane fragment. *The Journal of Cell Biology*, **186**, 589–600.

[10] Shattil, S.J., Kim, C., and Ginsberg, M.H. (2010) The final steps of integrin activation: the end game. *Nature Reviews: Molecular Cell Biology*, **11**, 288–300.

[11] Nishida, N., Xie, C., Shimaoka, M. *et al.* (2006) Activation of leukocyte β_2 integrins by conversion from bent to extended conformations. *Immunity*, **25**, 583–594.

[12] Calderwood, D.A., Zent, R., Grant, R. *et al.* (1999) The talin head domain binds to integrin β subunit cytoplasmic tails and regulates integrin activation. *Journal of Biological Chemistry*, **274**, 28071–28074.

[13] Cram, E.J. and Schwarzbauer, J.E. (2004) The talin wags the dog: new insights into integrin activation. *Trends in Cell Biology*, **14**, 55–57.

[14] Wegener, K., Partridge, A., Han, J. *et al.* (2007) Structural basis of integrin activation by talin. *Cell*, **128**, 171–182.

[15] Qin, J., Vinogradova, O., and Plow, E.F. (2004) Integrin bidirectional signaling: a molecular view. *PLoS Biology*, **2** (6), e169.

[16] Vinogradova, O., Velyvis, A., Velyviene, A. *et al.* (2002) A structural mechanism of integrin $\alpha_{IIb}\beta_3$ "inside-out" activation as regulated by its cytoplasmic face. *Cell*, **110**, 587–597.

[17] Anthis, N.J., Wegener, K.L., Ye, F. *et al.* (2009) The structure of an integrin/talin complex reveals the basis of inside-out signal transduction. *EMBO Journal*, **28**, 3623–3632.

[18] Evans, E.A. and Calderwood, D.A. (2007) Forces and bond dynamics in cell adhesion. *Science*, **316** (5828), 1148–1153.

[19] Alon, R. and Dustin, M.L. (2007) Force as a facilitator of integrin conformational changes during leukocyte arrest on blood vessels and antigen-presenting cells. *Immunity*, **26**, 17–27.

[20] Schwartz, M.A. (2010) Integrins and extracellular matrix in mechanotransduction. *Cold Spring Harbor Perspectives in Biology*, **2**, a005066.

[21] Saez, A., Buguin, A., Silberzan, P., and Ladoux, B. (2005) Is the mechanical activity of epithelial cells controlled by deformations or forces? *Biophysical Journal*, **89**, L52–L54.

[22] Friedland, J.C., Lee, M.H., and Boettiger, D. (2009) Mechanically activated integrin switch controls $\alpha_5\beta_1$ function. *Science*, **323**, 642–644.

[23] Wang, N., Butler, J., and Ingber, D. (1993) Mechanotransduction across the cell surface and through the cytoskeleton. *Science*, **260**, 1124–1127.

[24] Zhu, J., Luo, B.-H., Xiao, T. *et al.* (2008) Structure of a complete integrin ectodomain in a physiologic resting state and activation and deactivation by applied forces. *Molecular Cell*, **32**, 849–861.

[25] Harjanto, D. and Zaman, M.H. (2010) Matrix mechanics and receptor–ligand interactions in cell adhesion. *Organic & Biomolecular Chemistry*, **8**, 299–304.

[26] Patla, I., Volberg, T., Elad, N. *et al.* (2010) Dissecting the molecular architecture of integrin adhesion sites by cryo-electron tomography. *Nature Cell Biology*, **12**, 909–915.

[27] Cukierman, E., Pankov, R., and Yamada, K.M. (2002) Cell interactions with three-dimensional matrices. *Current Opinion in Cell Biology*, **14**, 633–639.

[28] Cavalcantiadam, E., Volberg, T., Micoulet, A. *et al.* (2007) Cell spreading and focal adhesion dynamics are regulated by spacing of integrin ligands. *Biophysical Journal*, **92**, 2964–2974.

[29] Curtis, A. and Wilkinson, C. (1997) Topographical control of cells. *Biomaterials*, **18**, 1573–1583.

[30] Balaban, N.Q., Schwarz, U.S., Riveline, D. *et al.* (2001) Force and focal adhesion assembly: a close relationship studied using elastic micropatterned substrates. *Nature Cell Biology*, **3**, 466–472.

[31] Berry, C.C., Campbell, G., Spadiccino, A. *et al.* (2004) The influence of microscale topography on fibroblast attachment and motility. *Biomaterials*, **25**, 5781–5788.

[32] Beningo, K.A., Dembo, M., and Wang, Y.-L. (2004) Responses of fibroblasts to anchorage of dorsal extracellular matrix receptors. *Proceedings of the National Academy of Sciences of the United States of America*, **101**, 18024–18029.

[33] Lo, S.H. (2006) Focal adhesions: what's new inside. *Developmental Biology*, **294**, 280–291.

[34] Zaidel-Bar, R., Itzkovitz, S., Ma'ayan, A. *et al.* (2007) Functional atlas of the integrin adhesome. *Nature Cell Biology*, **9**, 858–867.

[35] Zamir, E. and Geiger, B. (2001) Components of cell–matrix adhesions. *Journal of Cell Science*, **114**, 3577–3579.

[36] Zamir, E. and Geiger, B. (2001) Molecular complexity and dynamics of cell–matrix adhesions. *Journal of Cell Science*, **114**, 3583–3590.

[37] Arnaout, M.A., Goodman, S., and Xiong, J.-P. (2007) Structure and mechanics of integrin-based cell adhesion. *Current Opinion in Cell Biology*, **19**, 495–507.

[38] Orr, A.W., Ginsberg, M.H., Shattil, S.J. *et al.* (2006) Matrix-specific suppression of integrin activation in shear stress signaling. *Molecular Biology of the Cell*, **17**, 4686–4697.

[39] Ballestrem, C., Hinz, B., Imhof, B.A., and Wehrle-Haller, B. (2001) Marching at the front and dragging behind. *The Journal of Cell Biology*, **155**, 1319–1332.

[40] Zimerman, B., Volberg, T., and Geiger, B. (2004) Early molecular events in the assembly of the focal adhesion–stress fiber complex during fibroblast spreading. *Cell Motility and the Cytoskeleton*, **58**, 143–159.

[41] Zaidel-Bar, R., Ballestrem, C., Kam, Z., and Geiger, B. (2003) Early molecular events in the assembly of matrix adhesions at the leading edge of migrating cells. *Journal of Cell Science*, **116**, 4605–4613.

[42] Zamir, E., Katz, M., Posen, Y. *et al.* (2000) Dynamics and segregation of cell–matrix adhesions in cultured fibroblasts. *Nature Cell Biology*, **2**, 191–196.

[43] Katz, B.-Z., Zamir, E., Bershadsky, A. *et al.* (2000) Physical state of the extracellular matrix regulates the structure and molecular composition of cell–matrix adhesions. *Molecular Biology of the Cell*, **11**, 1047–1060.

[44] Sottile, J. and Hocking, D.C. (2002) Fibronectin polymerization regulates the composition and stability of extracellular matrix fibrils and cell–matrix adhesions. *Molecular Biology of the Cell*, **13**, 3546–3559.

[45] Marquis, M.-E., Lord, E., Bergeron, E. *et al.* (2009) Bone cells–biomaterials interactions. *Frontiers in Bioscience: A Journal and Virtual Library*, **14**, 1023–1067.

[46] Yamada, K.M., Pankov, R., and Cukierman, E. (2003) Dimensions and dynamics in integrin function. *Brazilian Journal of Medical and Biological Research*, **36**, 959–966.

[47] Galbraith, C.G., Yamada, K.M., and Sheetz, M.P. (2002) The relationship between force and focal complex development. *The Journal of Cell Biology*, **159**, 695–705.

[48] Riveline, D., Zamir, E., Balaban, N.Q. *et al.* (2001) Focal contacts as mechanosensors: externally applied local mechanical force induces growth of focal contacts by an mDia1-dependent and ROCK-independent mechanism. *The Journal of Cell Biology*, **153**, 1175–1186.

[49] Renkawitz, J., Schumann, K., Weber, M. *et al.* (2009) Adaptive force transmission in amoeboid cell migration. *Nature Cell Biology*, **11**, 1438–1443.

[50] Friedl, P. and Wolf, K. (2003) Tumour-cell invasion and migration: diversity and escape mechanisms. *Nature Reviews: Cancer*, **3**, 362–374.

[51] Shih, J. and Keller, R. (1992) Patterns of cell motility in the organizer and dorsal mesoderm of *Xenopus laevis*. *Development*, **116**, 915–930.

[52] Chien, S., Li, S., Shiu, Y.-T., and Li, Y.S. (2005) Molecular basis of mechanical modulation of endothelial cell migration. *Frontiers in Bioscience: A Journal and Virtual Library*, **10**, 1985–2000.

[53] Huber, F., Käs, J., and Stuhrmann, B. (2008) Growing actin networks form lamellipodium and lamellum by self-assembly. *Biophysical Journal*, **95**, 5508–5523.

[54] Ponti, A., Machacek, M., Gupton, S.L. *et al.* (2004) Two distinct actin networks drive the protrusion of migrating cells. *Science*, **305**, 1782–1786.

[55] Li, S., Guan, J.-L., and Chien, S. (2005) Biochemistry and biomechanics of cell motility. *Annual Review of Biomedical Engineering*, **7**, 105–150.

[56] Pollard, T.D. and Borisy, G.G. (2003) Cellular motility driven by assembly and disassembly of actin filaments. *Cell*, **112**, 453–465.

[57] Mogilner, A. and Oster, G. (1996) Cell motility driven by actin polymerization. *Biophysical Journal*, **71**, 3030–3045.

[58] Ananthakrishnan, R. (2007) The forces behind cell movement. *International Journal of Biological Sciences*, **3**, 303–317.

[59] Renkawitz, J. and Sixt, M. (2010) Mechanisms of force generation and force transmission during interstitial leukocyte migration. *EMBO Reports*, **11**, 744–750.

[60] Broussard, J.A., Webb, D.J., and Kaverina, I. (2008) Asymmetric focal adhesion disassembly in motile cells. *Current Opinion in Cell Biology*, **20**, 85–90.

[61] Guan, J.-L. (2010) Integrin signaling through FAK in the regulation of mammary stem cells and breast cancer. *IUBMB Life*, **62**, 268–276.

[62] Carter, S.B. (1967) Haptotaxis and the mechanism of cell motility. *Nature*, **213**, 256–260.

[63] Lo, C.M., Wang, H.B., Dembo, M., and Wang, Y.L. (2000) Cell movement is guided by the rigidity of the substrate. *Biophysical Journal*, **79**, 144–152.

[64] Li, S., Butler, P., Wang, Y. *et al.* (2002) The role of the dynamics of focal adhesion kinase in the mechanotaxis of endothelial cells. *Proceedings of the National Academy of Sciences of the United States of America*, **99**, 3546–3551.

[65] Devreotes, P.N. and Zigmond, S.H. (1988) Chemotaxis in eukaryotic cells: a focus on leukocytes and *Dictyostelium*. *Annual Review of Cell Biology*, **4**, 649–686.

[66] Lombardi, M.L., Knecht, D.A., Dembo, M., and Lee, J. (2007) Traction force microscopy in *Dictyostelium* reveals distinct roles for myosin II motor and actin-crosslinking activity in polarized cell movement. *Journal of Cell Science*, **120** (9), 1624–1634.

[67] Kamm, R., Lammerding, J., and Mofrad, M. (2010) Cellular nanomechanics, in *Springer Handbook of Nanotechnology*, 3rd edn. (ed. B. Bhushan), Springer, pp. 1171–1200.

[68] Cukierman, E., Pankov, R., Stevens, D.R., and Yamada, K.M. (2001) Taking cell–matrix adhesions to the third dimension. *Science*, **294**, 1708–1712.

[69] Fraley, S.I., Feng, Y., Krishnamurthy, R. *et al.* (2010) A distinctive role for focal adhesion proteins in three-dimensional cell motility. *Nature Cell Biology*, **12**, 598–604.

[70] Kubow, K.E. and Horwitz, A.R. (2011) Reducing background fluorescence reveals adhesions in 3D matrices. *Nature Cell Biology*, **13**, 3–5.

[71] Grinnell, F. (2003) Fibroblast biology in three-dimensional collagen matrices. *Trends in Cell Biology*, **13**, 264–269.

[72] Grinnell, F., Ho, C.-H., Tamariz, E. *et al.* (2003) Dendritic fibroblasts in three-dimensional collagen matrices. *Molecular Biology of the Cell*, **14**, 384–395.

[73] Friedl, P. and Wolf, K. (2009) Proteolytic interstitial cell migration: a five-step process. *Cancer Metastasis Reviews*, **28**, 129–135.

[74] Panková, K., Rösel, D., Novotný, M., and Brábek, J. (2010) The molecular mechanisms of transition between mesenchymal and amoeboid invasiveness in tumor cells. *Cellular and Molecular Life Sciences*, **67**, 63–71.

[75] Yoshida, K. and Soldati, T. (2006) Dissection of amoeboid movement into two mechanically distinct modes. *Journal of Cell Science*, **119**, 3833–3844.

[76] Zaman, M.H., Trapani, L.M., Sieminski, A.L. *et al.* (2006) Migration of tumor cells in 3D matrices is governed by matrix stiffness along with cell–matrix adhesion and proteolysis. *Proceedings of the National Academy of Sciences of the United States of America*, **103**, 10889–10894.

[77] Friedl, P. (2004) Prespecification and plasticity: shifting mechanisms of cell migration. *Current Opinion in Cell Biology*, **16**, 14–23.

[78] Friedl, P. and Bröcker, E.B. (2000) The biology of cell locomotion within three-dimensional extracellular matrix. *Cellular and Molecular Life Sciences*, **57**, 41–64.

[79] Zaman, M.H., Matsudaira, P., and Lauffenburger, D.A. (2007) Understanding effects of matrix protease and matrix organization on directional persistence and translational speed in three-dimensional cell migration. *Annals of Biomedical Engineering*, **35**, 91–100.

[80] Harley, B.A.C., Kim, H.-D., Zaman, M.H. *et al.* (2008) Microarchitecture of three-dimensional scaffolds influences cell migration behavior via junction interactions. *Biophysical Journal*, **95**, 4013–4024.

[81] Friedl, P., Noble, P.B., Walton, P.A. *et al.* (1995) Migration of coordinated cell clusters in mesenchymal and epithelial cancer explants *in vitro*. *Cancer Research*, **55**, 4557–4560.

[82] Gumbiner, B.M. (1996) Cell adhesion: the molecular basis of tissue architecture and morphogenesis. *Cell*, **84**, 345–357.

[83] Reinhart-King, C.A., Dembo, M., and Hammer, D.A. (2008) Cell–cell mechanical communication through compliant substrates. *Biophysical Journal*, **95**, 6044–6051.

[84] Kim, J.-H., Dooling, L.J., and Asthagiri, A.R. (2010) Intercellular mechanotransduction during multicellular morphodynamics. *Journal of the Royal Society, Interface*, **7** (Suppl. 3), S341–S350.

[85] Winer, J.P., Oake, S., and Janmey, P.A. (2009) Non-linear elasticity of extracellular matrices enables contractile cells to communicate local position and orientation. *PLoS ONE*, **4**, e6382.

[86] Stupack, D.G. and Cheresh, D.A. (2002) Get a ligand, get a life: integrins, signaling and cell survival. *Journal of Cell Science*, **115**, 3729–3738.

[87] Schwartz, M.A. and Assoian, R.K. (2001) Integrins and cell proliferation: regulation of cyclin-dependent kinases via cytoplasmic signaling pathways. *Journal of Cell Science*, **114**, 2553–2560.

[88] Peluso, J.J., Pappalardo, A., and Fernandez, G. (2001) E-cadherin-mediated cell contact prevents apoptosis of spontaneously immortalized granulosa cells by regulating Akt kinase activity. *Biology of Reproduction*, **64** (4), 1183–1190.

[89] Martin-Bermudo, M.D. (2000) Integrins modulate the Egfr signaling pathway to regulate tendon cell differentiation in the *Drosophila* embryo. *Development*, **127**, 2607–2615.

[90] Velleman, S.G. and McFarland, D.C. (2004) $\beta 1$ integrin mediation of myogenic differentiation: implications for satellite cell differentiation. *Poultry Science*, **83**, 245–252.

[91] Maitra, N., Flink, I.L., Bahl, J.J., and Morkin, E. (2000) Expression of α and β integrins during terminal differentiation of cardiomyocytes. *Cardiovascular Research*, **47**, 715–725.

[92] Adams, J.C. and Watt, F.M. (1993) Regulation of development and differentiation by the extracellular matrix. *Development*, **117**, 1183–1198.

[93] Engler, A., Sen, S., Sweeney, H., and Discher, D. (2006) Matrix elasticity directs stem cell lineage specification. *Cell*, **126**, 677–689.

[94] Chicurel, M.E., Singer, R.H., Meyer, C.J., and Ingber, D.E. (1998) Integrin binding and mechanical tension induce movement of mRNA and ribosomes to focal adhesions. *Nature*, **392**, 730–733.

[95] Shih, Y.-R.V., Tseng, K.-F., Lai, H.-Y. *et al.* (2011) Matrix stiffness regulation of integrin-mediated mechanotransduction during osteogenic differentiation of human mesenchymal stem cells. *Journal of Bone and Mineral Research: The Official Journal of the American Society for Bone and Mineral Research*, **26**, 730–738.

[96] Lindner, U., Kramer, J., Behrends, J. *et al.* (2010) Improved proliferation and differentiation capacity of human mesenchymal stromal cells cultured with basement-membrane extracellular matrix proteins. *Cytotherapy*, **12**, 992–1005.

[97] Solanki, A., Shah, S., Memoli, K.A. *et al.* (2010) Controlling differentiation of neural stem cells using extracellular matrix protein patterns. *Small*, **6**, 2509–2513.

[98] Pek, Y.S., Wan, A.C.A., and Ying, J.Y. (2010) The effect of matrix stiffness on mesenchymal stem cell differentiation in a 3D thixotropic gel. *Biomaterials*, **31** (3), 385–391.

[99] Kim, S., Schein, A.J., and Nadel, J.A. (2005) E-cadherin promotes EGFR-mediated cell differentiation and MUC5AC mucin expression in cultured human airway epithelial cells. *American Journal of Physiology: Lung Cellular and Molecular Physiology*, **289** (6), L1049–L1060.

2

Regulatory Mechanisms of Kinesin and Myosin Motor Proteins: Inspiration for Improved Control of Nanomachines

Sarah Rice

Northwestern University Medical School, USA

2.1 Introduction

The biological motors kinesin and myosin are highly efficient and effective nanomachines. These linear transporters move cargo along microtubules and actin, respectively, and small adaptations in their structure can adapt their basic core design to many different intracellular tasks. Their core mechanisms have remarkable similarities, and these have been mimicked in several recent efforts to develop useful synthetic and semi-synthetic nano-transporters. Recent work on kinesins and myosins has illuminated several regulatory mechanisms governing their activity. This review describes a very generalized mechanism of cytoskeletal motor action, then focuses on four common regulatory mechanisms that control the movement of kinesin and myosin motors: direct auto-regulatory control of the nucleotide-binding "switches"; phospho-regulation; binding of external, calcium-dependent regulators; and collective behavior resulting from mechanical coupling of several motors. All of these regulatory mechanisms can provide inspiration for designing improved control of synthetic motors.

2.2 Generalized Mechanism of Cytoskeletal Motors

The first mechanism for a molecular motor protein was proposed in 1954 by two groups studying the mechanism of muscle contraction [1,2]. Their main findings showed that thick

---

Nano and Cell Mechanics: Fundamentals and Frontiers, First Edition. Edited by Horacio D. Espinosa and Gang Bao.
© 2013 John Wiley & Sons, Ltd. Published 2013 by John Wiley & Sons, Ltd.

filaments, made of myosin-II, slid along thin filaments, made of actin, in an ATP-dependent manner. Subsequent X-ray diffraction work by H. Huxley's laboratory demonstrated that crossbridges exist between the thin and thick filaments, and these crossbridges tilt at two different angles relative to the thin filament axis [3]. These results together led these authors to propose that thick filament crossbridges undergo a cycle of ATP hydrolysis, coupled to thin filament binding and unbinding, as well as crossbridge tilting. Lymn and Taylor [4] subsequently worked out the correct coordination of these events: nucleotide-free myosin binds to ATP and dissociates from the actin filament. Upon ATP hydrolysis, myosin in solution undergoes an internal conformational change to adopt a "pre-powerstroke" conformation, and re-binds to the actin filament. Phosphate release stimulates the powerstroke, a conformational change of myosin-II while bound to actin. The powerstroke is a movement of a structural element within myosin-II called the "lever arm," and this lever arm movement corresponds to the crossbridge tilting observed by Huxley. After ADP release, ATP can re-bind and restart the cycle of activity. In the years since these papers described the mechanism of myosin-II, a wealth of supporting evidence has accumulated from a wide variety of approaches. This model remains a dominant paradigm not just for the myosin-II mechanism, but also as a generalized molecular motor mechanism (see Huxley [5] for an excellent historical review of this subject).

Kinesin family motors transport cargo generally toward the plus end of microtubules. While the exact coupling of their ATPase cycle to microtubule binding and internal conformational changes differs from myosin-II, the same general principle governs their activity. The canonical kinesin protein, kinesin-1, is ADP-bound in solution and releases ADP upon binding to the microtubule. When microtubule-bound kinesin-1 binds ATP, it undergoes a conformational change of an element called the "neck linker," which has a similar function to the myosin-II lever arm in generating a unidirectional conformational change on the partner filament. After ATP hydrolysis and phosphate release, ADP-bound kinesin-1 dissociates from the microtubule and can undergo another cycle of hydrolysis [6–8]. Kinesin-1 differs from myosin-II in that a dimer of two kinesin-1 enzymatic heads is catalytically coupled to function processively, taking hundreds of steps along a microtubule before dissociating from it. The coupling of the two enzymatic heads is mediated by tension between the two neck linkers that connect them [9–11]. This tension-mediated coupling presents an interesting twist to the original model of molecular motor motility: the cycle of ATP hydrolysis and filament binding feed forward to stimulate the neck linker conformational change at the correct point in the cycle and the neck linker conformation also feeds back as a gating mechanisms for the enzymatic activity of the heads.

By now, the molecular mechanisms for a large number of motors have been determined in detail. The recently described structure of dynein, the primary transporter to the microtubule minus end, is completely distinct from that of kinesin or myosin, but its cyclic mechanism of activity follows the same general principle described above for myosin-II [12–14]. In the zoo of molecular motors, there are processive myosins [15,16], nonprocessive kinesins [17,18], and motors such as myosin-VI and kinesin-5 that may behave primarily as cross-linkers for their partner filaments [19,20]. While their finely tuned properties are highly varied, all of these motors couple filament binding and unbinding to an ATP-dependent, unidirectional conformational change on the partner filament.

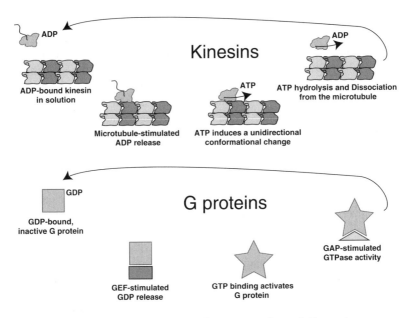

Figure 2.1   Cyclic activity of motor proteins and G proteins

Several structural and functional features of the enzymatic head domains of kinesin and myosin motors are extremely well conserved and, interestingly, they are also highly similar to the conserved GTPase superfamily (G proteins). Like the motor proteins described above, G proteins (e.g., the small signal transduction GTPase ras, the elongation GTPases EF-Tu and EF-G, and the SRP GTPases, such as Ffh and FtsY) undergo conformational cycles that are regulated by the binding and hydrolysis of nucleotide and that are modulated by interactions with other proteins [21]. A basic enzymatic cycle that describes the cyclic activity of both motor proteins and G proteins is outlined in Figure 2.1.

2.3   Switch I: A Controller of Motor Protein and G Protein Activation

Given the enzymatic similarities between motors and G proteins, it is not surprising that their structures are all remarkably similar. In fact, kinesins are structurally more similar to G proteins than they are to most other ATPases [21]. The heads of both kinesin motors and G proteins comprise four conserved motifs, termed 1–4, that define the nucleotide-binding site. Motifs 1, 3, and 4, are similar in both motor proteins and G proteins, and these are relatively static during the ATPase or GTPase cycle, while conserved motif 2 is dynamic. Each motif is coupled to adjacent "switch" regions. Switch I lies adjacent to motif 2, while switch II lies adjacent to motif 3. The switch regions are nucleotide and effector recognition elements that undergo structural rearrangements that transmit or modulate signals to and from the protein surface.

Switch I has a conserved function in all of these enzymes; it determines the rate of nucleotide release following effector binding. For kinesin, this is the rate of microtubule-stimulated ADP release; for myosin, it is the rate of actin-activated ADP and/or phosphate release; and for small GTPases, it is the rate of GDP release following binding of various GEFs (GTPase effector proteins). Upon effector binding, switch I moves toward the nucleotide pocket and stimulates nucleotide release, a movement termed the "closing" of switch I in the myosin and kinesin fields [22–24]. This is not simply a steric effect as the term "closing" implies; rather, the movement of switch I alters the set of hydrogen bonds holding the magnesium and ADP or GDP in the active site, weakening the nucleotide–enzyme interaction. While the structural changes seem subtle, the effect is not. For example, in kinesin-1, effector (microtubule) binding, coupled with switch I "closing," reduces its ADP affinity ∼1000-fold [25].

Switch-I-containing enzymes are essentially catalytically inactive until they bind their respective effectors and switch I stimulates nucleotide ejection. Switch I is a logical target for a regulatory mechanism, as it is a master activator for motor proteins and G proteins alike. Indeed, several regulatory mechanisms for all of these enzymes have been found to target switch I, either directly or indirectly. These mechanisms block switch I from stimulating nucleotide release, trapping their target enzyme in an inactive state.

The G proteins are a massive superfamily of proteins and, depending on their specific function, nature has produced a myriad of different regulatory mechanisms for them. Taken together from different GTPases, such regulatory interactions exploit almost every surface of the core GTPase fold. One class of G protein regulator is called a GDP dissociation inhibitor (GDI). GDIs inhibit the release of bound GDP from the nucleotide pocket. There are three main classes of GDIs. Two of these, RhoGDIs and RabGDIs, inhibit GDP release by interacting with switch I, switch II, and the magnesium ion in the nucleotide pocket to force these elements into a conformation that prevents the target G protein from releasing GDP. The GoLoco proteins, a third major class of GDIs, inhibits GDP release by positioning a specific arginine residue to stabilize the α/β-phosphates of a bound nucleotide, which also interferes with the ability of switch I to stimulate nucleotide release [26,27].

Kinesin-1 is a multifunctional cargo transporter in cells, carrying neuronal vesicles, mitochondria, peroxisomes, and RNA, among other cargoes. Roughly 90% of kinesin in cells is not bound to microtubules or actively transporting cargo, but rather is inactive in the cytoplasm [28,29]. In this inactive state, kinesin-1 bends at a hinge in the middle of the molecule to produce a "folded over" conformation in which the N-terminal enzymatic motor heads and the C-terminal tails interact directly [30–32]. This conformation is enzymatically inactive, having very tight ADP affinity and weak microtubule affinity [33].

Recent structural work has determined that the kinesin-1 tail interacts directly with the head. In addition, these studies have revealed that a critical regulatory sequence in the tail (QIAKPIRP) may stabilize the "open" or high-ADP affinity conformation of switch I and preventing kinesin-1 from binding to microtubules and becoming active [32]. There is an interesting parallel between tail-dependent kinesin-1 auto-regulation and the mechanism by which the Goloco proteins inhibit their G protein targets. Kinesin-1 tail-dependent auto-regulation strictly requires a conserved lysine within the QIAKPIRP sequence in the tail. When this residue is changed to alanine, the tail still interacts with the heads, but the regulatory activity of the tail appears to be completely abolished [29, 33–35]. A similar result has been found with RGS14, a Goloco inhibitor of the G protein Gαi1. Interestingly, an R516A mutation of the critical regulatory arginine in RGS14 results in a 10-fold reduction in GDI activity, but does not affect the RGS14- Gαi1 interaction [26].

The kinesin-1 structure and Gαi1 structure are similar enough that they can be aligned, and such alignments have shown that the critical arginine of RGS14 would be perfectly positioned to coordinate the α/β-phosphates of kinesin-1's bound ADP [35]. This result suggests that the "critical arginine" Goloco inhibitory mechanism and the "critical lysine" kinesin-1 inhibitory mechanism may be very similar, and that switch I movement, and hence target activation, may be blocked by both regulators. Kinesin-1 regulation is a very active area of research. Of note, at the time this review went to press a recent X-ray crystal structure of the kinesin-1 head-tail complex suggested an entirely different, switch I-independent "double lockdown" head-tail regulatory mechanism for the kinesin-1 dimer [36]. Further research is necessary to determine if, and when, switch I-targeted vs. "double lockdown" regulation occurs.

Switch I can be directly or allosterically frozen by small-molecule inhibitors of myosin-II, as well as the kinesin family members kinesin-5 and CENP-E. These inhibitors prevent the coordinated movements of switch I and switch II that enable all of these motors to engage their partner filaments and productively move through their respective ATPase cycles. The myosin-II inhibitor blebbistatin allosterically stabilizes the closed switch I/switch II, post-ATP hydrolysis conformation of myosin-II. In doing so, it prevents phosphate release, which precedes actin re-binding in myosin-II motors after their recovery stroke [37,38]. Allosteric inhibitors of kinesin-5 interfere with the normal movements of switch I to induce a conformation that is tightly ADP bound. Inhibited kinesin-5 motors can bind to microtubules, but they are not stimulated by microtubules to release ADP [39–41]. An inhibitor of another mitotic kinesin family member, CENP-E, binds in a structurally similar location to the allosteric kinesin-5 inhibitors; but instead of stabilizing an ADP-bound state, the CENP-E inhibitor stabilizes a tightly microtubule-bound ADP·phosphate state of the motor [42]. The only difference is that the CENP-E inhibitor stabilizes a "switched-on" conformation, while the kinesin-5 inhibitors stabilize a "switched-off" conformation of their respective motor targets. In all cases, switch I movement is disrupted by these inhibitors. Figure 2.2 shows kinesin-5 bound to the inhibitor monastrol (**a**, PDB#1Q0B [40]), myosin-II bound to blebbistatin (**b**, PDB#1YV3 [41]), and the two structures overlaid (**c**). Note that the inhibitors are positioned on either side of the switch I element (blue); for kinesin-5 the inhibitor is in front and for myosin-II the inhibitor is behind switch I.

These examples are just a few vignettes of regulatory and small-molecule inhibitory mechanisms targeting motor proteins. In G proteins, switch I can be a target for direct binding of an external regulatory partner. In kinesin-1, switch I may be a target for auto-regulation of the motor's enzymatic activity by its own C-terminal tail domain. The movement of switch I can also be allosterically blocked by small-molecule inhibitors of both kinesins and myosins. These examples are by no means comprehensive, but they illustrate the versatility of switch I as a regulatory target.

2.4   Calcium-Binding Regulators of Myosins and Kinesins

Several calcium-dependent regulators of myosins have been described in detail and are well understood. One broad class of these regulators consists of calmodulin or calmodulin-like domains that bind to several unconventional myosins. These proteins bind to their myosin motor targets in the lever arm region that amplifies the ångström-level

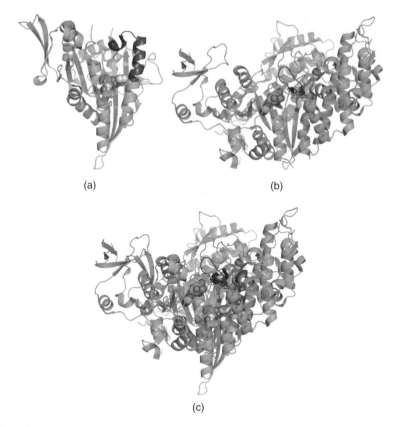

(a)   (b)

(c)

Figure 2.2 Structures of kinesin-5 and myosin-II bound to inhibitors. Refer to Plate 2 for the colored version

conformational changes taking place in the enzymatic myosin head into motor steps that can be as long as 36 nm [44].

Myosin-V is a processive motor that binds to six molecules of calmodulin in its lever arm. This gives the motor exceptionally long "legs," which it uses to take coordinated, processive, 36 nm steps along actin [15]. Two distinct but not mutually exclusive calcium-dependent regulatory mechanisms have been identified for myosin-V, both implicating these calmodulins. First, calcium binding can induce calmodulin molecules to dissociate from the myosin-V lever arm, rendering it flexible. This can shorten the myosin-V powerstroke and decouple the activity of the two heads by altering the flexibility of the connection between them. This regulatory mechanism appears to be shared by myosin-VI, which also uses calmodulin to stabilize its lever arm and coordinate the action of its two heads [45,46]. For myosin-V, however, a second, distinct effect of calcium binding to calmodulin has been identified. Similar to kinesin-1 described above, myosin-V can be auto-regulated by an interaction of its head and tail domains [47,48]. Part of this inter-action involves the hinge region between the myosin-V enzymatic head and calmodulin repeats, and calcium binding to calmodulin may weaken the myosin-V tail-head inter-action, effectively activating the molecule. These two effects are apparently competing:

calcium inhibits myosin-V activity by dissociating calmodulin from its lever arm, but activates it by dissociating the head–tail interaction. The key is the concentration of calmodulin: at low concentrations of calmodulin, calcium will induce slow, uncoordinated myosin-V movement, whereas at high concentrations of calmodulin, calcium will activate myosin-V for cargo movement. These coupled regulatory mechanisms can allow for fine control of myosin-V activity, simply by modulating calcium and calmodulin levels [49].

Calcium-dependent regulators of kinesin motors have also recently been described. A plant kinesin called kinesin calmodulin-binding protein (KCBP) is activated by calcium, and its mechanism has some similarities to those described above for calcium and calmodulin-dependent regulation of myosins. KCBP is a C-terminal kinesin; that is, its neck linker element that amplifies conformational changes within the enzymatic head into movement is C-terminal to its enzymatic head domain. KCBP also has a fascinating auto-regulatory element located at its N-terminus, before the motor domain. This element is termed the "neck mimic," because it binds to the KCBP head in place of the neck linker, displacing it, and then blocks microtubule binding by KCBP [50]. KIC (KCBP-interacting Ca^{2+} binding protein) activates KCBP by binding to the neck mimic and displacing it, so that the actual neck linker can amplify KCBP movement towards the microtubule

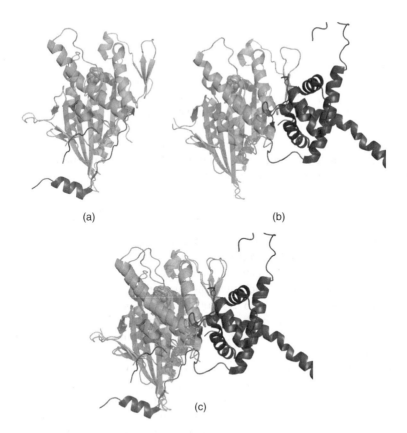

(a)   (b)

(c)

Figure 2.3   Regulation and calcium-dependent activation of KCBP. Refer to Plate 3 for the colored version

minus end [51]. Figure 2.3 shows KCBP with its neck mimic element in a regulatory conformation (**a**, PDB#3COB [52]; KCBP is cyan and the neck mimic is dark blue), then with KIC bound to the neck mimic (**b**, PDB ID# 3H4S [51]; KCBP is green, the neck mimic is dark blue, and KIC is red), and the two structures overlaid (**c**).

Another calcium-dependent kinesin regulatory protein has recently been discovered, although its mechanism of action is unknown. Kinesin-1 is a major mitochondrial transporter, and it is recruited to mitochondria by interacting with a calcium-binding outer mitochondrial membrane protein called Miro (*Mi*tochondrial *Rho* GTPase). Miro contains two GTPase domains flanking two Ca^{2+}-sensing EF hand domains. Calcium-dependent regulation of kinesin-based mitochondrial movement makes functional sense. As the ATP factory of the cell, mitochondria are preferentially localized to areas with high metabolic demand [53]. Inside neurons, these include presynaptic terminals and nodes of Ranvier, where ion pumps consume large quantities of ATP in their effort to restore the ionic environment following an action potential. The mobile versus stationary status of mitochondria is tightly coupled to intracellular calcium concentration $[Ca^{2+}]_i$: in low $[Ca^{2+}]_i$, mitochondria display bidirectional movement; in high $[Ca^{2+}]_i$, such as following depolarization, mitochondria are rendered immobile, strategically halted where they are needed most [54].

Recent work has shown that Ca^{2+}-dependent arrest of mitochondrial transport is mediated by the EF hands of Miro [55–57]. In neurons expressing a defective Miro that cannot bind calcium, Ca^{2+}-induced arrest of mitochondrial movement is effectively blocked and its interaction with kinesin-1 is altered. Conflicting evidence from these papers suggests two incompatible models: either calcium binding to Miro triggers the release of kinesin, or calcium-bound Miro itself binds kinesin and blocks microtubule binding [58]. The latter idea is intriguing by comparison to KCBP and KIC, as the mechanism of Miro-based inhibition of kinesin-1 could be analogous to the mechanism of KIC-based activation of KCBP. The existing data may be revealing what may be a fundamental design principle for calcium-dependent motor protein regulators: that, for reasons we do not yet understand, they tend to target the mechanical amplifier element. At present, this idea is highly speculative and more experimentation is required.

2.5   Phospho-Regulation of Kinesin and Myosin Motors

Phosphorylation is a ubiquitous biological regulatory mechanism, and nature has found several ways to exploit phospho-regulation to control myosin and kinesin motor proteins. Here, phospho-regulatory mechanisms are described that govern the activity and contractile action of myosin-II bipolar thick filaments, as well as the activation and movement of the mitotic kinesin, kinesin-5, to control the length of the mitotic spindle. Several distinct phospho-regulatory mechanisms act on both myosin-V and on kinesin-5 to fine-tune these motors to accomplish their respective tasks of muscle contraction and mitotic spindle elongation.

Smooth-muscle myosin motors are activated by phosphorylation of the regulatory light chain domain (RLC). Interestingly, this phosphorylation event is calcium dependent, and requires calmodulin. The target of this phospho-regulation, the RLC, binds to and stabilizes the lever arm of myosin-II, similar to the calmodulin domains on myosins V and VI. Calcium that enters a muscle cell binds to calmodulin, which then binds to and activates myosin light chain kinase (MLCK). MLCK phosphorylates the regulatory myosin light chain, which relieves auto-regulation of smooth muscle myosin to activate contraction [59,60].

A second myosin-II phospho-regulatory mechanism has been described for *Dictyostelium* myosin-II that does not involve the enzymatic heads at all. Myosin-II mediated contractility depends on the assembly of myosin-II molecules into highly regular bipolar thick filament structures composed of large numbers of myosin-II molecules joined at their C-terminal tails. When myosin heavy chain kinase phosphorylates three sites within *Dictyostelium* myosin-II tails, these thick filament structures fall apart [61,62]. Thus, phospho-regulation can also disrupt myosin-II by abolishing their structural coordination, independent of their enzymatic activity.

Kinesin-5 is in some ways functionally analogous to myosin-II, and its phospho-regulatory mechanisms bear some similarity as well. Kinesin-5 is a homotetrameric kinesin, consisting of two dimeric motor subunits joined at their tails. Like myosin-II, kinesin-5 functions as a bipolar, contractile structure. While myosin-II bipolar thick filaments move along actin crossbridges to drive muscle contraction, kinesin-5 homotetramers drive the separation of interpolar microtubules that control the length of the mitotic spindle.

Similar to myosin-II, kinesin-5 is phospho-regulated. Phospho-regulatory mechanisms have been described that affect the activity of its heads as well as its tails. The mitotic kinase mCDK phosphorylates a conserved threonine residue in the tail domain of all metazoan kinesin-5 motors [63]. This tail phosphorylation is required for kinesin-5 to localize to microtubules and set up the mitotic spindle during prophase [64]. The mechanism by which tail phosphorylation stimulates microtubule-binding is unknown, but *in vitro* studies by Cahu *et al.* [65] demonstrated that phosphorylation of this residue in a headless tetramer increases the affinity of the protein for isolated microtubules, suggesting that the tail may bind directly to the microtubule and stimulate kinesin-5 activation directly by tethering the kinesin-5 enzyme to its effector, in a phosphorylation-dependent manner.

Two different phospho-regulatory mechanisms have been described involving kinesin-5 heads: one in *Drosophila* and one in yeast. Garcia *et al.* [66] have suggested that the mitotic kinase Wee1 phosphorylates the enzymatic heads of the *Drosophila* kinesin-5 homologue, Klp61f. Meanwhile, Chee and Haase [67] described a conserved mCDK phosphorylation site in the heads of Cin8p and Kip1, the yeast kinesin-5 homologues. Intriguingly, these yeast kinesin-5 homologues are not phosphorylated in their tails by mCDK as the metazoan ones are.

The potential involvement of both Wee1 and M-Cdk in kinesin-5 regulation is intriguing, given that these two kinases play key competing roles in regulating cell division. M-Cdk is a serine–threonine kinase that initiates the G2 to M phase transition, and Wee1 phosphorylates and inactivates M-Cdk to prevent premature entry into mitosis [68]. To initiate the transition into M phase, the phosphatase Cdc25 reverses the effects of Wee1 phosphorylation, activating M-Cdk. When a small population of M-Cdk has been activated by this mechanism it feeds back positively on itself in two ways. M-Cdk phosphorylates the phosphatase Cdc25, activating Cdc25 to allow the dephosphorylation and activation of more M-Cdk. Active M-Cdk also phosphorylates Wee1, shutting down its kinase activity. Given this kinase feedback loop, it is tempting to speculate that Wee1 phosphorylation of

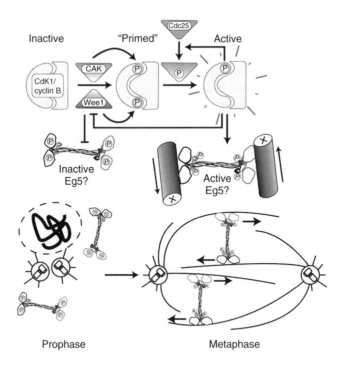

Figure 2.4   Possible counteracting activity of Wee1 and mCDK on kinesin-5

the enzymatic domain may inactivate kinesin-5 activity until the cell commits to mitosis, at which point M-Cdk will strongly activate kinesin-5, stimulating it to set up the mitotic spindle (Figure 2.4). Again, this is speculation and more experimentation is necessary to understand phosphoregulation of kinesin-5 motors.

2.6   Cooperative Action of Kinesin and Myosin Motors as a "Regulator"

Both of the phospho-regulated motors described above, myosin-II and kinesin-5, function in groups to assemble a large-scale contractile apparatus within the cell. Some of the phospho-regulatory mechanisms acting on these motors inhibit their function not by affecting their enzymatic activity, but rather by blocking their ability to assemble into functionally useful structures (bipolar thick filaments for myosin-II and homotetrameric, cross-linked motor-microtubule complexes for kinesin-5). Recent theoretical and experimental work in the motor protein field has identified several emergent properties of cooperating ensembles of motors. These properties can be exploited by regulators, such as the kinases that phosphorylate myosin-II and kinesin-5, for fine control of motor-dependent processes.

Intracellular cargoes are often simultaneously attached to multiple types of molecular motors. Intriguingly for several different types of microtubule-dependent cargoes, it has been shown that elimination of motors of one directionality (i.e., microtubule plus-end-directed kinesins or minus-end-directed dyneins) stops movement in both directions. This behavior has been observed for melanosomes [69], lipid droplets [70], peroxisomes [71,72], mitochondria [73], and RNA cargoes [74]. High-resolution tracking of peroxisome movement has shown that the transitions between plus-end-directed, kinesin-driven movement and minus-end-directed, dynein-driven movement are abrupt and discrete; that is, the motors take turns moving in one direction or the other, and do not compete in a "tug-of war" to move their cargoes [75]. Motors need not necessarily have opposite directionality to "sense" one another in this manner; intraflagellar transport (IFT) particles are moved by kinesin-II and Osm-3, which have the same directionality but velocities that differ by a factor of three. The two motors alternate control of IFT particles during different phases of transport in the flagellum [76]. These results together seem to indicate that different types of motors attached to cargoes are somehow coordinated so that they are able to sense each other's movement.

Both theoretical studies and experiments have explored the question of how different molecular motors coordinate their motility when transporting cargoes together. Theoretical treatments describing the movement of multiple motor systems using transition-rate models have largely recapitulated the behavior of experimental multiple-motor systems without invoking a specific coordinating element between motors. These treatments, as well as some simple experiments, have shown that, when molecular motors are coupled through attachment to a cargo, they can accelerate each others' dissociation from that cargo [72,77–80]. Under such conditions, small changes in the number of actively stepping motors can abruptly alter the motility behavior of a cargo.

The IFT system has certainly provided the most concrete example of how motor cooperativity can be exploited by a regulator. The BBS protein joins kinesin-II and Osm-3 on IFT particles. In the absence of intact BBS, the motility of IFT particles is altered because this coupling is lost. It is likely that other, similar "coupling regulator" proteins exist for biological motors, which may exploit the mechanical tension that can be generated between competing motors to alter the movement of an entire cargo ensemble [76].

2.7 Conclusion

The regulatory mechanisms described above are complex, and often seemingly conflicting. Switch I is a target of multiple, potentially competing regulators, calcium exerts both activating and inhibitory effects on the same motor protein, and phosphorylation by two competing kinases can control motor proteins. These mechanisms, as well as others not covered in this review, seem at first glance to be unnecessarily complex, literally turning our elegant biological motors into Rube Goldberg machines. In each case, however, the complexities of these regulatory pathways fine-tune motor activity for particular functions. It is anticipated that future developments of synthetic motors will create "nano-allosteric" molecular machines, synthetic motors that, like their biological counterparts, sacrifice simplicity for functional tenability. Tunable synthetic motors, like their biological counterparts, will have greater utility and adaptability for multiple applications.

References

[1] Huxley, A.F. and Niedergerke, R. (1954) Structural changes in muscle during contraction; interference microscopy of living muscle fibres. *Nature*, **173** (4412), 971–973.

[2] Huxley, H. and Hanson, J. (1954) Changes in the cross-striations of muscle during contraction and stretch and their structural interpretation. *Nature*, **173** (4412), 973–976.

[3] Huxley, H.E. (1969) The mechanism of muscular contraction. *Science*, **164** (886), 1356–1365.

[4] Lymn, R.W. and Taylor, E.W. (1971) Mechanism of adenosine triphosphate hydrolysis by actomyosin. *Biochemistry*, **10** (25), 4617–4624.

[5] Huxley, H.E. (2004) Fifty years of muscle and the sliding filament hypothesis. *European Journal of Biochemistry*, **271** (8), 1403–1415.

[6] Ma, Y.Z. and Taylor, E.W. (1997) Interacting head mechanism of microtubule-kinesin ATPase. *Journal of Biological Chemistry*, **272** (2), 724–730.

[7] Rice, S., Lin, A.W., Safer, D. *et al*. (1999) A structural change in the kinesin motor protein that drives motility. *Nature*, **402** (6763), 778–784.

[8] Vale, R.D. and Milligan, R.A. (2000) The way things move: looking under the hood of molecular motor proteins. *Science*, **288** (5463), 88–95.

[9] Hancock, W.O. and Howard, J. (1998) Processivity of the motor protein kinesin requires two heads. *Journal of Cell Biology*, **140** (6), 1395–1405.

[10] Hackney, D.D., Stock, M.F., Moore, J., and Patterson, R.A. (2003) Modulation of kinesin half-site ADP release and kinetic processivity by a spacer between the head groups. *Biochemistry*, **42** (41), 12011–12018.

[11] Guydosh, N.R. and Block, S.M. (2006) Backsteps induced by nucleotide analogs suggest the front head of kinesin is gated by strain. *Proceedings of the National Academy of Sciences of the United States of America*, **103** (21), 8054–8059.

[12] Burgess, S.A., Walker, M.L., Sakakibara, H. *et al*. (2003) Dynein structure and power stroke. *Nature*, **421** (6924), 715–718.

[13] Reck-Peterson, S.L., Yildiz, A., Carter, A.P. *et al*. (2006) Single-molecule analysis of dynein processivity and stepping behavior. *Cell*, **126** (2), 335–348.

[14] Carter, A.P., Cho, C., Jin, L., and Vale, R.D. (2011) Crystal structure of the dynein motor domain. *Science*, **331** (6021), 1159–1165.

[15] Rief, M., Rock, R.S., Mehta, A.D. *et al*. (2000) Myosin-V stepping kinetics: a molecular model for processivity. *Proceedings of the National Academy of Sciences of the United States of America*, **97** (17), 9482–9486.

[16] Rock, R.S., Rice, S.E., Wells, A.L. *et al*. (2001) Myosin VI is a processive motor with a large step size. *Proceedings of the National Academy of Sciences of the United States of America*, **98** (24), 13655–13659.

[17] Pechatnikova, E. and Taylor, E.W. (1999) Kinetics processivity and the direction of motion of Ncd. *Biophysical Journal*, **77** (2), 1003–1016.

[18] Foster, K.A. and Gilbert, S.P. (2000) Kinetic studies of dimeric Ncd: evidence that Ncd is not processive. *Biochemistry*, **39** (7), 1784–1791.

[19] Altman, D., Sweeney, H.L., and Spudich, J.A. (2004) The mechanism of myosin VI translocation and its load-induced anchoring. *Cell*, **116** (5), 737–749.

[20] Groen, A.C., Needleman, D., Brangwynne, C. *et al*. (2008) A novel small-molecule inhibitor reveals a possible role of kinesin-5 in anastral spindle-pole assembly. *Journal of Cell Science*, **121** (Pt 14), 2293–2300.

[21] Vale, R.D. (1996) Switches, latches, and amplifiers: common themes of G proteins and molecular motors. *Journal of Cell Biology*, **135** (2), 291–302.

[22] Yount, R.G., Lawson, D., and Rayment, I. (1995) Is myosin a "back door" enzyme? *Biophysical Journal*, **68** (4 Suppl), 44S–47S; discussion 47S–49S.

[23] Naber, N., Minehardt, T.J., Rice, S. *et al*. (2003) Closing of the nucleotide pocket of kinesin-family motors upon binding to microtubules. *Science*, **300** (5620), 798–801.

[24] Sindelar, C.V. and Downing, K.H. (2007) The beginning of kinesin's force-generating cycle visualized at 9-Å resolution. *Journal of Cell Biology*, **177** (3), 377–385.

[25] Ma, Y.Z. and Taylor, E.W. (1995) Mechanism of microtubule kinesin ATPase. *Biochemistry*, **34** (40), 13242–13251.

[26] Kimple, R.J., Kimple, M.E., Betts, L. *et al*. (2002) Structural determinants for GoLoco-induced inhibition of nucleotide release by Gα subunits. *Nature*, **416** (6883), 878–881.

[27] Willard, F.S., Kimple, R.J., and Siderovski, D.P. (2004) Return of the GDI: the GoLoco motif in cell division. *Annual Review of Biochemistry*, **73**, 925–951.

[28] Hollenbeck, P.J. (1989) The distribution, abundance and subcellular localization of kinesin. *Journal of Cell Biology*, **108** (6), 2335–2342.

[29] Cai, D., Hoppe, A.D., Swanson, J.A., and Verhey, K.J. (2007) Kinesin-1 structural organization and conformational changes revealed by FRET stoichiometry in live cells. *Journal of Cell Biology*, **176** (1), 51–63.

[30] Hirokawa, N., Pfister, K.K., Yorifuji, H. *et al.* (1989) Submolecular domains of bovine brain kinesin identified by electron microscopy and monoclonal antibody decoration. *Cell*, **56** (5), 867–878.

[31] Hackney, D.D., Levitt, J.D., and Suhan, J. (1992) Kinesin undergoes a 9 S to 6 S conformational transition. *Journal of Biological Chemistry*, **267** (12), 8696–8701.

[32] Dietrich, K.A., Sindelar, C.V., Brewer, P.D. *et al.* (2008) The kinesin-1 motor protein is regulated by a direct interaction of its head and tail. *Proceedings of the National Academy of Sciences of the United States of America*, **105** (26), 8938–8943.

[33] Hackney, D.D. and Stock, M.F. (2000) Kinesin's IAK tail domain inhibits initial microtubule-stimulated ADP release. *Nature Cell Biology*, **2** (5), 257–260.

[34] Hackney, D.D. and Stock, M.F. (2008) Kinesin tail domains and Mg^{2+} directly inhibit release of ADP from head domains in the absence of microtubules. *Biochemistry*, **47** (29), 7770–7778.

[35] Wong, Y.L., Dietrich, K.A., Naber, N. *et al.* (2009) The kinesin-1 tail conformationally restricts the nucleotide pocket. *Biophysical Journal*, **96** (7), 2799–2807.

[36] Kaan, H.Y., Hackney, D.D., and Kozielski, F. (2011) The structure of the kinesin-1 motor-tail complex reveals the mechanism of autoinhibition. *Science*, **333** (6044), 883–885.

[37] Kovács, M., Tóth, J., Hetényi, C. *et al.* (2004) Mechanism of blebbistatin inhibition of myosin II. *Journal of Biological Chemistry*, **279** (34), 35557–35563.

[38] Takacs, B., Billington, N., Gyimesi, M. *et al.* (2010) Myosin complexed with ADP and blebbistatin reversibly adopts a conformation resembling the start point of the working stroke. *Proceedings of the National Academy of Sciences of the United States of America*, **107** (15), 6799–6804.

[39] Maliga, Z., Kapoor, T.M., and Mitchison, T.J. (2002) Evidence that monastrol is an allosteric inhibitor of the mitotic kinesin Eg5. *Chemistry and Biology*, **9** (9), 989–996.

[40] Yan, Y., Sardana, V., Xu, B. *et al.* (2004) Inhibition of a mitotic motor protein: where, how, and conformational consequences. *Journal of Molecular Biology*, **335** (2), 547–554.

[41] Larson, A.G., Naber, N., Cooke, R. *et al.* (2010) The conserved L5 loop establishes the pre-powerstroke conformation of the kinesin-5 motor, eg5. *Biophysical Journal*, **98** (11), 2619–2627.

[42] Wood, K.W., Lad, L., Luo, L. *et al.* (2010) Antitumor activity of an allosteric inhibitor of centromere-associated protein-E. *Proceedings of the National Academy of Sciences of the United States of America*, **107** (13), 5839–5844.

[43] Allingham, J.S., Smith, R., and Rayment, I. (2005) The structural basis of blebbistatin inhibition and specificity for myosin II. *Nature Structural and Molecular Biology*, **12** (4), 378–379.

[44] Hodge, T. and Cope, M.J. (2000) A myosin family tree. *Journal of Cell Science*, **113** (Pt 19), 3353–3354.

[45] Morris, C.A., Wells, A.L., Yang, Z. *et al.* (2003) Calcium functionally uncouples the heads of myosin VI. *Journal of Biological Chemistry*, **278** (26), 23324–23330.

[46] Bahloul, A., Chevreux, G., Wells, A.L. *et al.* (2004) The unique insert in myosin VI is a structural calcium–calmodulin binding site. *Proceedings of the National Academy of Sciences of the United States of America*, **101** (14), 4787–4792.

[47] Liu, J., Taylor, D.W., Krementsova, E.B. *et al.* (2006) Three-dimensional structure of the myosin V inhibited state by cryoelectron tomography. *Nature*, **442** (7099), 208–211.

[48] Thirumurugan, K., Sakamoto, T., Hammer III,, J.A. *et al.* (2006) The cargo-binding domain regulates structure and activity of myosin 5. *Nature*, **442** (7099), 212–215.

[49] Krementsov, D.N., Krementsova, E.B., and Trybus, K.M. (2004) Myosin V: regulation by calcium, calmodulin, and the tail domain. *Journal of Cell Biology*, **164** (6), 877–886.

[50] Vinogradova, M.V., Reddy, V.S., Reddy, A.S. *et al.* (2004) Crystal structure of kinesin regulated by Ca^{2+}-calmodulin. *Journal of Biological Chemistry*, **279** (22), 23504–23509.

[51] Vinogradova, M.V., Malanina, G.G., Reddy, A.S. *et al.* (2009) Structure of the complex of a mitotic kinesin with its calcium binding regulator. *Proceedings of the National Academy of Sciences of the United States of America*, **106** (20), 8175–8179.

[52] Vinogradova, M.V., Malanina, G.G., Reddy, V.S. *et al*. (2008) Structural dynamics of the microtubule binding and regulatory elements in the kinesin-like calmodulin binding protein. *Journal of Structural Biology*, **163** (1), 76–83.

[53] Chang, D.T., Honick, A.S., and Reynolds, I.J. (2006) Mitochondrial trafficking to synapses in cultured primary cortical neurons. *Journal of Neuroscience*, **26** (26), 7035–7045.

[54] Yi, M., Weaver, D., and Hajnóczky, G. (2004) Control of mitochondrial motility and distribution by the calcium signal: a homeostatic circuit. *Journal of Cell Biology*, **167** (4), 661–672.

[55] Saotome, M., Safiulina, D., Szabadkai, G. *et al*. (2008) Bidirectional Ca^{2+}-dependent control of mitochondrial dynamics by the Miro GTPase. *Proceedings of the National Academy of Sciences of the United States of America*, **105** (52), 20728–20733.

[56] Macaskill, A.F., Rinholm, J.E., Twelvetrees, A.E. *et al*. (2009) Miro1 is a calcium sensor for glutamate receptor-dependent localization of mitochondria at synapses. *Neuron*, **61** (4), 541–555.

[57] Wang, X. and Schwarz, T.L. (2009) The mechanism of Ca^{2+}-dependent regulation of kinesin-mediated mitochondrial motility. *Cell*, **136** (1), 163–174.

[58] Cai, Q. and Sheng, Z.H. (2009) Moving or stopping mitochondria: Miro as a traffic cop by sensing calcium. *Neuron*, **61** (4), 493–496.

[59] Olney, J.J., Sellers, J.R., and Cremo, C.R. (1996) Structure and function of the 10 S conformation of smooth muscle myosin. *Journal of Biological Chemistry*, **271** (34), 20375–20384.

[60] Wendt, T., Taylor, D., Trybus, K.M., and Taylor, K. (2001) Three-dimensional image reconstruction of dephosphorylated smooth muscle heavy meromyosin reveals asymmetry in the interaction between myosin heads and placement of subfragment 2. *Proceedings of the National Academy of Sciences of the United States of America*, **98** (8), 4361–4366.

[61] Liang, W., Warrick, H.M., and Spudich, J.A. (1999) A structural model for phosphorylation control of *Dictyostelium* myosin II thick filament assembly. *Journal of Cell Biology*, **147** (5), 1039–1048.

[62] Hostetter, D., Rice, S., Dean, S. *et al*. (2004) *Dictyostelium* myosin bipolar thick filament formation: importance of charge and specific domains of the myosin rod. *PLoS Biology*, **2** (11), e356.

[63] Sawin, K.E., LeGuellec, K., Philippe, M., and Mitchison, T.J. (1992) Mitotic spindle organization by a plus-end-directed microtubule motor. *Nature*, **359** (6395), 540–543.

[64] Sharp, D.J., McDonald, K.L., Brown, H.M. *et al*. (1999) The bipolar kinesin, KLP61F, cross-links microtubules within interpolar microtubule bundles of *Drosophila* embryonic mitotic spindles. *Journal of Cell Biology*, **144** (1), 125–138.

[65] Cahu, J., Olichon, A., Hentrich, C. *et al*. (2008) Phosphorylation by Cdk1 increases the binding of Eg5 to microtubules in vitro and in *Xenopus* egg extract spindles. *PLoS ONE* **3**(12), e3936.

[66] Garcia, K., Stumpff, J., Duncan, T., and Su, T.T. (2009) Tyrosines in the kinesin-5 head domain are necessary for phosphorylation by Wee1 and for mitotic spindle integrity. *Current Biology*, **19** (19), 1670–1676.

[67] Chee, M.K. and Haase, S.B. (2010) B-cyclin/CDKs regulate mitotic spindle assembly by phosphorylating kinesins-5 in budding yeast. *PLoS Genetics*, **6** (5), e1000935.

[68] Parker, L.L. and Piwnica-Worms, H. (1992) Inactivation of the p34cdc2–cyclin B complex by the human WEE1 tyrosine kinase. *Science*, **257** (5078), 1955–1957.

[69] Gross, S.P., Tuma, M.C., Deacon, S.W. *et al*. (2002) Interactions and regulation of molecular motors in *Xenopus* melanophores. *Journal of Cell Biology*, **156** (5), 855–865.

[70] Gross, S.P., Welte, M.A., Block, S.M., and Wieschaus, E.F. (2002) Coordination of opposite-polarity microtubule motors. *Journal of Cell Biology*, **156** (4), 715–724.

[71] Kim, H., Ling, S.C., Rogers, G.C. *et al*. (2007) Microtubule binding by dynactin is required for microtubule organization but not cargo transport. *Journal of Cell Biology*, **176** (5), 641–651.

[72] Ally, S., Larson, A.G., Barlan, K. *et al*. (2009) Opposite-polarity motors activate one another to trigger cargo transport in live cells. *Journal of Cell Biology*, **187** (7), 1071–1082.

[73] Pilling, A.D., Horiuchi, D., Lively, C.M., and Saxton, W.M. (2006) Kinesin-1 and dynein are the primary motors for fast transport of mitochondria in *Drosophila* motor axons. *Molecular Biology of the Cell*, **17** (4), 2057–2068.

[74] Ling, S.C., Fahrner, P.S., Greenough, W.T., and Gelfand, V.I. (2004) Transport of *Drosophila* fragile X mental retardation protein-containing ribonucleoprotein granules by kinesin-1 and cytoplasmic dynein. *Proceedings of the National Academy of Sciences of the United States of America*, **101** (50), 17428–17433.

[75] Kural, C., Kim, H., Syed, S. *et al*. (2005) Kinesin and dynein move a peroxisome in vivo: a tug-of-war or coordinated movement? *Science*, **308** (5727), 1469–1472.

[76] Pan, X., Ou, G., Blacque, O.E. *et al*. (2006) Mechanism of transport of IFT particles in *C. elegans* cilia by the concerted action of kinesin-II and OSM-3 motors. *Journal of Cell Biology*, **174** (7), 1035–1045.

[77] Klumpp, S. and Lipowsky, R. (2005) Cooperative cargo transport by several molecular motors. *Proceedings of the National Academy of Sciences of the United States of America*, **102** (48), 17284–17289.

[78] Muller, M.J., Klumpp, S., and Lipowsky, R. (2008) Tug-of-war as a cooperative mechanism for bidirectional cargo transport by molecular motors. *Proceedings of the National Academy of Sciences of the United States of America*, **105** (12), 4609–4614.

[79] Larson, A.G., Landahl, E.C., and Rice, S.E. (2009) Mechanism of cooperative behaviour in systems of slow and fast molecular motors. *Physical Chemistry Chemical Physics*, **11** (24), 4890–4898.

[80] Jamison, D.K., Driver, J.W., Rogers, A.R. *et al*. (2010) Two kinesins transport cargo primarily via the action of one motor: implications for intracellular transport. *Biophysical Journal*, **99** (9), 2967–2977.

3

Neuromechanics: The Role of Tension in Neuronal Growth and Memory

Wylie W. Ahmed[1], Jagannathan Rajagopalan[2], Alireza Tofangchi[1], and Taher A. Saif[1]

[1]*University of Illinois, USA*
[2]*Arizona State University, USA*

3.1 Introduction

In recent years it has become increasingly evident that mechanical stimuli play an important role in the differentiation, growth, development, and motility of cells [1–6]. Cells sense and respond to cues from their mechanical microenvironment as well as externally applied mechanical stimuli. Most studies of cell mechanics have focused on cell types that are obviously subjected to mechanical forces during everyday function, such as: smooth muscle and endothelial cells, which experience strain as blood vessels expand and contract [7]; cardiac and skeletal muscle cells, which undergo contraction to pump blood and generate movement [8]; and epithelial cells throughout the body, such as in the lungs during breathing [9]. However, studies of mechanical forces during development and embryogenesis have highlighted their importance in all types of cells [10–15].

Growing experimental evidence suggests that mechanical tension plays a significant role in determining the growth, guidance, and function of neurons [16–19]. Recent developments in experimental techniques have made quantitative studies at the level of individual cells possible [20–22]. Here, we discuss cellular neuromechanics, which focuses on the mechanics of neurons at the cellular level and their function. The purpose of this chapter is to present a brief review of some significant studies supporting the role of mechanical tension in neuronal growth and its implications in memory. We begin with an introduction to the structure and function of the neuron (Sections 3.1.1–3.1.3). Then

Nano and Cell Mechanics: Fundamentals and Frontiers, First Edition. Edited by Horacio D. Espinosa and Gang Bao.
© 2013 John Wiley & Sons, Ltd. Published 2013 by John Wiley & Sons, Ltd.

we discuss experimental measurements (*in vitro* and *in vivo*) of neuronal force and its role in structural development (Section 3.2). Next, we explore the role of tension in neuronal function, including neurotransmission and vesicle dynamics (Section 3.3). In Section 3.4 we discuss models of the mechanical behavior of axons to interpret their behavior. Finally, we conclude the review by highlighting some unanswered questions in cellular neuromechanics (Section 3.5).

3.1.1 What is a Neuron?

Neurons are the basic communication element in the nervous system of nearly all animals. They communicate with other neurons and other types of cells via chemical and/or electrical signaling. For instance:

- our thoughts are processed by a complex network of neurons in our brains;
- our vision is mediated by light-sensitive optical neurons in our eyes;
- we move our bodies by actuating muscles signaled by motor neurons; and
- we are able to sense using sensory neurons wired throughout our bodies.

Many specialized types of neurons exist; however, most neurons share common biological structures [23]. The neuron is composed of four main parts: the soma (cell body), the dendrite, the axon, and the synapse, as shown in Figure 3.1. The soma of a neuron is very similar to that of other cells. The majority of the cell body is the nucleus, which holds the genetic information in the form of DNA. The cell body also contains the usual organelles, such as mitochondria, endoplasmic reticulum, ribosomes, the Golgi complex, and so on [24]. The dendrites of a neuron are a highly branched network of processes originating from the cell body. The branches contain many small protrusions, called dendritic spines, which function as locations for synaptic input from other cells. Through the dendritic network, a single neuron can receive signals from thousands of

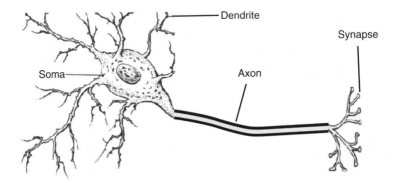

Figure 3.1 Schematic of a neuron showing the cell body (soma), dendrites, axon, and synapses. Neurons receive synaptic input from other cells through their dendrites and send outgoing signals along their axon to communicate with other cells via synapses. Image modified from www.mindcreators.com

cells simultaneously. The axon is a long, thin, tube-like structure that originates from the soma and can extend for meters to terminate at a synapse. They usually do not branch along their length (except very near their terminals) and maintain an approximately constant diameter. Axons are able to rapidly transmit electrical signals (action potentials) from soma to synapse and this process is crucial to neuronal signaling. Like most cells, neurons have an underlying support structure called the cytoskeleton consisting of actin, neurofilaments, and microtubules, as shown in Figure 3.2. Actin is an important structure known to regulate a variety of processes including the growing tips of axons, membrane shape, and dendritic spines [4,25–27]. Neurofilaments are the least understood cytoskeletal element, but it is known that their organization is related to disease pathogenesis [28]. Microtubules function as support structures as well as tracks for molecular-motor-based transport of organelles and vesicles [29,30]. Microtubules have a polar structure and their organization in neurons is specialized. Axons of vertebrate neurons exhibit "plus-end out" microtubules, meaning that the plus-end of the microtubule is pointing towards the distal end of the axon (synapse), and vertebrate dendrites usually have mixed polarity [31,32].

The synapse is a highly specialized structure, which serves as a communication element between cells (Figure 3.2). It is one of the most important elements of a neuron and its structure and function have been the focus of cellular neurobiology. There are two kinds

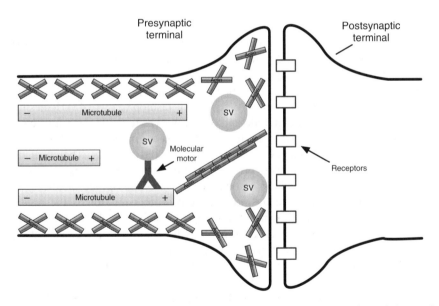

Figure 3.2 The subcellular structure of a neuron is composed of actin, microtubules, and neurofilaments (not shown). Microtubules serve as structural support and as the main tracks for molecular-motor-based transport of organelles and vesicles. Actin serves as structural support in advancing growth cones, as well as being involved in signaling processes, including tethering synaptic vesicles (SV) at the presynaptic terminal for neurotransmission. Synaptic vesicles are released from the presynaptic terminal and activate receptors on the postsynaptic cell for signal transmission

of synapses: chemical and electrical. Here we will focus on chemical synapses because they are the most well-studied. Chemical synapses are composed of distinct presynaptic and postsynaptic sides that are separated by a 20–30 nm gap called the synaptic cleft. The presynaptic terminal is the swollen axon terminal that contains neurotransmitter vesicles to be released to elicit a change in the postsynaptic cell. Understanding synaptic function is a key goal of modern neuroscience. Malfunctioning synapses are associated with a large number of debilitating diseases, such as Parkinson's disease, depression, and schizophrenia [33]. It is widely believed that changes in synaptic properties are a key factor in determining the plasticity of the nervous system and the processes of learning and memory [23,34–36].

3.1.2 How Does a Neuron Function?

The primary function of a neuron is to receive, integrate, and transmit signals. Supporting this function are the processes of neuronal transport, electrical signaling, and synaptic plasticity.

Neuronal transport involves the directed movement of organelles and vesicles throughout the intracellular space of the neuron. Since an axon can be over a meter in length, a protein or molecule produced in the cell body would require several days to diffuse to the synapse. To distribute components along axons and dendrites, neurons use a system of directed transport, which utilizes molecular motors to transport cargo along microtubules. Molecular motors are a large family of motor proteins that use ATP as an energy source to drive movement. The molecular motors that function as transporters on cytoskeletal elements are myosin, kinesin, and dynein. Myosin interacts with actin filaments to allow short-range transport and contractile force generation [37–40]. Kinesin and dynein are the primary molecular motors involved in neuronal transport and function by moving along microtubule structures in a polarity-dependent fashion [29,41–43]. Kinesin moves towards the plus-end of the microtubule and dynein moves towards the minus-end. Since the plus-ends of the microtubules point towards the distal end of the axon, kinesin transports molecules towards the synapse and dynein towards the cell body. Therefore, different types of cargoes, such as organelles, proteins, neurotransmitter vesicles, and so on, can be transported throughout the cell by associating with the appropriate motor.

One function that is crucial to the process of neurotransmission is the transport of vesicles containing neurotransmitters to the synapse, where they cluster at regions called active zones. These vesicles are called synaptic vesicles since they cluster at, and are released from, the presynaptic terminal. Synaptic vesicles are transported back and forth along the axon to maintain a steady pool of neurotransmitters ready to be released to communicate with the postsynaptic cell [44,45].

Neurons rapidly send messages down their axons. Like all other cells, the neuron is enclosed by a plasma membrane, which is an electrically insulating barrier that separates the inside of the neuron from the extracellular space. This plasma membrane prevents passive diffusion of charged ions in and out of the cell and maintains a potential difference across the membrane, which is important for signaling. Ion channels in the membrane have selective permeability to different ions, and this results in an unequal distribution of electrical charge (carried by ions) across the membrane. The primary ions that determine the membrane potential are K^+, Na^+, Cl^-, and organic ions (A^-). Of these ions, Na^+ and Cl^- are more concentrated outside the cell and K^+ and A^- are more concentrated inside. The membrane potential of a neuron at rest is dominated by K^+ channels, and thus the resting potential can be calculated to be -70 mV [36]. When a nerve cell is at rest, the steady influx of Na^+ is balanced by the steady efflux of K^+, so that the membrane potential is constant.

During signal transmission, a depolarizing stimulus above a certain threshold causes large and rapid change in membrane potential, resulting in the propagation of an action potential down the axon from the cell body to the axon terminal. The disturbance in membrane potential rapidly opens voltage-gated Na^+ channels, which increases membrane permeability to Na^+ and leads to a net influx of positive charge. As the positive charge increases, more Na^+ channels open and this positive feedback system quickly drives the system towards the Na^+ equilibrium potential of $+55$ mV (the peak of the action potential). The action potential is terminated by inactivation of Na^+ channels and the opening of voltage-gated K^+, leading to efflux of K^+. Thus, since voltage-gated channels of Na^+ open faster than K^+ channels do, a depolarization of the cell results in rapid depolarization and repolarization process known as the action potential [36]. The action potential is considered an "all-or-none" event, which means that the depolarizing stimulus must be above a certain threshold to elicit a response. However, the amplitude of the action potential is fixed and does not reflect the amplitude of the depolarizing stimuli.

The transmission of an action potential along an axon results in many processes, including the release of neurotransmitters at the synapse. When the neuronal membrane becomes depolarized, cytosolic Ca^{2+} levels increase due to influx through voltage-gated ion channels. The increased level of intracellular Ca^{2+} activates exocytosis of vesicles, which results in neurotransmitter release from the presynaptic terminal to the postsynaptic cell [46]. Additionally, action potentials can induce release of neuropeptides that diffuse into the extracellular environment for long-range communication with distant cells [47].

Synaptic plasticity is defined loosely as synaptic change as a function of activity. This synaptic change results in modulation of synaptic efficiency in terms of neurotransmission and is believed to be the basis of learning and memory [34]. When an organism develops, its genetic map ensures that there are fixed networks of neurons with invariant connections. Although the physical connections are predetermined, the precise strength of the synapses is not specified genetically. Rather, experiences tune the strength and effectiveness of the synapses [48]. One instance of synaptic plasticity is observed when the amount of neurotransmitter release in response to electrical stimulation changes due to activity. An example is when a neuron is stimulated, an action potential is elicited that leads to exocytic vesicle fusion and neurotransmitter release. A repetitive sequence of stimulation may result in increased sensitivity (potentiation) of the synapse; that is, the the synapse "remembers" its usage in the past. As a result, the synapse can release more neurotransmitters in response to the same stimulus. And vice versa: if the neuron becomes less sensitive (depression), then the number of neurotransmitters released will decrease. This seemingly simple change in sensitivity or the strength, known as synaptic plasticity, is mediated by changes in structural, biochemical, and electrical changes of the synapse [49–53]. The concept of synaptic plasticity, originally proposed by Ramon y Cajal during the early twentieth century [36], forms the basis of memory and learning, and hence is a hallmark of neuroscience. It involves the clustering of neurotransmitter vesicles at the presynaptic terminal, without which neurotransmission or change of synaptic strength would not be possible.

3.1.3 How Does a Neuron Grow?

Neurons are unique, since, unlike most cells, terminally differentiated neurons do not divide. Instead, many neurons will increase their size dramatically to accommodate development and growth of an organism. Axonal growth can be simplified into two stages: pathfinding and towed growth (Figure 3.3). During development, a neuron must extend an axon to reach its synaptic target, and this process is known as pathfinding. The axon terminal exhibits an enlarged and spread-out morphology, called the growth cone, which explores the extracellular environment in search for guidance cues to reach its postsynaptic partner [17]. As neuronal growth cones navigate, they apply traction forces to the underlying substrate to generate motion [54]. These traction forces applied by the growth cone generate a tension along the axon, which results in growth and elongation. By this method, the growth cone continuously elongates the axon by pulling until it reaches its final destination. Once the growth cone has reached its target, it structurally and biochemically differentiates into a synapse. After a synapse has formed (synaptogenesis), the axon is anchored between the cell body and the presynaptic terminal. As the embryo develops further, and increases its size, the synapse is stretched or towed outward, and the axon undergoes towed growth to accommodate expansion of the organism.

This dependence of extension on tension provides a plausible mechanism to regulate neuronal growth during development [55,56]. For example, if tension does not induce growth, then, as the organism grows, tension in the axon will continue to increase or

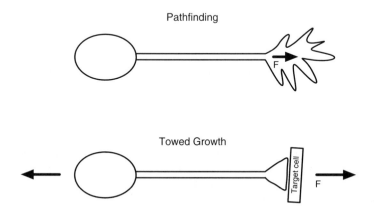

Figure 3.3 Neuronal growth can be divided into two stages. (1) Pathfinding occurs when the growth cone of a growing axon explores the extracellular environment in search for its postsynaptic partner. During this stage the growth cone applies traction forces to the underlying substrate to generate tension along the axon and stimulate elongation and growth. (2) Towed growth occurs after a synapse has formed (post-synaptogenesis), and the axon is stretched to accommodate expansion of the embryo during growth

the axon would thin significantly due to viscoelastic extension. Thus, it is reasonable to hypothesize that axons grow in response to tension; and in order to avoid uncontrolled growth, there must be a threshold above which growth is initiated. Similarly, axons may have a lower force threshold, below which axonal tension is generated. This lower force threshold would give rise to a rest tension in axons. These upper and lower force thresholds may regulate axonal elongation during growth of an organism and axonal retraction during specification of neural networks, respectively. In between the lower and upper thresholds there may exist a regime of passive axonal behavior. This regime could provide a buffer between stimulating force generation and relaxation in the axon induced by normal movement. Later, it will be shown that such a tension might also be essential for clustering of neurotransmitter vesicles at the synapse, and hence for their core functionality.

3.2 Tension in Neuronal Growth

The concept of mechanical tension in neuronal growth is not new. Qualitative observations were reported in 1945, when Weiss observed that nerve tissue advancing along grooves generated tension and distorted the nerve explants [57]. Decades later, Bray showed that growth cones of cultured neurons exert mechanical tension and that there was a correlation between the tension in a neurite and its diameter [58]. More importantly, these experiments showed that the direction of growth cone advance (and hence neurite growth) was dictated mainly by the tension existing in the neurite. This suggested that mechanical tension, and not the growth cone itself, was the primary determinant of neurite growth [58]. A subsequent study by Bray [59] showed that externally applied tension can indeed both initiate new neurites (from rounded cell bodies) and elongate existing neurites. Furthermore, the ultrastructure of the neurites initiated by external tension was identical to spontaneously generated (growth-cone-mediated) neurites, with abundant longitudinally aligned microtubules and neurofilaments. This observation was also confirmed in a later study by Zheng *et al.*, who showed that tension-initiated neurites exhibit rapid and normal microtubule assembly [60].

Based on these observations it was postulated that tension was a primary signal for neuronal growth during normal development, with the tension being supplied by the locomotory activity of the growth cone before synaptogenesis and later by the movement of the innervated tissues in a growing organism. While it was fairly obvious that the moving tissues would exert tension on the neurites, the question as to whether growth cones applied a similar pulling force on neurites was unclear. This question was clarified by Lamoureux *et al.* [54], who showed that an increase in neurite tension was directly correlated to growth cone advance and apparent neurite growth *in vitro*. In contrast, no increase in tension was seen in neurites with stationary growth cones. These studies established that mechanical tension is essential for neurite initiation and growth *in vitro* and, quite likely, *in vivo* as well.

3.2.1 In Vitro *Measurements of Tension in Neurons*

The first quantitative measurements of tension in cultured neurites were performed by Dennerll *et al.* [61], who examined the mechanical properties of PC-12 neurites.

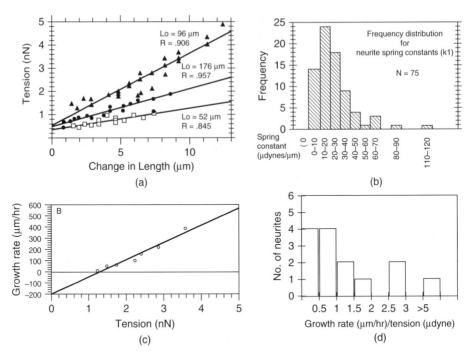

Figure 3.4   *In vitro* measurements of force in neurons. (a) Typical plots of neurite axial tension versus change in neurite length showing a linear force–displacement relationship. Each point represents a single "pull" and the *y*-intercept represents a rest tension. (b) Frequency distribution for neurite spring constants showing a large range of stiffness (note 1 μdyn = 10 pN). (c) Growth rate of a neurite as a function of applied tension. The zero-growth intercept shows the tension threshold to induce elongation. (d) Frequency distribution of tension sensitivity of neurite growth measured in terms of growth rate per unit tension [62]. Adapted from [61] and [62]

They applied force on the neurites with calibrated glass needles and measured their force–displacement characteristics. Their measurements showed a linear force–displacement relationship for small, rapid distensions and the presence of a positive rest tension (300–400 pN) in the neurites (Figure 3.4). In addition, their experiments revealed a complementary force interaction, with microtubules under compression resisting the tension exerted by the actin network. This interaction appeared to dictate neurite shape and size: an increase in the compressive load on microtubules led to their depolymerization and consequent retraction of the neurite. Along similar lines, it was suggested that a reduction in compressive load on microtubules (by shifting of tension to the substrate by an advancing growth cone) could lead to their assembly and, hence, neurite elongation.

In a subsequent study using similar techniques, Dennerll *et al.* uncovered three distinct phases of response to tension in PC-12 neurites and chick sensory neurons [55]. When the tension in the neurites was reduced below a lower threshold they actively contracted to restore tension, sometimes to a value above their resting tension. It has been suggested that the ability to actively develop tension and contract could provide a potential mechanism for *in vivo* axonal retraction from neuromuscular junctions [63–65], which results in a pattern of innervation where only one motor neuron synapses with each skeletal muscle fiber. On the other hand, when the applied tension exceeded an upper threshold (1 nN), the neurites grew in response, with the growth rate being proportional to the applied tension. For intermediate values of applied tension the neurites behaved as passive viscoelastic solids; that is, an initial elastic response followed by force relaxation to a steady-state value.

Following the work of Dennerll *et al.*, Zheng *et al.* examined the relationship between applied tension and neuronal growth rates in greater detail [62]. They applied force on chick sensory neurons in increasing steps, holding the force fixed for 30–60 min in each step. They found a linear relationship between applied force (in excess of a tension threshold) and growth rate (Figure 3.4) and a surprisingly high sensitivity ($150 \, \mu\text{m} \, \text{h}^{-1} \, \text{nN}^{-1}$) of growth rate to tension. They reasoned that the linear relationship between growth rate and tension provides a simple control mechanism for axons to accommodate tissue expansion in growing animals that consistently maintain a moderate rest tension on axons. Furthermore, the force required to initiate new neurites was found to be similar to the tension threshold for elongating neurites, suggesting that the basic mechanism underlying both initiation and elongation of neurites was similar.

3.2.2   In Vivo *Measurements of Tension in Neurons*

As described in Section 3.2.1, numerous studies have revealed the important role of mechanical tension in the initiation, development, elongation, and retraction of neurites *in vitro*. A similar role has been long suggested for mechanical forces *in vivo*. For example, Van Essen [66] hypothesized that tension in axons may underlie many aspects of morphogenesis of the brain, especially of the cortical regions of the brain. As a case in point, it was suggested that the folding of the cerebral cortex is due to the tension exerted by axons that connect relatively distant regions of the brain and that the folding minimizes the communication time between interconnected brain regions. While experiments examining the *in vivo* mechanical behavior of neurons have been scarce, two recent studies have provided evidence of the role of forces in neuronal development and function *in vivo*.

Experiments by Siechen *et al.* [18] (discussed in greater detail in Section 3.3) have shown that vesicle clustering in the presynaptic terminal of the neuromuscular junction in *Drosophila* embryos is dependent on mechanical tension in the axons. That study also provided preliminary evidence for the presence of a rest tension and viscoelastic behavior in *Drosophila* axons.

Following this work, Rajagopalan *et al.* examined the *in vivo* mechanical response of *Drosophila* neurons using high-resolution micromechanical force sensors (Figure 3.5) [67]. They found that:

1. *Drosophila* axons have a rest tension in the range 1–13 nN.
2. In response to fast deformation, axons behave like elastic springs, showing a linear force–deformation response, which is followed by force relaxation to a steady-state value after 15–30 min.

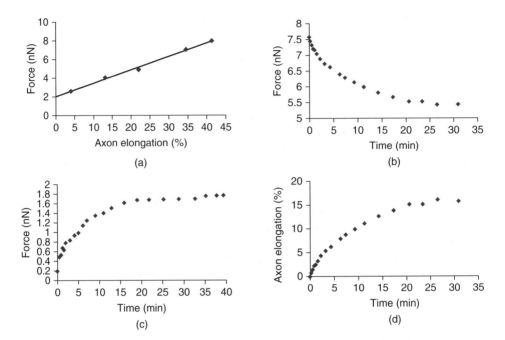

Figure 3.5 *In vivo* measurements of force in neurons. (a) Force–deformation response of an axon during fast stretch showing a linear response. Note the presence of a rest tension (∼2 nN) at zero elongation. (b) Force relaxation over time for the axon shown in (a). The force in the axon decays exponentially to a steady state over approximately 20 min. (c) Force buildup after unloading for axon shown in (a). The force increases exponentially to a value close to the initial rest tension over approximately 20 min. (d) Elongation of the axon during force decay shown in (b). Adapted from [67]

3. When the applied deformation is sufficiently large, the axons adopt a slack appearance upon force removal. However, the axons tauten and build up the tension, often to a level close to their rest tension, in a period of 15–60 min. In other words, the axons actively generate force and contract to restore tension.

These observations of *in vivo* neuronal mechanical behavior were remarkably similar to those from *in vitro* studies, suggesting that mechanical forces could prominently influence neuronal growth and function *in vivo*. Furthermore, Rajagopalan *et al.* [67] uncovered two other interesting facets of axonal response to applied tension. The axons that underwent large force relaxation (80–90%) became considerably longer and thinner during the process. More remarkably, axons that showed low relaxation (30% or less) became stronger (with their stiffness doubling in some cases) even though their length increased during force relaxation (Figure 3.6).

Both these observations, however, are consistent with the observations of Lamoureux *et al.*, who showed that *in vitro* axonal growth proceeds through a combination of steps: lengthening by viscoelastic stretching and intercalated addition of material [68]. In their study, during viscoelastic stretching, a noticeable thinning of the axons was observed,

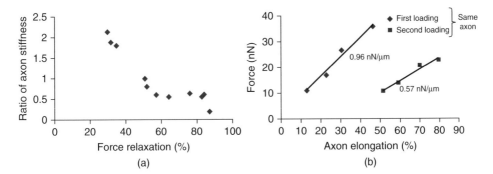

Figure 3.6 Stiffening of *in vivo* axons. (a) Ratio of the stiffness of the axons during the first and second loadings plotted as a function of their force relaxation after the first loading. Axons that show a large force relaxation show diminished stiffness during the second loading, and vice versa. (b) Force–deformation response for the first and second loadings of an axon that underwent a large force relaxation after the first loading. Adapted from [67]

but the axons eventually regained their thickness through material addition. On the other hand, material addition to the axon (and an increase in axon diameter) was also found to precede lengthening in both spontaneously growing (growth-cone-mediated) and towed axons. If this increase in axonal diameter is sufficiently large then the axon can become stiffer in spite of an increase in length as seen in the *in vivo* study of Rajagopalan *et al.* [67]. In other words, not only are the mechanical responses of axons *in vivo* and *in vitro* remarkably similar, but the mode through which tension induces axonal growth also appears identical. Thus, the force response and sensitivity seem to be highly conserved properties of neurons.

In this context, it is worth noting the observations of Pfister *et al.*, who found that integrated axon tracts could be stretched up to 8 mm per day to lengths over 10 cm without disruption [69]. During this elongation process the axons increased in diameter by 35% and maintained normal density of cytoskeletal structures and organelles, and also generated normal action potentials [70]. For a more comprehensive review of the role of force in axonal elongation, we suggest a recent article by Suter and Miller [71].

3.2.3 Role of Tension in Structural Development

As discussed in the previous sections, neurons actively regulate their tension both *in vivo* and *in vitro* and tension stimulates both the initiation and elongation of neurites. But in addition to regulating neurite initiation, mechanical tension has been shown to specify axonal fate. While a neuron can have many neurites emanating from the cell body, usually only one of them becomes an axon. Previous work on axonal specification had suggested that neurites need to attain a critical length before showing axonal character [72,73], implying that neurite length was the major determinant of axonal fate. However, Lamoureux *et al.* showed that by applying mechanical tension one can induce axonal characteristics even in minor neurites [74]. In other words, neurites under external tension would express axonal characteristics even when their length was lower than sibling

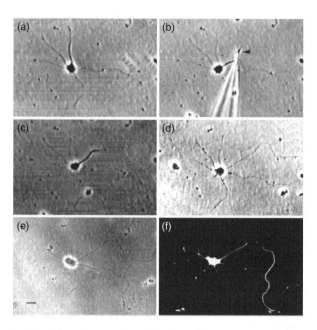

Figure 3.7 Neurite initiation and axonal specification. (a) A neuron with several long processes before manipulation. (b) A neurite is initiated by mechanical pulling with a glass needle (arrowhead). (c) The glass needle is removed. (d, e) The neurite initiated by mechanical tension continues to elongate. (f) The neurite was fixed and stained to show that only the neurite initiated by mechanical tension expresses axonal proteins. Adapted from [74]

neurites. Figure 3.7 shows a minor neurite being initiated by external tension from a neuron with many existing neurites. After external tension is removed, the neurite continues to elongate and expresses axonal characteristics, as shown by the fluorescent staining. This suggests that tension, rather than length, specifies axonal specification.

Apart from its role in the initiation and growth of neurites, tension appears to be capable of influencing the development of the neural system in two other important ways. The first concerns growth-cone guidance, the process through which developing axons find their synaptic targets. As mentioned earlier, axonal elongation in early development (before synaptogenesis) is mediated by growth-cone-generated tension. But tension also appears to influence growth-cone pathfinding. A simple and elegant study by Suter *et al.* showed that the direction of growth-cone pathfinding can be altered by applied tension [75]. They attached beads to cell surface receptors and showed that, by restraining bead motion, growth cones would generate tension in the direction of the bead, followed by cytoskeletal polymerization and growth-cone advance. Thus, tension could determine the direction of cytoskeletal polymerization and subsequent growth-cone pathfinding. Figure 3.8 shows time-lapse images of actin and microtubules responding to tension generated by bead manipulation. The second way in which tension could influence neural development is through the elimination of unnecessary axonal branches. As growth cones extend during the pathfinding phase, branches are often extended and retracted to explore the extracellular environment. Anava *et al.* discovered that tension along a growing axon

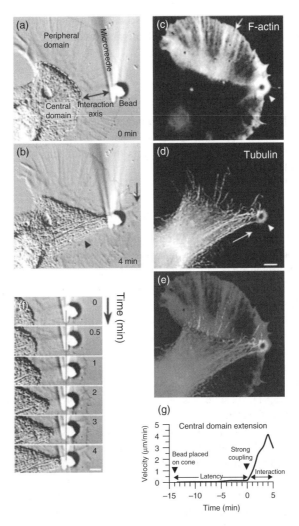

Figure 3.8 Tension can change the direction of growth-cone advance and cytoskeletal polymerization. (a, b) A glass pipette was used to restrain a bead attached to the cell surface and to generate a force on the growth cone. (c–e) Fluorescent staining shows actin accumulation around the bead as well as microtubule polymerization in the direction of the force. (f) Sequence of images showing the microtubule extension in the direction of force (line indicates the initial position of the needle). (g) Velocity of the extending central domain of microtubules towards the glass needle plotted as a function of time. Adapted from [75]

promotes stabilization of one set of axon branches while causing retraction or elimination of collaterals [76]. Thus, by the elimination of unnecessary axon collaterals, tension could influence the wiring of the neural network.

As an aside, it is worth noting that several other mechanical cues also influence neuronal development. For example, neurons show significantly higher neurite branching activity on softer polyacrylamide substrates that mimic the stiffness of brain tissue rather than stiff cell culture surfaces [77]. Growing neurons also use substrate stiffness as a guidance cue to avoid stiff substrates [78], a possible mechanism to avoid stiff scar tissue that develops as a result of injuries.

3.3 Tension in Neuron Function

It is clear that mechanical tension exists and is actively regulated in neurons both *in vivo* and *in vitro*. Now, the question that arises is: What is the purpose of mechanical tension in neurons? The following discussion highlights the role of tension in neurotransmission and vesicle dynamics.

3.3.1 Tension Increases Neurotransmission

Uncovering the role of mechanical tension in neuronal signaling is a topic in its infancy, despite the fact that the first observations were reported 60 years ago. In 1951, Fatt and Katz reported that stretching a muscle (and presumably its motor neurons) by 10–15% beyond its resting length resulted in a reversible increase of 2.5–3 times in the rate of spontaneous electrical potentials at the muscle endplate [79].

A later study conducted by Chen and Grinnell showed that stretching a whole frog muscle could more than double the release of neurotransmitters from motor nerve terminals [16]. They also showed that the stretch enhancement of neurotransmitter release was mediated by integrin, a well-known mechanosensor. Interestingly, they found that the stretch enhancement of neurotransmission bypassed the usual Ca^{2+} triggering step in vesicle fusion [80,81]. Vesicle fusion is usually triggered by an influx of extracellular Ca^{2+} ions through stretch-activated ion channels. Their experiments showed that stretch enhancement of neurotransmission is reduced, but still occurred in the absence of extracellular Ca^{2+} influx and even when internal stores of Ca^{2+} were buffered. These results suggest that Ca^{2+} plays a role in determining the amount of stretch enhancement but that it is not necessary for it to occur. Based on these findings, Chen and Grinnell concluded that second messenger pathways or chemical modification are not plausible mechanisms for stretch-enhanced neurotransmission [80]. Figure 3.9 shows the increased electrical firing frequency as a function of muscle stretch. In experiments with *Drosophila*, Grinnell *et al.* [81] showed that hypertonic swelling also induced enhancement of neurotransmission, which was strongly dependent on the amount of cAMP/PKA activity. They hypothesized that the cAMP/PKA cascade regulates the size of the vesicle pool available for release, and thus determines the amount of enhanced neurotransmission [81]. These studies suggest that mechanical tension can modulate neurotransmission on time scales of seconds.

3.3.2 Tension Affects Vesicle Dynamics

Since tension is involved in modulating neurotransmission, it must also affect vesicle dynamics. A recent study showed that synaptic vesicle clustering is dependent on mechanical tension within the axon [18]. Siechen *et al.* found that vesicles failed to cluster when tension was removed by axotomy. However, if tension is reapplied to the severed axon,

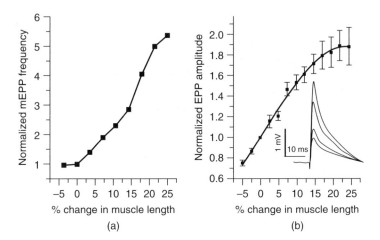

Figure 3.9   Stretch enhancement of neurotransmission in frog muscles. (a) Plot of mEPP (miniature end plate potential) frequency as a function of percent muscle stretch. (b) Plot of EPP (excitatory postsynaptic potential) as a function of percent muscle stretch. The inset shows the EPP amplitude at 95, 100, 110, and 120% of rest length. Adapted from [81]

vesicle clustering was restored. In addition, they showed that axons have a rest tension and that if this tension is perturbed then axons respond to maintain tension. They showed that if axonal tension is increased by external stretching, then the amount of synaptic vesicle accumulation at the presynaptic terminal increases by more than two times [18]. And as hypothesized by Grinnell *et al.* [81], this could lead to increased amount of neurotransmission. The study by Siechen *et al.* [18] suggests that mechanical tension is a necessary regulatory signal of presynaptic vesicle clustering, and thus is involved in the tuning of synaptic efficiency (i.e., synaptic plasticity).

To investigate the time evolution of tension-induced vesicle accumulation *in vivo*, Ahmed *et al.* stretched axons while observing by live imaging [82]. They found that vesicle accumulation increased by approximately 30% after approximately 5 min of mechanical stretch, as shown in Figure 3.10. This stretch enhancement of vesicle accumulation remained for at least 30 min after stretch was released (maximum time of experiment), indicating a persistent change. Experiments by Ahmed *et al.* [82] were conducted in Ca^{2+}- free medium, suggesting that influx of external Ca^{2+} is not necessary for stretch enhancement of synaptic vesicle accumulation. In comparison with Siechen *et al.* [18], the stretch-enhanced vesicle accumulation was attenuated in both time and amount, which may be explained by the absence of extracellular Ca^{2+} influx, as observed by Grinnell *et al.* [81]. Ahmed *et al.* [82] also found that vesicle accumulation is not affected when tension is removed from an intact axon. Since it is known that relaxed axons will rebuild their tension quickly [67], Ahmed *et al.* hypothesized that this rebuilding of tension does not allow vesicle accumulation to decrease [82]. This may be a protective mechanism to maintain vesicle clustering at synapses. These studies show that mechanical tension can modulate synaptic vesicle accumulation on time scales of minutes to hours.

If mechanical tension modulates vesicle clustering, then it must also affect local vesicle dynamics. Ahmed *et al.* found that reducing the mechanical tension in Aplysia neurites

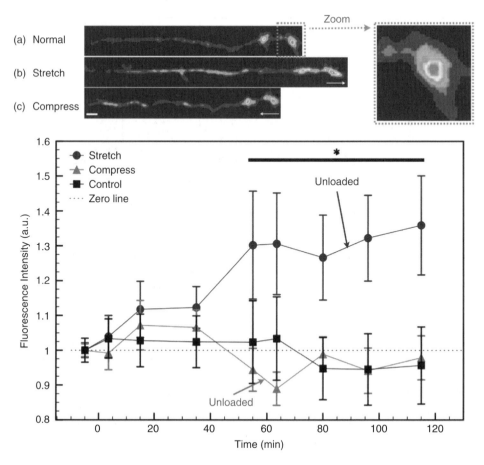

Figure 3.10 Stretch-induced accumulation of synaptic vesicles. (a) A control axon is shown on the PDMS surface. The inset on the right shows a magnified image of the presynaptic terminal where a 2.5 μm square region was used to quantify synaptic vesicle accumulation. (b) An axon is shown stretched by substrate deformation, notice the straightness of the axon (increased tension). (c) An axon is shown compressed by substrate deformation; notice the axon is squiggly (decreased tension) (scale bar: 5 μm). The plot shows the fluorescence intensity of GFP-tagged synaptic vesicles at the presynaptic terminal of the *Drosophila* neuromuscular junction as a function of time. Control samples show no significant change in synaptic vesicle accumulation. When axons are stretched, increased accumulation is observed after approximately 50 min and the effect persists for at least 30 min after stretch is removed. In compressed axons, no significant change occurs in accumulation during compression or after it is removed [82]

resulted in a significant decrease of vesicle motion [82]. Figure 3.11 shows that the range of vesicle motion and processivity decreases dramatically when neurite tension is decreased. This suggests that mechanical tension may also be necessary for normal vesicle motion.

When cells are stretched, it is known that actin polymerization increases [83,84]. This increase in actin polymerization due to stretching may also increase the amount of actin

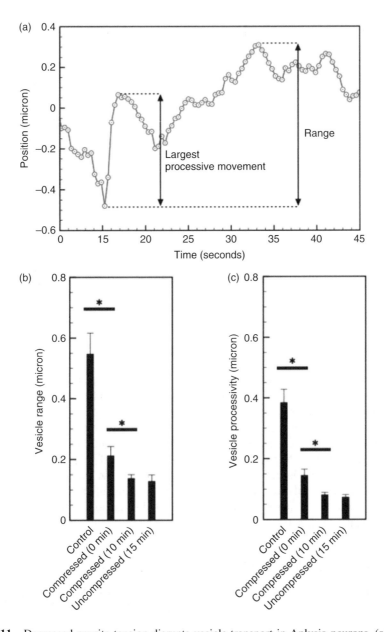

Figure 3.11 Decreased neurite tension disrupts vesicle transport in Aplysia neurons. (a) A representative plot showing the range and largest processive motion of a single vesicle. (b) Vesicle range of motion was approximately 550 nm in the control samples. Mechanical compression caused an immediate decrease to 200 nm and the range continued to decrease to 140 nm. (c) In control samples the largest processive motion of vesicles was approximately 380 nm. This decreased to 140 nm immediately after compression and continued to decrease to 80 nm. In both cases, the effect persists for over 15 min after compression was removed [82]

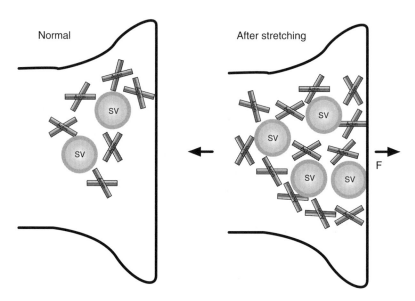

Figure 3.12 Tension is known to increase actin polymerization. Increased actin polymerization may lead to increased synaptic vesicle accumulation due to a larger number of actin–synapsin binding sites

at the synapse, which could increase the number of actin–synapsin binding sites [50, 51]. Therefore, assuming that axonal transport properties do not change, then, based on a higher number of binding sites for synaptic vesicles, the vesicle accumulation at the synapse could increase (Figure 3.12). This conceptual framework suggests that mechanical tension could play a role in synaptic plasticity and ultimately in learning and memory.

The emergent view is that axons actively maintain a rest tension, and such tension is essential to maintain vesicle transport and clustering at the synaptic terminal [18,82]. Synaptic vesicle accumulation is necessary for neurotransmitter release at a functional synapse [36]. Therefore, it seems natural to conclude that axonal tension is critical in maintaining neuronal function.

3.4   Modeling the Mechanical Behavior of Axons

It is clear from the above that *in vitro* neurons in culture show the following mechanical characteristics:

1. Axons actively maintain mechanical tension.
2. Axons behave as linear springs under fast stretch. Under a fixed stretch, the increased tension decays with time to a steady value.
3. When the tension is relaxed mechanically, axons shorten to rebuild the tension.
4. Axons can grow in response to mechanical stretch, which may serve as a mechanism of force relaxation under fixed stretch.

Dennerll et al. [55] postulated that there are two force thresholds for axons. When the upper threshold is exceeded due to stretching (mechanically or during development), the axon grows in length by synthesizing cytoskeletal structure. If the stretch is held fixed, this growth relaxes the tension. On the other hand, if the tension falls below the lower threshold, axons shorten by undoing the growth of the current cytoskeleton. In between the two thresholds, axons behave as a viscoelastic material. Thus, a mechanics model of the axon is represented as a combination of a "growth dashpot" (Figure 3.13) and a classical viscoelastic element. The growth dashpot is linear, in the sense that the rate of growth is linearly dependent on the applied tension above the upper threshold, as observed in experiments with chick sensory neurons [62]. The spring constants of the linear springs and the time constants of the dashpots of the model can be estimated from experiments.

It is hypothesized that molecular motors serve as force actuators in axons. A mechanical model that represents the motors and viscoelasticity of neurites was proposed by Bernal et al. [85] (Figure 3.14). Here, molecular motors are attached to parallel fibers (such as F-actin). The motors slide the fibers with respect to each other. If the fibers are anchored at both ends, then tensile force develops within the fibers, which results in a net tensile force in the neurite. The model does not require any remodeling of the cytoskeleton, either under stretch or during shortening.

Force evolution in a network of neurites of a single neuron was presented in a seminal paper by Bray [58]. Here, each neurite, mostly on a straight line, is considered as a unit vector pointing towards the tip of the neurite. The neurites orient themselves such that

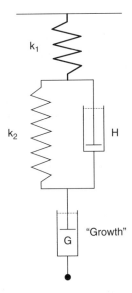

Figure 3.13 Mechanical model of neurite behavior with growth dashpot. The axon is represented as a combination of a "growth dashpot" and a classical viscoelastic element. The "growth dashpot" allows the axon to elongate when the force is above an upper threshold and shorten when the force is under a lower threshold. Adapted from [55]

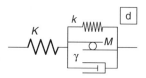

Figure 3.14   Mechanical model of neurite behavior with a molecular motor element. The axon is represented using the same mechanical model proposed in Dennerll *et al*. [55] with an additional element (*M*) representing the action of the molecular motors. Adapted from [85]

their vector sum is close to zero. This implies that the neurites possibly generate similar forces, and their evolution (growth, morphology, substrate attachment) is determined by force equilibrium. If this equilibrium is perturbed (e.g., by mechanical cutting of a neurite), then the network responds by reorienting and regenerating new neurites that also develop tension and maintain equilibrium. When a neurite branches into two new neurites, then the forces T in the parent neurite and those in the branches (t_1 and t_2) are related by a power-law relation: $T^n = t_1^n + t_2^n$, where $n = 1.6 \pm 0.09$.

Studies on the mechanical behavior of *in vivo* neurons are limited. The only report, to the best of our knowledge (described above [67]) shows that motor neurons that have formed neuromuscular junctions (i.e., post synaptogenesis) in embryonic fruit flies behave very similarly to those of *in-vitro*-cultured neurites. The significant features are: (a) force response of axons is linear with fast stretches; (b) axons maintain a rest tension actively. When stretched rapidly and the stretch is held fixed, axons relax tension to a steady value with time. When axon tension is released mechanically by bringing the neuromuscular junction closer to the central nervous system, the slack axons shorten with time linearly and regain tension in minutes. (c) Stiffness or the spring constant of the axons along the longitudinal direction changes after an applied stretch and after the tension relaxes to a steady value. When considering many experiments, for small relaxation (less than 30% of the force immediately after stretch), axon stiffness increases compared with the initial (unstretched) value. Stiffness progressively decreases with increasing force relaxation. When the relaxation is large (>80%), the stiffness is significantly lower than the initial value (Figure 3.6).

In order to interpret the above observations, we propose a simple mechanics model based on the hypothesis that axons have cytoskeletal fibers (actin, microtubule, and potentially other filaments), pairs of which are bridged by motor proteins. Various other proteins, such as tau, also bridge the fibers. The motors slide the fibers with respect to one another. When the fibers are restrained from motion (e.g., due to attachment with a substrate such as the neuromuscular junction), tensile force develops. As a result, tension generates in the axons macroscopically.

Consider such a pair of fibers that are pointing towards opposite directions. A motor is bridging the two. The heads of the motor can "walk" along either directions of the fibers by consuming ATP, but the activation barrier for taking a step along one direction is higher than in the other direction. Hence, it walks more frequently along the easier direction. Let $E_a \pm F_0 d_0$ be the activation barriers for walking along the harder and easier directions. Here, d_0 is the activation distance, F_0 is a characteristic force, often called the stall force [86]. $\pm F_0 d_0$ results in asymmetry in the energy landscape. When the fibers are free from any restraints, then the motor keeps walking by consuming ATP, and the constant average velocity of the fibers with respect to each other is

$$\dot{\gamma}_{sl} = \dot{\gamma}_0 \left\{ -\exp\left[\frac{-(E_a - F_0 d_0)}{kT}\right] + \exp\left[\frac{-(E_a + F_0 d_0)}{kT}\right] \right\} \qquad (3.1)$$

where $\dot{\gamma}_0$ is a constant. Note that $\dot{\gamma}_{sl}$ is positive towards the right in Figure 3.15. If a force F is applied at the ends of the fibers as shown in Figure 3.15, then the energy landscape will change. For $F < F_0$, sliding continues to be in the same direction as when $F = 0$. When $F > F_0$, the motion is reversed. In the presence of F, the rate of change of displacement of the fibers with respect to each other is given by

$$\dot{\gamma}_{sl} = \dot{\gamma}_0 \left\{ -\exp\left[\frac{-(E_a - F_0 d_0 + F d_0)}{kT}\right] + \exp\left[\frac{-(E_a + F_0 d_0 - F d_0)}{kT}\right] \right\} \qquad (3.2)$$

The displacement of the fibers with respect to each other has two components: an elastic component $\gamma_1 = F/G$, which accounts for the elastic deformation of the fibers and the motor due to F and the spring constant G, and a sliding component γ_{sl}. Thus, the net rate of displacement at the fiber ends $\dot{\gamma}$ is

$$\dot{\gamma} = \frac{\dot{F}}{G} + \dot{\gamma}_{sl} \qquad (3.3)$$

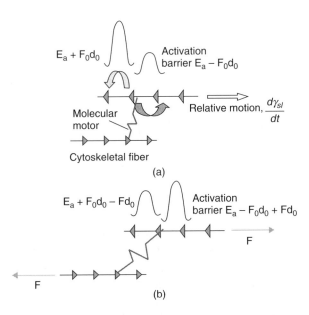

(a)

(b)

Figure 3.15 Mechanical model for a molecular motor bridging two fibers. (a) The molecular motor is biased to walk in one direction more often than the other. The activation barriers for walking along either direction is $E_a \pm F_0 d_0$, where F_0 is the characteristic stall force and d_0 is the activation distance. Thus, $\pm F_0 d_0$ results in asymmetry in the energy landscape. (b) When a force is applied, the energy landscape changes accordingly and may change the direction of sliding. For $F < F_0$ the sliding continues to be in the direction opposite of F and when $F > F_0$ the motion is reversed

Now consider a special case where the ends of the freely sliding fibers are suddenly brought to a stop by holding them at time $t = 0$. Then, $\dot{\gamma}$ is zero for all times $t > 0$; that is:

$$\dot{\gamma} = \frac{\dot{F}}{G} + \dot{\gamma}_{sl} = 0 \tag{3.4}$$

Owing to the anchorage at the ends (the ends are held fixed), the constraining force F increases and the energy landscape changes until the activation barriers for the motor for its forward and backward steps become equal at $F = F_0$. From this point onwards, $F = F_0$ remains steady.

In the second case, let the fibers be pulled suddenly by a force $F = F_* > F_0$ at time $t = 0$. The ends of the fibers are held fixed thereafter. Corresponding elastic displacement of the ends is F_*/G. Owing to the high force, the energy landscape for the motor becomes skewed, such that walking backwards becomes easier than walking forward. Thus, the fibers slide in the reverse direction compared with the case of $F = 0$. Thus, the force on the fibers relaxes with time until the force reaches a steady value $F = F_0$.

For

$$\frac{|F_0 - F|d_0}{kT} \ll 1$$

Equation (3.4) can be linearized, retaining only the first-order terms:

$$\dot{F} + \left[\frac{2G\dot{\gamma}_0 \exp(-E_a/kT)}{kT} \right](F - F_0) = 0 \tag{3.5}$$

Consider two initial conditions: first, when ends of the fibers are held fixed from $t = 0$ (i.e., $F(t = 0) = 0$); second, when the ends of the fibers are pulled by a force F_* at $t = 0$ and the stretch is held fixed (i.e., $F(0) = F_*$). Then the solutions for F for these two initial conditions are

$$F = F_0 \left[1 - \exp\left(\frac{-t}{\tau} \right) \right] \tag{3.6}$$

$$F = F_* \exp\left(\frac{-t}{\tau} \right) + F_0 \left[1 - \exp\left(\frac{-t}{\tau} \right) \right] \tag{3.7}$$

where

$$\frac{1}{\tau} = \frac{2G\dot{\gamma}_0 \exp(-E_a/kT)}{kT} \tag{3.8}$$

which predicts: (a) if the fiber ends are held fixed, tensile force will develop in the fibers reaching a steady value of F_0 with a time constant τ; (b) if the ends are pulled by a force $F_* > F_0$ and the ends are held fixed, the force will decay to a steady force F_0 with the same time constant τ. If no force or constraint is applied at the ends, the fibers will slide with respect to each other at a constant velocity. These three qualitative features were observed in our *in vivo* experiments. If the above model is applied macroscopically (i.e., F is the force on the axon, F_0 is the rest tension, and F_* is the force when a sudden stretch is applied on the axon), then τ is the time constant for force generation or relaxation. Figure 3.16 shows the force relaxation and force generation with time for the

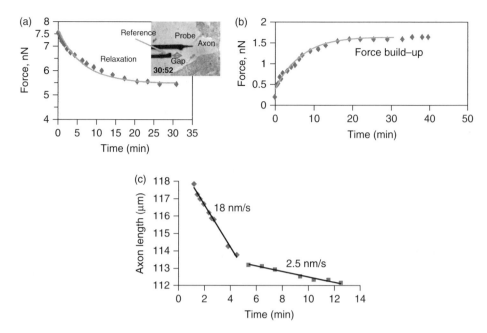

Figure 3.16 Force relaxation and generation in the *in vivo* axon. Force relaxation and generation have the same time constant of $\tau = 12.4$ min, as shown in the two upper plots. The lower plot shows that the axon shortens at a constant speed when the force is removed. The axon exhibits two distinct speeds during shortening

same axon. The solid line is an exponential fit to the data. Both the fits have the same time constant, $\tau = 12.4$ min. The rest tension of the axon was about 2 nN. After a stretch and force relaxation, the axon was slackened when the tension in the axon was reduced to zero. The axon then began to shorten its length and generate force until the force reached a steady value of about 1.8 nN. In a separate experiment to study axon shortening, the neuromuscular junction of an axon was brought close to the central nervous system to slacken the axon (Figure 3.16). The axon again shortened at a constant rate. However, the axon exhibits two distinct rates of shortening. Initially the rate is high, and then it abruptly decreases. The reason for two distinct rates of shortening is currently unknown.

The model above does not account for any change in stiffness of the axon after an applied stretch. Stiffness may increase due to alignment of fibers in a polymer. However, in an embryonic axon with diameter less than a micrometer and length of about $100\,\mu$, cytoskeletal structures, particularly microtubules and neurofilaments, are typically aligned along the longitudinal direction. Thus, there is not much room for further alignment. But how could stiffness be increased after stretch? Prior experiments show that *in vitro* axons can add cytoskeletal materials under stretch [68,69]. However, the mechanism by which axons initiate polymerization in response to tension or stretch remains elusive. A mechanism-based mechanics model can be developed when further biological insight is acquired on cytoskeletal growth and remodeling in axons due to forces.

3.5   Outlook

Here, we have discussed some important studies highlighting the role of mechanical tension in neuronal function. Neurons actively regulate tension along their axons, and this regulates growth, guidance, and signaling. Mechanical tension may serve as a regulatory signal to modulate both neurotransmission and vesicle clustering in functional synapses. There remain many unresolved issues in the field of cellular neuromechanics, including:

- What is the origin of force generation in axons?
- How does axonal tension lead to vesicle accumulation?
- What role does tension play in the plasticity of synapses (i.e., learning and the formation of memory)?

References

[1] Pelham, R.J. and Wang, Y.L. (1997) Cell locomotion and focal adhesions are regulated by substrate flexibility. *Proceedings of the National Academy of Sciences of the United States of America*, **94** (25), 13661–13665.

[2] Vogel, V. and Sheetz, M.P. (2006) Local force and geometry sensing regulate cell functions. *Nature Reviews Molecular Cell Biology*, **7** (4), 265–275.

[3] Engler, A.J., Sen, S., Sweeney, H.L., and Discher, D.E. (2006) Matrix elasticity directs stem cell lineage specification. *Cell*, **126** (4), 677–689, doi:10.1016/j.cell.2006.06.044.

[4] Lecuit, T. and Lenne, P.F. (2007) Cell surface mechanics and the control of cell shape, tissue patterns and morphogenesis. *Nature Reviews Molecular Cell Biology*, **8** (8), 633–644, doi:10.1038/nrm2222.

[5] Janmey, P.A. and McCulloch, C.A. (2007) Cell mechanics: integrating cell responses to mechanical stimuli. *Annual Review of Biomedical Engineering*, **9**, 1–34, doi:10.1146/annurev.bioeng.9.060906.151927.

[6] Fletcher, D.A. and Mullins, R.D. (2010) Cell mechanics and the cytoskeleton. *Nature*, **463** (7280), 485–492, doi:10.1038/nature08908.

[7] Furchgott, R. and Vanhoutte, P. (1989) Endothelium-derived relaxing and contracting factors. *The FASEB Journal: Official Publication of the Federation of American Societies for Experimental Biology*, **3** (9), 2007–2018.

[8] Huxley, A.F. and Simmons, R.M. (1971) Proposed mechanism of force generation in striated muscle. *Nature*, **233** (5321), 533–538.

[9] Wirtz, H. and Dobbs, L. (1990) Calcium mobilization and exocytosis after one mechanical stretch of lung epithelial cells. *Science*, **250** (4985), 1266–1269.

[10] Wozniak, M.A. and Chen, C.S. (2009) Mechanotransduction in development: a growing role for contractility. *Nature Reviews Molecular Cell Biology*, **10** (1), 34–43, doi:10.1038/nrm2592.

[11] Gorfinkiel, N., Blanchard, G., and Adams, R. (2009) Mechanical control of global cell behaviour during dorsal closure in *Drosophila*. *Development*, **136**, 1889–1898, http://dev.biologists.org/cgi /content/abstract/136/11/1889.

[12] Lecuit, T. (2008) "developmental mechanics": cellular patterns controlled by adhesion, cortical tension and cell division. *HFSP Journal*, **2** (2), 72–78, doi:10.2976/1.2896332.

[13] Rauzi, M., Verant, P., Lecuit, T., and Lenne, P.F. (2008) Nature and anisotropy of cortical forces orienting *Drosophila* tissue morphogenesis. *Nature Cell Biology*, **10** (12), 1401–1410, doi:10.1038/ncb1798.

[14] Martin, A.C., Kaschube, M., and Wieschaus, E.F. (2009) Pulsed contractions of an actin–myosin network drive apical constriction. *Nature*, **457** (7228), 495–499, doi:10.1038/nature07522.

[15] Thompson, D.W. (1917) *On Growth and Form,* University Press, Cambridge.

[16] Chen, B.M. and Grinnell, A.D. (1995) Integrins and modulation of transmitter release from motor nerve terminals by stretch. *Science*, **269** (5230), 1578–1580.

[17] Lowery, L.A. and Vactor, D.V. (2009) The trip of the tip: understanding the growth cone machinery. *Nature Reviews Molecular Cell Biology*, **10** (5), 332–343, doi:10.1038/nrm2679.

[18] Siechen, S., Yang, S., Chiba, A., and Saif, M.T.A. (2009) Mechanical tension contributes to clustering of neurotransmitter vesicles at presynaptic terminals. *Proceedings of the National Academy of Sciences of the United States of America*, **106** (31), 12611–12616, doi:10.1073/pnas.0901867106.

[19] Franze, K. and Guck, J. (2010) The biophysics of neuronal growth. *Reports on Progress in Physics*, **73**, 094601.

[20] Bao, G. and Suresh, S. (2003) Cell and molecular mechanics of biological materials. *Nature Materials*, **2** (11), 715–725, doi:10.1038/nmat1001.

[21] Van Vliet, K.J., Bao, G., and Suresh, S. (2003) The biomechanics toolbox: experimental approaches for living cells and biomolecules. *Acta Materialia*, **51** (19), 5881–5905.

[22] Rajagopalan, J. and Saif, M. (2011) MEMS sensors and microsystems for cell mechanobiology. *Journal of Micromechanics and Microengineering*, **21**, 054002.

[23] Levitan, I. and Kaczmarek, L. (2002) *The Neuron: Cell and Molecular Biology*, Oxford Univerity Press.

[24] Alberts, B., Johnson, A., Lewis, J. *et al*. (2002) *Molecular Biology of The Cell*, 4th edn., Garland Science, New York.

[25] Fischer, M., Kaech, S., Knutti, D., and Matus, A. (1998) Rapid actin-based plasticity in dendritic spines. *Neuron*, **20** (5), 847–854.

[26] Dent, E.W. and Gertler, F.B. (2003) Cytoskeletal dynamics and transport in growth cone motility and axon guidance. *Neuron*, **40** (2), 209–227.

[27] Letourneau, P. (2009) Actin in axons: stable scaffolds and dynamic filaments. *Results and Problems in Cell Differentiation*, **48**, 65–90, doi:10.1007/400_2009_15.

[28] Lee, M.K. and Cleveland, D.W. (1996) Neuronal intermediate filaments. *Annual Review of Neuroscience*, **19**, 187–217, doi:10.1146/annurev.ne.19.030196.001155.

[29] Hirokawa, N. and Takemura, R. (2005) Molecular motors and mechanisms of directional transport in neurons. *Nature Reviews Neuroscience*, **6** (3), 201–214, doi:10.1038/nrn1624.

[30] Kulic, I.M., Brown, A.E.X., Kim, H. *et al*. (2008) The role of microtubule movement in bidirectional organelle transport. *Proceedings of the National Academy of Sciences of the United States of America*, **105** (29), 10011–10016, doi:10.1073/pnas.0800031105, http://www.pnas.org/content/105/29/10011.

[31] Rasband, M.N. (2010) The axon initial segment and the maintenance of neuronal polarity. *Nature Reviews Neuroscience*, **11** (8), 552–562, doi:10.1038/nrn2852.

[32] Barnes, A.P. and Polleux, F. (2009) Establishment of axon–dendrite polarity in developing neurons. *Annual Review of Neuroscience*, **32**, 347–381, doi:10.1146/annurev.neuro.31.060407.125536.

[33] De Vos, K.J., Grierson, A.J., Ackerley, S., and Miller, C.C.J. (2008) Role of axonal transport in neurodegenerative diseases. *Annual Review of Neuroscience*, **31**, 151–173, doi:10.1146 /annurev.neuro.31.061307.090711.

[34] Hawkins, R.D., Kandel, E.R., and Siegelbaum, S.A. (1993) Learning to modulate transmitter release: themes and variations in synaptic plasticity. *Annual Review of Neuroscience*, **16**, 625–665, doi:10.1146/annurev.ne.16.030193.003205.

[35] Martin, S.J., Grimwood, P.D., and Morris, R.G. (2000) Synaptic plasticity and memory: an evaluation of the hypothesis. *Annual Review of Neuroscience*, **23**, 649–711, doi:10.1146/annurev.neuro.23.1.649.

[36] Kandel, E., Schwartz, J., and Jessell, T. (eds.) (2000) *Principles of Neural Science*, 4th edn., McGraw-Hill, New York.

[37] DePina, A.S. and Langford, G.M. (1999) Vesicle transport: the role of actin filaments and myosin motors. *Microscopy Research and Technique*, **47** (2), 93–106, doi:10.1002/(SICI)1097-0029(19991015)47 :2<93::AID-JEMT2>3.0.CO;2-P.

[38] Bridgman, P.C. (2004) Myosin-dependent transport in neurons. *Journal of Neurobiology*, **58** (2), 164–174, doi:10.1002/neu.10320.

[39] Bridgman, P.C. (2009) Myosin motor proteins in the cell biology of axons and other neuronal compartments. *Results and Problems in Cell Differentiation*, **48**, 91–105, doi:10.1007/400_2009_10.

[40] Lodish, H., Berk, A., Matsudaira, P. *et al*. (2003) *Molecular Cell Biology*, 5th edn., W.H. Freeman.

[41] Hirokawa, N., Nitta, R., and Okada, Y. (2009) The mechanisms of kinesin motor motility: lessons from the monomeric motor KIF1A. *Nature Reviews Molecular Cell Biology*, **10** (12), 877–884, doi:10.1038/nrm2807.

[42] Verhey, K.J. and Hammond, J.W. (2009) Traffic control: regulation of kinesin motors. *Nature Reviews Molecular Cell Biology*, **10** (11), 765–777, doi:10.1038/nrm2782.

[43] Kardon, J.R. and Vale, R.D. (2009) Regulators of the cytoplasmic dynein motor. *Nature Reviews Molecular Cell Biology*, **10** (12), 854–865, doi:10.1038/nrm2804.

[44] Takamori, S., Holt, M., Stenius, K. *et al.* (2006) Molecular anatomy of a trafficking organelle. *Cell*, **127** (4), 831–846, doi:10.1016/j.cell.2006.10.030.

[45] Kidokoro, Y., Kuromi, H., Delgado, R. *et al.* (2004) Synaptic vesicle pools and plasticity of synaptic transmission at the *Drosophila* synapse. *Brain Research Reviews*, **47** (1–3), 18–32, doi:10.1016 /j.brainresrev.2004.05.004.

[46] Kelly, R.B. (1993) Storage and release of neurotransmitters. *Cell*, **72** (Suppl), 43–53.

[47] Hökfelt, T., Broberger, C., Xu, Z.Q. *et al.* (2000) Neuropeptides – an overview. *Neuropharmacology*, **39** (8), 1337–1356.

[48] Kandel, E.R. (2001) The molecular biology of memory storage: a dialogue between genes and synapses. *Science*, **294** (5544), 1030–1038, doi:10.1126/science.1067020.

[49] Holtmaat, A. and Svoboda, K. (2009) Experience-dependent structural synaptic plasticity in the mammalian brain. *Nature Reviews Neuroscience*, **10** (9), 647–658, doi:10.1038/nrn2699.

[50] Cingolani, L.A. and Goda, Y. (2008) Actin in action: the interplay between the actin cytoskeleton and synaptic efficacy. *Nature Reviews Neuroscience*, **9** (5), 344–356, doi:10.1038/nrn2373.

[51] Dillon, C. and Goda, Y. (2005) The actin cytoskeleton: integrating form and function at the synapse. *Annual Review of Neuroscience*, **28** (1), 25–55, doi:10.1146/neuro.2005.28.issue-1.

[52] Dityatev, A., Schachner, M., and Sonderegger, P. (2010) The dual role of the extracellular matrix in synaptic plasticity and homeostasis. *Nature Reviews Neuroscience*, **11** (11), 735–746, doi:10.1038/nrn2898.

[53] Siegelbaum, S.A. and Kandel, E.R. (1991) Learning-related synaptic plasticity: LTP and LTD. *Current Opinion in Neurobiology*, **1** (1), 113–120.

[54] Lamoureux, P., Buxbaum, R.E., and Heidemann, S.R. (1989) Direct evidence that growth cones pull. *Nature*, **340** (6229), 159–162, doi:10.1038/340159a0.

[55] Dennerll, T.J., Lamoureux, P., Buxbaum, R.E., and Heidemann, S.R. (1989) The cytomechanics of axonal elongation and retraction. *The Journal of Cell Biology*, **109** (6 Pt 1), 3073–3083, http://www .jcb.org/cgi/reprint/109/6/3073.

[56] Heidemann, S.R. and Buxbaum, R.E. (1990) Tension as a regulator and integrator of axonal growth. *Cell Motility and the Cytoskeleton*, **17** (1), 6–10, doi:10.1002/cm.970170103.

[57] Weiss, P. (1945) Experiments on cell and axon orientation in vitro; the role of colloidal exudates in tissue organization. *The Journal of Experimental Zoology*, **100**, 353–386.

[58] Bray, D. (1979) Mechanical tension produced by nerve cells in tissue culture. *Journal of Cell Science*, **37**, 391–410.

[59] Bray, D. (1984) Axonal growth in response to experimentally applied mechanical tension. *Developmental Biology*, **102** (2), 379–389.

[60] Zheng, J., Buxbaum, R.E., and Heidemann, S.R. (1993) Investigation of microtubule assembly and organization accompanying tension-induced neurite initiation. *Journal of Cell Science*, **104** (Pt 4), 1239–1250.

[61] Dennerll, T.J., Joshi, H.C., Steel, V.L. *et al.* (1988) Tension and compression in the cytoskeleton of PC-12 neurites. II: Quantitative measurements. *The Journal of Cell Biology*, **107** (2), 665–674.

[62] Zheng, J., Lamoureux, P., Santiago, V. *et al.* (1991) Tensile regulation of axonal elongation and initiation. *The Journal of Neuroscience: The Official Journal of the Society for Neuroscience*, **11** (4), 1117–1125.

[63] Korneliussen, H. and Jansen, J.K. (1976) Morphological aspects of the elimination of polyneuronal innervation of skeletal muscle fibres in newborn rats. *Journal of Neurocytology*, **5** (8), 591–604.

[64] Bixby, J.L. (1981) Ultrastructural observations on synapse elimination in neonatal rabbit skeletal muscle. *Journal of Neurocytology*, **10** (1), 81–100.

[65] Morrison-Graham, K. (1983) An anatomical and electrophysiological study of synapse elimination at the developing frog neuromuscular junction. *Developmental Biology*, **99** (2), 298–311.

[66] Van Essen, D.C. (1997) A tension-based theory of morphogenesis and compact wiring in the central nervous system. *Nature*, **385** (6614), 313–318, doi:10.1038/385313a0.

[67] Rajagopalan, J., Tofangchi, A., and Saif, T. (2010) *Drosophila* neurons actively regulate axonal tension in vivo. *Biophysical Journal*, **99** (10), 3208–3215, doi:10.1016/j.bpj.2010.09.029.

[68] Lamoureux, P., Heidemann, S.R., Martzke, N.R., and Miller, K.E. (2010) Growth and elongation within and along the axon. *Developmental Neurobiology*, **70** (3), 135–149, doi:10.1002/dneu.20764.

[69] Pfister, B.J., Iwata, A., Meaney, D.F., and Smith, D.H. (2004) Extreme stretch growth of integrated axons. *The Journal of Adhesion Neuroscience: The Official Journal of the Society for Neuroscience*, **24** (36), 7978–7983, doi:10.1523/JNEUROSCI.1974-04.2004.

[70] Pfister, B.J., Bonislawski, D.P., Smith, D.H., and Cohen, A.S. (2006) Stretch-grown axons retain the ability to transmit active electrical signals. *FEBS Letters*, **580** (14), 3525–3531, doi:10.1016/j.febslet.2006.05.030.

[71] Suter, D.M. and Miller, K.E. (2011) The emerging role of forces in axonal elongation. *Progress in Neurobiology*, **94**, 91–101, doi:10.1016/j.pneurobio.2011.04.002.

[72] Dotti, C.G. and Banker, G.A. (1987) Experimentally induced alteration in the polarity of developing neurons. *Nature*, **330** (6145), 254–256, doi:10.1038/330254a0.

[73] Goslin, K. and Banker, G. (1989) Experimental observations on the development of polarity by hippocampal neurons in culture. *The Journal of Cell Biology*, **108** (4), 1507–1516.

[74] Lamoureux, P., Ruthel, G., Buxbaum, R.E., and Heidemann, S.R. (2002) Mechanical tension can specify axonal fate in hippocampal neurons. *The Journal of Cell Biology*, **159** (3), 499–508, doi:10.1083/jcb.200207174.

[75] Suter, D.M., Errante, L.D., Belotserkovsky, V., and Forscher, P. (1998) The Ig superfamily cell adhesion molecule, apCAM, mediates growth cone steering by substrate–cytoskeletal coupling. *The Journal of Cell Biology*, **141** (1), 227–240.

[76] Anava, S., Greenbaum, A., Ben Jacob, E. *et al.* (2009) The regulative role of neurite mechanical tension in network development. *Biophysical Journal*, **96** (4), 1661–1670, doi:10.1016/j.bpj.2008.10.058.

[77] Flanagan, L.A., Ju, Y.E., Marg, B. *et al.* (2002) Neurite branching on deformable substrates. *Neuroreport*, **13** (18), 2411–2415, doi:10.1097/01.wnr.0000048003.96487.97.

[78] Franze, K., Gerdelmann, J., Weick, M. *et al.* (2009) Neurite branch retraction is caused by a threshold-dependent mechanical impact. *Biophysical Journal*, **97** (7), 1883–1890, doi:10.1016/j.bpj.2009.07.033.

[79] Fatt, P. and Katz, B. (1952) Spontaneous subthreshold activity at motor nerve endings. *The Journal of Physiology*, **117** (1), 109–128.

[80] Chen, B.M. and Grinnell, A.D. (1997) Kinetics, Ca^{2+} dependence, and biophysical properties of integrin-mediated mechanical modulation of transmitter release from frog motor nerve terminals. *The Journal of Neuroscience: The Official Journal of the Society for Neuroscience*, **17** (3), 904–916.

[81] Grinnell, A.D., Chen, B.M., Kashani, A. *et al.* (2003) The role of integrins in the modulation of neurotransmitter release from motor nerve terminals by stretch and hypertonicity. *Journal of Neurocytology*, **32** (5-8), 489–503, doi:10.1023/B:NEUR.0000020606.58265.b5.

[82] Ahmed, W., Li, T., Rubakhin, S. *et al.* (2012) Mechanical tension modulates local and global vesicle dynamics in neurons. *Cellular and Molecular Bioengineering*, **5**, 115–164.

[83] Pender, N. and McCulloch, C.A. (1991) Quantitation of actin polymerization in two human fibroblast sub-types responding to mechanical stretching. *Journal of Cell Science*, **100** (Pt 1), 187–193.

[84] Ahmed, W.W., Kural, M.H., and Saif, T.A. (2010) A novel platform for *in situ* investigation of cells and tissues under mechanical strain. *Acta Biomaterialia*, **6** (8), 2979–2990, doi:10.1016/j.actbio.2010.02.035.

[85] Bernal, R., Pullarkat, P.A., and Melo, F. (2007) Mechanical properties of axons. *Physical Review Letters*, **99** (1), 018301, http://scitation.aip.org/getabs/servlet/GetabsServlet?prog=normal&id=PRLTAO000099000001018301000001&idtype=cvips&gifs=yes.

[86] Howard, J. (2009) Mechanical signaling in networks of motor and cytoskeletal proteins. *Annual Review of Biophysics*, **38**, 217–234, doi:10.1146/annurev.biophys.050708.133732.

Part Two

Nanoscale Phenomena

4

Fundamentals of Roughness-Induced Superhydrophobicity

Neelesh A. Patankar
Northwestern University, USA

4.1 Background and Motivation

Hydrophobic or water-hating surfaces are generally regarded to be those on which a water drop, when deposited, would have a contact angle of greater than 90° (Figure 4.1); that is, water does not like to spread on the surface. The liquid need not necessarily be water; rather, this type of phenomenon is possible for many different liquid–solid combinations. The term oleophobic is used to specifically imply surfaces that hate oil. Generic terms such as omniphobic, lyophobic, and hygrophobic have been used in the past to imply liquid-hating characteristics. Omniphobic implies "hating everything," lyophobic means resistance of a substance to dissolve in a liquid, while hygrophobic means "moisture hating" per typical usage of the word "hygro." Moisture is generally associated with water. Perhaps a generic term such as rheustophobic, where rheusto is a Greek term for fluid, should be used to imply liquid-hating surfaces. Throughout this chapter, the term hydrophobic will be used, since it is a term that is widely used. However, the discussion and the concepts are not confined to water or any particular type of liquid. The fundamental ideas are applicable to any liquid–solid combination.

It has been known that textured surfaces made from hydrophobic materials amplify the water-hating behavior [2–4]. For example, the apparent contact angle of a drop of water on a textured surface, made from hydrophobic material, is larger (i.e., the drop beads up more) than the contact angle of a drop on a flat surface made from the same hydrophobic material (see drop on a lotus leaf in Figure 4.1). Such surfaces have been commonly referred to as superhydrophobic surfaces (the apparent contact angle of the drop is typically greater than 150°). Two models for the apparent contact angle of a drop on a textured surface – the Cassie–Baxter (CB) model [2,3] and the Wenzel model [4] – corresponding to two

Nano and Cell Mechanics: Fundamentals and Frontiers, First Edition. Edited by Horacio D. Espinosa and Gang Bao.
© 2013 John Wiley & Sons, Ltd. Published 2013 by John Wiley & Sons, Ltd.

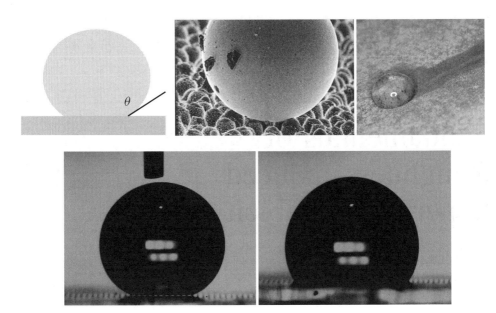

Figure 4.1 *Top left*: Angle θ is defined as the contact angle of a drop on a surface. $\theta > 90°$ is regarded as a hydrophobic, and $\theta < 90°$ is regarded as a hydrophilic contact. *Top middle*: SEM image of a drop on a lotus leaf. *Top right*: Self-cleaning behavior due to a drop rolling off on a lotus leaf. Top figures (Middle & Right) – Reprinted with permissions from [1]. © 1997 Springer. *Bottom left*: A drop on a textured surface in a Cassie–Baxter state [2,3]. The presence of air in the roughness grooves is apparent from windows of light underneath the drop. *Bottom right*: A drop on a textured surface in a Wenzel state [4]. The liquid has impaled the roughness grooves; thus, no light is visible underneath the drop. Bottom figures (Left & Right) – Reprinted with permission from [5]. © 2003 American Chemical Society

different configurations at the liquid–solid interface (Figure 4.1), have been known for more than 60 years. Some of the early applications of this phenomenon were in the textile industry.

The above phenomenon has been observed in nature. Water drops deposited on "super-hydrophobic" or water-hating leaves, e.g. lotus, tend to bead up (Figure 4.1). Additionally, the water drops roll off easily from these surfaces and clean the surface in the process (Figure 4.1). In the mid 1990s, Barthlott and Neinhuis [1] took scanning electron microscope (SEM) images of the surface of lotus and other leaves. They found that the surface of lotus and similar leaves had a hierarchical roughness structure. It was observed that drops deposited on lotus leaves settle on top of the peaks of the double roughness structure [1]; that is, the drops are in the CB state [3] (see Figure 4.1). Since there are air pockets below the drop, it can move easily and clean the surface in the process. Barthlott and Neinhuis named this the lotus effect [1].

Low resistance to flow and self-cleaning capability made lotus-type surfaces ideal for technological applications. Given the advances in micro- and nano-fabrication technologies in recent decades, there has been a worldwide interest in artificially fabricating such surfaces. Research over the past decade has subsequently unveiled a wide variety of additional applications; for example, substrates for efficient dropwise condensation and heat transfer, nucleate boiling heat transfer, desalination, deicing, efficient drop shedding in steam turbines and fuel cells, textiles, oil–water separation, reduced biofouling, biofilm growth and tissue engineering, and substrates to study cell mechanics [6–14].

The performance in the above applications relies greatly on the wetting state of liquid on rough surfaces. In one of the states, the liquid resides on top of roughness features; that is, in a CB state (Figure 4.1) [3]. A state in which the liquid impales the roughness grooves represents another commonly observed scenario called the Wenzel state (Figure 4.1) [4]. Depending on the application, one or the other state is desired. By the early 2000s, it was realized that roughness geometry is crucial to determine which of the two states is observed experimentally. There was a need to develop a theoretical understanding of this phenomenon. This chapter summarizes the fundamental theory underlying this phenomenon and how it can be used to design appropriate textures. Thus, micro/nanoscale texturing together with surface chemistry, made possible by nanotechnology, can help develop better materials for many engineering and bioengineering applications.

4.2 Thermodynamic Analysis: Classical Problem (Hydrophobic to Superhydrophobic)

One of the most studied configurations in roughness-induced superhydrophobicity has been a drop deposited on a textured surface in an open environment of air [15]. There are two types of situations that have been considered. The first situation is the one where the surface material is hydrophobic and roughness is used to obtain superhydrophobic behavior. In the second situation the surface material is hydrophilic (contact angle <90°), but roughness is used to obtain hydrophobic or superhydrophobic behavior. The first case will be considered in this section and the second case will be presented in Section 4.3. Both these scenarios will be referred to as the classical problem. Another important scenario that has not been widely studied, but is technologically important nonetheless, will be discussed in Section 4.4.

Typically, three key features are sought in the classical configuration of a drop deposited on a surface in ambient air (especially to enable the low-drag scenario where the drop moves easily on the surface):

1. The liquid–solid contact has the CB state; that is, the liquid does not occupy the grooves of the roughness geometry (see Figure 4.1). The liquid–air interface hanging between the roughness protrusions provides low resistance to liquid flow next to the solid, thus effectively reducing the drag experienced by the liquid [16].
2. The area of footprint of the drop on the surface is minimized and the apparent contact angle is large.
3. The contact angle hysteresis is low. One manifestation of this is the fact that the drop on a tilted surface does not roll off until the surface has a critical angle of inclination. This happens because the contact angles at the advancing and receding fronts are unequal. The greater the difference between the advancing and receding angles, the greater is the surface tilt required to move the drop (due to its own weight) and the greater is the hysteresis. Low-hysteresis surfaces are desired for high mobility of a drop on a surface.

It is noted that the term superhydrophobicity has been used variously in literature. It is most commonly used to imply feature (1) above or feature (2) and (3) above. It is now known that feature (1) is one essential ingredient to obtain feature (3) – a drop in a Wenzel state where the liquid impales the roughness grooves exhibits very high hysteresis [17,18]. Thus, in this work, a desirable superhydrophobic surface is regarded to be one that has all three of the features above. Finding appropriate roughness geometries that enable the three features above is essential. In this section, a general thermodynamic framework presented in an earlier study [19] is summarized. It can be used to gain insights into the design of surface texture for each of the features above.

4.2.1 Problem Formulation

Consider a static or equilibrium angle θ_e of the substrate material which is assumed to be hydrophobic; that is, $\theta_e > 90°$. Consider a liquid drop deposited on a rough substrate with a pillar-type roughness geometry as shown in Figure 4.2. Analysis of a pore-type geometry will be discussed in Section 4.3. The drop size is considered large compared with the length scale of the roughness. Assume that the ambient air is at constant temperature T_0 and pressure P_0. Let the air phase be denoted by α. Let the liquid drop phase be denoted by β. The different interfaces between the solid, liquid, and air will be collectively denoted by phase σ, which will be assumed to be sharp. Stable equilibrium states of this system can be obtained by minimizing the availability, which is the relevant energy function [19,20]

$$U^\sigma + \sum_\varphi (U^\varphi + P_0 V^\varphi - T_0 S^\varphi) \tag{4.1}$$

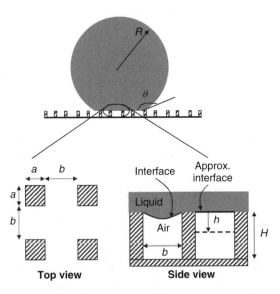

Figure 4.2 Schematic of a drop on a pillar-type roughness geometry [19]. A square pillar geometry is considered simply as a representative case; similar analysis can be done for any other geometric configurations. Reprinted with permission from [19]. © 2010 American Chemical Society

where

$$U^\varphi = T^\varphi S^\varphi - P^\varphi V^\varphi + N^\varphi \mu^\varphi \tag{4.2}$$

In Equations (4.1) and (4.2), superscripts φ denote phases α or β. The summation with respect to φ in Equation (4.1) represents addition of the quantity in the parentheses for phases α and β. U^φ is the total internal energy, V^φ is the total volume, S^φ is the total entropy and μ^φ is the chemical potential of phase ϕ. T^φ and P^φ are the temperature and pressure of phase φ. N^φ is the number of moles of phase φ. U^σ is the total surface energy of all the interfaces. It is obtained by summing up the surface tension (energy) times the area of all the interfaces. Contact-angle measurements are assumed to be made on time scales much shorter than the evaporation time scale. Therefore, in Equation (4.1), it is assumed that each of the phases α and β is a single component and that the mass of each individual phase remains constant. This assumption implies that chemical equilibrium (i.e., equating the chemical potentials of all component species in the phases) is not imposed across liquid–air interfaces.

Using Equations (4.2) in (4.1), the stable equilibrium states are obtained by minimizing

$$U^\sigma + \sum_\varphi [(P_0 - P^\varphi)V^\varphi + (T^\varphi - T_0)S^\varphi + N^\varphi \mu^\varphi] \tag{4.3}$$

The gravitational potential is ignored by assuming that the drop sizes are less than the capillary length scale (for water it is $\sim$2.7 mm). By imposing thermal equilibrium, each phase is assumed to be at temperature T_0. Thus, the temperature terms in Equation (4.3) make no contribution. Since N^α is assumed constant, the corresponding chemical potential term has no effect on the minimization of availability. Phase β is considered incompressible and of constant mass. Consequently, its pressure–volume and chemical potential terms can be reformulated in terms of a Lagrange multiplier corresponding to the volume constraint. The availability A to be minimized becomes

$$A = (P_0 - P^\beta)(V^\beta - V_0^\beta) + U^\sigma \tag{4.4}$$

which is the Gibbs free energy of the system. The first term is due to the volume constraint on phase β. The volume V^β of phase β will be equal to its fixed volume V_0^β after the minimization. The corresponding Lagrange multiplier is $P_0 - P^\beta$, where P^β is the mechanical pressure of phase β; the thermodynamic pressure is undefined for an incompressible fluid.

Consider the drop in a CB state with the liquid–air interface hanging between the pillar tops as shown in Figure 4.2. The hanging liquid–air interface is curved. For the purposes of quantifying its surface energy it will be assumed to be flat (see Figure 4.2), since the actual shape is not known unless a more complex problem is solved. Upon substituting the expressions for surface energy in Equation (4.4), the availability A_C corresponding to the CB state is given by [19]

$$A_C = (P_0 - P^\beta)(V^\beta - V_0^\beta) + \sigma_{lg}\underbrace{[2\pi R^2(1 - \cos\theta)]}_{S_{cap}} - \sigma_{lg}\underbrace{(\pi R^2 \sin^2\theta)}_{A_{base}} \cos\theta_r^c + \sigma_{sg}A_{tot}$$

$$\tag{4.5}$$

where R is the radius of the drop (Figure 4.2), S_{cap} is the area of the spherical cap of the drop, θ is the apparent contact angle, A_{base} is the base area of the drop projected on the horizontal plane, A_{tot} is the total solid–air area of the dry surface, and σ_{sg} is the solid–air surface energy. θ_r^c is the apparent contact angle based on the CB formula [2]: $\cos \theta_r^c = \varphi \cos \theta_e + \phi - 1$, where ϕ is the area fraction of the pillar tops in the horizontal plane. The third and fourth terms on the right-hand side of Equation (4.5) denote the changes in energy of an initially dry substrate that is wetted in the region corresponding to A_{base}. The question of whether the availability A_C of the CB state corresponds to an equilibrium state or not and, if so, whether it is the global minimum or not is explored below.

Consider that the interface at the top of the pillars, in the CB state, impales the grooves to some depth h (Figure 4.2). The availability A of this general state can be written as [19]

$$A = (P_0 - P^\beta)(V^\beta - V_0^\beta - V_{imp}) + (P_0 - P^\beta)V_{imp} + \sigma_{lg} 2\pi R^2 (1 - \cos \theta)$$
$$- \pi R^2 \sin^2 \theta \sigma_{lg} \cos \theta_r^c - (r - 1)\pi R^2 \sin^2 \theta \sigma_{lg} \cos \theta_e \qquad (4.6)$$

where V_{imp} is the volume of the fluid that has impaled the grooves. $(P_0 - P^\beta)V_{imp}$ is added and subtracted because it is the pressure–volume energy associated with the liquid that has impaled the roughness. The area of the pillar sides relative to the horizontal area is given by $r - 1 = 4ah/(a + b)^2$, where a, b, and h are as defined in Figure 4.2. The term $\sigma_{sg} A_{tot}$ is not included since it is constant and plays no role in the minimization. Minimizing the availability in Equation (4.6) will give the stable equilibrium states of the system.

The volume of the liquid phase V^β is given by

$$V^\beta[R, \theta, h] = \underbrace{(\pi R^3/3)(2 - 3\cos \theta + \cos^3 \theta)}_{V_{up}} + \underbrace{h(1 - \phi)\pi R^2 \sin^2 \theta}_{V_{imp}} \qquad (4.7)$$

where V_{up} is the volume of the drop shape above the substrate. The square bracket denotes "function of" and the expression for the volume V_{imp} of the liquid that has impaled the roughness is also indicated in Equation (4.7). The availability to be minimized can be written as [19]

$$A[P^\beta, R, \theta, h] = (P_0 - P^\beta)(V_{up}[R, \theta] - V_0^\beta) + \sigma_{lg} S_{cap}[R, \theta] - \sigma_{lg} A_{base}[R, \theta]$$
$$\times \left[\cos \theta_r^c + \frac{4ah}{(a + b)^2} \cos \theta_e + h(1 - \phi)\frac{P^\beta - P_0}{\sigma_{lg}} \right] \qquad (4.8)$$

where S_{cap} and A_{base} are as defined in Equation (4.5), and expressions for $r-1$ and V_{imp} have been substituted. Equation (4.8) shows that the availability should be minimized with respect to four parameters: P^β, R, θ, and h. The macroscopic parameters of the drop are represented by P^β, R, and θ, whereas h represents the microscopic parameter that specifies the location of the liquid–air interface in the roughness geometry.

The term multiplying A_{base} in Equation (4.8) represents the surface energy per unit area associated with the wetting of the rough substrate. Thus, it represents the energy of the microscopic state of the substrate. The others terms quantify the energy of the macroscopic drop. The microscopic state can be analyzed independent of the macroscopic problem resulting in a decoupling in the energy analysis [19].

In the next three subsections the thermodynamic formulation above will be used to obtain insights into each of the three features (1)–(3) listed at the beginning of Section 4.2. This will be done by considering the four questions listed below:

1. How do we obtain the CB state (Section 4.2.2)?
2. How do we predict transition to the Wenzel state so that it can be avoided (Section 4.2.3)?
3. How does the microscopic state affect the apparent contact angle of the drop (Section 4.2.4)?
4. How do we model contact-angle hysteresis (Section 4.2.5)?

4.2.2 The Cassie–Baxter State

Consider a drop on top of a rough substrate or a liquid pool/bath on top of a rough substrate. The liquid–solid contact area will be in the CB state if a liquid–air interface can hang between the pillar tops; that is, if the CB state is a stable equilibrium configuration. The related analysis presented next is thus relevant not only to a "drop on a surface" configuration, but also for the "liquid bath on a surface" configuration.

Equation (4.8) implies that the energy E_{lg} per unit area, nondimensionalized by σ_{lg}, at the liquid–solid contact (i.e., the microscopic state) is given by

$$E_{lg} = \underbrace{-\cos\theta_r^c}_{\text{Cassie}} + \underbrace{\frac{-4ah}{(a+b)^2}\cos\theta_e}_{\text{wetting}} + \underbrace{h(1-\phi)\frac{P_0 - P^\beta}{\sigma_{lg}}}_{\text{pressure–work}} \tag{4.9}$$

where the first term in Equation (4.9) corresponds to the effective interfacial energy in the CB state, the second term corresponds to the wetting of roughness grooves if the liquid–air interface impaled by a distance h (see Figure 4.2), and the third term is related to the work done by pressure during impalement. A schematic plot of Equation (4.9) as a function of h is shown in Figure 4.3. When the slope is positive it implies that the CB state is in stable equilibrium; that is, the liquid–air interface can remain pinned at the top of the pillars (no transition scenario in Figure 4.3) [19]. It follows from Equation (4.9) that this is possible when [19]

$$\underbrace{h(1-\phi)(a+b)^2(P^\beta - P_0)}_{\text{pressure–work}} < \underbrace{-4ah\sigma_{lg}\cos\theta_e}_{\text{wetting-energy}} \tag{4.10}$$

that is, when the work done by pressure during impalement (left-hand side) is not sufficient to overcome the energy required to wet the roughness grooves during impalement (right-hand side). Thus, if the roughness is such that, for all the operating pressures $P^\beta - P_0$, Equation (4.9) is satisfied, then the CB state will be feasible and transition to the Wenzel state has to overcome an energy barrier (Figure 4.3).

Figure 4.3 shows that when the liquid–air interface proceeds to the bottom of the grooves at depth H, the nondimensional energy per unit area E_{lg} reaches a maximum value followed by a decrease by an amount equal to $(1-\phi)(1+\cos\theta_e)$ corresponding to the wetting of the bottom of the roughness groove. This results in a border minimum

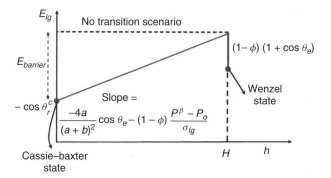

Figure 4.3 A schematic of the nondimensionalized effective liquid–solid contact energy E_{lg} as a function of the degree of invasion h of the liquid–air interface into the roughness groove

which represents the energy corresponding to the Wenzel state [21]. Thus, the CB and Wenzel states are separated by an energy barrier at a given pressure (Figure 4.3) [19]. If the CB state is desirable, then robust surfaces would be those for which the CB state is lower in energy compared with the Wenzel state, as shown in the scenario sketched in Figure 4.3. This implies that the energy barrier $E_{\text{b,W–C}} = (1 - \phi)(1 + \cos\theta_e)$ for the Wenzel-to-CB transition would be smaller than

$$E_{\text{b,W–C}} = \frac{-4aH}{(a+b)^2}\cos\theta_e + H(1-\phi)\frac{P_0 - P^\beta}{\sigma_{\text{lg}}}$$

for the CB-to-Wenzel transition at a given pressure. Indeed, these barriers will depend on the pressure. As the pressure increases, the transition to the Wenzel state becomes increasingly favorable until a pressure where Equation (4.10) is no longer satisfied. At that pressure, the CB state is no longer an equilibrium state and cannot be sustained (equivalently, the slope of the line sketched in Figure 4.3 will be negative); this is equivalent to the de-pinning of the liquid–air interface from the top of the pillars [19]. A good design strategy is to ensure that a sufficiently large energy barrier exists at the largest operating pressure.

When the roughness texture is at the nanoscale, the energy barriers, at a given pressure, are important to determine the probability of transition (according to the equipartition theorem) from one state to another due to thermal fluctuations [22]. Similar to the Arrhenius equation for reaction rates, the transition rates between CB and Wenzel states would depend on the energy barriers:

$$k_{\text{C–W}} \propto e^{-A(E_{\text{b,C–W}}/k_{\text{B}}T)} \qquad \text{and} \qquad k_{\text{W–C}} \propto e^{-A(E_{\text{b,W–C}}/k_{\text{B}}T)} \qquad (4.11)$$

where $k_{\text{C–W}}$ and $k_{\text{W–C}}$ are the rate constants corresponding to CB-to-Wenzel and Wenzel-to-CB transitions, respectively, k_{B} is the Boltzmann constant, and $A = \sigma_{\text{lg}} \times$ (scale for area) is a dimensional constant. Correspondingly, a first-order model leads to the ratio of probability of CB and Wenzel states:

$$\frac{p_{\text{C}}}{p_{\text{W}}} \propto \exp\left(A\frac{E_{\text{b,C–W}} - E_{\text{b,W–C}}}{k_{\text{B}}T}\right) \qquad (4.12)$$

where p_C and p_W are the probabilities of the CB and Wenzel states, respectively. For $p_C \gg p_W$, it is desirable to have $E_{b,C-W} \gg E_{b,W-C}$. This is feasible when the CB state is lower in energy than the Wenzel state.

Designing Surface Roughness for the Cassie–Baxter State

The above discussion implies that, ideally, the roughness should be designed such that:

1. The CB state is lower in energy than the Wenzel state.
2. The effective surface energy in the CB state, which is equal to $-\cos\theta_r^c$ ($=-\phi\cos\theta_e - \phi + 1$), is large, so that liquid deposited on the surface will not prefer to spread on it.

It follows from the above formulation that this can be accomplished by the following choices, among others:

1. Surface roughness that results in small values of liquid–solid contact area; that is, ϕ is small. This is evident in the success of small cross-sectional area pillar geometries in producing superhydrophobic surfaces [23,24].
2. Pillar geometries with large H so that transition to the Wenzel state is minimized [23,24]. These are surfaces on which even if drops are deposited with high velocity, they will typically bounce back (if they do not break up) instead of transitioning to the Wenzel state.
3. Maximizing θ_e. This can be accomplished in two ways. Use inherently high contact-angle hydrophobic materials (using material chemistry) or use double/multiple roughness geometries so that the effective θ_e for the coarsest roughness scale is relatively high [25]. This approach has also successfully led to superhydrophobicity in nature (e.g., lotus [1]) and on fabricated surfaces [26].

4.2.3 Predicting Transition from Cassie–Baxter to Wenzel State

The above formulation helps to consolidate various mechanisms of transition from the CB to Wenzel state [19]. There are at least three basic modes that have been proposed:

1. De-pinning of the liquid–air interface hanging between pillars, leading to the invasion of the liquid into the roughness grooves (Figure 4.4) [27–29].
2. Transition by way of the liquid–air interface touching the bottom of the roughness groove due to its sag [27–30].
3. Transition triggered by de-pinning at the contact line in a drop on a surface configuration [35,36].

The first two mechanisms (Figure 4.4), discussed below, are relevant in the "drop on a surface" configuration as well as the "liquid pool/bath on a surface" configuration. The third, not elaborated here, is relevant in the "drop on a surface" configuration [35,36].

The liquid–air interface hanging between pillars is curved due to the pressure difference $P^\beta - P_0$ across it [27–29]. If the hanging interface is such that it cannot remain pinned at

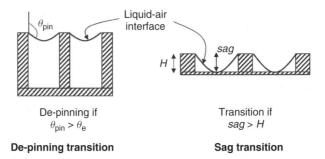

Figure 4.4 Side views of the transition from the CB to Wenzel state due to de-pinning and sag mechanisms [27–34]. A pillar-type roughness geometry is considered. Reprinted with permission from [19]. © 2010 American Chemical Society

the pillar tops, then it proceeds downward into the roughness grooves and fully wets the surface. Lack of pinning occurs if the contact angle formed by the liquid–air interface is greater than the maximum contact angle that can be sustained at a corner [37,38]. This is the de-pinning mechanism. In the sag-based mechanism, if the curved liquid–air interface is such that it touches the bottom of the roughness groove, then transition occurs [27,29].

Transition Criteria Based on Geometric Considerations

Formulas based on geometric considerations have been proposed to identify the criteria for de-pinning or sag-based transitions [27–30,32]. To obtain an expression for the de-pinning transition based on geometric considerations, consider a spherical liquid–air interface hanging between diagonally opposite pillars and impose the condition that the interface will impale the grooves if the contact angle at the pillar tops is greater than the equilibrium contact angle. The corresponding condition for de-pinning or impalement is given by [27–30]

$$P^\beta - P_0 > \frac{-4\sigma_{lg}\cos\theta_e}{\sqrt{2}b} \tag{4.13}$$

Another condition can be derived if the interface is considered as hanging between adjacent pillars. In that case the de-pinning condition is given by

$$P^\beta - P_0 > \frac{-4\sigma_{lg}\cos\theta_e}{2b} \tag{4.14}$$

The criteria above are based on approximations, since they are derived by assuming a shape of the liquid–air interface which is not known exactly unless the complex shape of the interface is solved.

The sag criterion for transition from the CB to Wenzel state is given by [27,29]

$$H < R\left(1 - \sqrt{1 - \frac{b^2}{2R^2}}\right) \tag{4.15}$$

where R is the radius of curvature based on the pressure; that is, $R = 2\sigma_{lg}/(P^\beta - P_0)$.

Transition Criterion Based on Energy Formulation

In the approximate energy formulation, presented in previous sections, the CB state is not sustained if it is no longer an equilibrium state. In this case the liquid–air interface cannot remain hanging at the top of the pillars and, therefore, proceeds into the roughness grooves to wet it. Thus, it follows from Equations (4.9) and (4.10) that transition from the CB to Wenzel state will occur if

$$P^\beta - P_0 > \frac{-4\sigma_{lg} \cos \theta_e}{2b \left(1 + \frac{b}{2a}\right)} \tag{4.16}$$

for a pillar-type geometry. This criterion for transition can be extended to pillars of any cross-section. This can be done by generalizing Equations (4.9) and (4.10), which imply that transition will occur when the work done by pressure is greater than the energy required to wet the sides of the pillars; that is, when [39]

$$\underbrace{(P^\beta - P_0)A_p h}_{\text{pressure–work}} > \underbrace{-\sigma_{lg} P_p \cos \theta_e h}_{\text{wetting-energy}} \tag{4.17}$$

where A_p is the projected area (on the horizontal plane) of a unit cell (of the periodic pillar structure) upon which the pressure difference acts during impalement of the liquid–air interface [39]. P_p is the perimeter cross-section of a single pillar. Thus, the generalized energy-based criterion for transition becomes

$$(P^\beta - P_0) > \frac{-\sigma_{lg} P_p \cos \theta_e}{A_p} \tag{4.18}$$

for an array of pillars of any cross-section. Equation (4.18) leads to Equation (4.16) if expressions for A_p and P_p for a square pillar geometry are substituted.

The criteria in Equations (4.16) or (4.18) are derived solely based on energy considerations. The local geometry of the liquid–air interface is not considered in deriving the de-pinning or sag transition condition, as has been previously reported.

Correlation between Energy-Based and Geometry-Based Transition Criteria

The correlation between the energy-based criterion and the geometrically derived de-pinning and sag transition conditions can be understood as follows [39].

Let $R = 2\sigma_{lg}/(P^\beta - P_0)$ be the mean radius of curvature of the liquid–air interface corresponding to the pressure drop $P^\beta - P_0$ across it. Let there be a length scale R_{crit} defined as

$$R_{crit} = \frac{-2A_p}{P_p \cos \theta_e} \tag{4.19}$$

Equation (4.18) implies that transition will occur if $R < R_{crit}$. Thus, R_{crit} is an average measure of the radius of curvature of the liquid meniscus required to impale the roughness grooves. As stated above, this radius of curvature is not derived based on local geometric considerations, but follows from the overall energetics of transition.

For square cross-section pillars $R_{crit} = -2b[1 + (b/2a)]/2 \cos \theta_e$. It can be verified that this critical radius of curvature of the liquid–air interface scales in accordance with the geometrically derived critical radius of curvature for de-pinning transition in Equations (4.13) and (4.14), $R_{crit} \sim -b/\cos\theta_e$, when the post spacing b is small compared with the cross-sectional size a of the pillar [39]. Note that the geometrically derived de-pinning condition is also an approximation.

When the post spacing b is large, the critical radius of curvature for transition $R_{crit} \sim -b^2/2a\cos\theta_e$ according to the energy-based criteria in Equations (4.16) and (4.18) [39]. For large b/a this estimate is typically greater than $-b/\cos\theta_e$, which is the geometrically derived critical radius of curvature for de-pinning transition. Thus, the energy-based criterion is a more restrictive estimate for wetting transition at large b/a. Additionally, at large b/a it is the sag transition that is more restrictive than the de-pinning transition. The critical radius of curvature, based on geometric considerations, for sag transition is b^2/H according to Equation (4.15). Thus, if $H \sim a$, it turns out that at large b/a the energy-based critical radius of curvature, Equation (4.19), has the same scaling as that due to sag transition, Equation (4.15); that is, $R_{crit} \sim -b^2/2a\cos\theta_e \sim -b^2/H$. This is convenient, since de-pinning transition is relevant at small b, whereas sag transition is relevant at large b. Therefore, the energy-based criterion can conveniently cover the limits of de-pinning and sag transitions when $H \sim a$. Thus, a single expression can be used to analyze transition instead of differentiating between de-pinning and sag-based criteria [39]. This is further corroborated by the agreement of experimental data with the energy-based criterion over a range of pillar spacings reported elsewhere [39].

Pressure Sources for Transition

Equation (4.18) is the primary equation to predict transition from the CB to Wenzel state for surface textures with an array of pillars. The pressure drop $P^\beta - P_0$ drives the transition. This pressure drop can be created in a variety of ways. Accordingly, $P^\beta - P_0$ will scale differently, leading to a rich variety of phenomena. Some (but not all) of the pressure sources studied experimentally and compared with theoretical predictions of transition are listed below.

1. Laplace pressure of a drop [27–30,39]. In this case, $P^\beta - P_0 \sim 1/R_{drop}$, where R_{drop} is the radius of curvature of the drop. This leads to an interesting behavior where smaller drops transition to the Wenzel state. If a drop evaporates starting with a CB state, then as the drop radius becomes smaller than R_{crit} it transitions to a Wenzel state. Typical drop length scales are smaller than the capillary length scale; hence, the effect of gravity is not very important.
2. Pressure due to squeezing of drops or pressurizing the liquid [26].
3. Dynamic pressure during drop impact [30,31,33,34]. Drops impacting a textured surface can bounce off the surface if they do not transition to the Wenzel state. Transition to the Wenzel state can be predicted based on Equation (4.18) with $P^\beta - P_0 \sim \rho V_d^2$, where ρ is the liquid density and V_d is the velocity of the drop at impact. This scaling for pressure arises from the convective inertia term in the fluid equation – also termed the dynamic pressure scaling.
4. Pressure due to transient or oscillatory motion [40]. In this case, $P^\beta - P_0 \sim \rho(dV_d/dt) \sim \rho\omega^2$, where ω is the frequency of oscillation.

5. Water hammer pressure [30,39]. In this case, rapid changes in velocity, which can occur during droplet impact or even during droplet deposition [39], can lead to pressure spikes. The pressure in the drop then scales as $P^\beta - P_0 \sim \rho(dV_d/dt) \sim \rho c_s V_d$, where c_s is the speed of sound in the liquid. Since c_s is typically large in liquids, this can lead to huge pressure spikes similar to water hammer in pipelines and lead to transition to a Wenzel state even when the dynamic pressure ρV_d^2 is not sufficient to induce transition [30].

4.2.4 The Apparent Contact Angle of the Drop

Equation (4.8) shows that the availability is linear with respect to h under the approximations considered. As discussed above, the possible stable equilibrium values of h are 0 and H, where H is the pillar height. Once h is known, A can be extremized with respect to P^β, R, and θ [41]. Setting $\partial A/\partial \theta = 0$, $\partial A/\partial P^\beta = 0$, and $\partial A/\partial R = 0$ gives equilibrium conditions. For example, it follows that, at equilibrium, the apparent contact angle θ is given by

$$\cos \theta = \cos \theta_r^c = \varphi \cos \theta_e + \phi - 1 \quad \text{for } h = 0 \qquad (4.20)$$

which is the CB formula. Additionally, the extremization conditions imply that $V^\beta[R, \theta, h] = V_0^\beta$ and $P^\beta - P_0 = 2\sigma_{lg}/R$. These two equations and the apparent contact-angle formula, Equation (4.20), can be solved for θ, R, and P^β for a known value of h. It can be shown that this equilibrium solution is in fact the stable equilibrium with respect to θ, R, and P^β [21,42].

As seen above, $h = H$ is also a possible stable equilibrium solution corresponding to the Wenzel state. In this case, an additional term equal to $-\sigma_{lg}A_{base}[R, \theta](1 - \phi)(1 + \cos \theta_e)$ must be added to the availability function to account for the wetting of the bottom of the grooves. Upon minimizing the availability function in Equation (4.8) with this change, the apparent contact angle is given by [19]

$$\cos \theta = \frac{4aH}{(a + b)^2} \cos \theta_e + 2(\phi - 1)\frac{H}{R} \quad \text{for } h = H \qquad (4.21)$$

As expected, the first term on the right-hand side of Equation (4.21) is the same as the Wenzel formula [4]. The term involving H/R in Equation (4.21), where R is the drop radius, is an additional term associated with the pressure–volume energy of the liquid that has impaled the roughness. It is negligible for large drops or small pillar heights. Physically, it implies the tendency of the liquid at high pressure to spread laterally through the rough substrate. Consequently, its effect is to reduce the apparent contact angle. This term is equivalent to a negative line tension τ given by $\tau = -\sigma_{lg}(1 - \phi)H \sin \theta$. It is noted, however, that there is an opposing effect that is not resolved in the approximate formulation above. While accounting for the surface energy associated with the substrate in a Wenzel state, the interfacial energy of the liquid front within the roughness is not included in the theory or in the Wenzel formula. This contribution to the energy will scale as $\sigma_{lg}H/R \sin \theta$ and it will oppose spreading due to pressure; that is, the corresponding line tension will be positive. This contribution to the apparent contact angle is expected to reduce the effect of the H/R term in Equation (4.21). Appropriate modeling of the line-tension-type contributions is essential to describe the lateral invasion of the liquid within the grooves. There could be a different final static equilibrium of the liquid front *within* the roughness grooves (e.g., circular, square) depending on the dynamic process during lateral invasion of the grooves by the liquid [43,44]. This issue is not considered above. A circular front was assumed similar to the assumption inherent in the CB and Wenzel models.

Comments on the Energy Minimization Approach Versus the "Contact-Line Approach"

The theoretically determined apparent contact angle, discussed above, is based on minimization over a coarser parameter space due to the approximations involved. For example, the drop shape is assumed spherical and the contact line is assumed circular, when in reality the actual contact line is distorted and not perfectly circular. The primary question is whether this theoretically determined approximate model for the apparent contact angle is in agreement with experimental observations. Consequently, the validity of the theory underlying the CB and Wenzel formulas has been questioned and debated in the literature [45–47]. An "alternate" view has been proposed, which states that it is the contact-line configuration that determines the apparent contact angle [45,47,48]. This has led to the so-called divide between the contact line versus the contact area/area fraction (note that area fraction arises in the CB formula) approaches to understand the apparent contact angles of drops on rough surfaces. This issue is discussed below.

A fundamental principle to find a static equilibrium state of a drop is energy minimization (which reduces to surface energy minimization in certain scenarios) [20]. Thus, if the energy minimization problem is solved with full resolution of the surface geometry, then it will lead to variety of equilibrium drop states, one of which will be the lowest energy state [49,50]. Such simulations using Surface Evolver [51] have been reported in the past [50]. Other methods for full-resolution computations have also been reported [52]. It has been shown theoretically that if the drop size is large compared with the length scale of the roughness features, then the CB and Wenzel formulas are good approximations to the computed lowest energy drops corresponding to their respective wetting states at the liquid–solid interface [50]. The drop shapes obtained by solving the energy minimization problem with full resolution of the surface geometry are consistent with experimental observations. The full-resolution calculations give *the contact line as a part of the solution of the energy minimization process*. Thus, the contact-line approach is not an alternate point of view. Its effect is inherently present in the full-resolution energy minimization formulations.

An argument used by some against the energy-minimization-based approach is also the fact that in reality there is force balance at the contact line [45,47], and that, historically, Young did not consider this problem in terms of surface energy. Thus, it is claimed that force balance at contact lines is the fundamental approach. However, force balance (or the equilibrium problem) can be equivalently formulated in terms of an energy function – a minimum of which gives the force balance or the equilibrium state. Thus, force balance at the contact line is resolved in the energy-minimization-based approaches – these are not independent methodologies.

As discussed above, the CB or Wenzel formulas are not based on full-resolution energy minimization. They are obtained based on an approach that searches for a minimum energy state in a *coarser parameter space*; hence, it is an approximate approach. Therefore, these formulas represent an approximate contact line which does not include all the distortions of the actual contact line. These formulas represent an average (or homogenized) measure of the minimum energy state experienced by the contact line [53]. Thus, the area fraction in, for example, the CB formula for the apparent contact angle is used to estimate the homogenized surface energy *in the neighborhood* of the contact line [54], contrary to counter-examples depicted in some papers [45,47]. Note the extremization discussed before Equation (4.20).

There can be disagreements between the approximate theoretical formulas for the apparent contact angles and the measured contact angles. Such disagreements do not implicate the energy minimization approach, but rather highlight the limitations caused by the approximations. In spite of the disagreements, these formulas provide useful guidance to design rough superhydrophobic surfaces, since they provide a measure of the ground state of the liquid–surface contact [15]. If one raises the ground-state energy, then the surface would tend to be superhydrophobic.

In light of the potential disagreement between approximate theoretical formulas and experiments, it is still necessary to enquire whether the approximate formulas are practically useful or not. This issue is considered next, in Section 4.2.5, in the context of contact-angle hysteresis.

4.2.5   Modeling Hysteresis

The contact angle of a drop on a surface that is observed in experiments depends on how the drop is formed. If it is formed by increasing the drop volume then the observed angle (of a static drop) is called the advancing angle θ_{adv} and if the drop is formed by decreasing the drop volume then the observed angle (again, of the static drop) is called the receding angle θ_{rec}. This is typically true of drops on all materials – not just on textured surfaces. θ_{adv} and θ_{rec} also depend on velocity of the moving contact line. However, in this discussion, only the static case is considered.

The difference between θ_{adv} and θ_{rec}, where $\theta_{adv} > \theta_{rec}$, is a measure of the contact-angle hysteresis. The theoretically predicted CB and Wenzel angles are typically between the advancing and receding angles of drops on rough surfaces (in the respective configurations at the liquid–solid interface). It is known that drops in the Wenzel state exhibit large hysteresis [17,18]. For this reason they do not move easily on surfaces. Thus, the Cassie state is a preferred state in many applications and it is considered in the discussion that follows.

Three questions will be considered in this section:

1. How does the CB formula compare with the experimentally observed advancing and receding contact angles? This will help determine the utility of the CB formula.
2. What is the origin of the deviation of the CB angle from the experimentally observed angles?
3. What predictive models could be developed to quantify the deviation of the CB angles from the experimentally observed angle?

Comparison of the Cassie–Baxter Angle with Experimental Data

Kwon *et al.* [53] compared two sets of experimental data for advancing and receding contact angles to the predictions from the CB formula. Pillar- and hole-type rough surfaces were considered (Figure 4.5). A range of roughness geometries were fabricated by varying the spacing between the pillars and holes in the respective cases. The CB angles were predicted according to the following equation:

$$\cos\theta_{Cass} = \phi\cos\theta_{(A/R)} + \phi - 1 \tag{4.22}$$

where θ_{Cass} is the predicted CB angle. θ_A is the advancing angle and θ_R is the receding angle of the material used to make the textured surface. If θ_A is used in Equation (4.22), then the corresponding value of θ_{Cass} is the advancing angle on the textured surface that is predicted by the CB formula. This was compared (see Figure 4.5) with the experimentally measured advancing angle on the textured surface at different values of ϕ. A similar comparison was made for the receding angle (see Figure 4.5). Thus, two fundamentally different kinds of roughness features were considered: a pillar-type geometry that results in a discontinuous contact line and a hole-type roughness that results in a continuous contact line. Figure 4.5 shows that in all cases considered – except for the advancing front on pillar-type roughness, which has strong pinning and, therefore, a constant contact angle – the apparent contact angles follow the overall trend of the CB formula. Replotting data from other work for pillar-type geometry in the same way led to similar conclusions [55]. Deviations from the CB formula did not significantly compromise the primary trend except for one case (advancing angle on pillars). Thus, disregarding the CB formula as irrelevant is not warranted in general.

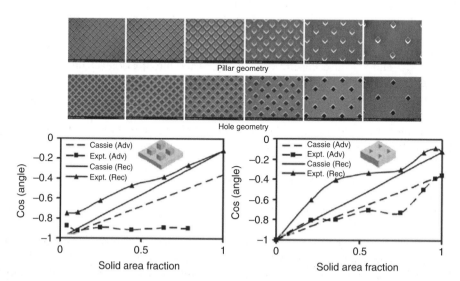

Figure 4.5 Pillar (top) and hole (middle) geometries for contact-angle measurements. Comparison of experimentally measured contact angles with CB formula for pillar (bottom left) and hole geometries (bottom right). Reprinted with permission from [53]. © 2010 American Chemical Society

Reason for Deviation from the Cassie–Baxter Angle

In the energy minimization framework, the deviation of the CB angle from the experimental angle can be understood by noting two examples:

1. The actual surface energy scanned by the contact line in the neighborhood of its equilibrium state could be different from that implied by the effective surface energy modeled in the CB formula (see Sections 4.2.2 and 4.2.3). Consider a pillar geometry with the contact-line location as shown in Figure 4.6. In the case depicted, the solid area fraction (0.5) scanned by the contact line around its equilibrium state is different from the average area fraction (0.25) of the substrate. Thus, the apparent contact angle is found to be in agreement with a modified CB formula in which the locally scanned area fraction (0.5) is used instead of the average area fraction (0.25) [56].

2. It is known that sharp edges or changes in surface properties can lead to pinning of the contact line [37]. This can cause additional distortion of the contact line, which in turn distorts the liquid–air interface, leading to a different apparent angle [37,53]. Consider Figure 4.7, where advancing and receding fronts on a flat surface with checkerboard pattern of contact angle of the surface are simulated [53]. In each of the advancing and receding cases, the contact-line configurations A and E correspond to the CB state where the apparent contact angle is in reasonable agreement with the CB angle. When the advancing front pins at the edge of the more hydrophobic patch on the checkerboard pattern, the microscopic state of the contact line changes (see contact lines B and C in Figure 4.7) from the CB state to a pinned state. Correspondingly, the apparent angle deviates away from the CB angle. When the contact line de-pins (transition of the contact line from C to D to E in Figure 4.7) the C–B state of the contact line is recovered and the apparent contact angle drops back to the C–B angle within the margin of the numerical error. Similar behavior is observed in case of the receding front in Figure 4.7.

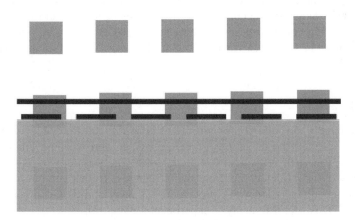

Figure 4.6 Top view of a receding contact line (solid to dotted line) on the top of a pillar geometry. The spacing between the pillars is equal to the width of each pillar. Thus, the solid area fraction scanned by the receding contact line is 0.5, as opposed to a mean solid area fraction of 0.25 used in the CB formula

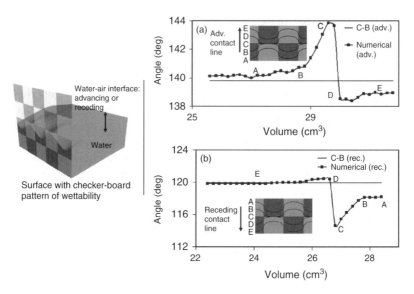

Figure 4.7 Simulation of advancing and receding fronts on a checkerboard pattern of contact angle (left). Calculation of the advancing (right top) and receding (right bottom) angles, and the corresponding shapes of the contact line (inset). Reprinted with permission from [53]. © 2010 American Chemical Society

The above results also show that there are "ideal" CB states of the contact line that result in an apparent angle that is nearly equal to the CB angle. In this ideal CB state the contact line need not be without deformation – it can be wavy, as seen in Figure 4.7 – dispelling the notion that contact-line deformations cause deviations from the CB angle. In fact, in the pinned states shown by configuration C in Figure 4.7, the contact line is straight; even so, the apparent contact angle deviates from the CB state. Thus, it is not the absolute deformation of the contact line that causes deviation from the CB angle, but the distortion of the contact line away from the ideal CB state. This distortion can be caused by a different local area fraction [56] in the neighborhood of the contact line or by pinning [53].

Modeling Hysteresis

This is still a largely unresolved problem. It has been difficult to identify generic models that accurately capture the deviations of the advancing and receding contact angles on rough surfaces from their respective CB predictions. One way forward is to quantify the experimentally observed deviations from the CB angles. To that end, an *ad hoc* general-ization of the theoretical framework provided by Joanny and de Gennes [57] (hereafter referred to as JD) is useful to analyze experimental data [55].

JD developed a theory for chemically or physically heterogeneous surfaces. A physical heterogeneity or defect would be a roughness feature on a base substrate. They assumed that the defects were dilute; that is, their number density is low, so that their effect is additive. They also assumed that the defects were strong, so that the contact line can be effectively pinned, which plays an important role in causing hysteresis.

Under the aforementioned assumptions, JD showed that the advancing contact angle θ_{adv} deviates from the "equilibrium" contact angle θ_E on a surface according to

$$\gamma_{lg}\left(\cos\theta_{adv} - \cos\theta_E\right) = -E_{adv} \tag{4.23}$$

where γ_{lg} is the liquid–air surface tension and E_{adv} is the energy dissipated per unit area by an advancing contact line. The equilibrium contact angle is defined as the apparent contact angle at the contact line in the presence of thermal agitation or mechanical vibrations. This would be the most stable or ground state(s) of the liquid–solid contact. Thus, θ_E could be regarded as the contact angle according to CB or Wenzel formulas depending on the nature of the wetting configuration on the rough surface. This is consistent with the derivations of JD for dilute defects. Similarly, JD showed that the receding contact angle θ_{rec} deviates from θ_E because of the energy per unit area E_{rec} dissipated by a receding contact line:

$$\gamma_{lg}(\cos\theta_{rec} - \cos\theta_E) = E_{rec} \tag{4.24}$$

They obtained expressions for E_{adv} and E_{rec} assuming dilute defects. Equations (4.23) and (4.24) imply that

$$\gamma_{lg}(\cos\theta_{rec} - \cos\theta_{adv}) = E_{rec} + E_{adv} = W_d \tag{4.25}$$

where W_d is the total energy dissipated per unit area around a hysteresis cycle.

Equations (4.23)–(4.25) were derived rigorously by JD for dilute defects. Additionally, the heterogeneous surfaces were assumed to be made of materials that do not exhibit hysteresis. To analyze realistic scenarios, both these assumptions can be restrictive. In typical realistic scenarios the area fractions of roughness features are not low and the materials used to make rough surfaces have unequal advancing and receding contact angles even when they are "flat." However, an *ad hoc* conceptual generalization of the formulation by JD could be useful as an analysis tool [55]. Consider drops in CB states. Consider surfaces with pillar-type roughness features. Consider that the surface is made of a material with an advancing angle θ_A and a receding angle θ_R. Let θ_{EA} be the CB prediction for the advancing contact angle on the rough surface, where $\cos\theta_{EA} = \phi\cos\theta_A + \phi - 1$. Similarly, $\cos\theta_{ER} = \phi\cos\theta_R + \phi - 1$ gives θ_{ER} as the CB prediction for the receding contact angle on the rough surface. *Ad hoc* generalization of Equations (4.23) and (4.24) gives

$$\left.\begin{aligned}\gamma_{lg}(\cos\theta_{adv} - \cos\theta_{EA}) &= -E_{adv}\\ \gamma_{lg}(\cos\theta_{rec} - \cos\theta_{ER}) &= E_{rec}\end{aligned}\right\} \tag{4.26}$$

where E_{adv} and E_{rec} are the energies associated with additional dissipation by advancing and receding contact lines due to the presence of the roughness. A generalized form of Equation (4.25) becomes

$$\gamma_{lg}(\cos\theta_{rec} - \cos\theta_{adv}) = \gamma_{lg}\phi(\cos\theta_R - \cos\theta_A) + E_{adv} + E_{rec} \tag{4.27}$$

where $\gamma_{lg}\phi(\cos\theta_R - \cos\theta_A)$ represents the energy per unit area dissipated due to hysteresis of the surface material and $E_{adv} + E_{rec}$ represents the additional energy dissipated in the hysteresis cycle due to the presence of surface roughness.

Information about E_{adv} and E_{rec} could be obtained from experimental data [55]. It is nontrivial to obtain simple models for E_{adv} and E_{rec} for nondilute area fractions of the pillars. Some mechanisms that influence the values of E_{adv} and E_{rec} have been successfully presented, such as pinning on pillar geometry during advancing [48,55,58], a sharp rise in E_{rec} at low area fractions due to pinning of the contact line on a single pillar during receding [55,58–60], and the effect of local area fraction near the contact line [56]. However, no single model has emerged that quantitatively resolves the key mechanisms underlying the all the trends [55]. Many of the present models can be useful to set meaningful bounds on expected values, or to estimate the primary trends, of advancing and receding angles. However, better models are still required.

4.3 Thermodynamic Analysis: Classical Problem (Hydrophilic to Superhydrophobic)

The above section discusses a scenario where rough surfaces made from hydrophobic materials (contact angles $>90°$) show enhanced hydrophobicity or superhydrophobicity. If, instead, the surface material is hydrophilic (contact angle $<90°$), then the liquid typically prefers to occupy the roughness grooves of a textured surface. This is because it is energetically favorable to wet the surface; this is also termed the wicking effect. This results in superhydrophilic textured surfaces. To obtain hydrophobic or superhydrophobic textured surfaces made from hydrophilic materials, it is essential to sustain air pockets in the roughness grooves. Two approaches have thus far been proposed and demonstrated:

1. Re-entrant surface geometries like, for example, a nail-type geometry can pin a liquid–air interface between the heads of the nails (the local hydrophilic contact angle is satisfied due to the re-entrant shape) [61–64]. This can help sustain air pockets underneath droplets, thus resulting in hydrophobicity.
2. Cavity geometry can trap air inside the pores (since the air in the pore is not part of the ambient atmosphere, unlike the air surrounding pillars in pillar-type geometries), thus maintaining air pockets underneath droplets to result in hydrophobic behavior [41,65].

The design process according to the first approach, listed above, is discussed elsewhere by others [63,64]. The thermodynamic framework underlying the second approach is briefly discussed below, the details of which are provided elsewhere [41].

The problem of drops on top of rough surfaces with pillars or protrusions is different from the one in which rough surfaces have cavities. In case of rough substrates (made of hydrophilic materials) with pillars, a Cassie–Baxter state will typically transition to a Wenzel state by simply displacing the air *which is part of the ambient*, thus trapping of air is not possible. In case of cavities, it is of interest to enquire the conditions under which the trapping of air is feasible.

Consider a liquid drop deposited on a rough substrate with cavities. Consider the drop size to be much larger than the length scale of the cavities (typically more than 10 times). Assume that the ambient air is at constant temperature T_0 and pressure P_0. Let this phase be denoted by α. Let the liquid drop phase be denoted by β. The air in any cavity i will be considered a separate phase γ_i. Additionally, there are interfaces between the solid, liquid, and air. These interfaces will be collectively denoted as phase σ and will be assumed to be sharp. Similar to the analysis in Section 4.2, the stable equilibrium states of this system can be obtained by minimizing the availability function in Equations (4.1) and (4.2). Superscripts ϕ in Equations (4.1) and (4.2) denote phases α, β, or γ_i.

If it is assumed that the contact angle measurements are made on time scales much shorter than the evaporation time scale, then each of the phases α, β, and γ_i are considered to be single components and the mass of each individual phase is constant. This assumption implies that chemical equilibrium (i.e., equating the chemical potentials of all component species in the phases) is not imposed across interfaces of different phases. As such, the equilibrium state sought is a quasi-equilibrium state. If the drop is in dry air, then it will eventually evaporate, which is the true equilibrium state. However, before substantial evaporation, the state of the drop is well approximated by the quasi-equilibrium state. Once again, small drop sizes are assumed, so that the gravitational potential is not accounted in the availability function.

The availability function in Equation (4.1) becomes [41]

$$A_{cav} = (P_0 - P^\beta)(V^\beta - V_0^\beta) + \sigma_{lg}[2\pi R^2(1 - \cos\theta)] - \sigma_{lg}(\pi R^2 \sin^2\theta)\cos\theta_r^c$$

$$+ I\left[(P_0 - P^\gamma)V^\gamma + P_0 V_{cav}\ln\frac{P^\gamma}{P_0} + \Delta U_\gamma^\sigma\right] \tag{4.28}$$

where thermal equilibrium and a constant volume of the liquid phase is assumed. Equation (4.28) is applicable for hydrophobic or hydrophilic materials. Minimizing the above function will give the equilibrium states of the system. It is assumed that each cavity underneath the drop is in the same state. Thus, the summation over the cavities is replaced by I, the number of cavities under the drop. Superscript/subscript γ denotes any cavity under the drop. Similar to the problem discussed in Section 4.2, except for the first two terms, all other terms correspond to the "state" at the contact of the drop with the textured surface. Additionally, the problem can be similarly solved by first minimizing the availability function with respect to the "microscopic" parameters and then with respect to macroscopic parameters P^β, R, and θ [41]. Minimizing with respect to the microscopic parameters leads to an understanding of the state of the liquid–air interface within the cavity. Minimizing with respect to the macroscopic parameters gives the apparent contact angle of the drop in its ground state – the CB formula is recovered when air is trapped in the CB state, as shown in Figure 4.8 [41].

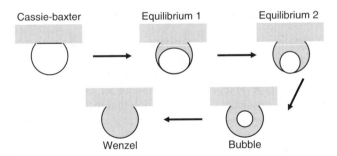

Figure 4.8 Schematics of the different equilibrium states of the liquid–air interface within a spherical cavity made of hydrophilic material [41]. Reprinted with permission from [41]. © 2009 Taylor and Francis Group

Minimization of the availability function reveals various equilibrium states within the cavity. For example, Figure 4.8 shows a particular scenario where there are five equilibrium states in a spherical cavity made of hydrophilic material. In this case, the CB, equilibrium 2, and Wenzel states are stable equilibria [41]. The Wenzel state is the lowest energy state if the surface material is hydrophilic; however, transition to the Wenzel state from the CB state can be prevented due to intermediate energy barrier states. The number of equilibrium states can vary depending on the parameters. One of the important parameters is the ratio of the pressure force to the surface tension force P_σ:

$$P_\sigma = \frac{P_0 R_{\text{cav}}}{2\sigma_{\text{lg}}} \tag{4.29}$$

where R_{cav} is the radius of the spherical cavity. Note that P_0 is the initial pressure of air in the cavity. As the liquid–air interface invades the cavity the air pressure increases due to compression. The greater the pressure force relative to the surface tension force, the greater is the energy barrier for transition to the Wenzel state. This is because pressure resists the invasion of the liquid–air interface into the cavity. Larger P_σ helps sustain the CB state on rough surfaces made from hydrophilic materials. If P_σ is small, then the surface tension force dominates and the equilibrium state inside the cavity is such that a significant portion of the wall of the cavity is wetted. This, in turn, would increase hysteresis. Hysteresis could be reduced by choosing $P_\sigma > 1$. Thus, larger values of P_σ are desirable for superhydrophobic properties. The value of P_σ can be set by choosing an appropriate value of R_{cav}. $P_\sigma \gg 1$ implies $R_{\text{cav}} \gg 2\sigma_{\text{lg}}/P_0$. For the case of water–air interface and P_0 at atmospheric condition, the constraint on the cavity radius becomes $R_{\text{cav}} \gg 1.43\,\mu\text{m}$. It is also noted that the surface tension force should dominate the gravity force, else the cavities can be filled up due to gravity. This condition is satisfied if $R_{\text{cav}} \ll a_{\text{cap}}$, the capillary length. For the water–air case, $a_{\text{cap}} = 2.72$ mm. Therefore, the ideal values of the cavity radius are given by $1.43\,\mu\text{m} \ll R_{\text{cav}} \ll 2.72$ mm.

While the above analysis is useful to understand the state of trapped air in the cavity, it should be noted that the air can dissolve out of the cavity if the liquid is not saturated with air. Thus, on a long time scale the air pockets may not be sustained. The next section addresses this by considering a new approach based on stabilizing the vapor of the liquid itself in the cavities or pores.

4.4 Thermodynamic Analysis: Vapor Stabilization

Previous sections detailed the theory underlying superhydrophobic surfaces that rely on the presence of an air–like fluid in the grooves of roughness geometries. In pillar-type geometries the air in the roughness grooves is assumed to be part of the ambient environment, whereas in cavity-type geometries the ambient air is assumed to initially occupy the pores. The roughness geometry is designed such that the impalement of the roughness grooves by the liquid, and the consequent transition to a wetted or Wenzel state, is energetically unfavorable. This approach works well when droplets are deposited on substrates in an open system. However, if such substrates form the walls of enclosed fluidic networks then the air in the roughness grooves is no longer connected to the ambient. In this case the air could be dissolved into the liquid, depending on the degree of saturation of air in the liquid. If the air does dissolve out, then a transition to a wetted state is possible. Thus, such substrates may not be robust for many technologically relevant scenarios.

Patankar [66,67] considered whether it is possible to design textured surfaces that do not rely on the presence of air in the roughness grooves. To that end, whether the roughness features can be designed in such a way that the vapor phase of the liquid itself occupies the roughness grooves and exists in equilibrium with the liquid was explored. This approach is summarized below. Such vapor-stabilizing substrates could be crucial for robustness in practical applications such as low drag [66] or efficient nucleate boiling [67].

Some background is provided first before considering a specific example problem to illustrate the concept. Figure 4.9 shows a pressure–temperature (P–T) diagram for liquid–vapor phase change. Below the critical point and above the triple point, the phase change line is given by a coexistence or phase-change curve. This curve is also referred to as the binodal line. The stable liquid phase is above this curve and the stable vapor phase is below it. On either side of the coexistence curve are the spinodal curves. Besides the usual stable phases, metastable vapor and liquid phases can also exist in the regions between the spinodal and coexistence curves, as shown in Figure 4.9. As shown in Figure 4.9, typically, capillary evaporation involves phase change of a stable liquid (A) to a metastable vapor (B). This phase change is caused by small-scale pores that stabilize the metastable state. Boiling involves phase change of a metastable liquid (E) to a stable vapor (D). Similarly, capillary condensation is usually a stable vapor (D) to a metastable liquid (E) transition, whereas condensation in heat exchangers is usually a metastable vapor (B) to a stable liquid (A) transition. During equilibrium phase change the liquid and vapor chemical potentials are the same. The liquid and vapor phases, in chemical equilibrium but at different pressures, can coexist in mechanical equilibrium if they are separated by a curved liquid–vapor interface. This phenomenon is also referred to as the Kelvin effect. The relationship between the pressures of the liquid and the vapor during phase change is obtained by equating the chemical potentials or, equivalently, the fugacities (which is the tendency to escape from one phase to another) at that temperature. Assuming ideal

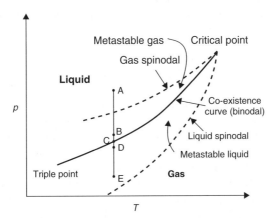

Figure 4.9 A schematic phase diagram for liquid–vapor phase transition. Reprinted with permission from [67]. © 2010 Royal Society of Chemistry

gas for the vapor phase, the equation is [20]

$$p_v = p_{sat} \exp\left[\frac{V_l}{R_U T}(p_l - p_{sat})\right] \qquad (4.30)$$

where p_l and p_v are the liquid and vapor pressures during phase transition. For example, during the boiling phase transition shown in Figure 4.9, $p_l = p_E$ and $p_v = p_D$. p_{sat} is the saturation pressure (i.e., the pressure on the coexistence curve at temperature T – point C in Figure 4.9) and V_1 is the molar volume of the liquid.

To illustrate the concept of vapor stabilization, consider a textured surface with cylindrical pores that is immersed in a liquid. Let the surface material be hydrophobic with a contact angle of 120°. Robust design should require that the cavity geometry be such that it stabilizes the vapor phase even at ambient conditions. Thus, assume that there are no trapped gases in the cavities. The goal is to seek the pore diameter and aspect ratio that will stablize the vapor phase inside it.

Let the surface with cylindrical cavities be immersed in a liquid that is connected to a constant-temperature and -pressure reservoir; that is, consider an isobaric–isothermal ensemble for a closed system. The temperature and chemical potentials are assumed to be uniform everywhere, implying thermal and chemical equilibrium. Let the reference state be the one where the liquid fills the cavity and no vaporization has occurred. All energies will be compared with this reference state.

Now consider a state where some liquid (A in Figure 4.9) has vaporized to a metastable state (B in Figure 4.9) at the same temperature (i.e., at thermal equilibrium). Assuming chemical equilibrium, the difference in liquid and vapor pressures can be calculated using Equation (4.30). For water at standard ambient atmospheric conditions $T = 298$ K and $p_l = p_A = 1$ atm (101.3 kPa), $p_{sat} = p_0 = 3.17$ kPa and (at A) $V_1 = 1.8 \times 10^{-5} \, m^3 \, mol^{-1}$. Using these values and $R_U = 8.314$ J K^{-1} mol^{-1} in Equation (4.30), the pressure of the metastable vapor (B) at equilibrium with the liquid is found to be nearly equal to p_{sat} (i.e. p_0) at point 0. The liquid at high pressure and the vapor at low pressure will be in mechanical equilibrium via an interface that will curve into the cavity as shown in Figure 4.10. The radius of curvature R_{co} of this interface at ambient conditions is given by the Young–Laplace equation:

$$R_{co} = \frac{2\sigma_{lv}}{p_A - p_B} \qquad (4.31)$$

where σ_{lv} is the liquid–vapor surface tension (assumed constant for simplicity). Using $\sigma_{lv} = 72$ mN m^{-1} for water and the equilibrium pressure difference between the liquid and vapor phases at ambient conditions, $R_{co} = 1.47$ μm. Mechanical equilibrium also requires that the contact angle of the liquid–vapor interface with the solid wall should be equal to the equilibrium contact angle (assumed to be 120° in this discussion) inside the cavity. In general, this condition will be satisfied at the corner at the top of the cavity (Figure 4.10) if the material inside the cavity is hydrophobic [37].

For robust vapor-stabilizing cavities, it should be ensured that the CB state with the liquid–vapor interface hanging at the top of the cavity is a stable equilibrium state and that it is at lower energy than the Wenzel state, where the cavity is completely filled with the liquid. This can be ensured by choosing an appropriate diameter and aspect ratio of

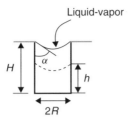

Figure 4.10 Schematic of a liquid–vapor interface hanging at the top of a cylindrical cavity. The liquid outside the cavity is in equilibrium with its own vapor phase inside the cavity. Dotted curve shows an intermediate state of the liquid–vapor interface during impalement into the cavity. Reprinted with permission from [67]. © 2010 Royal Society of Chemistry

the cylindrical cavity. To enable this, the availability A of a vaporized state relative to the reference state is defined by [67]

$$A = S_{sv}\sigma_{lv}\cos\theta_c + S_{lv}\sigma_{lv} - (p_v - p_l)V_v \qquad (4.32)$$

where S_{sv} and S_{lv} are the solid–vapor and liquid–vapor contact areas, respectively, and only one cylindrical cavity is considered for simplicity. V_v is the volume of the vapor phase and $\theta_c = 120°$ is the equilibrium contact angle of the material inside the cavity. $p_l - p_v = p_A - p_B$ is known for equilibrium at ambient conditions and can be represented in terms of R_{co} (Equation (4.31)). The availability can be calculated for different values of the interface location h (see Figure 4.10). Using Equation (4.31) in Equation (4.32), substituting expressions for S_{sv} and S_{lv}, and nondimensionalizing gives [67]

$$A' = \frac{A}{\pi R_{co}^2\sigma_{lv}} = 2\frac{h}{R_{co}}\left(\frac{R}{R_{co}} + \cos\theta_c\right)\frac{R}{R_{co}} + \left(\frac{R}{R_{co}}\right)^2\cos\theta_c + \frac{2}{3}(1 - \sin^3\alpha) \quad (4.33)$$

where R is the radius of the cylindrical cavity, h is the height to which the liquid–vapor interface has moved up in the cavity of height H (see Figure 4.10), and α is as defined in Figure 4.10. Geometric considerations give $\cos\alpha = R/R_{co}$. R and H are to be determined, while all other parameters are known. The smallest value of h possible for the chosen configuration of the liquid–vapor interface is $h_{min} = R_{co}(1 - \sin\alpha)$ and the maximum value is H. Below h_{min} the liquid–vapor interface will touch the bottom of the cavity and other configurations need to be explored. Equation (4.33) implies that the availability will decrease with increasing vaporization, i.e. with increasing h, only if

$$\frac{R}{R_{co}} < -\cos\theta_c \qquad (4.34)$$

This is desirable because that will stabilize the equilibrium state where the liquid–vapor interface hangs at the top of the cavity. For $\theta_c = 120°$, this implies $R/R_{co} < 1/2$. The radius of the cavity can be determined based on this result. In this discussion $R/R_{co} = 1/3$ will be chosen; that is, $R = 490$ nm.

Figure 4.11 shows the variation of availability vs. h/R_{co} according to Equation 4.31 with $\theta_c = 120°$, $R/R_{co} = 1/3$, and $\alpha = \cos^{-1}(R/R_{co}) = 70.53°$. When there is no vaporization in the cavity $A' = 0$. At $h = h_{min}$, A' is positive, beyond which it decreases as h increases.

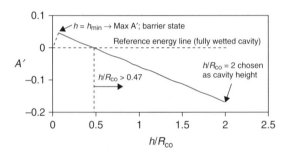

Figure 4.11 Plot of availability as a function of the position of the liquid–vapor interface (Equation (4.33)). Reprinted with permission from [67]. © 2010 Royal Society of Chemistry

Thus, $h = h_{min}$ is a measure of the energy barrier between the fully wetted state of the cavity and a vaporized state with the liquid–vapor interface at the top of the cavity.

Figure 4.11 shows that at $h/R_{co} = 0.47$, A' decreases to zero. This implies that, if the height of the cavity $H/R_{co} < 0.47$, the equilibrium state with the liquid–vapor interface at the top of the cavity will be at higher energy compared with the fully wetted state. It will still be a local minimum, but not the global minimum. For robustness, one could choose a cavity height $H/R_{co} > 0.47$. In that case the equilibrium state at the top of the cavity will be at lower energy than the fully wetted state. When the surface is initially immersed in the liquid bath it is the equilibrium state at the top of the cavity that will be encountered first. Additionally, if this state is at a lower energy compared with a fully wetted cavity, then the energy barrier to transition to the wetted state will be larger. The likelihood of transition to a wetted state will be low and vapor-filled cavities will be sustained at ambient conditions. Thus, the deeper the cavities the better it is. This helps to determine the second geometric parameter that was sought.

Thus, the above example worked out by Patankar [67] shows that the geometry of the surface texture can be chosen so that the vapor phase of the liquid is stabilized in roughness cavities. A similar approach can be used for different types of geometries and optimal textures could be searched for a particular application. This concept can be generalized to other phase transitions as well, thus allowing phase manipulation at the solid interface using surface textures and material chemistry (e.g., the material contact angle).

4.5 Applications and Future Challenges

Research through the past decade has shown that superhydrophobic surfaces can be effective in a variety of applications, including low-drag surfaces, self-cleaning surfaces, efficient dropwise condensation and heat transfer, nucleate boiling heat transfer, desalination, deicing, efficient drop shedding in steam turbines and fuel cells, textiles, oil–water separation, reduced biofouling, biofilm growth and tissue engineering, and substrates to study cell mechanics, among others [6–14].

There is worldwide interest in this area owing to such wide ranging applications. However, several challenges remain on the fundamental research side, as well as in enabling a mature technology that can be translated into industrial applications. Some of the key challenges are summarized below:

1. *Vapor stabilization*. As stated in the previous section, several applications require that superhydrophobicity should not rely on the presence of air; rather, an approach based on vapor stabilization may be warranted. While the theoretical underpinning has been proposed [66,67], it is essential to demonstrate efficacy of this concept in realistic application scenarios, such as in drag reduction on submerged bodies, enhancement of nucleate boiling heat transfer, and so on.

2. *Molecular/nanoscale effects*. A better understanding of nanoscale effects will help in better design strategies that take advantage of material chemistry as well as surface textures. For example, nanopores could resist the invasion of liquids due to molecular scale effects even if the bulk material itself may have hydrophilic properties (i.e., liquid would invade a larger pore made of same material) [68]. This effect could be used to develop vapor-stabilizing surfaces from hydrophilic materials. This is important for industrial applications, where the materials are often metals and ceramics which have hydrophilic properties.

3. *Wear resistance*. Many microfabricated textured surfaces do not have wear resistance that is essential to withstand industrial scale applications. More research on wear resistance materials that are also superhydrophobic is essential.

4. *Bio-fouling or bio-film growth*. An understanding of how surface texture affects the growth of biomaterial is not fully developed [13,14,69]. A theoretical framework to help design textured surfaces either for enhanced bio-film growth or reducing bio-fouling is essential. This is relevant in applications such as fouling reduction on low-drag surfaces, biofilm growth for energy harvesting, tissue engineering, and studying cell mechanics, among others.

5. *Mass production*. Many applications require a large surface area of textured surface; for example, drag-reduction surfaces. Thus, there is an order of magnitude difference between the roughness length scale and the length scale of the surface size. Highly scalable manufacturing techniques such as spray coating [70] are essential to enable industrial-scale application of textured surfaces. New manufacturing protocols need to be explored in general.

Acknowledgments

NAP acknowledges support from the Initiative for Sustainability and Energy at Northwestern (ISEN) University.

The author acknowledges support from the Initiative for Sustainability and Energy at Northwestern (ISEN) University. Northwestern (ISEN) University. Section 4.2.1 draws from [19], Section 4.2.2 from [39] and and Section 4.2.3 from [53], and reprinted with permission from the American Chemical Society. Some texts in Section 4.3 draws from [41], reprinted with permission from Taylor and Francis. Lastly, some texts in Section 4.4 draws from [67], reprinted with permission from the Royal Society of Chemistry.

References

[1] Barthlott, W. and Neinhuis, C. (1997) Purity of the sacred lotus, or escape from contamination in biological surfaces. *Planta*, **202** (1), 1–8.

[2] Cassie, A.B.D. (1948) Contact angles. *Discussions of the Faraday Society*, **3**, 11–16.

[3] Cassie, A.B.D. and Baxter, S. (1944) Wettability of porous surfaces. *Transactions of the Faraday Society*, **40**, 546–551.

[4] Wenzel, R.N. (1949) Surface roughness and contact angle. *Journal of Physical and Colloid Chemistry*, **53** (9), 1466–1467.

[5] He, B., Patankar, N.A. and Lee, J. (2003) Multiple Equilibrium Droplet Shapes and Design Criterion for Rough Hydrophobic Surfaces. *Langmuir*, **19** (12), pp. 4999–5003. Copyright 2003 American Chemical Society.

[6] Narhe, R.D. and Beysens, D.A. (2004) Nucleation and growth on a superhydrophobic grooved surface. *Physical Review Letters*, **93** (7), 076103.

[7] Varanasi, K.K., Hsu, M., Bhate, N. *et al*. (2009) Spatial control in the heterogeneous nucleation of water. *Applied Physics Letters*, **95** (9), 094101.

[8] Chen, R., Lu, M.C., Srinivasan, V. *et al*. (2009) Nanowires for enhanced boiling heat transfer. *Nano Letters*, **9** (2), 548–553.

[9] Li, C., Wang, Z., Wang, P.I., *et al*. (2008) Nanostructured copper interfaces for enhanced boiling. *Small*, **4** (8), 1084–1088.

[10] Mishchenko, L., Hatton, B., Bahadur, V. *et al*. (2010) Design of ice-free nanostructured surfaces based on repulsion of impacting water droplets. *ACS Nano*, **4** (12), 7699–7707.

[11] Meuler, A.J., Smith, J.D., Varanasi, K.K. *et al*. (2010) Relationships between water wettability and ice adhesion. *ACS Applied Materials & Interfaces*, **2** (11), 3100–3110.

[12] Varanasi, K.K., Deng, T., Smith, J.D. *et al*. (2010) Frost formation and ice adhesion on superhydrophobic surfaces. *Applied Physics Letters*, **97** (23), 234102.

[13] Salta, M., Wharton, J.A., Stoodley, P. *et al*. (2010) Designing biomimetic antifouling surfaces. *Philosophical Transactions of the Royal Society A: Mathematical Physical and Engineering Sciences*, **368** (1929), 4729–4754.

[14] Scardino, A.J. and de Nys, R. (2011) Mini review: Biomimetic models and bioinspired surfaces for fouling control. *Biofouling*, **27** (1), 73–86.

[15] Quere, D. (2008) Wetting and roughness. *Annual Review of Materials Research*, **38**, 71–99.

[16] Rothstein, J.P. (2010) Slip on superhydrophobic surfaces. *Annual Review of Fluid Mechanics*, **42**, 89–109.

[17] He, B., Lee, J., and Patankar, N.A. (2004) Contact angle hysteresis on rough hydrophobic surfaces. *Colloids and Surfaces A: Physicochemical and Engineering Aspects*, **248** (1–3), 101–104.

[18] Lafuma, A. and Quere, D. (2003) Superhydrophobic states. *Nature Materials*, **2** (7), 457–460.

[19] Patankar, N.A. (2010) Consolidation of hydrophobic transition criteria by using an approximate energy minimization approach. *Langmuir*, **26** (11), 8941–8945.

[20] Modell, M. and Reid, R.C. (1983) *Thermodynamics and Its Applications*, 2nd edn., Prentice-Hall, Englewoods Cliffs, NJ.

[21] Marmur, A. (2003) Wetting on hydrophobic rough surfaces: to be heterogeneous or not to be? *Langmuir*, **19** (20), 8343–8348.

[22] Koishi, T., Yasuoka, K., Fujikawa, S. *et al*. (2009) Coexistence and transition between Cassie and Wenzel state on pillared hydrophobic surface. *Proceedings of the National Academy of Sciences of the United States of America*, **106** (21), 8435–8440.

[23] Lee, C., Choi, C.H., and Kim, C.J. (2008) Structured surfaces for a giant liquid slip. *Physical Review Letters*, **101** (6), 064501.

[24] Lau, K.K.S., Bico, J., Teo, K.B.K, *et al*. (2003) Superhydrophobic carbon nanotube forests. *Nano Letters*, **3** (12), 1701–1705.

[25] Patankar, N.A., (2004) Mimicking the lotus effect: influence of double roughness structures and slender pillars. *Langmuir*, **20** (19), 8209–8213.

[26] Kwon, Y., Patankar, N., Choi, J., and Lee, J. (2009) Design of surface hierarchy for extreme hydrophobicity. *Langmuir*, **25** (11), 6129–6136.

[27] Jung, Y.C. and Bhushan, B. (2007) Wetting transition of water droplets on superhydrophobic patterned surfaces. *Scripta Materialia*, **57** (12), 1057–1060.

[28] Kusumaatmaja, H., Blow, M.L., Dupuis, A., and Yeomans, J.M. (2008) The collapse transition on superhydrophobic surfaces. *Europhysics Letters*, **81** (3), 36003.

[29] Reyssat, M., Yeomans, J.M., and Quere, D. (2008) *Impalement of fakir drops. Europhysics Letters*, **81** (2), 26006.

[30] Deng, T., Varanasi, K.K., Hsu, M. *et al*. (2009) Nonwetting of impinging droplets on textured surfaces. *Applied Physics Letters*, **94** (13), 133109.

[31] Bormashenko, E., Pogreb, R., Whyman, G., and Erlich, M. (2007) Resonance Cassie–Wenzel wetting transition for horizontally vibrated drops deposited on a rough surface. *Langmuir*, **23** (24), 12217–12221.

[32] Bormashenko, E., Pogreb, R., Whyman, G., and Erlich, M. (2007) Cassie–Wenzel wetting transition in vibrating drops deposited on rough surfaces: is the dynamic Cassie–Wenzel wetting transition a 2D or 1D affair? *Langmuir*, **23** (12), 6501–6503.

[33] Bartolo, D., Bouamrirene, F., Verneuil, E. *et al*. (2006) Bouncing or sticky droplets: Impalement transitions on superhydrophobic micropatterned surfaces. *Europhysics Letters*, **74** (2), 299–305.

[34] Extrand, C.W. (2004) Criteria for ultralyophobic surfaces. *Langmuir*, **20** (12), 5013–5018.

[35] Jung, Y.C. and Bhushan, B. (2008) Dynamic effects of bouncing water droplets on superhydrophobic surfaces. *Langmuir*, **24** (12), 6262–6269.

[36] Reyssat, M., Pepin, A., Marty, F. *et al*. (2006) Bouncing transitions on microtextured materials. *Europhysics Letters*, **74** (2), 306–312.

[37] Oliver, J.F., Huh, C., and Mason, S.G. (1977) Resistance to spreading of liquids by sharp edges. *Journal of Colloid and Interface Science*, **59** (3), 568–581.

[38] Patankar, N.A. (2004) Transition between superhydrophobic states on rough surfaces. *Langmuir*, **20** (17), 7097–7102.

[39] Kwon, H.-M., Paxson, A.T., Varanasi, K.K., and Patankar, N.A. (2011) Rapid deceleration driven wetting transition during pendant drop deposition on superhydrophobic surfaces. *Physical Review Letters*, **106**, 036102.

[40] Jung, Y.C. and Bhushan, B. (2009) Dynamic effects induced transition of droplets on biomimetic superhydrophobic surfaces. *Langmuir*, **25** (16), 9208–9218.

[41] Patankar, N.A. (2009) Hydrophobicity of surfaces with cavities: making hydrophobic substrates from hydrophilic materials? *Journal of Adhesion Science and Technology*, **23** (3), 413–433.

[42] Patankar, N.A. (2003) On the modeling of hydrophobic contact angles on rough surfaces. *Langmuir*, **19** (4), 1249–1253.

[43] Courbin, L., Denieul, E., Dressaire, E. *et al*. (2007) Imbibition by polygonal spreading on microdecorated surfaces. *Nature Materials*, **6** (9), 661–664.

[44] Sbragaglia, M., Peters, A.M., Pirat, C. *et al*. (2007) Spontaneous breakdown of superhydrophobicity. *Physical Review Letters*, **99** (15), 156001.

[45] Gao, L.C. and McCarthy, T.J. (2007) How Wenzel and Cassie were wrong. *Langmuir*, **23** (7), 3762–3765.

[46] Marmur, A. and Bittoun, E. (2009) When Wenzel and Cassie are right: reconciling local and global considerations. *Langmuir*, **25** (3), 1277–1281.

[47] Gao, L.C. and McCarthy, T.J. (2009) An attempt to correct the faulty intuition perpetuated by the Wenzel and Cassie "laws". *Langmuir*, **25** (13), 7249–7255.

[48] Extrand, C.W. (2002) Model for contact angles and hysteresis on rough and ultraphobic surfaces. *Langmuir*, **18** (21), 7991–7999.

[49] Johnson, R.E. and Dettre, R.H. (1964) Contact angle hysteresis. 3. Study of an idealized heterogeneous surface. *Journal of Physical Chemistry*, **68** (7), 1744–1750.

[50] Marmur, A. (2009) Solid-surface characterization by wetting. *Annual Review of Materials Research*, **39**, 473–489.

[51] Brakke, K.A. (1992) The Surface Evolver, http://www.cs.berkeley.edu/~sequin/CS284/TEXT/brakke.pdf.

[52] Dupuis, A. and Yeomans, J.M. (2005) Modeling droplets on superhydrophobic surfaces: equilibrium states and transitions. *Langmuir*, **21** (6), 2624–2629.

[53] Kwon, Y., Choi, S., Anantharaju, N. *et al*. (2010) Is the Cassie–Baxter formula relevant? *Langmuir*, **26** (22), 17528–17531.

[54] Panchagnula, M.V. and Vedantam, S. (2007) Comment on how Wenzel and Cassie were wrong by Gao and McCarthy. *Langmuir*, **23** (26), 13242.

[55] Patankar, N.A. (2010) Hysteresis with regard to Cassie and Wenzel states on superhydrophobic surfaces. *Langmuir*, **26** (10), 7498–7503.

[56] Choi, W., Tuteja, A., Mabry, J.M. *et al*. (2009) A modified Cassie–Baxter relationship to explain contact angle hysteresis and anisotropy on non-wetting textured surfaces. *Journal of Colloid and Interface Science*, **339** (1), 208–216.

[57] Joanny, J.F. and de Gennes, P.G. (1984) A model for contact-angle hysteresis. *Journal of Chemical Physics*, **81** (1), 552–562.

[58] Gao, L.C. and McCarthy, T.J. (2006) Contact angle hysteresis explained. *Langmuir*, **22** (14), 6234–6237.

[59] Dorrer, C. and Ruhe, J. (2006) Advancing and receding motion of droplets on ultrahydrophobic post surfaces. *Langmuir*, **22** (18), 7652–7657.

[60] Reyssat, M. and Quere, D. (2009) Contact angle hysteresis generated by strong dilute defects. *Journal of Physical Chemistry B*, **113** (12), 3906–3909.

[61] Ahuja, A., Taylor, J.A., Lifton, V. *et al*. (2008) Nanonails: a simple geometrical approach to electrically tunable superlyophobic surfaces. *Langmuir*, **24** (1), 9–14.

[62] Herminghaus, S. (2000) Roughness-induced non-wetting. *Europhysics Letters*, **52** (2), 165–170.

[63] Tuteja, A., Choi, W., Ma, M.L. *et al*. (2007) Designing superoleophobic surfaces. *Science*, **318** (5856), 1618–1622.

[64] Tuteja, A., Choi, W., Mabry, J.M. *et al*. (2008) Robust omniphobic surfaces. *Proceedings of the National Academy of Sciences of the United States of America*, **105** (47), 18200–18205.

[65] Abdelsalam, M.E., Bartlett, P.N., Kelf, T., and Baumberg, J. (2005) Wetting of regularly structured gold surfaces. *Langmuir*, **21** (5), 1753–1757.

[66] Patankar, N.A. (2010) Vapor stabilizing substrates for superhydrophobicity and superslip. *Langmuir*, **26** (11), 8783–8786.

[67] Patankar, N.A. (2010) Supernucleating surfaces for nucleate boiling and dropwise condensation heat transfer. *Soft Matter*, **6** (8), 1613–1620.

[68] Qiao, Y., Liu, L., and Chen, X. (2009) Pressurized liquid in nanopores: a modified Laplace–Young equation. *Nano Letters*, **9** (3), 984–988.

[69] Scardino, A.J., Zhang, H., Cookson, D.J. *et al*. (2009) The role of nano-roughness in antifouling. *Biofouling*, **25** (8), 757–767.

[70] Tiwari, M.K., Bayer, I.S., Jursich, G.M. *et al*. (2010) Highly liquid-repellent, large-area, nanostructured poly(vinylidene fluoride)/poly(ethyl 2-cyanoacrylate) composite coatings: particle filler effects. *ACS Applied Materials & Interfaces*, **2** (4), 1114–1119.

5

Multiscale Experimental Mechanics of Hierarchical Carbon-Based Materials[1]

Horacio D. Espinosa[1], Tobin Filleter[2], and Mohammad Naraghi[1,3]

[1] *Northwestern University, USA*
[2] *University of Toronto, Canada*
[3] *Texas A&M University, USA*

5.1 Introduction

Nature utilizes self-organized hierarchical mechanical structures, which exhibit extraordinary stiffness, strength and toughness with low weight, to accomplish a vast number of functions. These structures range from internal components such as mammalian bone and tendons, to protective elements such as seashell nacre, to external survival tools such as spider silks. Research has shown that the superior mechanical behavior of all of these materials relies intimately on their hierarchical structures [1–8]. New research efforts are increasingly being developed to apply these lessons from nature toward the development of artificial materials that can achieve a similar marriage of strength and toughness, such as synthetic composites that emulate nacre and bioinspired fibers that emulate spider silk [9,10].

In particular, the application of crystalline one-dimensional (1D) and two-dimensional (2D) carbon-based materials, such as carbon nanotubes (CNTs) and graphene, as building blocks in artificial hierarchical structures has received a great deal of attention. The unique sp^2 in-plane carbon bonding found in CNTs and graphene sheets yields some of the stiffest and strongest materials known to man. Recent experiments on both CNTs and graphene

[1] This chapter is based on a contribution recently published in *Advanced Materials*. H.D. Espinosa, T. Filleter, and M. Naraghi, Multiscale experimental mechanics of hierarchical carbon-based materials. *Advanced Materials*, Wiley–VCH, doi: 10.1002/adma.201104850. Published 11 May 2012.

Nano and Cell Mechanics: Fundamentals and Frontiers, First Edition. Edited by Horacio D. Espinosa and Gang Bao.
© 2013 John Wiley & Sons, Ltd. Published 2013 by John Wiley & Sons, Ltd.

have confirmed theoretical predictions, and demonstrated elastic moduli of up to 1 TPa and strengths of over 100 GPa [11–13]. These proven intrinsic mechanical properties have sparked the questions of whether these materials can be organized in hierarchical structures to achieve performance even superior to naturally occurring materials. In the case of CNTs, their 1D nature is envisioned to act as a building block in engineered fibers akin to the mineralized collagen fibrils found in tendon. Alternatively, the 2D nature of graphene sheets can be used to mimic the aragonite platelets found in the layered structure of nacre [2]. While the intrinsic properties of these carbon-based building blocks are well understood, a great deal of work is required to understand and engineer how they can be effectively assembled to emulate the complex hierarchical structures of natural materials.

One of the fundamental challenges to assemble carbon-based materials into high-performance hierarchical structures is engineering their interfacial interactions across multiple scales. In the natural case of tendons, the structure is organized in seven hierarchical levels with different interaction mechanisms acting at each level (see Figure 5.1a) [14,15]. At the first level, tropocollagen molecules consist of three polypeptides arranged in a triple helix bonded together laterally via hydrogen bonding (H-bonding). Upon tensile loading, the H-bonds rupture in a reversible process, yielding a mechanism allowing the collagen fibrils to deform up to 50% prior to failure while dissipating energy [1,16]. At the third level of hierarchy, collagen fibrils are organized into arrays connected by a biopolymer phase, which provides additional energy dissipation mechanisms. These different interfacial interactions at multiple length scales collectively contribute to the macroscopic toughness of tendons. Recent modeling shows that these interfacial interactions follow length scales associated with the interface geometry and chemical bonds [17]. A similarly complex hierarchical structure with varying interfacial interactions is also found in spider silk and nacre. An overarching mechanism that is common between these natural materials is a tolerance to flaws. In silk, it has been shown that macroscopic fibers and bundles can have high strength and toughness as a result of the delocalization of stress concentrators at scales of ~100 nm [8,10]. In nacre, the brick-and-mortar-like microstructure allows for damage propagation over millimeter length scales, dramatically increasing toughness [9]. Similarly, engineering interfacial interactions across hierarchical length scales and the utilization of toughening mechanisms is required in order to develop tough macroscopic materials out of CNTs (see Figure 5.1b) and graphene building blocks, which are otherwise inherently brittle. To address this challenge, there is a great need for characterization and testing of mechanical properties and shear interactions across multiple length scales. This need has sparked a new multiscale approach to understanding the mechanics of carbon-based materials which merges established nanomechanical testing (such as atomic force microscopy (AFM), *in-situ* micro-electrical–mechanical systems (MEMS)-based testing, *in-situ* Raman spectroscopy testing) and classical micromechanical testing methods. Only through application of such methods, and addressing each length scale in the hierarchical structure, can novel carbon-based materials be designed to emulate and surpass the properties of natural materials.

The envisioned carbon-based materials have the potential to make a significant impact, in particular on aerospace and military applications, which utilize fiber-reinforced structural composites and armor structures. Hierarchical CNT-based fibers are emerging as a class of material proposed to replace more traditional high-strength carbon fibers in composite materials [18–23]. In addition to applications requiring high performance in

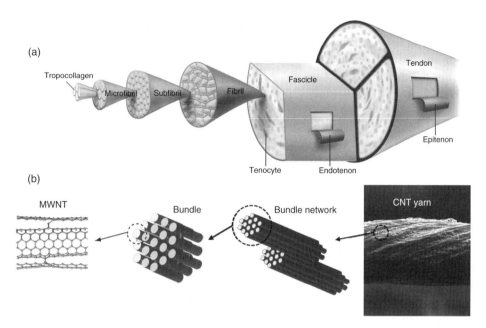

Figure 5.1 (a) Schematic representation of the multiple levels of the hierarchical structure found in natural tendons. (b) SEM image of a carbon nanotube yarn including a schematic representation of the hierarchical structure that makes up its structure. Reprinted with permission from [14]. © 2008 American Society for Clinical Investigation

mechanical properties, CNT fibers and sheets are also being applied as templates to develop multifunctional materials [24] and temperature-invariant artificial muscles [25]. For example, Lima *et al.* have recently developed a bi-scrolling method of fabricating composite yarns based on the spinning method of CNTs from forests, in which drawn sheets of CNTs are used as hosts of otherwise unspinnable powders and granulates to develop multifunctional yarns suitable for high-tech applications, such as superconductors and flexible battery cathodes [24].

Macroscopic graphene composites and graphene-oxide-based materials are also beginning to emerge [26,27]; however, initial implementations are yet to approach the intrinsic superior mechanical properties of the building block constituents. The gap between the mechanical properties of the composites and their nanometer-size building blocks further emphasizes the great need for a better multiscale understanding of how the hierarchical structures and interactions at multiple levels within the materials play a role in the macroscopic behavior. We envision that this will ultimately be achieved in great part by the development and application of unique experimental tools to enable the understanding of deformation mechanisms at each length scale.

5.2 Multiscale Experimental Tools

Elucidating a complete understanding of the mechanical properties and behavior of carbon-based composite materials requires a diverse toolbox of techniques which not only can resolve the forces acting at each hierarchical level, but also reveal deformation and failure of the structures at each scale. While it has become common place to identify strain and failure at macro- and micro-scopic scales through optical techniques, it has only been in the last few decades that novel techniques have been developed to characterize mechanical behavior at the nanoscale. This has been, in the most part, enabled by the application of two revolutionary approaches to the study of mechanics: AFM and *in-situ* scanning and transmission electron microscopy (*in-situ* SEM and TEM). Together, these techniques have addressed the most fundamental requirements for studying mechanics of nanoscale constituents such as CNTs and graphene sheets; namely, (1) the capability to measure forces in the range of pico- to nano-newtons and (2) real-space imaging resolution extending down to individual atoms. Although these approaches have already yielded a great deal of understanding of individual CNT and graphene properties, they are only beginning to address characterization at higher scales in carbon-based materials, such as the shear interactions between adjacent nanoscale constituents. This level in the hierarchy of carbon-based composites remains one of the biggest bottlenecks impeding the development of high-performance macroscopic materials. This has initiated the application of additional experimental techniques, such as Raman spectroscopy and nanoindentation, to characterize interactions and local mechanical behavior of the constituents within larger networks. When coupled with more traditional micromechanical testing methods, this combination of techniques will be essential in moving forward the frontiers of multiscale experiments on carbon-based materials. As an example, Figure 5.2 outlines the different hierarchical levels in CNT yarns and details the above-mentioned experimental tools as well as modeling techniques which can be applied at each length scale.

5.2.1 Revealing Atomic-Level Mechanics: In-Situ TEM Methods

TEM has become one of the most powerful tools to characterize the real-space atomic-level structure of a wide variety of materials. With application to low-dimension carbon materials it has been particularly useful, most notably facilitating the discovery of CNTs by Ijima [31] and revealing the nature of defects in CNTs and graphene through direct atomic-resolution imaging [12,32]. The later accomplishment epitomizes one of the greatest strengths of TEM: local ångström-level spatial resolution, which can be achieved due to the relatively small wavelengths of high-energy electrons used to probe the sample. In addition, such atomic-scale imaging is not limited to surfaces (information related to the real space) and can be combined with electron diffraction (information related to the reciprocal space), giving TEM a very unique capability. These strengths of TEM coupled with the need to understand the mechanical behavior of emerging nanomaterials initiated a new field in mechanics that aimed at conducting mechanical testing of nanostructures with near-real-time observation of their deformation and failure: *in-situ* TEM mechanical testing. Initial work in the field utilized tensile straining holders and nanomanipulators, providing important initial findings on dislocation-based plasticity and failure mechanisms in a variety of materials [33]. In the case of CNTs, nanomanipulation *in-situ* TEM allowed visualization of the sword-in-sheath failure mechanism in which adjacent multiwall carbon nanotube (MWNT) shells slide with respect to one another due to weak shear interactions (see Figure 5.3a) [12,28], demonstrating one of the central limitations in scaling individual CNT shell mechanical properties to macroscopic materials.

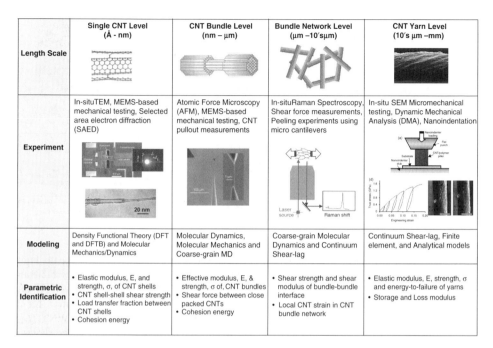

	Single CNT Level (Å - nm)	CNT Bundle Level (nm – μm)	Bundle Network Level (μm –10's μm)	CNT Yarn Level (10's μm –mm)
Length Scale				
Experiment	In-situ TEM, MEMS-based mechanical testing, Selected area electron diffraction (SAED)	Atomic Force Microscopy (AFM), MEMS-based mechanical testing, CNT pullout measurements	In-situ Raman Spectroscopy, Shear force measurements, Peeling experiments using micro cantilevers	In-situ SEM Micromechanical testing, Dynamic Mechanical Analysis (DMA), Nanoindentation
Modeling	Density Functional Theory (DFT and DFTB) and Molecular Mechanics/Dynamics	Molecular Dynamics, Molecular Mechanics and Coarse-grain MD	Coarse-grain Molecular Dynamics and Continuum Shear-lag	Continuum Shear-lag, Finite element, and Analytical models
Parametric Identification	• Elastic modulus, E, and strength, σ, of CNT shells • CNT shell-shell shear strength • Load transfer fraction between CNT shells • Cohesion energy	• Effective modulus, E, & strength, σ of, CNT bundles • Shear force between close packed CNTs • Cohesion energy	• Shear strength and shear modulus of bundle-bundle interface • Local CNT strain in CNT bundle network	• Elastic modulus, E, strength, σ and energy-to-failure of yarns • Storage and Loss modulus

Figure 5.2   Multiscale experimental and theoretical modeling approaches applied to study hierarchical carbon-nanotube-based fiber materials. *Single* CNT *level* adapted with permission from [28], © 2000 AAAS and [12], © 2008 Nature Publishing Group 2008. CNT *bundle level* adapted with permission from [22]. © 2010 American Chemical Society. *Bundle network level*, adapted from [29]. CNT *yarn level* adapted with permission from [22], © 2010 American Chemical Society and adapted from [30]

While these techniques have been effective in visualizing mechanical deformation and failure, they are limited in that they do not allow for the simultaneous measurement of forces. This limitation was overcome by the use of MEMS technologies [35–37]. In its most advanced version, thermal actuation was employed to achieve nanometer displacements and differential capacitance to measure load with nanonewton resolution [35–37] (see Figure 5.3b). Other implementations consisted of *in-situ* electrostatic force actuators and three-plate capacitive displacement-sensing technologies initially developed for nanoindentation applications. In the case of *in-situ* nanoindentation, one direction of particular interest was the study of plasticity in metals using *in-situ* TEM nanoindentation [38,39]. Through the direct visualization of dislocation dynamics in metallic structures, these studies demonstrated nanoscale mechanical phenomena such as dislocation starvation [39] and grain-boundary strengthening [38]. A detailed account of progress utilizing the above-mentioned techniques has been previously given by Legros *et al.* [33]. Although *in-situ* nanoindentation has shed a great deal of light onto nanomechanical size effects, the interpretation of indentation testing is indirect compared with uniaxial testing methods and is not suitable for characterizing 1D nanostructures such as CNTs. In addition, traditional *in-situ* TEM straining holders do not provide the force-sensing capabilities required to fully characterize their mechanical behavior.

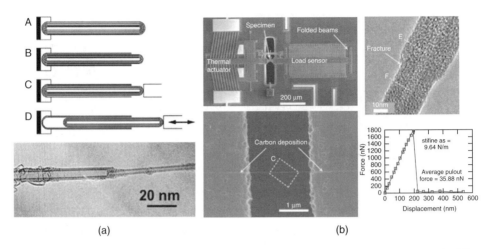

(a)   (b)

Figure 5.3 (a) *In-situ* TEM nanomanipulation of an MWNT revealing sword-in-sheath failure in which inner CNT shells pull out with respect to outer shells. (*Top*) Schematic representation of the MWNT manipulation. (*Bottom*) High-resolution TEM (HRTEM) image of an MWNT after manipulation. Reprinted with permission from [28]. © 2000 The American Association for the Advancement of Science. (b) MEMS-based *in-situ* TEM uniaxial tensile testing of an MWNT. (*Left*) The top SEM image shows an MEMS-based tensile testing device and the bottom image shows an MWNT suspending the device. Reprinted with permission from [34] © 2009 Springer. (*Right*) The top shows an HRTEM image of a MWNT after tensile testing. Reprinted with permission from [12] © 2008 Nature Publishing Group. The bottom shows a force vs displacement curve for MWNT sword-in-sheath failure. Reprinted with permission from [34] © 2000 The American Association for the Advancement of Science

The challenge of conducting uniaxial mechanical testing of nanostructures in *in-situ* TEM with simultaneous load sensing has been addressed by several approaches, including AFM and MEMS. Because of the advantages in using MEMS platforms for *in-situ* TEM, our discussion will first be focused on MEMS-based testing technologies. AFM techniques are also reviewed in the following section owing to their application to nanomaterials despite their indirect nature and the need for models to extract properties. *In-situ* uniaxial TEM MEMS testing requires addressing many challenges, including specimen handling and loading, force sensing, and displacement or strain sensing. Prior to testing, nanometer-scale specimens must be attached to the MEMS testing device. One common method is performed via a combination of nanomanipulation and bonding techniques utilizing piezoactuation and special glues or electron-induced deposition [12,23,40] *in-situ* optical or scanning electron microscopes. Loading of the specimen can be achieved by either piezo [35], thermal [12,40,41], or electrostatic actuation [37,42–44]. Force measurements are typically achieved by calibrated electrostatic comb drives [45] in force-controlled experiments and by deflection measurements of a force sensor beam with calibrated stiffness in displacement-controlled experiments, either through direct displacement imaging [46] or via capacitive sensing [41]. Finally, specimen displacement is determined by direct TEM imaging, while atomic-level strain can be measured from selected-area electron diffraction (SAED) of the specimen [40], a technique which exploits the atomic-scale spatial resolution achievable in TEM (see Figure 5.3b). For a more detailed review of *in-situ* MEMS-based mechanical testing, see the review articles of Haque and coworkers [35,47].

5.2.2 Measuring Ultralow Forces: AFM Methods

One of the most fundamental requirements towards understanding the mechanical behavior of carbon-based materials at the atomic level and micrometer scales, as discussed above, is high force resolution measurement. This requirement has been addressed by quantitative AFM, which allows force measurements in the nano- to pico-newton regime. This capability has become one of the most widely applied mechanical testing techniques for nanostructures in general, and has allowed mechanical characterization of individual carbon nanostructures such as CNTs, graphene, and graphene oxide (GO) sheets. In particular, AFM has been applied in a variety of implementations, including membrane deflection [13,48], as an extension of the membrane deflection experiments first developed in the study of elasticity and size-scale plasticity in sub-micrometer thin films [49–51], and in frictional [52,53] studies of graphene and GO as well as tensile [11,22,54] and bending studies [55] of CNTs and CNT bundles. In the case of graphene, AFM-based membrane deflection was essential in confirming the extraordinary in plane strength ($\sim$100 GPa) and modulus ($\sim$1 TPa) of single-layer sheets, which had been previously predicted by quantum mechanics simulations (see Figure 5.4) [13]. This experimental demonstration helped to spur the incorporation of graphene in hierarchical composite materials as nature's strongest and stiffest building block.

The unique force sensitivity of AFM relies on high-resolution deflection sensing of a micro-fabricated cantilever beam engineered to have well-defined spring constants. In AFM, the deflection sensing is commonly achieved via an optical laser beam deflection

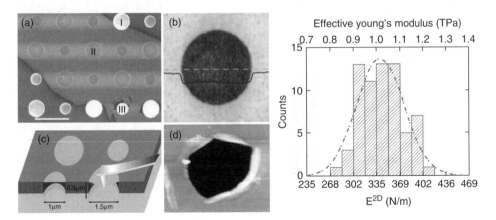

Figure 5.4   AFM indentation of suspended graphene sheets. (a, b) SEM and AFM images of a suspended graphene sheet. (c) Schematic representation of an AFM indentation experiment of a graphene sheet. (d) AFM image of a fractured graphene membrane. (*Right*) Histogram of the experimentally measured elastic modulus of graphene sheets revealing stiffness of $\sim$1 TPa. Reprinted with permission from [13]. © 2008 American Association for the Advancement of Science

scheme, but other techniques, such as piezo-resistive detection, have also been applied. For a detailed description of both the modes of operation and quantitative analysis of force measurements using AFM, refer to Meyer *et al*. [56].

5.2.3   *Investigating Shear Interactions: In-Situ SEM/AFM Methods*

Despite the remarkable mechanical properties of CNTs and graphene, the strengths and stiffnesses of their nanocomposite films and fibers are only a small fraction of the corresponding intrinsic properties, pointing to the inefficient load transfer between neighboring CNTs and graphene sheets. Moreover, owing to the weak interactions between shells of, for example, as-produced MWNTs, only the outermost shell will participate in the load transfer, substantially reducing the effective strength and modulus of MWNTs. Therefore, a key element to the design of CNT- and graphene-based nanocomposites with high mechanical properties is to better understand and engineer the shear interactions between adjacent CNTs and graphene sheets and shells of MWNTs.

As discussed in Section 5.2.1, the first experimental evidence of the sliding of the inners shells of MWNTs inside outer shells was revealed through electron microscopy visualization [28]. While these early experiments pointed to a nearly wear-free shear interactions between adjacent shells of MWNTs, they did not allow the quantification of forces involved. To achieve this, *in-situ* SEM experiments, which utilized an AFM cantilever as the load sensor with the direct visualization of electron microscopy, were developed by Yu *et al*. [57]. Their experiments included the loading of the outer shell of MWNTs in tension, followed by its rupture and the controlled pull-out of the inner shells, as shown in Figure 5.5a, in a so-called sword-in-sheath failure. Owing to the

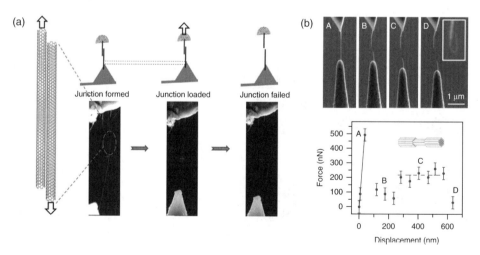

Figure 5.5   *In-situ* SEM shear testing of CNT interfaces. (a) SEM images and schematics of the formation, shear testing, and failure of a junction of two MWNTs using an Si cantilever load sensor . Reprinted with permission from [17]. © 2012 American Chemical Society. (b) SEM images and force displacement curve measured for the pullout of inner DWNTs from a DWNT bundle. Reprinted with permission from [60]. © 2012 American Chemical Society

point-to-point resolution of the SEM, correlation between the number of failed shells and measured force was not possible. This was first accomplished by Espinosa and coworkers using MEMS technology *in-situ* TEM [12]. In their work, the peak load and post-failure load–displacement curve was measured and correlated to the failure of a single CNT shell in an MWNT. Furthermore, the modulus and strength predicted by quantum mechanics for a single CNT shell was achieved, for the first time, on pristine arc-discharge CNTs.

Shear force measurement between CNTs has also been obtained by incorporating AFM [59,60]. Using AFM, Kaplan-Ashiri *et al.* investigated, both experimentally and theoretically, the stiffness of interlayer shear sliding of MWNTs by loading individual MWNTs in three-point bending, causing small shear strains between the shells of MWNTs [59]. Another type of AFM-based study of the shear interactions between CNTs was performed by mounting an MWNT on the AFM cantilever tip and scanning across a suspended single-wall carbon nanotube (SWNT) in tapping mode [60]. The variations of the AFM cantilever vibration parameters were used to estimate the frictional force and the shear strength of the contact between the two tubes.

Recently, experimental schemes coupled with theoretical models and simulations have been employed to investigate the shear interactions, in a different level of the hierarchy of CNT materials, between shells of contacting CNTs and their bundles, aiming at developing a fundamental understanding of the effect of surface functionalization and CNT surface quality on the CNT–CNT interactions [17,61,62] (see Figure 5.5).

In addition to nanoscale experiments aimed at investigating the interactions between graphitic shells in MWNTs, macroscale experiments have also been devised and implemented to indirectly measure the shear interactions between MWNTs [63,64]. These experiments typically have a higher throughput than the nanoscale experiments presented earlier in this section, mainly due to the size of the sample (which facilitates its preparation) and the higher forces involved. However, the larger scale experiments can only quantify the average response of the MWNT sliding, without being able to elucidate the details of their atomic-scale sliding.

5.2.4 Collective and Local Behavior: Micromechanical Testing Methods

Owing to their remarkable mechanical properties, since their discovery, CNTs have been considered as one of the promising fillers for nanocomposites, adding strength and stiffness to different matrices. In general, based on relative mass/volume CNT content, CNT-based nanocomposites can be divided into two groups: nanocomposites with low and high CNT ratio. The demarcation line between the low and high CNT content can be considered the percolation threshold of CNT concentration, above which a continuous network of CNTs will develop in the composite facilitating the transfer of mechanical load, electric charges, and heat. The latter type of CNT nanocomposite (containing high concentration of CNT) has become an active field of research, especially in the last decade, and is the main focus of this section. More general reviews of CNT composites, which focus on low CNT-ratio composites, and their processing and mechanical properties, can be found elsewhere [65,66]. As pointed out in the previous section, by enhancing the shear interactions between CNTs in *high-CNT*-content nanocomposites, the remarkable mechanical properties of CNTs at the nanoscale should be achieved at the macroscale. In addition,

higher contents of CNTs in nanocomposites can enhance both their electrical properties and thermal stability. Several methods have been commonly used to investigate the
mechanical behavior of CNT-based composites, including dynamic mechanical analysis
(DMA), micro-tension tests on CNT-based composites and yarns, and *in-situ* mechanical
characterizations inside analytical chambers such as a scanning electron microscope or
Raman spectrometer.

DMA is typically used to investigate the viscoelastic response of CNT–polymer composites, and it can be used to investigate the interactions between CNTs and polymer
chains [67–71]. In DMA, a sinusoidal stress field is applied to the sample and the strain
response is monitored. From the analysis of the DMA experiments, the dynamic moduli of the CNT–polymer system with two major components are extracted: the storage
modulus and the loss modulus, both functions of the strain oscillation frequency. The
former modulus is a measure of the mechanical energy stored in the system and the latter
represents the energy loss in the system, mostly in the form of heat. DMA experiments on
CNT composites are typically carried out in a temperature-sweep mode to reveal information about the glass transition temperature T_g of the system and its variations with
CNT–polymer interactions. Generally, the dependence of T_g on CNT content contains
information about the nature of interactions between CNTs and the polymer matrix.

Tension tests have also been widely used to measure the mechanical behavior of
CNT composites, (see Figure 5.6a–d). Unlike DMA experiments, which reveal only the
viscoelastic mechanical properties of the composites, in tension tests other material properties, such as strength and energy to failure, have also been reported [29,67,73–76].
These parameters are all of significant interest to the design of high-performance composites and yarns [22]. In addition, the statistical analysis of fracture parameters such as
toughness, strength, and ductility can be instrumental in obtaining an average distribution
of defects in the samples [20]. The shape of the stress–strain curves, obtained in tension
tests, provides evidence about the deformation mechanisms in CNT composites and the
ensemble behavior of the CNT–polymer chain interactions [73]. Tension tests on CNT
composites carried out in, for example, *in-situ* SEM and Raman analytical chambers, can
also be used to reveal the microscopic mechanisms of deformation of CNT composites.
For example, Espinosa and coworkers investigated the effect of lateral contraction and
twisting of CNT yarns inside SEM chambers [22]. Moreover, mechanical testing in an
in-situ Raman spectrometer has been used to assess the local strain distribution in mechanically loaded CNT composites [29,77]. In Raman spectroscopy, the sample is excited by
a monochromatic light source and the inelastic scattering of light is collected. The shift
in the frequency of the scattered light compared with that of the incident light is a characteristic of the modes of vibration of the sample and atomic spacing. For instance, in the
Raman spectrum of CNTs, the so-called radial breathing mode (RBM) peaks, the graphite
(G) peak and the disorder (D) peak are among the most commonly studied signatures.
The RBM modes (scattering shift of $<300\,\mathrm{cm}^{-1}$) reflect the vibration of carbon atoms of
CNTs in the radial direction, and are characteristic of the diameter of CNTs. The G peak
(scattering shift at $\sim1600\,\mathrm{cm}^{-1}$) results from the in-plane vibration of carbon–carbon
bonds, and reflects the graphitic nature of CNTs, while the D peak (scattering shift at
$\sim1300\,\mathrm{cm}^{-1}$) is the defect-induced peak [78,79]. In *in-situ* Raman experiments, straining of CNTs will result in the change of the carbon–carbon interaction energy and the
frequencies of their different modes of vibration, inducing a change in the Raman signal

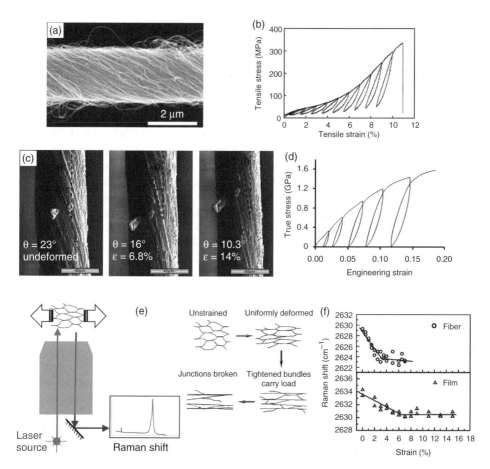

Figure 5.6 Micromechanical testing methods applied to CNT yarns. (a) SEM image of the CNT yarns spun from a forest of CNTs and (b) its mechanical behavior measured in tension. Reprinted with permission from [72]. © 2004 American Association for the Advancement of Science. (c) Three snapshots of the tension test on a CNT yarn, tested by *in-situ* SEM. Reprinted with permission from [22]. © 2010 American Chemical Society The Poisson ratios of the yarns are ~1 as measured as the ratio of the axial strain to the relative change in diameter of the yarn. (d) The mechanical behavior of the yarn shown in (c). (e) Schematics of the *in-situ* Raman tension tests on CNT yarns fabricated from forests of CNTs, and the corresponding changes in the location of the G′ peak as a function of strain. Adapted from [29]

shifts. This shift in the Raman signal can be collected, for instance during a tension-test experiment on a CNT composite, to estimate the average strain in CNTs under different loading conditions [29,77].

Another type of experiment used to investigate the mechanical properties of CNT-based materials is nanoindentation [80–85]. In this type of experiment, an indenter is pushed against the sample surface and the mechanical interactions between the two are extracted and used to calculate the sample modulus and hardness at the location of the indentation.

Moreover, by modifying the sample boundary conditions and utilizing proper nanoindenter tip geometry, a nanoindentation setup can be used to perform bending and compression experiments on composites [30,82]. In a nanoindentation experiment, the modulus is calculated by measuring the unloading stiffness of the nanoindenter, and hardness is measured as the ratio of the maximum load to the residual indentation area [86]. The main advantage of nanoindentation over other mechanical characterization methods is that it does not require extensive sample preparation, which becomes specifically beneficial for nanoscale samples such as electrospun nanofibers containing CNTs [84]. Moreover, it allows for the measurement of local mechanical properties of the samples. In the case of composites with nanofiller reinforcement, such as CNT-based composites, together with rule-of-mixture estimates of elastic moduli, nanoindentation becomes an enabling tool for correlating CNT dispersion inhomogeneities with mechanical properties [85]. The main disadvantage of nanoindentation is that, owing to its indirect nature, a model for the mechanical behavior of the sample and tip is required to extract the sample's material properties.

5.3   Hierarchical Carbon-Based Materials

In the following we will review progress in the development and understanding of hierarchical carbon-based materials, which has been achieved through the application of the multiscale experimental techniques discussed in the previous section. In particular, the focus will be on materials that utilize a high density of CNTs, graphene, or GO as building blocks. Such materials, which are based primarily on constituents with a high level of order, are envisioned to allow for optimal properties similar to those found in nature. A focus on CNTs and graphene is motivated by their extraordinary intrinsic mechanical properties, which have recently been confirmed [12,13]. For a detailed review on the intrinsic mechanical properties of individual CNTs and graphene sheets, refer to the review article by Kis and Zettl [87].

5.3.1   Weak Shear Interactions between Adjacent Graphitic Layers

One consequence of developing hierarchical materials with a high density of highly ordered graphitic layers (graphene sheets and CNT shells) is the corresponding high density of direct interactions between adjacent graphitic layers. As discussed previously, this poses one of the greatest challenges to scaling up the exceptional mechanical properties of CNTs and graphene to macroscopic materials, as these interfaces tend to have inherently weak interactions dominated by van der Waals (vdW) forces. Examples of these interfaces include shell–shell interactions in MWNTs, tube–tube interactions in close-packed CNT bundles, and sheet–sheet interactions in few-layer graphene. These weak interfacial interactions lead to low strength shear transfer between adjacent layers, which significantly limits the effective mechanical performance of the macroscopic constituents. As pointed out in Section 5.2.3, the investigation of binding energy and load transfer between such graphitic layers can provide us with deeper insight into the mechanical behavior of carbon-based composites, and potentially opens new horizons in designing CNT- and graphene-based composites with enhanced mechanical performance.

One of the first direct demonstrations of easy shear between individual adjacent graphitic layers was through *in-situ* TEM axial stretching of MWNTs [28]. Cumings and Zettl visualized the pull-out of the inner shells of an MWNT from the outmost shell of MWNTs, using *in-situ* TEM, by welding a probe to the inner shells and pulling the inner shells out. Figure 5.3 shows the inner shells of an MWNT being pulled out of an outer tube of shells in a reversible process. Without measuring the pull-out force experimentally, they estimated the vdW interlayer force to be in the order of 9 nN, corresponding to 2.3×10^{-14} N per atom. They demonstrated that the pulled-out inner shells tend to be pulled back in, due to the vdW interactions between the inner and outer shells, without any detectable damage to the atomic structure of CNTs, pointing to nearly perfect sliding surfaces with minimal wear. Subsequent pull-out studies, which utilized an AFM force sensor in *in-situ* TEM, estimated that the interlayer shear was dominated by vdW interactions [88]. In that study the interlayer shear strength was estimated to be <0.05 MPa, with an interlayer cohesion energy of 33 meV/atom, which was in good agreement with estimates determined from an energy analysis of collapsed MWNTs [89].

The interactions between shells of an MWNT were also quantified by Yu *et al*. [57]. They used an AFM cantilever as the load sensor mounted on a manipulator inside an scanning electron microscope. One end of an MWNT was attached to the AFM cantilever tip, while the other end was attached to a substrate, both by carbon deposition. In the experiments, the outer shell of an MWNT was pulled in tension. First, the outer shell of the MWNT ruptured, which was followed by controlled pull-out of the inner shells. The pull-out force measured by Yu *et al*. was described as being composed of two components: the shear interaction between shells, which scales with the overlap (embedded) length; and a second interaction, induced, for instance, by the capillary effect and the interactions by dangling bonds at the two ends of the shells, which scales only with the circumference of the inner tube. The static shear strength was estimated to be 0.08–0.3 MPa, and the combined capillary and edge-effect force was measured to be 80–150 nN, dominating the experimentally observed forces. In that study, it was assumed that the shear transfer takes places along the whole contact length between the inner and outer shells. Therefore, the active overlap length could be overestimated, and the reported shear strength represents a lower bound. However, given the finite stiffness of the CNTs, the shear forces could be distributed over a shorter length, leaving the rest of the CNTs traction free. This could in part explain the wide range of measured shear strength between CNT shells presented by Yu *et al*. [57]. While such nanoscale experiments can quantitatively capture shear sliding between layers of MWNTs, owing to the limited resolution of SEM in measuring the interlayer sliding they are not capable of capturing the interlayer sliding shear modulus.

Kaplan-Ashiri *et al*. investigated, both experimentally and theoretically, the stiffness of interlayer shear sliding of MWNTs, by loading individual MWNTs in bending using AFM to achieve sub-nanometer deflection resolution [59]. Their theoretical analysis, using the density-function-based tight binding (DFTB) method, pointed to an interlayer shear modulus of about 2 GPa. The relatively low value of the interlayer shear modulus compared with the axial modulus of CNTs (the ratio of shear modulus to axial modulus of CNTs is ~0.02 compared with a value of 0.3 for an isotropic material) resulted in considerable shear sliding between shells of the MWNTs in the AFM bending experiment, such that as much as 25% of the lateral deflection of CNTs was due to elastic shear deformation. For comparison, the shear deformations would contribute to less than 1% of the total lateral deformation of an isotropic material with similar dimensions. Locascio *et al*. applied an *in-situ* TEM method to measure the shear sliding between the inner and outer shells of an MWNT [33]. They measured an average post failure force of 35 nN required to pull out the 11 inner shells of a 14 nm diameter MWNT with respect to the outer shell (see Figure 5.3b).

CNT bundles have also been shown to exhibit weak interaction between adjacent outer CNT shells within close-packed bundles. *In-situ* SEM tensile experiments conducted on SWNTs and double-wall nanotubes (DWNTs) revealed a similar sword-in-sheath failure mechanism in which inner CNTs within bundles pulled out with respect to an outer shell of CNTs [22,54]. Yang *et al*. experimentally investigated the friction between SWNTs by loading an SWNT bundle in tension [63]. Their SWNT bundles were relatively long ($\sim$3 mm) and were grown in a chemical-vapour deposition (CVD) reactor. During the tensile loading, they monitored both the elastic behavior and the inelastic behavior of the yarns due to SWNTs sliding on each other. They estimated the cohesive energy per unit area of the SWNTs to be in the range 0.1–$0.6\,\mathrm{J\,m^{-2}}$, by normalizing the friction energy (dissipated during the plastic deformation) by the change in the contact area between the bundles. While this method can be used to obtain an average value for the cohesive energy between SWNT studies, it suffers from uncertainties in the estimation of the edge effect and the true contact area between SWNTs. More recently, a novel method of *in-situ* SEM peeling between two SWNT bundles was used to measure an adhesion energy of 0.12–$0.16\,\mathrm{nJ\,m^{-1}}$ [90]; however, the width of the contact between the two bundles was not reported, therefore, the cohesion energy per unit area is unknown, making comparison with other studies difficult. The nature of shear interactions within DWNT bundles has also been recently investigated via an experimental–computational approach which compared *in-situ* SEM pull-out tests with molecular mechanics (MM) and density functional theory (DFT) simulations [58]. In that study, Filleter *et al*. measured a normalized pullout force of $1.7 \pm 1.0\,\mathrm{nN/CNT}$–CNT interaction for sliding of a smaller inner bundle of DWNTs out of a larger outer shell of DWNTs (see Figure 5.5b). Through comparison with MM and DFT simulations of sliding between adjacent CNTs in bundles it was identified that factors contributing to the pull-out force included the creation of new CNT surfaces, carbonyl functional groups terminating the free ends, corrugation of the CNT–CNT interaction, and polygonalization of the CNTs in the bundle. In addition, a top-down analysis of the experimental results revealed that greater than one-half of the pull-out force was due to dissipative forces. This finding of behavior at the CNT bundle level differs significantly from the behavior of pull-out in individual MWNTs, for which dissipation is found to be negligible [28].

Bhushan *et al*. investigated the interactions between an individual SWNT and MWNTs in ambient conditions using AFM [60]. The former was mounted over a trench on a substrate, while the latter were mounted at the tip of an AFM cantilever. The cantilever was scanned over the trench in tapping mode to induce contact between the SWNT and MWNTs, as shown in Figure 5.5b. The maximum deflection of the cantilever as the MWNT was pulled away from the SWNT was used to estimate the work of adhesion between the two tubes as $0.03\,\mathrm{J\,m^{-2}}$. This value is significantly lower than previous measurements of the graphite surface energy [91], potentially due to the uncertainties in the contact area between the two tubes. Therefore, it should be considered as a lower bound for the work of adhesion in graphite in ambient conditions. In addition, the attenuation of the amplitude of vibration of the cantilever, upon the formation of contact between two CNTs, was used to estimate the power loss due to friction, from which the friction force and the shear strength of the contact between the two tubes was estimated. The shear strength was estimated to be $\sim$4 MPa, which is higher than that measured in vacuum [57], most likely due to the presence of thin layers of water molecules strengthening the junctions between the two tubes in ambient conditions.

Peeling experiments have also been used to investigate the interactions between graphitic surfaces. Ishakawa *et al.* used a self-detecting microcantilever inside an SEM chamber to measure the interactions between an MWNT and cleaved graphitic surfaces in pure peeling mode (mode I) with negligible net shear stress component on the interface [92]. They observed discrete jumps in the adhesive force with the peak values in the range of a few tens of nanonewtons, corresponding to adhesive energies of as high as 78 keV, with the number of discontinuous jumps decreasing as the length of the CNTs decreased, due to the enhanced bending stiffness of the MWNTs.

Wei *et al.* developed a continuum based shear-lag model which successfully predicted the saturation regime in the shear force as a function of overlap length for SEM experiments of shear between MWNTs, the details of which are shown in Figure 5.5a [17]. In that study the data from shear experiments performed on unfunctionalized MWNTs fell into the region in between two theoretical curves with shear strengths of 30 MPa and 60 MPa for arm-chair and zig-zag tubes, respectively. This suggested that the shear between two MWNTs is dependent on chirality, which was also verified by atomistic calculations. Furthermore, through their model, they demonstrated that CNT alignment, although required, is not the sufficient condition for optimal mechanical performance of CNT-based yarns. Rather, the average overlap lengths between CNTs needs to be chosen properly, based on the mechanical properties of the constituents and the shear interactions between CNTs, to achieve the highest mechanical performance, such as the highest elastic energy density and full utilization of the CNT strength. Likewise, to achieve ductility and associated high failure energies, spreading of the sliding and delay of deformation localization are needed [9]. It is to be noted, however, that the complexities and limitations related to fabrication of CNT yarns with a prescribed architecture (characterized by parameters such as overlap length and alignment) remain to be addressed. Similar experimental–analytical approaches can be implemented to realize the role of CNT surface functional groups and cross-linking chemistries on the enhancement of CNT interface interactions.

In addition to microscopic studies of shear within CNTs and bundles, weak interlayer binding and shear between adjacent graphene layers has also been extensively studied indirectly through a variety of techniques, including macroscopic friction testing of graphite [93]. It is generally accepted that graphite acts as a good solid lubricant due to shearing of adjacent graphitic planes. More recently, nanometer-scale friction measurements on graphite and graphene have confirmed ultralow frictional properties [52,94]. In addition, several experimental techniques, including heat of wetting experiments [95] and molecule desorption studies [96], have measured the cohesion energy of graphite to be 0.26–$0.37\,\mathrm{J\,m^{-2}}$, similar to the reported measurements on CNTs and CNT bundles. A summary of both the shear strengths and cohesion energies measured for the variety of interfacial systems discussed in this section is found in Figure 5.2.

5.3.2 Cross-linking Adjacent Graphitic Layers

One approach to address the limitation of weak interlayer shear between graphitic layers, as discussed in the previous section, is the introduction of cross-linking bonds between adjacent sheets. This approach inherently requires modifications to the in-plane sp^2 bonding within the graphitic sheet. In the case of CNTs, one method to achieve this is via high-energy electron irradiation at energies above the knock-on requirement to displace or remove carbon atoms and create covalent cross-linking defects; in the case of graphene, chemical surface modifications have been applied to create GO sheets in which functional surface groups are utilized to enhance interlayer linking.

Irradiation by high-energy particles, such as ions, electrons, or neutrons, to induce modifications to carbon nanostructures has been predicted and demonstrated to result in the creation of vacancy and interstitial defects that can bridge adjacent atomic layers [97,98]. This has proven to be a unique method to greatly enhance the effective properties of both MWNTs and CNT bundles. Quantum mechanics calculations have demonstrated that irradiation leads to the formation of divacancy, interstitial, and Frenkel pair defects (see Figure 5.7a), resulting in covalent bonds being formed between adjacent graphitic sheets [98]. MM simulations have shown that only a low density of such defects ($\sim$0.2–0.4 Å^{-1}) is required to approach theoretical limits of load transfer between adjacent graphitic sheets, such as inner and outer shells in DWNTs (see Figure 5.7b) [12,34]. This approach towards achieving efficient load transfer facilitated by cross-linking defects has been applied to address load transfer at multiple lengths scales within CNT-based materials. In the majority of high-density CNT-based materials, the first two hierarchical length

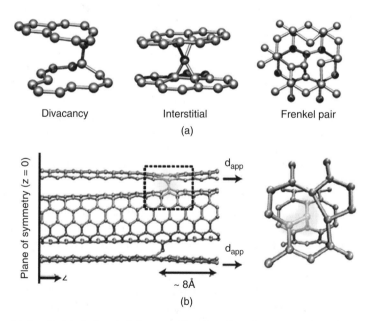

Figure 5.7 (a) Irradiation-induced defects predicted to form between adjacent atomic layers in graphite. Adapted with permission from [99]. © 2011 American institute of Physics. (b) MM simulation of load transfer for Frenkel pair defects between the outer and inner shells of a DWNT. Adapted with permission from [34], © 2009 Springer

scales are shell–shell interactions within individual MWNTs and tube–tube interactions between adjacent outer shells, both of which have been demonstrated to be successfully cross-linked by irradiation.

At the first hierarchical level, Peng *et al.* demonstrated the effects of electron irradiation cross-linking through *in-situ* TEM tensile testing experiments conducted on irradiated MWNTs [12]. In that study, high-resolution TEM imaging revealed that, in the absence of irradiation, only the outer shell of the MWNT carried an appreciable load until failure, after which it slid with respect to the inner shell by the previously documented sword-in-sheath failure mechanism. At higher irradiation levels MWNTs were found to fail across multiple shells, as load transfer to inner shells was increased due to cross-links. Analysis of the true stress acting on the shells carrying load revealed that, at low levels of irradiation, the outer few CNT shells had a remarkable failure stress and modulus of $\sim$100 GPa and $\sim$1 TPa, respectively, whereas increased irradiation led to a reduction in true strength and modulus to 35 GPa and 590 GPa, respectively [12]. This demonstrates the inherent tradeoff between enhancing the effective mechanical properties of the MWNT while reducing the intrinsic mechanical properties of individual CNT shells through the creation of defects.

At the tube–tube level, Kis *et al.* applied an AFM-based deflection method to investigate the effects of electron-irradiation-induced cross-linking on enhancing the bending modulus of SWNT bundles (see Figure 5.8f and g) [55]. Here, it was demonstrated that low doses ($\sim$5 × 10^{20} e^{-} cm^{-2}) of electron irradiation yielded an increase in the effective bending modulus up to $\sim$750 GPa [55]. Again, higher irradiation doses significantly reduced the modulus, in this case to as low as $\sim$100 GPa (at doses >40 × 10^{20} e^{-} cm^{-2}). The AFM-based method, however, did not allow for a determination of the strength of the

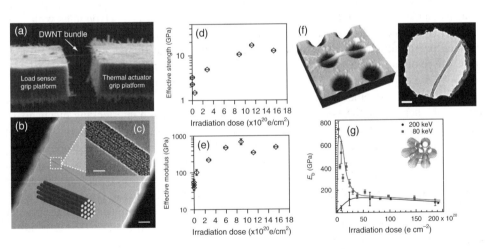

Figure 5.8 Irradiation-induced covalent cross-linking enhancements in DWNT and SWNT bundles. (a–c) *In-situ* TEM tensile testing of DWNT bundles. (d, e) Enhancements in effective strength and modulus for DWNT bundles as a function of irradiation dose. Adapted from [23]. (f) AFM-based deflection experiments of SWNT bundles. Reprinted with permission from [55] © 2004 Nature Publishing Group. (g) Enhancements in effective bending modulus for SWNT bundles as a function of irradiation dose [55]. Adapted by permission from [55]. © 2004 Nature Publishing Group.

irradiated SWNT bundles. More recently, Filleter *et al.* conducted *in-situ* TEM studies on DWNT bundles exposed to high-energy electron irradiation [23]. In this case, both levels of hierarchy, inter-tube shell–shell interactions and inter-bundle tube–tube interactions, are present and cross-linked via irradiation [23]. Here, it was found that both the effective strength and stiffness of the DWNT bundles was increased by irradiation up to 17 GPa and 693 GPa, respectively (see Figure 5.8a–e). HRTEM images of the DWNT bundles revealed that, in the case of minimally irradiated bundles ($0.5 \times 10^{20} e^- \, cm^{-2}$), the outer tubes within the bundles failed during the tensile test and slid with respect to the inner bundle of DWNTs, akin to the sword-in-sheath failure observed for MWNT shells. However, at optimal irradiation doses (~ 9–$11 \times 10^{20} e^- \, cm^{-2}$) the bundles were found to fail across the entire cross-section of shells and tubes, confirming effective load transfer to the entire inner core of material. The potential benefits of irradiation-induced cross-linking have also been investigated via molecular dynamics (MD) simulations of CNT bundles. Cornwell and Welch recently conducted simulations of SWNTs cross-linked by interstitial carbon atoms which predicted significantly increased load transfer leading to strengths of up to 60 GPa for SWNT bundles with optimal cross-linking density and overlap geometry [99]. Although all of these studies have demonstrated that irradiation-induced cross-linking is effective at improving load transfer within CNTs and bundles, they have also identified that there is a limit in the achievable mechanical properties, as the complexity of the material is increased due to the inherent introduction of structural defects into the material.

Another approach that has been implemented recently to develop high-performance CNT yarns and to enhance the shear interactions between CNTs is the *in-situ* CVD functionalization of tubes (DWNT bundles) [22]. Unlike conventional methods for the functionalization of CNTs, this method, which is based on the development of polymer radicals inside the CVD reactor, together with the formation of DWNT bundles with very low defect density, allows for the covalent functionalization of DWNT bundles at the existing defect sites with active polymer species such as substituted acrylic acid groups, therefore maintaining the integrity of the structure of the bundles. The nanoscale experiments on the effect of *in-situ* CVD functionalization of the tubes on the shear interactions between DWNTs together with the multiscale simulations of the shear experiments have pointed to shear strengths as high as 300 MPa, mostly due to the interlocking mechanisms between the polymer chains of the shearing bundles and the high degree of alignment of polymer chains achieved during the process of shearing [61]. These strong interactions have led to unconventional strength and energy to failure of as high as 1.5 GPa and 100 J g^{-1}, respectively [22]. Coarse-grain molecular modeling is emerging as an attractive technique that enables studying the effects of polymer cross-linking in large CNT bundles and fibers. Through coarse-grain modeling, Bratzel *et al.* have demonstrated that polymer cross-links with optimized concentration and length can lead to increases of fourfold in strength and fivefold in toughness for cross-linked CNT bundles [100].

In the case of graphene, a similar challenge exists in improving load transfer from one graphene sheet to the next. At the core of the challenge is the lack of reactive chemical handles on the basal planes. In addition to electron-irradiation-induced cross-linking, another attractive method to address this limitation, which has been applied to graphene, is the oxidation of graphite to produce GO sheets, which bear oxygen functional groups on their basal planes and edges. These functional groups then act as chemical cross-linking sites between adjacent sheets. Despite the similarities between the atomic structures of CNTs and graphene, their responses to functionalization processes may be very different. On the one hand, graphene will likely start to react with functional groups at the edges and the defects where dangling bonds are present. On the other hand, the surface curvature of CNTs and the associated bending energy stored in C–C bonds facilitates the chemical reaction of C atoms with functional groups. In the case of GO, similar limitations on the mechanical properties exist to those demonstrated for irradiation-induced cross-linking of CNTs: the oxidation process inherently leads to the formation of voids in the graphene lattice, which reduces the strength and stiffness. AFM-based experiments, similar to those used to demonstrate the exceptional strength and stiffness of pristine graphene sheets [13], have demonstrated that by oxidizing graphene to create GO the elastic modulus is reduced from ~1 TPa to 208 GPa [48]. It should also be noted that further reduction of GO, which attempts to return the structure to a pristine state, has yielded a material with a stiffness of 250 GPa; hence, there is no recovery of the exceptional mechanical properties of pristine graphene [101]. Although no quantitative study of the strength of GO has been conducted to date, the reduction in stiffness and the observed defect structures observed via HRTEM imaging of GO [102] suggest a substantial reduction in the strength of GO. If we assume a correlation between the reduction of modulus and the induced defect density in the graphitic structures, the higher reduction in modulus observed for GO compared with irradiated CNTs would suggest strengths of less than 17–35 GPa for GO [12,23]. This is one area in which novel fabrication methods are needed to reduce both the size and density of defects in oxidized graphene sheets while maintaining enough functional groups to facilitate effective cross-linking handles. To this effect, initial studies on tuning the C/O ratio of GO are just beginning to emerge [103].

Despite the intrinsic reductions in the mechanical properties of oxidized graphene sheets, the effective benefits of cross-linking adjacent layers of GO has shown some promise towards macroscopic graphene-based materials. In its uncrosslinked state, GO paper already exhibits mechanical properties superior to many technologically relevant paper materials, such as bucky paper, flexible graphite, and vermiculite. However, when compared with the intrinsic behavior of graphene, it exhibits mechanical properties that are orders of magnitude lower; that is, a modulus of ~32 GPa and a strength of ~80 MPa [27]. These reductions are mainly influenced by characteristics of the hierarchical structure, such as the discontinuous layering and inherent waviness, as well as weak interlayer shear interactions. Initial chemical methods to enhance the layer–layer shear interactions have included introducing divalent ions and polyallylamine crosslinks [104,105]. While these reports have shown some improvements in the macroscopic mechanical properties of the paper materials, the enhancements have been only minor, and the underlying mechanisms remain to be studied via experiments and simulations at smaller length scales.

5.3.3 Local Mechanical Properties of CNT/Graphene Composites

One of the great challenges in developing advanced carbon-based composites is bridging an understanding between the mechanics of the nanoscale constituents and macroscopic materials. In the previous two sections, progress in understanding the mechanical behavior of CNT and graphene building blocks and how two elements (e.g., two adjacent CNTs or G/GO nanoparticles) interact was discussed. The next level of hierarchy within a carbon-based material that requires a more fundamental understanding is small networks composed of many of these building blocks. To investigate this length scale, experimental techniques are required that both probe local mechanical properties on volumes of the order of a few hundred cubic nanometers and measure average local behavior of many interacting nanoscale constituents. One notable example of the latter is measuring the average strain in individual CNTs in a loaded composite.

The local mechanical response of CNTs within larger networks has been investigated by utilizing mechanical testing in *in-situ* Raman spectroscopy. Given the sensitivity of the location of peaks of the Raman spectra of CNTs to the applied strain, tension tests in *in-situ* Raman spectrometry have also been used to assess the efficiency of load transfer between CNTs in a composite. In such experiments, an average local strain on CNTs can be estimated as a function of the shift in Raman peaks and compared with the global strain on the sample [29,77]. Ma *et al.* investigated the shifts in the G peak shape and location to identify the micromechanical mechanisms of deformation of CNT fibers and films [29]. At sufficiently low strains, they observed a downshift in the G peak, potentially due the straining of individual CNTs and the resulting weakening of C−C bonds. However, the rate of shift relative to the composite average strain was a small fraction of the value corresponding to strained individual CNTs [106], indicating that most of the macroscale strain was due to the change of the shape of the CNT network through CNTs rotation towards the stretching direction, rather than their stretching, to accommodate the global strain. At higher strains, the location of the G peak became stationary, pointing to a change in the deformation mechanism from CNT network stretching to network rupture and stress redistribution, such that the average strain in individual CNTs remained nearly unchanged. Moreover, the straining of the samples resulted in G peak broadening, attributed to nonuniform strain in different regions of the sample.

Another type of experiment that has been applied to the investigation of the local mechanical properties of CNT/graphene-based materials is nanoindentation [80–83], which is capable of measuring material properties, such as modulus and hardness, on localized volumes confined to between tens and hundreds of cubic nanometers. This capability of nanoindentation has been specifically useful in the assessment of the quality (uniformity) of dispersion of CNTs in CNT composites [85]. For instance, Bakshi *et al.* utilized the distribution in CNT dispersion (measured by image processing of the composite surfaces) together with the Halpin–Tsai relation to successfully explain the scatter observed in the composite moduli measured by nanoindentation [85]. Therefore, a reverse method in which the local modulus measurements of a composite by nanoindentation are used to quantitatively express the dispersion of CNTs in composites is imaginable.

AFM-based methods have also been used to perform nanoindentation experiments on CNT composites [84,85]. The sharpness of AFM cantilever tips, in the range of few tens of nanometers, allows for the indentation of nanometer-scale samples. The tilt of the AFM cantilever should either be included in the data analysis to estimate the normal indentation force from the cantilever deflection, or a tilt compensation scheme should be experimentally implemented [107]. In addition, for uncompensated experiments, given the inherent tilt of the AFM cantilevers with respect to the sample surface, pushing the AFM cantilever tip down on the sample will generate some forward motion of the tip, further complicating the data analysis. Therefore, AFM-based indentations should, in general, only be carried out to induce relatively shallow indents, such that reliable loading–unloading curves can be obtained and to avoid excessive material pile-up, which can complicate the stress field.

In the application of nanoindentation to investigate the mechanical behavior of composites, care should be taken in choosing the proper type of nanoindenter tip and the boundary conditions of the sample to distinguish between different modes of sample loading, namely indentation and sample bending, and proper models should be incorporated to extract the mechanical properties of the sample.

An example of different loading modes using a nanoindenter is the work by Lee and Cui, in which they used a nanoindenter to load CNT-based nanocomposite films both in indentation (collapsed samples, sitting on a substrate) and bending (a free-standing film gripped on two edges) modes [82]. The geometry of the nanoindenter tip provides a helpful guideline to distinguish between the two cases: a flat tip, wider than the width of a free-standing sample, favors the bending experiment, while a sharp tip, with a tip radius sufficiently less than the width of the sample, on a collapsed sample provides a stress field closer to an ideal indentation experiment.

In addition to nanoindentation and bending experiments, a nanoindenter can be used to measure the mechanical properties of CNT composites in compression, by incorporating a flat punch as the nanoindenter tip. For instance, Garcia et al. used a nanoindenter with flat punch to investigate the compressive modulus of CNT epoxy nanocomposites with CNTs oriented parallel to the loading direction [30]. A schematic of their setup is shown in Figure 5.9. They reported a substantial increase in compressive modulus of the composites from 3.7 GPa to 11.8 GPa by adding 2 vol.% of CNTs. Compression tests of multilayer CNT–polymer arrays have also been conducted recently by Misra and coworkers [108,109]. These multilayer structures were found to sustain large compressive deformations and exhibit high energy absorption compared with synthetic materials of similar density.

5.3.4 High Volume Fraction CNT Fibers and Composites

One of the focuses of research in the past decade has been the development of CNT-based composites and fibers with high volume fractions of CNTs. In these novel materials, unlike in conventional CNT composites, owing to the high concentration of CNTs, which is well above the percolation threshold, the mechanical load is transferred through a continuous network of CNTs. Therefore, the mechanical properties of the composite are substantially controlled by the average interactions between adjacent CNTs. This is in contrast to conventional composites, where the interaction between CNTs and the matrix controls the overall mechanical behavior of the composite. Several experimental techniques have been commonly used to investigate the ensemble behavior of CNT composites and fibers with high CNT concentration, such as DMA and tension tests. As will be pointed out in this section, in addition to mechanical characterization, proper analysis of the data can provide us with information about the nature of CNT–CNT interactions.

As shown in Section 5.2.4, DMA is commonly used to investigate the viscoelastic response of CNT–polymer composites. Generally, it is accepted that an increase in the storage modulus indicates the reinforcing effect of CNTs in a polymer matrix [67–71]. This is based on the fact that strong interactions between CNTs prevent slippage between

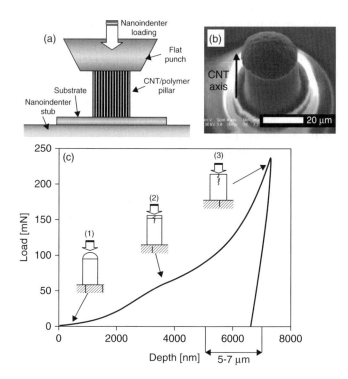

Figure 5.9 Schematics of (a) the nanocompression experiment on vertically aligned CNT–epoxy composites and (b) CNT–epoxy nanopillar composite. (c) Plot of typical axial compressive force as a function of the indentation depth. Adapted from [30]

CNTs and the matrix, further enhancing the elastic energy storage in the sample at sufficiently low strains. Moreover, the variations of the T_g, measured as the temperature corresponding to the maximum loss modulus with CNT content, contains valuable information about the ensemble interactions between CNTs and the polymer matrix. For instance, the increase in T_g with CNT content is considered as an indication of the loss of mobility of polymer chains in the presence of CNTs [67,71]. Despite the hindrance effect of CNTs on chain mobility, Hwang *et al.* reported a reduction in T_g as a function of CNTs in poly(methyl methacrylate) (PMMA) up to 20 wt% CNTs, when CNTs themselves were grafted with PMMA [69]. The reduction in T_g with CNT content was attributed to the plasticizing effect of the functional groups in the aforementioned matrix. Moreover, they observed a second peak in the loss modulus at ∼9 wt% of CNTs, suggestive of nonuniform dispersion of CNTs in the composites for 9 wt% CNT content and above, which further facilitates the polymer chain motion.

In addition to the storage modulus and T_g, the shape of the dynamic moduli curves as a function of temperature provides additional clues about the nature of the interactions between CNTs and polymers. For instance, Shaffer and Windle reported a broadening of loss modulus curves by increasing the CNT contents in CNT–polyvinylalcohol (PVA) to ∼60 wt%, which was attributed to the loss of mobility of the unconstrained portions of the PVA chains by adding CNTs [70].

Despite the information they reveal about the material stiffness and the nature of inter-actions between CNTs and polymers, DMA experiments can only capture the small-strain mechanical response of the sample. Another class of experiments, used to capture both the small deformation (such as modulus) and large deformation (such as strength and tough-ness) mechanical properties of the CNT composites, is tension tests. Similar to DMA experiments, in tension tests it is commonly accepted that an increase in the modulus or strength of CNT composites with the addition of CNTs or with proper functionaliza-tion of CNTs points to enhanced interaction between CNTs and the matrix due to proper functionalization of CNTs or their uniform dispersion [67,74,75].

In addition, given the low ductility of typical CNT yarns, the statistical analysis of the mechanical properties, such as strength and modulus, can be instrumental in assessing the average distribution of defects in the as-produced sample. For instance, Windle and coworkers measured the strength of *in-situ* CVD-fabricated CNT yarns with varying gage lengths ranging from 1 to 20 mm [19,20] to assess the defect density in the samples. At all gage lengths, they observed a peak in strength at around $1\,GPa\,g^{-1}\,cm^{-3}$ (0.5–1.5 GPa), while at sufficiently low gage length (1–2 mm) another peak appeared at $\sim 5\,GPa\,g^{-1}\,cm^{-3}$ ($\sim 9\,GPa$); see Figure 5.10. The variation of strength with gage length and the peak observed at the smallest gage length pointed to the presence of defects in the sample that

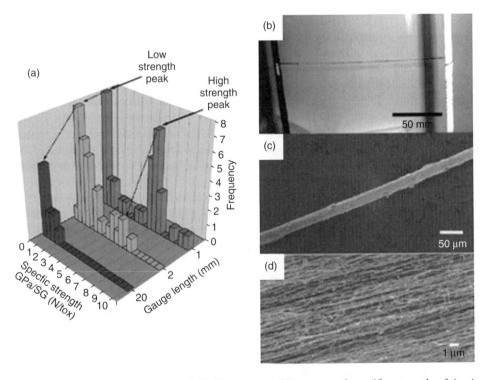

Figure 5.10   Micro-tensile testing of CNT yarns. (a) Histogram of specific strength of *in-situ* CVD spun CNT yarns for multiple gage lengths. Reprinted with permission from [20]. © 2007 The American Association for the Advancement of Science. (b–d) Optical and SEM images of *in-situ* CVD spun yarns at different length scales. Reprinted with permission from [19]. © 2004 AAAS

were, on average, on the order of 1–2 mm apart from one another, which controlled the mechanical response of samples with sufficiently large gage length. Moreover, the good correlation between the modulus and strength in all the samples eliminated the possibility of the presence of flaws in the sample that would introduce stress concentrations, as the adverse effect of such flaws on strength would be more significant than on the modulus. From these findings they concluded that the defects in the samples were likely due to a lower densification in some parts of the fibers, which affects the modulus and strength similarly.

Baughman and coworkers pointed to the crucial role of twist in the mechanical behavior of yarns of CNTs, drawn from CVD-grown forests of CNTs, to induce transverse forces between constituents to bind CNTs together [72]. As shown by polarized Raman spectroscopy, this method of drawing of CNTs from CNT forests induces partial alignment between CNTs in the as-drawn sheets of CNTs [110]. A combination of CNT alignment and enhanced interactions between CNTs due to twist resulted in yarns with strength close to $\sim$1 GPa (specific strength of $\sim$0.6 GPa g^{-1} cm^{-3}) and toughness as high as $\sim$30 J g^{-1} measured in tension, which is comparable to high-performance fibers [111]. The role of lateral interactions through polymer coatings has also been revealed by Espinosa and coworkers. They found that yarns with a very high volume fraction of DWNT incorporating polymer intermediaries, primarily composed of substituted acrylates formed during CVD growth, exhibited specific strengths in excess of 1 GPa g^{-1} cm^{-3} and toughness as high as $\sim$100 J g^{-1} [22]. These yarns were among the first to achieve simultaneously high specific strengths (higher than the ones previously reported for CNTs) and energies to failure similar to one of nature's toughest materials, spider silk (see Figure 5.11).

The significant Poisson effect observed in the yarns has been successfully exploited to develop mechanical strain sensors with sensitivities of as high as 0.5 μV/$\mu\varepsilon$. The basis of the operation of these electromechanical sensors is the reduction in the volume of yarns upon axial loading, and the consequent reduction in their electric charge capacity [112]. However, it is to be pointed out that the twist has an adverse effect on the modulus of yarns, resulting in yarns with moduli substantially lower than the modulus of their constituents (CNTs), due to the loading of CNTs in an inclined direction with respect to the yarn axis [113].

Tension tests on CNT composites can also been carried out in *in-situ* analytical chambers, such as with SEM, to visualize the deformation mechanisms of CNT composites. Naraghi *et al.* investigated the mechanical behavior of twisted CNT yarns *in-situ* SEM [22]. They observed significantly high lateral contraction of the yarns in response to axial loading and Poisson ratios of as high as unity, mostly due to the twisted structure of the yarns, similar in nature to the giant Poisson effect observed for axially loaded CNT yarns spun from CVD-grown forests of CNTs [18,72,111]. As explained by Baughman and coworkers [72] and experimentally demonstrated by Naraghi *et al.* [22], the giant Poisson effect is rooted in the twisted morphology of the yarn and the consequent "unwinding" of its helical structure upon axial loading. This effect can result in volume loss of the sample and its lateral contraction during axial loading, thus enhancing the interactions between CNTs. Together with enhanced alignment of CNTs during axial loading, the enhanced load transfer between CNTs can account for the increase of elastic modulus of the yarns with plastic strain [29,72]. Through their *in-situ* SEM experiments on yarns of DWNT

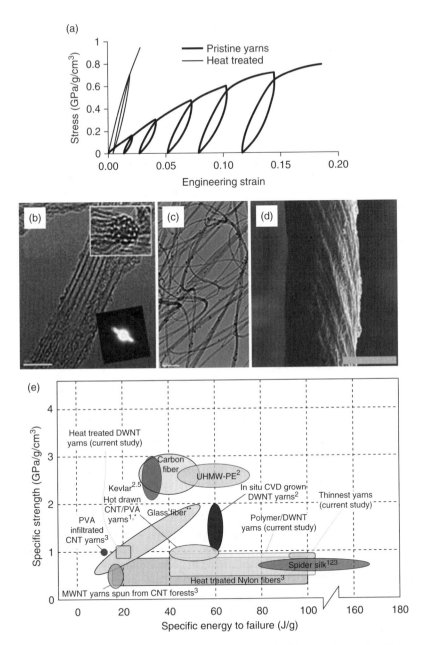

Figure 5.11 (a) Stress–strain curve of DWNT yarns with and without polymer crosslinking. (b–d) SEM images of DWNT–polymer yarn at different hierarchical length scales. (e) Ashby-style plot of specific strength as a function of specific energy-to-failure for advanced yarns and fibers [22]. Reprinted with permission from [22]. © 2010 American Chemical Society

bundles with an inherent polymer coating developed in the CVD reactor, Naraghi *et al.* demonstrated that the yarns accommodate the applied deformation by both reorientation of the network of CNTs (due to the yarn twisted morphology) and the network stretching, with similar contributions from each component [22]. Moreover, by coupling macroscale tension experiments on yarns with *in-situ* SEM tension tests on DWNT bundles, they demonstrated that only a small portion of the total energy given to the system by the tension test apparatus is stored in CNTs, while most of it is dissipated by the shear interactions between CNTs and the inherent polymer coating. Continuum and analytical models can also be applied to predict macroscopic fiber and yarn mechanical properties based on experimental data and microscopic simulations. Vilatela *et al.* recently developed a model of the strength of aligned CNT yarns and identified optimization parameters of the subunits of the yarns which included CNT length and interlayer shear strength [113]. In particular, it was predicted that, by using large-diameter CNTs with few walls, the degree of contact between adjacent CNTs could be increased, leading to higher strength.

In addition to the measurement of mechanical properties, the shape of stress–strain curves obtained in tension tests can give insights into deformation mechanisms involved in CNT composites and the ensemble behavior of the CNT–polymer chain interactions. For instance, the lack of plastic deformation and the "wave-like pattern" of the curves, in layer-by-layer deposited CNT composites with as high as 50 wt% CNTs, was suggestive of the uncoiling of the entangled polymer chains, followed by their rupture [73].

5.4 Concluding Remarks

The emergence of hierarchical design in carbon-based composites is showing great promise in the development of mechanically superior and multifunctional materials. Progress in this field relies heavily on understanding both the mechanical behavior of the constituents and interactions between them at multiple length scales, a challenge that has been addressed in part by multiscale experiments. At the nanoscale, the application of AFM and *in-situ* TEM methods have been used to measure the intrinsic mechanical properties of promising building blocks, most notably the strength and stiffness of CNT shells [11,12] and graphene sheets [13], which are now known to a high degree of certainty. The same, however, cannot be said for the mechanical interactions between constituents such as adjacent CNTs or graphene sheets. While experimental techniques have provided a great deal of insight into the nature of these shear interactions, there remain large variations in reported measurements, one example being the shear strength measured for adjacent CNT shells (see Table 5.1). One approach to engineering these interfaces through cross-linking that has proven to be a promising direction is particle irradiation. Irradiation-induced cross-linking between adjacent tubes and shells of CNTs has been demonstrated to improve effective mechanical properties, such as strength and stiffness, by orders of magnitude [12,23,55]; however, this approach has not yet been demonstrated in macroscopic fibers or yarns owing to difficulties in uniformly irradiating large quantities of materials. Despite the lack of a complete understanding of the behavior at such intermediate length scales, a great deal of progress has already been made in developing macroscopic materials for end-user applications. CNT-based fibers and yarns, which have benefited from engineered interfacial interactions and geometrical parameters such as alignment and twist, have emerged that exhibit mechanical properties superior to

Table 5.1 Experimentally determined interfacial properties of graphitic materials

Material interface	Shear strength (MPa)	Cohesion energy ($J\,m^{-2}$)	Environment	Ref
MWNT shells	0.05	0.198	TEM/high vacuum	[88]
MWNT shells	0.08–0.3	–	SEM/high vacuum	[57]
Bundled SWNTs	–	0.1–0.6	Air	[63]
SWNT/MWNT	4	0.03	Air	[60]
Collapsed MWNT shells	–	0.21	TEM/high vacuum	[89]
Graphite layers	–	0.26	Air	[95]
Graphite/aromatics	–	0.37	Air	[114]

more commonly available materials such as steels and Kevlar [20,22,72]. In addition, the multifunctionality of macroscopic fibers that incorporate CNTs and graphene sheets has emerged recently, which addresses the growing need for novel technologies with energy applications. Initial demonstrations include yarns of superconductors, lithium-ion battery materials, and titanium dioxide for photocatalysis [24].

The next significant leaps in understanding of hierarchical carbon-based materials can benefit greatly from the coupling of multiscale experimental tools with computation approaches. Multiscale simulations on the class of materials discussed in this chapter are beginning to fill in some of the gaps in our understanding of their mechanical behavior. In the case of natural materials, such as tendon and spider silk, multiscale simulations have been instrumental in elucidating how the hierarchical structures and interactions between constituents contribute to the impressive macroscopic properties [1,3]. Moreover, a great deal of activity is now focused on applying these lessons of nature to materials engineered out of carbon constituents.

The modeling of carbon-based hierarchical materials requires the development and coupling of models across a number of length scales. At the nanoscale, DFT simulations have predicted the intrinsic mechanical properties of CNTs [115,116] and graphene [117] and can be extended to characterize the interaction of individual cross-linking elements. At intermediate length scales, MM and MD simulations can be applied to investigate load transfer between adjacent graphitic sheets in larger systems. In particular, MM and MD studies of CNT bundles have provided insights into the nature of shear interactions and load transfer mechanisms between adjacent CNTs within bundles [58,118]. In this case, the MM potentials used (e.g., modified second-generation Tersoff–Brenner [119]) can be directly validated by comparison with DFT calculations [12]. Similar studies also include simulations that incorporate many cross-linking elements, such as covalent cross-linking defects that occur as a result of electron irradiation [109]. For MD simulations, the application of ReaxFF reactive force fields has proven to be powerful in developing an understanding of carbon materials [120–122]. These have extended from studies of the catalytic formation of CNTs [121,122], to the fracture properties of individual graphene sheets [120]. On the other hand, networks of many interacting CNTs and graphene sheets, such as cross-linked bundles of tens to hundreds of CNTs, are typically too large and computationally costly to be simulated by traditional MM or MD; therefore, simulation approaches that use coarse grain elements can be applied [100]. Here, individual mechanical properties of the elements, as well as interaction parameters such as cohesive energies between adjacent constituents, are input from independent DFT and MM/MD calculations. Finally, continuum and analytical models can predict macroscopic fiber and yarn properties. Examples are the recent work of Espinosa and coworkers [17] and Vilatela *et al.* [113], in which the mechanical properties of highly aligned CNT yarns were modeled based on experimental data and MD simulations. Using their model, Vilatela *et al.* identified critical steps in the enhancement of the mechanical behavior of CNT yarns, for instance by utilizing thick CNTs with few shells that can "autocollapse," increasing the contact area between adjacent CNTs.

A combination of existing multiscale experimental and theoretical tools alone, however, will not be sufficient to fully explore the potential of hierarchical carbon materials. Novel tools and materials also need to be explored and developed. For example the *in-situ* electron microscopy testing capabilities that have been discussed earlier are typically quasi-static and restricted to extremely low temporal resolution due to limitations in image capturing. However, many applications, in particular the development of CNT/graphene-based bulletproof armors, will require a better understanding of dynamic failure properties. Furthermore, we would expect large strain rate dependence of CNT-based fibers and yarns. One approach to investigate such dependencies at the nanoscale would be the application of dynamic TEM to study deformation and failure in carbon-based materials. The viability and potential technical capabilities of dynamic TEM have been discussed recently [123], and it has already been applied to reveal the dynamic propagation of reaction fronts in layered materials with nanosecond resolution [124]. However, its application to atomic-scale studies of CNT- and graphene-based materials will require further improvements in overcoming the space charge effects, which limit the spatial resolution in current dynamic TEM experiments. Moreover, force and strain measurement techniques in *in-situ* TEM and SEM on nanoscale samples, which are mostly based on the deflection measurement of cantilevers, should be augmented, for instance by all-electronic *in-situ* SEM/TEM measurements such as the ones reported in Refs. [35–37]. By using such improved systems, load and strain resolution can be substantially enhanced, and stiff load sensors can be incorporated, for instance, in an attempt to stably capture weak interactions between CNTs (see Section 5.3.1).

Along with advancing novel methods of measuring material properties of CNT/graphene hierarchical materials, researchers must also continue advancing the study of both complementary and alternative materials. One example is the development of CNT composites incorporating strong materials such as Kevlar, which can form strong π-stacking interactions with unfunctionalized tubes. This approach has recently been demonstrated to produce CNT-reinforced Kevlar composites with strength as high as 6 GPa and modulus as high as 200 GPa [125]. However, these values of mechanical properties, while remarkable, are still a small fraction of the properties of CNTs, thus demanding future research in this area. Another material system that has attracted the attention of many researchers is non-carbon-based building blocks, such as boron nitride nanotubes and sheets. Boron nitride, also known as "white graphene," has a similar 2D hexagonal structure to graphene and has attracted a great deal of attention recently for a number of reasons, including its remarkably high mechanical properties (strength $\sim$30 GPa and modulus $\sim$900 GPa [126]), comparable to those of CNTs, and superior thermal stability.

References

[1] Buehler, M.J. (2006) Nature designs tough collagen: explaining the nanostructure of collagen fibrils. *Proceedings of the National Academy of Sciences of the United States of America*, **103**, 12285–12290.

[2] Espinosa, H.D., Rim, J.E., Barthelat, F., and Buehler, M.J. (2009) Merger of structure and material in nacre and bone – perspectives on *de novo* biomimetic materials. *Progress in Materials Science*, **54**, 1059–1100.

[3] Keten, S., Xu, Z., Ihle, B., and Buehler, M.J. (2010) Nanoconfinement controls stiffness, strength and mechanical toughness of -sheet crystals in silk. *Nature Materials*, **9**, 359–367.

[4] Meyers, M., Chen, P., Lin, Y., and Seki, Y. (2008) Biological materials: structure and mechanical properties. *Progress in Materials Science*, **53**, 1–206.

[5] Omenetto, F.G. and Kaplan, D.L. (2010) New opportunities for an ancient material. *Science*, **329** (5991), 528–531.

[6] Wegst, U. and Ashby, M. (2004) The mechanical efficiency of natural materials. *Philosophical Magazine*, **84** (21), 2167–2181.

[7] Nova, A., Keten, S., Pugno, N.M. *et al.* (2010) Molecular and nanostructural mechanisms of deformation, strength and toughness of spider silk fibrils. *Nano Letters*, **10** (7), 2626–2634.

[8] Giesa, T., Arslan, M., Pugno, N.M., and Buehler, M.J. (2011) Nanoconfinement of spider silk fibrils begets superior strength, extensibility, and toughness. *Nano Letters*, **11** (11), 5038–5046.

[9] Espinosa, H.D., Juster, A.L., Latourte, F.J. *et al.* (2011) Tablet-level origin of toughening in abalone shells and translation to synthetic composite materials. *Nature Communications*, **2**, 173.

[10] Elices, M., Guinea, G.V., Plaza, G.R. *et al.* (2011) Bioinspired fibers follow the track of natural spider silk. *Macromolecules*, **44** (5), 1166–1176.

[11] Yu, M.F., Lourie, O., Dyer, M.J. *et al.* (2000) Strength and breaking mechanism of multiwalled carbon nanotubes under tensile load. *Science*, **287** (5453), 637–640.

[12] Peng, B., Locascio, M., Zapol, P. *et al.* (2008) Measurements of near-ultimate strength for multiwalled carbon nanotubes and irradiation-induced crosslinking improvements. *Nature Nanotechnology*, **3** (10), 626–631.

[13] Lee, C., Wei, X., Kysar, J.W., and Hone, J. (2008) Measurement of the elastic properties and intrinsic strength of monolayer graphene. *Science*, **321** (5887), 385–388.

[14] Aslan, H., Kimelman-Bleich, N., Pelled, G., and Gazit, D. (2008) Molecular targets for tendon neoformation. *Journal of Clinical Investigation*, **118** (2), 439–444.

[15] Gautieri, A., Vesentini, S., Redaelli, A., and Buehler, M.J. (2011) Hierarchical structure and nanomechanics of collagen microfibrils from the atomistic scale up. *Nano Letters*, **11** (2), 757–766.

[16] Sun, Y.L., Luo, Z.P., Fertala, A., and An, K.N. (2004) Stretching type II collagen with optical tweezers. *Journal of Biomechanics*, **37** (11), 1665–1669.

[17] Wei, X., Naraghi, M., and Espinosa, H.D. (2012) Optimal length scales emerging from shear load transfer in natural materials: application to carbon-based nanocomposite design. *ACS Nano*, **6** (3), 2333–2344.

[18] Dalton, A.B., Collins, S., Munoz, E. *et al.* (2003) Super-tough carbon-nanotube fibres – these extraordinary composite fibres can be woven into electronic textiles. *Nature*, **423** (6941), 703–703.

[19] Li, Y.-L., Kinloch, I., and Windle, A. (2004) Direct spinning of carbon nanotube fibers from chemical vapor deposition synthesis. *Science*, **304**, 276–278.

[20] Koziol, K., Vilatela, J., Moisala, A. *et al.* (2007) High-performance carbon nanotube fiber. *Science*, **318** (5858), 1892–1895.

[21] Motta, M., Moisala, A., Kinloch, I.A., and Windle, A.H. (2007) High performance fibres from 'dog bone' carbon nanotubes. *Advanced Materials*, **19**, 3721–3726.

[22] Naraghi, M., Filleter, T., Moravsky, A. *et al.* (2010) A multi-scale study of high performance DWNT–polymer fibers. *ACS Nano*, **4** (11), 6463–6476.

[23] Filleter, T., Bernal, R., Li, S., and Espinosa, H.D. (2011) Ultrahigh strength and stiffness in cross-linked hierarchical carbon nanotube bundles. *Advanced Materials*, **23**, 2855–2860.

[24] Lima, M.D., Fang, S.L., Lepro, X. *et al.* (2011) Biscrolling nanotube sheets and functional guests into yarns. *Science*, **331** (6013), 51–55.

[25] Aliev, A.E., Oh, J.Y., Kozlov, M.E. *et al.* (2009) Giant-stroke, superelastic carbon nanotube aerogel muscles. *Science*, **323** (5921), 1575–1578.

[26] Stankovich, S., Dikin, D.A., Dommett, G.H. *et al.* (2006) Graphene based composite materials. *Nature*, **442**, 282–286.

[27] Dikin, D.A., Stankovich, S., Zimney, E.J. *et al.* (2007) Preparation and characterization of graphene oxide paper. *Nature*, **448**, 457–460.

[28] Cumings, J. and Zettl, A. (2000) Low-friction nanoscale linear bearing realized from multiwall carbon nanotubes. *Science*, **289** (5479), 602–604.

[29] Ma, W.J., Liu, L.Q., Yang, R. *et al.* (2009) Monitoring a micromechanical process in macroscale carbon nanotube films and fibers. *Advanced Materials*, **21** (5), 603–608.

[30] Garcia, E.J., Hart, A.J., Wardle, B.L., and Slocum, A.H. (2007) Fabrication and nanocompression testing of aligned carbon-nanotube–polymer nanocomposites. *Advanced Materials*, **19** (16), 2151–2156.

[31] Iijima, S. (1991) Helical microtubules of graphitic carbon. *Nature*, **354** (6348), 56–58.

[32] Meyer, J.C., Kisielowski, C., Erni, R. *et al.* (2008) Direct imaging of lattice atoms and topological defects in graphene membranes. *Nano Letters*, **8**, 3582–3586.

[33] Legros, M., Gianola, D.S., and Motz, C. (2010) Quantitative *in situ* mechanical testing in electron microscopes. *MRS Bulletin*, **35** (5), 354–360.

[34] Locascio, M., Peng, B., Zapol, P. *et al.* (2009) Tailoring the load carrying capacity of MWCNTs through inter-shell atomic bridging. *Experimental Mechanics*, **49** (2), 169–182.

[35] Haque, M.A. and Saif, M.T.A. (2003) A review of MEMS-based microscale and nanoscale tensile and bending testing. *Experimental Mechanics*, **43** (3), 248–255.

[36] Zhu, Y., Moldovan, N., and Espinosa, H.D. (2005) A microelectromechanical load sensor for *in situ* electron and X-ray microscopy tensile testing of nanostructures. *Applied Physics Letters*, **86** (1), 013506.

[37] Zhu, Y., Corigliano, A., and Espinosa, H.D. (2006) A thermal actuator for nanoscale *in situ* microscopy testing: design and characterization. *Journal of Micromechanics and Microengineering*, **16** (2), 242–253.

[38] Minor, A.M., Asif, S.A.S., Shan, Z.W. *et al.* (2006) A new view of the onset of plasticity during the nanoindentation of aluminum. *Nature Materials*, **5**, 697–702.

[39] Shan, Z.W., Mishra, R.K., Asif, S.A.S. *et al.* (2008) Mechanical annealing and source-limited deformation in submicrometer-diameter Ni crystals. *Nature Materials*, **7**, 115–120.

[40] Agrawal, R., Peng, B., Gdoutos, E.E., and Espinosa, H.D. (2008) Elasticity size effects in ZnO nanowires – a combined experimental–computational approach. *Nano Letters*, **8** (11), 3668–3674.

[41] Zhu, Y. and Espinosa, H.D. (2005) An electromechanical material testing system for *in situ* electron microscopy and applications. *Proceedings of the National Academy of Sciences of the United States of America*, **102** (41), 14503–14508.

[42] Osterberg, P.M. and Senturia, S.D. (1997) M-TEST: a test chip for MEMS material property measurement using electrostatically actuated test structures. *Journal of Microelectromechanical Systems*, **6** (2), 107–118.

[43] Ballarini, R., Mullen, R.L., Yin, Y. *et al.* (1997) The fracture toughness of polysilicon microdevices: a first report. *Journal of Materials Research*, **12** (4), 915–922.

[44] De Boer, M.P., Jensen, B.D., and Bitsie, F. (1999) A small area *in-situ* MEMS test structure to measure fracture strength by electrostatic probing, in *Materials and Device Characterization in Micromachining II*, Proceedings of SPIE, vol. 3875, SPIE, Bellingham, WA, pp. 97–103.

[45] Haque, M.A. and Saif, M.T.A. (2001) Microscale materials testing using MEMS actuators. *Journal of Microelectromechanical Systems*, **10** (1), 146–152.

[46] Haque, M.A. and Saif, M.T.A. (2005) *In situ* tensile testing of nanoscale freestanding thin films inside a transmission electron microscope. *Journal of Materials Research*, **20** (7), 1769–1777.

[47] Haque, M.A., Espinosa, H.D., and Lee, H.J. (2010) MEMS for *in situ* testing-handling, actuation, loading, and displacement measurements. *MRS Bulletin*, **35** (5), 375–381.

[48] Suk, J.W., Piner, R.D., An, J., and Ruoff, R.S. (2010) Mechanical properties of monolayer graphene oxide. *ACS Nano*, **4**, 6557–6564.

[49] Espinosa, H.D. and Prorok, B.C. (2003) Size effects on the mechanical behavior of gold thin films. *Journal of Materials Science*, **38** (20), 4125–4128.

[50] Espinosa, H.D., Prorok, B.C., and Peng, B. (2004) Plasticity size effects in free-standing submicron polycrystalline FCC films subjected to pure tension. *Journal of the Mechanics and Physics of Solids*, **52** (3), 667–689.

[51] Espinosa, H.D., Prorok, B.C., and Fischer, M. (2003) A methodology for determining mechanical properties of freestanding thin films and MEMS materials. *Journal of the Mechanics and Physics of Solids*, **51** (1), 47–67.

[52] Filleter, T., McChesney, J.L., Bostwick, A. *et al*. (2009) Friction and dissipation in epitaxial graphene films. *Physical Review Letters*, **102** (8), 086102.

[53] Li, Q.Y., Lee, C., Carpick, R.W., and Hone, J. (2010) Substrate effect on thickness-dependent friction on graphene. *Physica Status Solidi B: Basic Solid State Physics*, **247** (11–12), 2909–2914.

[54] Yu, M.F., Files, B., Arepalli, S., and Ruoff, R. (2000) Tensile loading of ropes of single wall carbon nanotubes and their mechanical properties. *Physical Review Letters*, **84** (24), 5552–5555.

[55] Kis, A., Csanyi, G., Salvetat, J.P. *et al*. (2004) Reinforcement of single-walled carbon nanotube bundles by intertube bridging. *Nature Materials*, **3** (3), 153–157.

[56] Meyer, E., Hug, H., and Bennewitz, R. (eds.) 2004 *Scanning Probe Microscopy: The Lab on a Tip*, Springer-Verlag, Berlin.

[57] Yu, M.F., Yakobson, B.I., and Ruoff, R.S. (2000) Controlled sliding and pullout of nested shells in individual multiwalled carbon nanotubes. *Journal of Physical Chemistry B*, **104** (37), 8764–8767.

[58] Filleter, T., Yockel, S., Naraghi, M. *et al*. (2012) Experimental–computational study of shear interactions within double-walled carbon nanotube bundles. *Nano Letters*, **12** (2), 732–742.

[59] Kaplan-Ashiri, I., Cohen, S.R., Apter, N. *et al*. (2007) Microscopic investigation of shear in multiwalled nanotube deformation. *Journal of Physical Chemistry C*, **111** (24), 8432–8436.

[60] Bhushan, B., Ling, X., Jungen, A., and Hierold, C. (2008) Adhesion and friction of a multiwalled carbon nanotube sliding against single-walled carbon nanotube. *Physical Review B*, **77** (16), 165428.

[61] Naraghi, M., Bratzel, G., Filleter, T. *et al*. (2012) Experimental–computational studies on the shear interactions between functionalized DWNT bundles. In review.

[62] Suekane, O., Nagataki, A., Mori, H., and Nakayama, Y. (2008) Static friction force of carbon nanotube surfaces. *Applied Physics Express*, **1** (6), 064001.

[63] Yang, T.Y., Zhou, Z.R., Fan, H., and Liao, K. (2008) Experimental estimation of friction energy within a bundle of single-walled carbon nanotubes. *Applied Physics Letters*, **93** (4), 041914.

[64] Syue, S.H., Lu, S.Y., Hsu, W.K., and Shih, H.C. (2006) Internanotube friction. *Applied Physics Letters*, **89** (16), 163115.

[65] Coleman, J.N., Khan, U., Blau, W.J., and Gun'ko, Y.K. (2006) Small but strong: a review of the mechanical properties of carbon nanotube–polymer composites. *Carbon*, **44** (9), 1624–1652.

[66] Spitalsky, Z., Tasis, D., Papagelis, K., and Galiotis, C. (2010) Carbon nanotube–polymer composites: chemistry, processing, mechanical and electrical properties. *Progress in Polymer Science*, **35** (3), 357–401.

[67] Yang, J.W., Hu, J.H., Wang, C.C. *et al*. (2004) Fabrication and characterization of soluble multi-walled carbon nanotubes reinforced P(MMA–co–EMA) composites. *Macromolecular Materials and Engineering*, **289** (9), 828–832.

[68] Wang, Z., Liang, Z.Y., Wang, B. *et al*. (2004) Processing and property investigation of single-walled carbon nanotube (SWNT) buckypaper/epoxy resin matrix nanocomposites. *Composites Part A: Applied Science and Manufacturing*, **35** (10), 1225–1232.

[69] Hwang, G.L., Shieh, Y.T., and Hwang, K.C. (2004) Efficient load transfer to polymer-grafted multiwalled carbon nanotubes in polymer composites. *Advanced Functional Materials*, **14** (5), 487–491.

[70] Shaffer, M.S.P. and Windle, A.H. (1999) Fabrication and characterization of carbon nanotube/poly(vinyl alcohol) composites. *Advanced Materials*, **11** (11), 937–941.

[71] Shieh, Y.T., Liu, G.L., Twu, Y.K. *et al*. (2010) Effects of carbon nanotubes on dynamic mechanical property, thermal property, and crystal structure of poly(L-lactic acid). *Journal of Polymer Science Part B: Polymer Physics*, **48** (2), 145–152.

[72] Zhang, M., Atkinson, K.R., and Baughman, R.H. (2004) Multifunctional carbon nanotube yarns by downsizing an ancient technology. *Science*, **306** (5700), 1358–1361.

[73] Mamedov, A.A., Kotov, N.A., Prato, M. *et al*. (2002) Molecular design of strong single-wall carbon nanotube/polyelectrolyte multilayer composites. *Nature Materials*, **1** (3), 190–194.

[74] Sahoo, N.G., Cheng, H.K.F., Cai, J.W. *et al*. (2009) Improvement of mechanical and thermal properties of carbon nanotube composites through nanotube functionalization and processing methods. *Materials Chemistry and Physics*, **117** (1), 313–320.

[75] Yang, K., Gu, M.Y., Guo, Y.P. *et al*. (2009) Effects of carbon nanotube functionalization on the mechanical and thermal properties of epoxy composites. *Carbon*, **47** (7), 1723–1737.

[76] Liu, L.Q., Tasis, D., Prato, M., and Wagner, H.D. (2007) Tensile mechanics of electrospun multiwalled nanotube/poly(methyl methacrylate) nanofibers. *Advanced Materials*, **19** (9), 1228–1233.

[77] Cooper, C.A., Young, R.J., and Halsall, M. (2001) Investigation into the deformation of carbon nanotubes and their composites through the use of Raman spectroscopy. *Composites Part A: Applied Science and Manufacturing*, **32** (3–4), 401–411.

[78] Mu, M.F., Osswald, S., Gogotsi, Y., and Winey, K.I. (2009) An *in situ* Raman spectroscopy study of stress transfer between carbon nanotubes and polymer. *Nanotechnology*, **20** (33), 335703.

[79] Dresselhaus, M.S., Dresselhaus, G., Saito, R., and Jorio, A. (2005) Raman spectroscopy of carbon nanotubes. *Physics Reports: Review Section of Physics Letters*, **409** (2), 47–99.

[80] Liu, T.X., Phang, I.Y., Shen, L. *et al*. (2004) Morphology and mechanical properties of multiwalled carbon nanotubes reinforced nylon-6 composites. *Macromolecules*, **37** (19), 7214–7222.

[81] Li, X.F., Lau, K.T., and Yin, Y.S. (2008) Mechanical properties of epoxy-based composites using coiled carbon nanotubes. *Composites Science and Technology*, **68** (14), 2876–2881.

[82] Lee, D. and Cui, T.H. (2011) Suspended carbon nanotube nanocomposite beams with a high mechanical strength via layer-by-layer nano-self-assembly. *Nanotechnology*, **22** (16), 165601.

[83] Minary-Jolandan, M. and Yu, M.F. (2008) Reversible radial deformation up to the complete flattening of carbon nanotubes in nanoindentation. *Journal of Applied Physics*, **103** (7), 073516.

[84] Mayor, M., Weber, H.B., Reichert, J. *et al*. (2003) Electric current through a molecular rod – relevance of the position of the anchor groups. *Angewandte Chemie International Edition*, **42**, 5834–5838.

[85] Bakshi, S.R., Batista, R.G., and Agarwal, A. (2009) Quantification of carbon nanotube distribution and property correlation in nanocomposites. *Composites Part A: Applied Science and Manufacturing*, **40** (8), 1311–1318.

[86] Pharr, G.M., Oliver, W.C., and Brotzen, F.R. (1992) On the generality of the relationship among contact stiffness, contact area, and elastic-modulus during indentation. *Journal of Materials Research*, **7** (3), 613–617.

[87] Kis, A. and Zettl, A. (2008) Nanomechanics of carbon nanotubes. *Philosophical Transactions of the Royal Society A: Mathematical Physical and Engineering Sciences*, **366** (1870), 1591–1611.

[88] Kis, A., Jensen, K., Aloni, S. *et al*. (2006) Interlayer forces and ultralow sliding friction in multiwalled carbon nanotubes. *Physical Review Letters*, **97** (2), 025501.

[89] Chopra, N.G., Benedict, L.X., Crespi, V.H. *et al*. (1995) Fully collapsed carbon nanotubes. *Nature*, **377** (6545), 135–138.

[90] Ke, C.H., Zheng, M., Zhou, G.W. *et al*. (2010) Mechanical peeling of free-standing single-walled carbon-nanotube bundles. *Small*, **6** (3), 438–445.

[91] Good, R.J., Girifalco, L.A., and Kraus, G. (1958) A theory for estimation of interfacial energies. 2. Application to surface thermodynamics of Teflon and graphite. *Journal of Physical Chemistry*, **62** (11), 1418–1421.

[92] Ishikawa, M., Harada, R., Sasaki, N., and Miura, K. (2008) Visualization of nanoscale peeling of carbon nanotube on graphite. *Applied Physics Letters*, **93** (8), 083122.

[93] Bhushan, B. (2002) *Introduction to Tribology*, John Wiley & Sons.

[94] Dienwiebel, M., Verhoeven, G.S., Pradeep, N. *et al*. (2004) Superlubricity of graphite. *Physical Review Letters*, **92** (12), 126101.

[95] Girifalco, L.A. and Lad, R.A. (1956) Energy of cohesion, compressibility, and the potential energy functions of the graphite system. *Journal of Chemical Physics*, **25** (4), 693–697.

[96] Zacharia, R., Ulbricht, H., and Hertel, T. (2004) Interlayer cohesive energy of graphite from thermal desorption of polyaromatic hydrocarbons. *Physical Review B*, **69** (15), 155406.

[97] Banhart, F. (1999) Irradiation effects in carbon nanostructures. *Reports on Progress in Physics*, **62** (8), 1181–1221.

[98] Telling, R.H., Ewels, C.P., El Barbary, A.A., and Heggie, M.I. (2003) Wigner defects bridge the graphite gap. *Nature Materials*, **2** (5), 333–337.

[99] Cornwell, C.F. and Welch, C.R. (2011) Very-high-strength (60-GPa) carbon nanotube fiber design based on molecular dynamics simulations. *Journal of Chemical Physics*, **134** (20), 204708.

[100] Bratzel, G.H., Cranford, S.W., Espinosa, H., and Buehler, M.J. (2010) Bioinspired noncovalently crosslinked "fuzzy" carbon nanotube bundles with superior toughness and strength. *Journal of Materials Chemistry*, **20** (46), 10465–10474.

[101] Gómez-Navarro, C., Burghard, M., and Kern, K. (2008) Elastic properties of chemically derived single graphene sheets. *Nano Letters*, **8**, 2045–2049.

[102] Erickson, K., Erni, R., Lee, Z. *et al*. (2010) Determination of the local chemical structure of graphene oxide and reduced graphene oxide. *Advanced Materials*, **22** (40), 4467–4472.

[103] Compton, O.C., Jain, B., Dikin, D.A. *et al.* (2011) Chemically active reduced graphene oxide with tunable C/O ratios. *ACS Nano*, **5** (6), 4380–4391.

[104] Park, S., Lee, K.S., Bozoklu, G. *et al.* (2008) Graphene oxide papers modified by divalent ions – enhancing mechanical properties via chemical cross-linking. *ACS Nano*, **2** (3), 572–578.

[105] Park, S., Dikin, D.A., Nguyen, S.T., and Ruoff, R.S. (2009) Graphene oxide sheets chemically cross-linked by polyallylamine. *Journal of Materials Chemistry C: Nanomaterials and Interfaces*, **113**, 15801–15804.

[106] Cronin, S.B., Swan, A.K., Unlu, M.S. *et al.* (2005) Resonant Raman spectroscopy of individual metallic and semiconducting single-wall carbon nanotubes under uniaxial strain. *Physical Review B*, **72** (3), 035425.

[107] Cannara, R.J., Brukman, M.J., and Carpick, R.W. (2005) Cantilever tilt compensation for variable-load atomic force microscopy. *Review of Scientific Instruments*, **76** (5), 053706.

[108] Misra, A., Raney, J.R., De Nardo, L. *et al.* (2011) Synthesis and characterization of carbon nanotube–polymer multilayer structures. *ACS Nano*, **5** (10), 7713–7721.

[109] Misra, A., Raney, J.R., Craig, A.E., and Daraio, C. (2011) Effect of density variation and non-covalent functionalization on the compressive behavior of carbon nanotube arrays. *Nanotechnology*, **22** (42), 425705.

[110] Zhang, M., Fang, S.L., Zakhidov, A.A. *et al.* (2005) Strong, transparent, multifunctional, carbon nanotube sheets. *Science*, **309** (5738), 1215–1219.

[111] Atkinson, K.R., Hawkins, S.C., Huynh, C. *et al.* (2007) Multifunctional carbon nanotube yarns and transparent sheets: fabrication, properties, and applications. *Physica B: Condensed Matter*, **394** (2), 339–343.

[112] Mirfakhrai, T., Oh, J., Kozlov, M.E. *et al.* (2011) Mechanoelectrical force sensors using twisted yarns of carbon nanotubes. *IEEE–ASME Transactions on Mechatronics*, **16** (1), 90–97.

[113] Vilatela, J.J., Elliott, J.A., and Windle, A.H. (2011) A model for the strength of yarn-like carbon nanotube fibers. *ACS Nano*, **5** (3), 1921–1927.

[114] Zacharia, R., Ulbricht, H., and Hertel, T. (2004) Interlayer cohesive energy of graphite from thermal desorption of polyaromatic hydrocarbons. *Physical Review B*, **69**, 155406.

[115] Ogata, S. and Shibutani, Y. (2003) Ideal tensile strength and band gap of single-walled carbon nanotubes. *Physical Review B*, **68** (16), 165409.

[116] Ozaki, T., Iwasa, Y., and Mitani, T. (2000) Stiffness of single-walled carbon nanotubes under large strain. *Physical Review Letters*, **84**, 1712–1715.

[117] Liu, F., Ming, P.M., and Li, J. (2007) *Ab initio* calculation of ideal strength and phonon instability of graphene under tension. *Physical Review B*, **76** (6), 064120.

[118] Qian, D., Liu, W.K., and Ruoff, R.S. (2003) Load transfer mechanism in carbon nanotube ropes. *Composites Science and Technology*, **63**, 1561–1569.

[119] Brenner, D.W., Shenderova, O.A., Harrison, J.A. *et al.* (2002) A second-generation reactive empirical bond order (REBO) potential energy expression for hydrocarbons. *Journal of Physics: Condensed Matter*, **14** (4), 783–802.

[120] Sen, D., Novoselov, K.S., Reis, P.M., and Buehler, M.J. (2010) Tearing graphene sheets from adhesive substrates produces tapered nanoribbons. *Small*, **6** (10), 1108–1116.

[121] Neyts, E.C., Shibuta, Y., van Duin, A.C.T., and Bogaerts, A. (2010) Catalyzed growth of carbon nanotube with definable chirality by hybrid molecular dynamics-force biased Monte Carlo simulations. *ACS Nano*, **4** (11), 6665–6672.

[122] Nielson, K.D., van Duin, A.C.T., Oxgaard, J. *et al.* (2005) Development of the ReaxFF reactive force field for describing transition metal catalyzed reactions, with application to the initial stages of the catalytic formation of carbon nanotubes. *Journal of Physical Chemistry A*, **109** (3), 493–499.

[123] Armstrong, M.R., Boyden, K., Browning, N.D. *et al.* (2007) Practical considerations for high spatial and temporal resolution dynamic transmission electron microscopy. *Ultramicroscopy*, **107**, 356–367.

[124] Kim, J.S., LaGrange, T., Reed, B.W. *et al.* (2008) Imaging of transient structures using nanosecond *in situ* TEM. *Science*, **321** (5895), 1472–1475.

[125] O'Connor, I., Hayden, H., Coleman, J.N., and Gun'ko, Y.K. (2009) High-strength, high-toughness composite fibers by swelling Kevlar in nanotube suspensions. *Small*, **5** (4), 466–469.

[126] Wei, X.L., Wang, M.S., Bando, Y., and Golberg, D. (2010) Tensile tests on individual multiwalled boron nitride nanotubes. *Advanced Materials*, **22** (43), 4895–4899.

6

Mechanics of Nanotwinned Hierarchical Metals

Xiaoyan Li and Huajian Gao
Brown University, USA

6.1 Introduction and Overview

Owing to their unique microstructures and enhanced mechanical/physical properties, nanotwinned materials have recently attracted intense interest from mechanics, materials science, and physics communities. These materials are characterized by a large number of coherent, nanometer-spaced, twin boundaries (TBs) embedded in micrometer- or sub-micrometer-sized grains. Experimental studies have shown that nanotwinned metals possess ultra-high strength [1,2] and hardness [3–5], as well as good tensile ductility [2,6,7] and strain hardening [2,6,8]. Nanotwinned bulk metals exhibit an electrical conductivity comparable to that of their conventional coarse-grained counterparts, but much higher than that of nanocrystalline metals [1]. Nanotwins cause substantial changes in electronic and optical properties of semiconductor nanowires [9–12]. These properties of nanotwinned materials are suggesting promising applications in microelectromechanical systems, micro-/nano-devices, quantum optics, biological sensing and high-performance structural applications.

A vast majority of the experimental, theoretical, and computational studies on the underlying deformation mechanisms that govern the mechanical responses of nanotwinned metals have been conducted during the last 5 years. From the perspectives of both scientific curiosity and technological promise, it has become increasingly important and necessary to develop an in-depth understanding of how the intrinsic microstructures of nanotwinned metals (e.g., the sub-grain twin structures) influence their mechanical behaviors and deformation mechanisms. Here, we present a brief overview of some of the recent studies on the microstructural characteristics and mechanical properties of nanotwinned materials, especially nanotwinned metals. Part of the discussion will be focused on the effects of intrinsic length-scales (e.g., TB spacing) in these materials on the deformation behaviors/mechanisms.

Nano and Cell Mechanics: Fundamentals and Frontiers, First Edition. Edited by Horacio D. Espinosa and Gang Bao.
© 2013 John Wiley & Sons, Ltd. Published 2013 by John Wiley & Sons, Ltd.

6.1.1 Nanotwinned Materials

In general, nanotwinned materials are crystalline materials in which a high density of nanoscale twin lamellae (i.e., with TB spacing below 100 nm) are embedded in the interior of a network of equiaxed or columnar grains. In each grain, a large number of coherent TBs are aligned parallel to each other and divide the grain interior into two kinds of lamellar structures: matrix and twin. The lattice structures of the matrix and twin lamellae retain a mirror symmetry with respect to the TB. In crystallography, a TB is a special type of planar defect with energy lower than other high-angle grain boundaries (GBs). In nanotwinned materials, TBs are highly organized to realize the regular arrangement of nanoscale twin lamellae inside the grains, which significantly alters the mechanical and physical properties of the materials [13]. In this sense, nanotwinned materials can be regarded as a class of hierarchical nanostructured material, where the twin lamellae act as a sub-grain microstructure. The stability of nanotwinned materials is closely related to TB and stacking fault energies. Generally, the lower the TB and stacking fault energies are, the more stable are the corresponding nanotwinned materials and the higher is the density of twins. Table 6.1 summarizes TB and stacking fault energies of selected materials.

Nanotwinned Metals

In the mid-1970s, Mertz and coworkers [19,20] fabricated fine-grained Cu and Ni foils via sputter deposition and introduced a high density of twins inside columnar grains; however, the TB spacing in such specimens is randomly distributed. During the last decade, Lu and coworkers [1,2] were able to synthesize ultrafine-grained Cu samples with sub-grain twin lamellae by means of electrodeposition, with TB spacing controlled down to tens of nanometers via changing deposition rates. Zhang and coworkers [3–5] produced epitaxial nanotwinned Cu and Ag thin films with thickness on the order of $\sim$1 μm using sputter deposition. Idrissi *et al.* [8] synthesized nanotwinned Pd ultrathin and ultrafine films via chemical evaporation at high temperature, with thicknesses ranging from 80 to 200 nm, and average grain size around 30 nm. More recently, a variety of nanotwinned metallic nanowires (such as Ag [21], Au [22,23], and Cu [24]) have been successfully produced by electrodeposition or electrochemical synthesis. In these nanotwinned nanowires, highly dense twins normal to the growth axis were introduced with TB spacing as small as 10 nm and even less. Progress has also been made on the

Table 6.1 TB and stacking fault energies of selected materials

Materials	TB energy γ_{TB} (mJ m^{-2})	Stacking fault energy γ_{SF} (mJ m^{-2})	Ref.
Ag	8	16	[14]
Au	15	32	[14]
Cu	24	45	[14]
Ni	43	125	[14]
Pd	97	180	[15]
CuAl	5–20	7–41	[16]
TiAl	<60	67–97	[17,18]

fabrication of nanotwinned metallic nanopillars. Greer and coworkers [25,26] achieved the synthesis of nanotwinned Cu nanopillars by precisely controlling the conditions under which Cu is electroplated into patterned polymethylmethacrylate templates. The diameters of these nanotwinned pillars range from 50 to 500 nm, with TB spacing below 5 nm. Remarkably, they could also control the orientation of TBs to create single-crystalline nanopillars with TBs inclined to the growth direction. For nanotwinned metals (in various forms, including bulk, thin film, nanowire, and nanopillars) with face-centered cubic (fcc) lattice structures, much research activity has been focused on the investigation of twin size and orientation effects on mechanical properties, such as yield strength, ductility, strain hardening, and fracture toughness.

Nanotwinned Semiconductor Nanowires

During the past decade, semiconductor nanowires containing a high density of twins, such as SiC [27], ZnSe [28], ZnS [29], and many III–V types [10,11], have been made using the vapour–liquid–solid (VLS) method. In these nanotwinned nanowires, TBs are usually periodically arranged and normal to the growth direction, resulting in zigzag micro-faceting of nanowire surfaces. Microscopically, each surface is a {111} side facet of the "truncated" octahedron. Such full control of defect structures in semiconductor nanowires leads to engineering of band gaps and novel electronic behaviors in nanotwinned nanowires. Interestingly, it is found that the twinning periodicity (i.e., TB spacing) is to some extent dependent on the diameter of individual nanowires [28]. At present, a critical research topic in this area is the formation mechanism of periodic twin structures. A commonly accepted point of view is that the regular arrangement of twins is associated with a balance of energies associated with solid/liquid surfaces, the olid–liquid–TB interface, and edges of TBs [30–32]. It has been suggested that the twin formation might be related to fluctuations between diffusion rate inside the catalytic droplet and the growth rate on the liquid–solid interface [29].

Twinned Alloys

Twin lamellae were incorporated into two-phase TiAl alloys during 1990s, generating the so-called poly-synthetically twinned (PST) crystals [33–35]. The flow behaviors of PST alloys were found to be strongly dependent on the orientation (i.e., the angle between TBs and the loading axis) of the lamellar structures [33–40], which has been associated with the activation of dislocations with different slip modes. To date, many studies have been performed on the plastic anisotropy of PST alloys under tension, compression, and creep.

Currently, most twinned materials have been synthesized via the formation of growth twins. A recent study showed that twinned CuAl alloys can be produced by annealing commercial alloys [41]. It was observed that a large number of micrometer-sized twins were formed during the annealing process and the fraction of annealing twins increases as the Al content rises. It has also been reported that nanoscale twin/matrix lamellae could be introduced via dynamic plastic deformation into Cu–4.5 wt% Al alloys with an average grain size of about 200 μm [42]. In the course of this treatment, deformation twinning occurred at strain levels ∼0.15 within many grains with preferred orientations. The formation of annealing and deformation twins is due to very low TB and stacking fault energies of CuAl alloys (see Table 6.1), which has resulted in a dependence of annealing twins on the Al content.

6.1.2 Numerical Modeling of Nanotwinned Metals

With rapid developments in high-performance computation, numerical modeling is becoming a reliable/indispensable tool of scientific investigation, with an increasingly important role in complementing theory and experiment.

During the last 5 years, finite-element and molecular dynamics (MD) simulations have been extensively used to model nanotwinned metals. Dislocation dynamics (DD), a mesoscopic modeling method in between length- and time-scales typically associated with continuum and atomistic simulations, has often been used to model grain-interior plasticity induced by collective motion of dislocations. However, at present, DD is not mature enough to handle intergranular plasticity, and a major challenge in DD simulations is how to describe the interaction between two-dimensional interfaces (i.e., free surface, GBs, and TBs) and dislocations. A recent attempt has been made on using DD to simulate dislocation motion in multilayer films (such as confined layer slip and pile-up at interfaces) based on a set of line integral field equations [43].

Finite-Element Method

In simulating plastic deformation in polycrystalline nanotwinned metals via the continuum finite-element method, important progress has been made on (1) adopting crystal plasticity constitutive models containing an intrinsic length scale and (2) treating TB as a strengthening zone of finite thickness.

The post-mortem transmission electron microscopy (TEM) observations showed plenty of dislocations remaining on or piling up against TBs. Motivated by such observations, Suresh and coworkers [44] proposed a TB-affected zone (TBAZ) model to capture the effects of TBs on plastic deformation of nanotwinned Cu. It is assumed that dislocation-based activities in the vicinity of TBs are the dominant plastic deformation mechanism [44]. In this model, TBAZs spanning 14–20 lattice parameters (5.1–7.2 nm) across TBs are endowed a special slip geometry that involves soft (shear parallel to TB) and hard (shear cross TB) modes of deformation. Combining the TBAZ model and rate-dependent crystal plasticity, Suresh and coworkers implemented two-dimensional finite-element simulations of nanotwinned Cu under uniaxial tension, and obtained results in agreement with the experimental results of Lu *et al.* [2]. Jerusalem *et al.* [45] extended the TBAZ model to three dimensions and elaborated the orientation-dependent dislocation blockage and absorption at TBs by modifying the relevant model parameters. Their simulations captured some of the three-dimensional features of plastic deformation in nanotwinned Cu and provided insights into the relevant failure mechanisms.

To model the strain hardening of TBs, Shen *et al.* [46] introduced a constitutive equation in which the flow stress to operate deformation in twins is expressed as a power-law function of the TB spacing. At the same time, they adopted crystal plasticity with grain-size-dependent flow stress to model the plastic deformation of the grain interior. Their simulations coupled both grain and twin size effects, with results showing the concentration of deformation near TBs. Wu *et al.* [47] used cohesive elements to model sliding and separation of TBs and regular GBs, and introduced a conventional theory of mechanism-based strain-gradient plasticity for the grain interior. It is noted that TBs have an intrinsic length scale determined by the cohesive constitutive relation used in the model.

Molecular Dynamics Simulations

Over the last decade, MD simulations have been widely recognized as a valuable tool to investigate mechanical, physical, and chemical behaviors of nanostructured materials. Generally, MD simulation methods of a many-body system can be divided into two classes: one is the classical MD (CMD) based on Newton's equations, where the interatomic interaction forces are described by a set of empirical potentials; another is the so-called *ab-initio* MD, which combines traditional CMD with electronic-structure calculations from quantum mechanics theories.

MD simulations are subject to inherent limitations in spatial and temporal scales. At present, CMD can only track the deformation of a number of grains (containing tens of millions of atoms) and nearly instantaneous (up to a few nanoseconds) loading. However, the rapidly advancing supercomputing technology has been continuously reducing the gap in length- and time-scales between CMD simulations and real experiments. Owing to their atomic-level resolutions, CMD simulations have become the most broadly used method to study deformation and failure mechanisms in nanostructured materials, especially nanocrystalline metals. A number of comprehensive review articles about CMD simulations of nanocrystalline metals can be found in the recent literature [48–50]. Compared with experimental conditions, CMD simulations are still limited to idealized model microstructures and extremely high deformation rates. However, the importance and value of CMD simulations lie in their ability to reveal deformation mechanisms that are often inaccessible to experiments.

During the last 5 years, a number of CMD simulations have been performed to study mechanical responses of nanotwinned metals subjected to different external loadings, such as nanotwinned bulk metals under uniaxial tension [51–59] or nanoindentation [60–63], nanotwinned nanowires under uniaxial tension [64–73], and nanotwinned nanopillars under compression [74,75]. At the same time, CMD simulations have been used to explore the underlying microscopic mechanisms of plastic deformation in nanotwinned metals, focusing on the interaction between dislocations and TBs [76–86]. Some results from these atomistic studies will be highlighted in Section 6.3 in comparison with relevant experimental observations.

So far, most CMD simulations of nanotwinned bulk metals have used exclusively columnar and quasi-three-dimensional (Q3D) simulation geometries based on periodic boundaries conditions. Such geometries cause severe reductions in the available slip systems of fcc metals (only three slip systems activated in Q3D geometry), as well as the degrees of freedom for GB configurations, which might suppress some important mechanisms (such as nucleation and interaction of dislocation loops) or promote artificial phenomena [56,59]. In this sense, fully three-dimensional simulations may be necessary for CMD simulations of the deformation mechanisms in nanotwinned metals.

For polycrystalline nanotwinned metals, the initial samples used in the simulations are usually free of dislocations. In such samples, massive nucleation of dislocations at the end of elastic deformation typically results in a sharp stress drop in the stress–strain curve, from a peak stress related to dislocation nucleation to a characteristic lower level associated with the operative deformation mechanisms. From this view, it will be more accurate and meaningful to define the flow/yield stress as the average stress after initial yielding, rather than the peak stress. In the subsequent section, we will compare the existing CMD simulations with experimental data.

6.2 Microstructural Characterization and Mechanical Properties of Nanotwinned Materials

In this section we will briefly review the microstructural characterizations of nanotwinned materials, as well as the principal mechanical properties of nanotwinned metals such as yield strength, ductility, strain hardening, strain-rate sensitivity, fracture toughness, and fatigue properties. First, let us consider the crystallographic structure of TBs, especially the coherent TBs (CTBs).

6.2.1 Structure of Coherent Twin Boundary

Crystallographically, a CTB is described as one of $\Sigma 3$ coincidence-site-lattice boundaries with a $\langle 011 \rangle$ misorientation axis, which belongs to a group of tilt GBs. For fcc metals, a $\Sigma 3$ GB can be obtained by rotating one monocrystal (the matrix domain) with respect to another (the twin domain) by an angle of 70.53° around a $\langle 011 \rangle$ axis [87]. The geometry of $\Sigma 3$ GB is characterized by a number of mutually perpendicular low-index lattice planes for both neighboring grains: {111}, {112}, and {011} [87]. As depicted in Figure 6.1, a

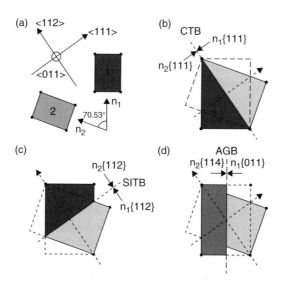

Figure 6.1 Geometry of bicrystals with a $\Sigma 3\langle 011 \rangle$ grain boundary. (a) $\Sigma 3$ misorientation for fcc elementary cells 1 and 2; (b) coherent twin boundary (CTB); (c) symmetrical incoherent twin boundary (SITB); (d) asymmetrical grain boundary (AGB) with an inclination angle of 35.26°. Figure replotted from [87]

CTB is parallel to a {111} plane; a symmetrical incoherent TB (SITB) is perpendicular to a {111} plane but parallel to a {112} plane; between CTB and SITB are asymmetrical GBs (AGBs) that can be identified by the inclination angle of the GB plane with respect to {111}. Figure 6.2 shows the atomic configuration of a CTB in fcc metals. It can be seen that a CTB is a highly symmetrical interface separating matrix and twin domains, so that atoms at both sides of the CTB are in perfect mirror symmetry. More details about the geometries of TBs in fcc, body-centered cubic (bcc), and hexagonal close-packed (hcp) metals can be found in the literature [88].

A special twin structure, the fivefold twin, has also been found in small particles/clusters and thin films [89], and has recently been observed in nanocrystalline fcc metals and alloys [90–94]. The fivefold twin structure consists of five tetrahedral subunits joined together on adjacent bounding faces, all of which are CTBs. Recent experimental observations [93] and CMD simulations [95,96] have revealed that the fivefold twin might have originated from sequential twinning involving multiple emissions of partial dislocations from GBs.

6.2.2 Microstructures of Nanotwinned Materials

Microstructures of Nanotwinned Metals

The fcc polycrystalline nanotwinned metals exhibit an inherent structural hierarchy: a high density of nanoscale twins are embedded in micrometer- or submicrometer-sized grains and act as a substructure that refines the grain interior. Such hierarchical structures play an important role in determining properties and functions of nanotwinned metals. TBs not only obstruct the motion of dislocations, but can also accommodate large plastic deformation via interfacial migration. Meanwhile, as a second-level structure, the refining twins can store plenty of dislocations and confine the length of dislocation segments, which substantially affects the capability of efficient dislocation storage in the first-level structure.

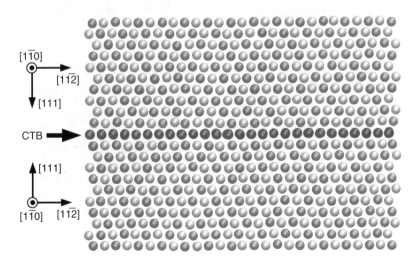

Figure 6.2 Illustration of a CTB in fcc metals. White and gray atoms are on adjacent {110} planes

Owing to various differences in synthesis methods, there have been two distinct grain structures in nanotwinned metals: one is the equiaxed grain structure where TBs are inclined to GBs, while another is the columnar-grained structure where TBs are nearly perpendicular to GBs along the columnar axis, as shown in the TEM images in Figure 6.3a and b, respectively. In nanotwinned metals with equiaxed grains, most GBs are of a high-angle type with mixed features of tilting and twisting. In nanotwinned metals with columnar grains, the GBs are primarily of the twisting type. It can be expected that these different microstructures would have significant effects on their respective mechanical behaviors.

Microstructures of Nanotwinned Nanowires

In nanotwinned semiconductor nanowires prepared by the VLS method, TBs are periodically aligned and perpendicular to the $\langle 111 \rangle$-oriented growth direction, resulting in a bamboo appearance in TEM images. More interestingly, each twin or matrix is a perfect truncated octahedron, and each facet of the octahedron is a {111} plane [30]. Adjacent octahedra are joined together along a {111} plane, with one of them rotated by exactly $60°$ around their shared $\langle 111 \rangle$ axis [30], creating a matrix-twin pair separated by a CTB. Multiple matrix-twin pairs are connected to each other, forming a twinned nanowire with a zigzag configuration. Such fascinating geometry is predicted to give rise to new and potentially useful electronic properties, including the formation of mini-bands and opening of band gaps [32]. The periodic twin structures with faceted surfaces may also enhance phonon scattering, which is relevant for thermoelectric applications [32].

In nanotwinned metallic nanowires, TBs still remain normal to the $\langle 111 \rangle$-oriented growth direction, like nanotwinned semiconductor nanowires. However, it appears that there are no ordered micro-facets on free surfaces of nanotwinned metallic nanowires,

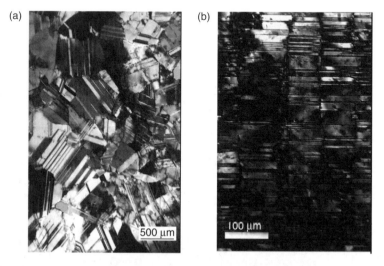

(a)   (b)

500 μm   100 μm

Figure 6.3   Microstructures of nanotwinned metals. (a) A bright-field TEM image of nanotwinned bulk Cu with equiaxed grains. (b) Cross-section TEM image of nanotwinned Ag film with columnar grains. Reprinted from [1,5] with permission from AAAS and Elsevier, respectively

presumably because these free surfaces have developed oxide layers during/after synthesis/processing. It was recently found that the regular arrangement of TBs can significantly improve the mechanical strength of nanotwinned metallic nanowires without altering their electrical properties [22,24]. Nanotwinned metallic nanowires are of great interest for applications in micro- and nano-electromechanical systems, where excellent mechanical and electrical properties are simultaneously required.

6.2.3 Mechanical and Physical Properties of Nanotwinned Metals

Yield Strength

Mechanical properties of nanotwinned metals have been characterized by uniaxial tension/compression tests and micro- or nanoindentation. Most existing data were obtained for fcc metals (e.g., Cu, Au, Ag, Ni, and Pd). It has been found experimentally that the yield strength of nanotwinned metals depends strongly on the intrinsic and extrinsic length-scales of the material. For nanotwinned polycrystals, the yield strength depends on two characteristic size scales; that is, TB spacing and grain size. For nanotwinned single crystals, the sample dimension (e.g., diameter of nanowires/nanopillars) and TB spacing become two crucial factors to determine the strength. Figure 6.4a and b show the typical stress–strain curves from a bulk sample of nanotwinned Cu under uniaxial tension [1] and nanotwinned Cu nanopillars under uniaxial compression [26], respectively. Nanotwinned metals exhibit higher yield strength than their TB-free counterparts. As shown in Figure 6.4a, the strength of nanotwinned Cu bulk samples can reach 1 GPa, which is more than double that of the nanocrystalline Cu and one order of magnitude higher than that of coarse-grained Cu samples. Recent experiments [26] also reported that the

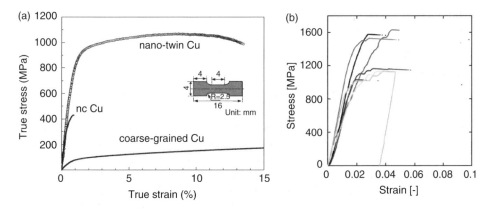

Figure 6.4 Stress–strain curves of nanotwinned metals. (a) A typical tensile stress–strain curve for as-deposited nanotwinned Cu (with average grain size d of 500 nm and twin boundary spacing λ of 15 nm) in comparison with that for a coarse-grained polycrystalline Cu ($d > 100$ μm) and nanocrystalline (nc) Cu ($d \approx 30$ nm). (b) Selected true stress–strain curves for nanotwinned Cu single-crystalline nanopillars (with diameter of 500 nm and twin boundary spacing λ of 8.8 nm). Different curves represent fluctuations among different tests on similar samples. Reprinted from [1,26] with permission from AAAS and American Chemical Society, respectively

compressive strength of nanotwinned Cu pillars with diameter of 500 nm can reach three times that of single-crystalline pillars with the same diameter.

For polycrystalline nanotwinned metals with TB spacing beyond a critical value, the yield strength increases as the TB spacing decreases, following the well-known Hall–Petch relation:

$$\sigma_y = \sigma_0 + k\lambda^{-1/2} \tag{6.1}$$

where σ_y is the yield strength, λ is the TB spacing or twin thickness, k is a material constant, and σ_0 is a friction stress. The data from experimental studies on twin-size dependence of yield strength in nanotwinned Cu polycrystals, as summarized in Figure 6.5a, agree well with the Hall–Petch law. Notably, two sets of data were obtained from nanoindentations of nanotwinned Cu thin films with columnar grains [3,4], with yield strength taken to be approximately one-third of the hardness. The strengthening behavior indicated in Figure 6.5a provides strong evidence of TBs blocking dislocation motion, as will be discussed again in Section 6.3.

So far, there are no systematic experimental studies on the effects of twin size on the strength of nanotwinned metallic nanowires and nanopillars, partly because of the technical difficulties in processing nanotwinned low-dimensional metals and manipulating them in tensile or compressive tests. A number of MD simulations [64–75] have been performed to investigate twin-size effects in metallic nanowires and nanopillars with TBs perpendicular to the growth axis. Figure 6.5b shows some recent data from MD simulations of nanotwinned Au nanowires and nanopillars. There is a clear increase in the yield strength of nanotwinned single crystals as the TB spacing is reduced. However, such strengthening does not follow the conventional Hall–Petch relation with one length variable, in that the strength here relies also on the extrinsic dimension; that is, the sample diameter. MD simulations revealed that the strengthening in nanotwinned nanowires arises from complicated interactions between dislocations and TBs. Deng and Sansoz [69] found that the yield strength of nanotwinned Au nanowires and nanopillars is proportional to the inverse of the TB spacing; that is:

$$\sigma_y = \sigma_0 + \frac{f(d)}{\lambda} \tag{6.2}$$

as shown in Figure 6.5b. In Equation (6.2), σ_0 represents the yield strength in the absence of TBs and $f(d)$ is a function of sample diameter d. This function represents the constraint of sample dimension on the curved dislocation loops emitted from free surfaces [69].

When the TB spacing falls below a critical value, the yield strength of nanotwinned metals decreases as the TB spacing is further reduced, reminiscent of the (somewhat controversial) "inverse Hall–Petch effect" in nanocrystalline metals. Such a softening phenomenon was observed for the first time during uniaxial tensile testing [2] of nanotwinned Cu with equiaxed grains, and subsequently in MD simulations of equiaxial-grained nanotwinned Cu [58] and Pd [59]. To account for the observed strength softening, there have been extensive experimental, theoretical, and computational studies showing that the softening is induced by TB migration via easy-slip of twinning partial dislocations along the TBs, resulting in narrowing of twin domains.

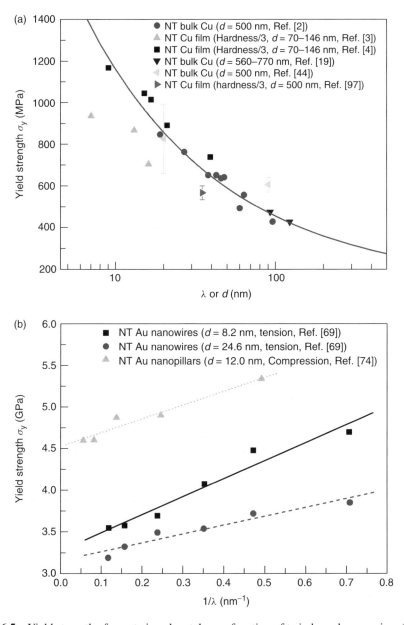

Figure 6.5 Yield strength of nanotwinned metals as a function of twin boundary spacing. (a) Yield strength of polycrystalline nanotwinned (NT) Cu. Data are from experiments (uniaxial tension or nanoindentation) [2–4,19,44,97]. Solid line represents the Hall–Petch relation. (b) Yield strength of nanotwinned Au nanowires and nanopillars. Data are from classic MD simulations of nanotwinned Au nanowires and nanopillars where twin boundaries are normal to the growth direction [69,74]

Ductility

Ductility is usually defined as the limit of strain for a material to sustain homogeneous plastic deformation before localized deformation occurs in the form of shear banding or necking. Recent decades have witnessed numerous ultrafine-grained and nanocrystalline metals with ultrahigh strengths. However, as the grain size is reduced to sub-micrometer and nanometer ranges, the ultrahigh strength achieved is typically accompanied by a great reduction in ductility compared with the conventional coarse-grained metals. A survey of published papers indicated that most nanocrystalline metals exhibit a tensile elongation to failure below $\sim$10% [97]. Early studies attributed the limited ductility of nanocrystalline metals to some kinds of flaws from processing and tensile instability [49]. However, recent *in-situ* TEM studies of nanocrystalline Al films showed that the limited ductility of nanocrystalline metals is due to the lack of viable deformation mechanisms, rather than flaw sensitivity and/or tensile instability [98,99]. In fact, the extreme stress homogenization near a notch tip due to localized grain rotation/rearrangements resulted in flaw-insensitive fracture and absence of stress concentration in the nominally brittle nanocrystalline materials [98,99].

Recent advancements of nanotwinned metals [1,2,6] demonstrated that the introduction of a high density of nanoscale twins into ultrafine-grained metals not only leads to substantial improvements in yield strength, but also imparts good ductility. An early study [6] reported that nanotwinned Cu with average grain size 200–300 nm has an elongation to failure $\sim$30% at 77 K, which is close to the tensile ductility (40–60%) of coarse-grained Cu. More recent investigations [2] of nanotwinned Cu with mean grain size $\sim$500 nm showed a pronounced rise in tensile ductility with decreasing TB spacing at room temperature, in contrast to the reduced ductility with grain refinement in nanocrystalline metals, as illustrated in Figure 6.6. In nanocrystalline metals, dislocations are nucleated from GBs and travel through the small grains to sustain plastic flow, but they can be quickly absorbed by the opposite GBs, which lower the dislocation density and prevent the accumulation of large strains. In the absence of very active GB-mediated deformation mechanisms, the ductility of nanocrystalline metals would decrease as the grain size is reduced. In nanotwinned metals, in contrast, the nanoscale twin lamellae provide ample room for dislocation nucleation and storage to sustain large plastic strains during deformation, resulting in a continuous rise in ductility with narrowing twins.

Strain Hardening

For metals undergoing homogeneous plastic deformation, the stress–strain behavior can usually be approximated as a power law:

$$\sigma = K_1 + K_2 \varepsilon^n \tag{6.3}$$

where K_1 represents the initial yield stress, K_2 is the strengthening coefficient, and the exponent n characterizes work hardening. Most coarse-grained metals have an n value between 0.1 and 0.5. Figure 6.7 shows a comparison of size effects on the strain-hardening exponent n between nanotwinned Cu and its twin-free counterparts [2]. As the grain size decreases, the strain-hardening coefficient of regular Cu decreases, especially for nanocrystalline metals, until it becomes nearly undetectable when the grain size goes down

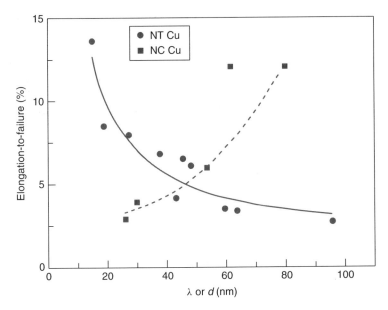

Figure 6.6 Elongation-to-failure as a function of mean twin boundary spacing for nanotwinned (NT) Cu with average grain size of 500 nm, in comparison with the grain size dependence of elongation-to-failure for nanocrystalline (NC) Cu. The solid and dashed lines serve only as visual guides. Figure replotted from [13]

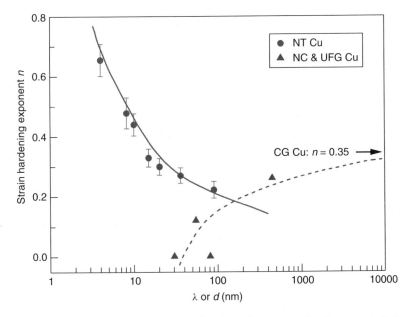

Figure 6.7 Strain-hardening exponent n as a function of mean spacing between twin boundaries for nanotwinned (NT) Cu with average grain size of 500 nm, in comparison with the grain size dependence of strain-hardening exponent of ultrafine-grained (UFG) and nanocrystalline (NC) Cu. The solid and dashed lines serve only as visual guides. Figure and data replotted from [2]

to tens of nanometers. However, for nanotwinned bulk Cu, the strain hardening monotonically increases with the reduction in TB spacing. The low strain-hardening behavior of nanocrystalline metals can be attributed to the saturation of effective dislocation density during plastic deformation, which is caused by dynamic recovery or dislocation annihilation in small grains [49]. While the detailed mechanism of strain hardening in nanotwinned metals is still being debated, it may be attributed to the extraordinary ability of CTBs to store a high density of geometrically necessary dislocations during deformation.

Strain-Rate Sensitivity

It is well known that the mechanical responses of metals can strongly depend on the applied strain rate. The strain-rate sensitivity is usually characterized by a nondimensional exponent m defined as

$$m = \frac{1}{\sigma} \frac{\partial \sigma}{\partial \ln \dot{\varepsilon}} \tag{6.4}$$

where σ is the flow/yield stress and is the strain rate. For ductile metals under uniaxial tension or compression, the strain-rate sensitivity can also be related to the so-called activation volume V [100]:

$$V = \sqrt{3} k_{\rm B} T \frac{\partial \ln \dot{\varepsilon}}{\partial \sigma} \tag{6.5}$$

where $k_{\rm B}$ is the Boltzmann constant and T is the absolute temperature. The activation volume represents the rate of decrease of the activation enthalpy with respect to flow stress at constant temperature [100]. Combining Equations (6.4) and (6.5) indicates

$$m = \frac{\sqrt{3} k_{\rm B} T}{\sigma V} \tag{6.6}$$

The strain-rate sensitivity index m and the activation volume V are often used to evaluate the sensitivity of flow stress to loading rate and to probe/identify the active deformation mechanisms. Typical values of m and V for coarse-grained, nanocrystalline, and nanotwinned bulk Cu are summarized in Table 6.2. Figure 6.8 shows a comparison of strain-rate sensitivity between nanocrystalline and nanotwinned bulk Cu [101]. When the average grain size is fixed, nanotwinned bulk Cu has higher strain-rate sensitivity than its twin-free counterpart. Recently, a systematic investigation [100] of rate sensitivity of

Table 6.2 Typical values of m and V for coarse-grained (CG), nanocrystalline (NC), and nanotwinned (NT) bulk Cu (b is the magnitude of Burgers vector of dislocation)

Material	Grain size	Mechanism	Strain-rate sensitivity m	Activation volume V
CG Cu	>1000 nm	Forest hardening	0–0.007	100–1000 b^3 [49]
NC Cu	10–100 nm	Dislocation activity and GB-mediated process	0.025–0.06 [49]	10–200 b^3
Creep tests for NC Cu	10–100 nm	GB-diffusion-mediated process	0.5–1	1–10 b^3 [49]
NT Cu	~500 nm	Interfacial plasticity	0.025–0.036 [100]	12–22 b^3 [100]

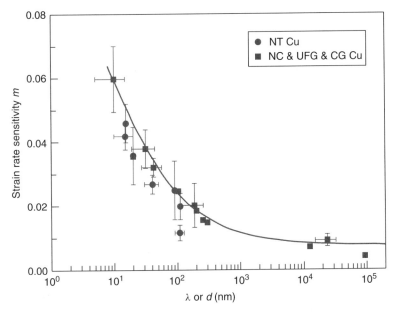

Figure 6.8 Strain-rate sensitivity as a function of twin boundary spacing for nanotwinned (NT) Cu, in comparison with that of coarse-grained (CG), ultrafine-grained (UFG), and nanocrystalline (NC) Cu as a function of grain size. The solid line serves as a visual guide. Figure and data replotted from [101]

nanotwinned bulk Cu via nanoindentation revealed that the interaction between CTBs and dislocations is responsible for elevated strain-rate sensitivity of nanotwinned bulk Cu.

Fracture Toughness and Fatigue Properties

To date, there have been only a few reports on the fracture and fatigue properties of nanotwinned metals. Qin and coworkers [102,103] measured the fracture toughness of nanotwinned bulk Cu processed by dynamic plastic deformation using three-point bending tests. They found that the fracture toughness increased with increases in volume fraction of nanoscale twins, and that nanotwinned samples exhibit dimpled fracture surfaces. It was also shown that the size and shape of dimples are closely correlated with the fraction of nanoscale twins. These phenomena may be attributed to the structural anisotropy of nanoscale deformation twins that are very slender in shape. The twin lamellar structures were found to be effective in energy absorption and arresting crack propagation during fracture [103]. Singh *et al.* [104] performed tensile fatigue tests on pre-cracked nanotwinned bulk Cu prepared by electrodeposition and determined fracture toughness using load–displacement curves. It was concluded that decreasing TB spacing (i.e., increasing volume fraction of twins) leads to improved fracture toughness. Table 6.3 summarizes the fracture toughness of nanotwinned, ultrafine-grained, and coarse-grained bulk Cu measured in recent experiments.

Shute *et al.* [105] examined the effects of 35 nm thickness twins on fatigue crack growth under cyclic loading in columnar-grained nanotwinned bulk Cu processed by magnetron

Table 6.3 Fracture toughness of nanotwinned (NT), ultrafine-grained (UFG), and coarse-grained (CG) bulk Cu

Specimen	Mean grain size d (nm)	Mean TB spacing λ (nm)	Fracture toughness K_{IC} (MPa m$^{1/2}$)	Test method
NT Cu	–	47	17.0	3-point bending [103]
NT Cu	60	49	24.4	3-point bending [103]
NT Cu	55	46	27.3	3-point bending [103]
CG Cu	–	–	9.4	3-point bending [103]
NT Cu	450	85	17.5	Fatigue tension [104]
NT Cu	450	32	22.3	Fatigue tension [104]
UFG Cu	450	–	14.9	Fatigue tension [104]

sputter deposition. They compared the $S-N$ curves (stress amplitude versus cycles-to-failure) of nanotwinned Cu with those of ultrafine-grained and coarse-grained Cu (see Figure 6.9a). It can be seen that both nanotwinned and ultrafine-grained Cu showed a markedly improved fatigue life compared with coarse-grained Cu, even in the high-cycle regime. However, it was quite unexpected that the $S-N$ curves for nanotwinned Cu are quite similar to those of ultrafine-grained Cu, as these two materials have drastically different microstructures and characteristic features on fracture surfaces. They argued that this similarity might be a coincidence, the interpretation being that cracks in both materials initiated from the intersections of certain soft domains with surfaces [105]. Dao and coworkers [104] investigated the fatigue properties of nanotwinned and ultrafine-grained bulk Cu with equiaxed grains (with mean grain size of ∼450 nm) synthesized via pulsed electrodeposition and provided a quantitative comparison of fatigue crack growth rate $\log(da/dN)$ versus $\log \Delta K$, as shown in Figure 6.9b. It was found that the higher the twin density is, the smaller the fracture crack growth rate is at any given value of ΔK.

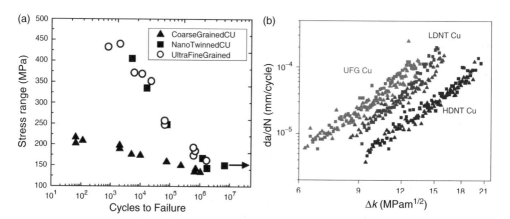

Figure 6.9 Fatigue properties of nanotwinned Cu. (a) $S-N$ curves of nanotwinned (twin boundary spacing 35 nm), ultrafine-grained and coarse-grained Cu. (b) Variation of da/dN versus $\log \Delta K$ for ultrafine-grained, low-twin-density nanotwinned (LDNT; twin boundary spacing about 85 nm), and high-twin-density nanotwinned (HDNT; twin boundary spacing about 32 nm) Cu with fixed grain size of about 450 nm. Reprinted from [104,105] with permission from Elsevier

This result is at variance with that obtained by Shute *et al.* [105], which might be due to microstructural differences (such as grain configuration and GB misorientation) of the specimen from different processing methods. This issue needs to be further clarified in order to facilitate the design of optimum microstructures of nanotwinned metals with enhanced fatigue properties.

Electrical Conductivity and Electromigration Resistance

For metals, GBs are generally considered to be deleterious to the electrical properties of materials, because GBs as planar defects can substantially scatter conducting electrons, which can become a primary mechanism as the sample size or grain size decreases [106]. By means of pulsed electrodeposition, Lu *et al.* [1] synthesized nanotwinned Cu with an average grain size of 450 nm and a mean TB spacing of 15 nm. They found a surprising phenomenon that nanotwinned Cu exhibits mechanical strength an order of magnitude higher than that of coarse-grained Cu, while at the same time retaining an electrical conductivity comparable to that of pure coarse-grained Cu. This combined property between mechanical strength and electrical conductivity originates from the ability of CTBs to obstruct dislocation motion effectively without scattering conducting electrons. A recent study estimated the electron scattering coefficient of CTBs to be $(1.5–5.7) \times 10^{-7}\,\mu\Omega\,cm^2$, which is an order of magnitude lower than that of other high-angle GBs [4]. This study also found that epitaxial nanotwinned Cu films with TB spacing of 7–16 nm exhibit a high ratio of strength-to-electrical resistivity $\sim$400 MPa $\mu\Omega^{-1}\,cm^{-1}$ [4]. Owing to a rare combination of ultrahigh strength, good ductility and low electrical resistivity, nanotwinned Cu shows high promise as interconnect components for the next-generation nano-electronic devices.

Using *in-situ* ultrahigh-vacuum and high-resolution TEM, Chen *et al.* [107] observed electromigration-induced atomic diffusion in the triple junctions of CTBs and GBs in nanotwinned Cu films, and found electromigration-induced void growth rate in nanotwinned Cu to be 10 times lower than that of twin-free Cu. TBs can impede current-driven atomic transport, which may suppress electromigration-induced failure of Cu line in integrated circuits [108]. Therefore, nanotwinned Cu can be expected to significantly improve the reliability of interconnects in next-generation electronic devices.

6.3 Deformation Mechanisms in Nanotwinned Metals

As described in Section 6.2, in comparison with twin-free nanocrystalline or ultrafine-grained metals, nanotwinned metals exhibit an unusual combination of ultrahigh strength, good tensile ductility, enhanced strain hardening, high strain-rate sensitivity, remarkable fracture toughness, and fatigue resistance. At present, the underlying deformation mechanisms responsible for these greatly enhanced mechanical properties are being subjected to intense experimental, theoretical, and computational studies aimed to understand the specific deformation behaviors in nanotwinned metals and to provide guidance for material processing and structural optimization. In this section we review some of the fundamental mechanisms in dislocation–TB interactions, as well as the strengthening and softening mechanisms responsible for the observed optimal strength of nanotwinned metals. This section will conclude with discussions on the mechanisms related to the fracture of nanotwinned metals.

6.3.1 Interaction between Dislocations and Twin Boundaries

CTBs are known to have a profound influence on the mechanical behaviors of metals, especially yielding and work hardening, because they can act as orientation-dependent barriers to obstruct dislocation motion and to arrest dislocations. Although the interaction between dislocations and CTBs has been studied for over half a century [88,109–113], the detailed reactions of dislocations at CTBs are still not fully understood. The recent studies of nanotwinned metals have generated renewed interest in the dislocation–TB interaction in fcc metals.

When a dislocation approaches a CTB with orientation inclined to the slip direction, there exist in general three types of reactions between the incident dislocation and the CTB: (1) full absorption of the dislocation by the CTB; (2) decomposition of the dislocation into a residual dislocation remaining on the CTB and a transmission dislocation crossing the CTB into the adjacent domain; (3) transfer of the dislocation through the CTB via cross-slip or other reaction mechanisms. The occurrence of these reactions depends strongly on the characteristics of the incident dislocation and the applied stress.

For fcc metals, the dislocation–TB interaction can be easily understood with the assistance of double Thompson tetrahedron, as shown in Figure 6.10. From *in-situ* TEM tensile experiments on nanotwinned Cu, Wang and Sui [114] observed extended dislocations passing through a CTB, leaving partial dislocations on the CTB. In the process of dislocation transmission through a CTB, the incident dislocation experiences shrinkage, combination and re-dissociation, which can be described by the following reactions:

$$DA \rightarrow D\gamma + \gamma A$$

$$D\gamma + \gamma A \rightarrow DA$$

$$DA \rightarrow BD' + \delta C \rightarrow B\gamma' + \gamma'D' + \delta C \text{ (partial } \delta C \text{ remains on TB)}$$

These reactions indicate that an extended dislocation would pass through a CTB through an indirect process with residual dislocations on the CTB.

Using atomistic simulations, Jin and coworkers [81,82] investigated the interaction between isolated screw and non-screw dislocations and a CTB in fcc metals, and quanti-

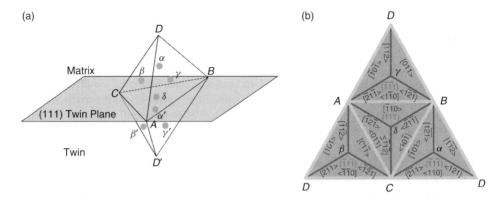

Figure 6.10 Double Thompson tetrahedron representation: (a) perspective illustration; (b) plane illustration with Burgers vectors denotation

tatively measured the critical stress for the dislocation to overcome the energy barrier of a CTB. For example, for Cu the critical shear stress of a screw dislocation is about 465–510 MPa, while for non-screw dislocations it needs to be larger than 1.2 GPa. Chassagne *et al.* [84] performed simulations similar to those of Jin and coworkers work on the screw dislocation–TB reaction and found that the reaction stress is associated with a parameter $\gamma_{SF}/\mu b_p$ (where μ is the shear modulus and b_p is the Burger vector of the Shockley partial), which is close to 400 MPa for Cu. Recent *in-situ* TEM straining experiments on Cu [84] showed screw dislocations crossing a CTB after substantial dislocation pile-ups had been established on the incident side. This observation implies that dislocation transmission through TBs may be the result of a collective process, where the local stress can be substantially magnified by dislocation pile-up [84].

Zhu *et al.* [115] systematically summarized dislocation–TB reactions in fcc metals and calculated the energy barriers for all the reactions based on the elastic theory of dislocations. Ezaz *et al.* [83] analyzed the energetic of dislocation–TB reactions using generalized planar stacking-fault energy curves calculated from atomistic simulations. Table 6.4 gives a summary of all the possible dislocation–TB reactions in fcc metals [115].

Other than their roles as dislocation barriers and traps, CTBs can also act as dislocation sources after they lose coherency during plastic deformation. In the process of *in-situ* TEM tensile experiments on electrodeposited nanotwinned Cu, Wang *et al.* [116] observed atomic steps at TBs associated with the dissociation of perfect dislocations into sessile Frank partial dislocations on TBs. Such steps can serve as dislocation sources to relieve stress concentration. A similar phenomenon has also been observed from straining experiments on ultrafine-grained Cu with low density of TBs [117,118].

6.3.2 Strengthening and Softening Mechanisms in Nanotwinned Metals

As introduced in "Microstructures of nanotwinned metals" in Section 6.2.2, uniaxial tensile tests [2] of nanotwinned Cu showed that there exists a critical twin thickness for the maximum yield strength; that is, the strength of nanotwinned Cu first increases with decreasing TB spacing and then decreases once the TB spacing is reduced below the critical value. The former is a size-strengthening effect which has been experimentally and computationally demonstrated to be attributed to TB blockage of dislocation motion, while the latter is a size-softening effect which is still under investigation. Here, we review some of the recent studies (including experiment, theory, and modeling) aimed at understanding the underlying mechanisms responsible for the observed strengthening and softening in nanotwinned bulk metals. At the same time, we will provide a comparison of Hall–Petch strengthening between nanotwinned and nanocrystalline metals, and also briefly introduce the deformation mechanisms of nanotwinned nanowires and nanopillars from CMD simulations.

Deformation Mechanisms in Nanotwinned Bulk Metals

The experiments [2–4,8] and CMD simulations [57–59] have shown that the size-strengthening behavior of nanotwinned bulk metals follows the Hall–Petch relation, as described by Equation (6.1). In polycrystalline metals, the Hall–Petch strengthening is known to arise from dislocation pile-up at GBs. For nanotwinned metals, this can be

Table 6.4 Summary of dislocation–TB reactions and their energy barriers in fcc metals (twin plane is ABC plane) [115]

Reaction description	Equation	Isotropic energy barriers[a]
30° partial, Bα		
Cross-slip onto TB	Bα → Bδ + δα	$E + 2.0F$
Stair-rod dislocation dissociation	δα → δB + Bα	$3.5E + 5.5F$
Transmit across TB	Bα → Bα′ + α′α	$2.7E + 4.1F$
90° partial, Dα		
Cross-slip onto TB	Dα → Aδ	0
Transmit across TB	Dα → δα + δα′ + α′D′	$2.0E + 4.0F$
	Dα → δα + δβ′ + β′D′	$2.0E + 4.0F$
	Dα → δα + δγ′ + γ′D′	$2.0E + 4.0F$
	Dα → 4/9Aδ + α′D′	$0.6E + 1.4F$
	Dα → 2/9δ + β′D′	$0.1E + 0.2F$
	Dα → 2/9δC + γ′D′	$0.1E + 0.2F$
Cross-slip of perfect screw dislocation, BC		0
No dislocation reaction needed		
Perfect 60° dislocation		
Cross-slip onto TB	BD → BC + CD	$6.0E + 7.2F$
Transmit across TB	BD → 2Bδ + D′B	$4.5E + 7.5F$
	BD → δC + D′A	$3.0E + 3.9$
	BD → δA + D′C	$3.0E + 3.9F$
30° leading partial cross-slip onto TB and trailing	D′δ → D′α′ + α′δ	$2.0E - 0.3F$
90° partial transmit across TB		
30° leading partial transmit across TB first and the 90° trailing partial transmit cross TB second	Dα + αα′ → 2α′δ + αD′δ	$4.0E$
90° leading partial and 30° trailing partial transmit across TB sequentially	Bα + αδ + α′δ → Bα′ + 2α′δ	$3.4E + 2.7F$
	Bα + αδ + β′δ → Bα′ + α′δ + β′δ	$2.8E + 3.7F$
	Bα + αδ + γ′δ → Bα′ + α′δ + γ′δ	$2.8E + 3.7F$

[a] $E = \{\mu a^2/[72\pi(1 - \nu)]\}\ln(2^{1/2}d/a)$ and $F = \mu a^2/[72\pi(1 - \nu)]$, where μ is the shear modulus, ν is Poisson's ratio, a is the lattice constant, and d the grain size.

attributed to CTBs acting as strong barriers to dislocation motion or nanoscale twins limiting the lengths of mobile dislocation segments. Gao and coworkers [58] performed large-scale three-dimensional CMD simulations of equiaxial-grained nanotwinned Cu under uniaxial tension, with mean grain size of 10–20 nm and TB spacing from 0.63 to 6.25 nm. They observed that, when TB spacing is greater than a critical value, a large number of dislocations nucleate from GBs and glide on the slip planes inclined to the TBs. In this case, the plastic deformation is dominated by dislocations cutting across TBs, as illustrated in Figure 6.11a. In the traditional Hall–Petch law $\sigma = \sigma_0 + k/d^{1/2}$, the slope k reflects the efficacy of GBs or TBs as dislocation obstacles, while the intercept

(a)

(b)

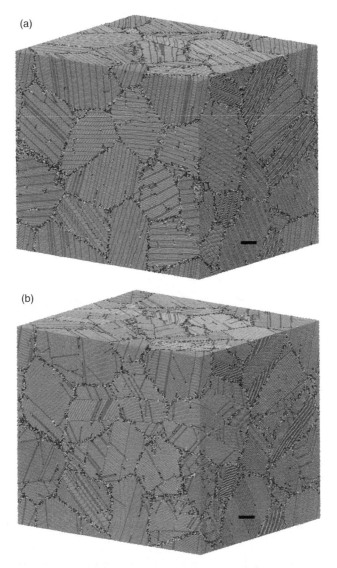

Figure 6.11 Simulated deformation patterns in nanotwinned samples with mean grain size $d = 20$ nm at 10% strain. (Scale bars, 5 nm.) Samples with twin boundary spacing (a) $\lambda = 1.25$ nm and (b) $\lambda = 6.25$ nm. In (a), plastic deformation is dominated by partial dislocations gliding parallel to twin planes, whereas in (b) dislocations cutting across twin planes is the controlling deformation mechanism. Figure replotted from [58]

σ_0 represents the lattice frictional stress required to move dislocations. To estimate these characteristic parameters (slope and intercept), we fit the Hall–Petch relation based on the previous experimental and computational results about nanotwinned and nanocrystalline metals. The slopes and intercepts obtained are given in Table 6.5, and all data are plotted in Figure 6.12. The two fitting parameters from experiments of nanotwinned Cu are very

Table 6.5 Fitting parameters of Hall–Petch relation based on experimental and computational results for nanocrystalline (NC) and nanotwinned (NT) Cu

Data source	σ_0 (GPa)	k (GPa nm$^{1/2}$)
NT Cu (exp.)	0.1278	3.266
NC Cu (exp.)	0.1210	3.366
NT Cu (CMD)	1.0528	3.085
NC Cu (CMD)	1.2689	4.067

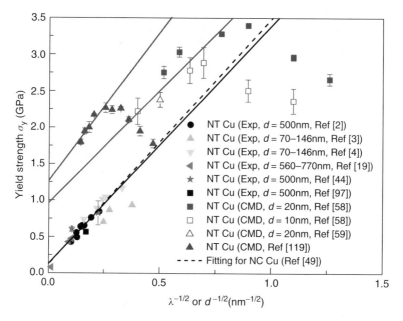

Figure 6.12 Comparison of Hall–Petch relationship between nanotwinned (NT) and nanocrystalline (NC) Cu based on experimental and computational results [2–4,19,44,49,58,59,97,119]

close to those obtained from nanocrystalline Cu. This implies that the size-strengthening in nanotwinned Cu indeed follows the Hall–Petch law, in a similar way to nanocrystalline Cu. The material constants (i.e., slope) from CMD simulations agree with those from experiments. However, the stresses (i.e., intercept) from CMD simulations are one order of magnitude higher than those from experiments. Such high resistance for dislocations in CMD simulations is due to high strain rate, small length scale, and high-energy configurations. Recent CMD simulations have shown that the stresses required for screw and non-screw dislocations transmission through TBs are about 0.5 GPa and 1.2 GPa, respectively. The value for non-screw dislocations is close to the present fitting parameters.

Using atomistic reaction pathway calculations (i.e., via free-end nudged elastic band method to determine the minimum energy paths), Zhu *et al.* [76] revealed that the mechanistic origin of high rate sensitivity and good tensile ductility in nanotwinned metals lies with the slip transfer reactions between dislocations and CTBs. They also pointed out that the slip transfer reactions mediated by TBs dominate plastic deformation as the rate-controlling mechanisms [76], consequently resulting in small activation volume in nanotwinned metals. Concomitantly, it was found that, in the slip transfer reactions, small activation volume arises because only a small group of atoms is involved during the cross-slip process [50]. In CMD simulations of nanotwinned Cu, Zheng *et al.* [79] observed complete transmission of screw dislocations across TBs via cross-slip, along with incomplete transmission of non-screw dislocations and transmission-induced jog formation. These mechanisms are operative in maintaining the high strength and good ductility of nanotwinned metals. Wu *et al.* [78] reported two new dislocation reactions with CTBs from atomistic simulations: the first one involves a 60° dislocation interacting with a CTB, creating a $\{001\}\langle 110\rangle$ Lomer dislocation, which, in turn, dissociated into Shockley, stair-rod, and Frank partials; the second one has a 30° partial slipping transfer across a CTB, generating three new Shockley partials. The former reaction leaves immobile dislocations on TBs, which can effectively pin or block further dislocation motion, while the latter reaction induces a high population of mobile dislocations on TBs, which facilitate further plastic flow.

Among studies on the stability of growth twins in nanotwinned metal, early experimental observations [85] showed rapid migration of TBs due to collective glide of multiple twinning partial dislocations, which can induce narrowing or annihilation of thin twins, corresponding to de-twinning. A recent experiment [120] found that the shear stress required to move partials on a TB can be as low as 10 MPa for fcc metals. This implies that it can be quite easy for dislocations to glide on or parallel to TBs.

Using large-scale atomistic simulations, Gao and coworkers [58] have systemically investigated the twin-size effect in nanotwinned Cu. When TB spacing is sufficiently small, it was observed that a large number of partial dislocation loops nucleate from the intersections between TBs and GBs, and glide on TBs, as shown in Figure 6.13. Such a mechanism prevails in nearly all grains, hence dominating the plastic deformation. As revealed in Figure 6.11, there exists a transition in deformation mechanism, from the classical Hall–Petch type of strengthening due to dislocation pile-up and cutting through twin planes to a dislocation–nucleation-controlled softening mechanism with TB migration resulting from nucleation and motion of partial dislocations parallel to the twin planes, as TB spacing is reduced. Based on the insights gained from CMD simulations, Gao and coworkers [58] have developed a kinetic model of dislocation nucleation to account for the observed strength softening, This model yields a scaling law between the shear strength of nanotwinned fcc metals and characteristic dimensions (mean grain size d and TB spacing λ) of the microstructure as follows:

$$\tau = \frac{\Delta U}{S V^*} - \frac{k_{\mathrm{B}} T}{S V^*}\ln\left(\frac{d}{\lambda}\frac{\nu_{\mathrm{D}}}{\dot{\varepsilon}}\right) \qquad (6.7)$$

where ΔU is the activation energy, S is a factor related to local stress concentration and geometry, V^* is the activation volume, k_{B} and T are Boltzmann's constant and temperature, ν_{D} is the Debye frequency, and $\dot{\varepsilon}$ is the macroscopic strain rate. The estimation of yield strength, based on this scaling law, is in excellent agreement with the experimental measurements and simulation results, as shown in Figure 6.14, which summarizes all the data about nanotwinned Cu. Recently, Wei [121] expressed thermally activated dislocation

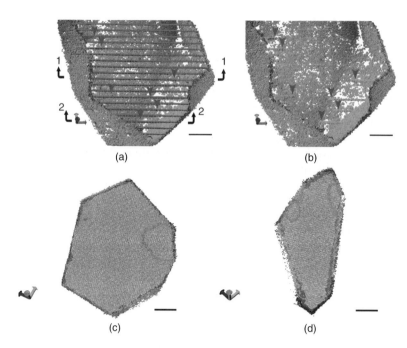

Figure 6.13 Dislocation nucleation in a representative grain in nanotwinned Cu with grain size of 20 nm and twin boundary spacing of 1.25 nm. Atomic views of dislocation structures (pointed by gray arrows). (a) Twin boundaries are painted in red. (b) Twin boundaries are invisible. (c, d) Top views of the cross-section marked by 1 and 2, respectively, in (a). The scale bars are 5 nm. Refer to Plate 4 for the colored version

nucleation frequency in a more accurate manner, and extended the above kinetic model to the following equation:

$$\frac{\tau}{\phi} = \tau_0 - \frac{k_B T}{V^*} \ln\left(\frac{\beta\pi}{2} \frac{v_D}{\dot{\varepsilon}} \frac{d}{\lambda} \sqrt{\frac{\Omega_0}{V^*}}\right) \tag{6.8}$$

where τ is the critical resolved shear stress, τ_0 is the friction stress, ϕ is a factor related to the stress concentration and crystallographic texture, β is a geometric coefficient, and Ω_0 is the atomic volume. Combining this with Equation (6.4), Wei also obtained the strain-rate sensitivity as a function of mean grain size and TB spacing:

$$\frac{1}{m} = \frac{\tau_0 V^*}{k_B T} - \ln\left(\frac{\beta\pi}{2} \frac{v_D}{\dot{\gamma}} \frac{d}{\lambda} \sqrt{\frac{\Omega_0}{V^*}}\right) \tag{6.9}$$

where $\dot{\gamma}$ is the shear strain rate. This expression shows qualitative agreement with experimental observations that the rate sensitivity m increases as the TB spacing λ decreases.

Considering de-twinning as collective motion of twinning partials [85], Wei [121,122] treated nanoscale twins as Eshelby inclusions with an eigenstrain e^T, regarded de-twinning as a phase transformation of inclusions, and derived an energetic model as follows:

$$\tau = \tau_0 + \frac{\alpha\pi}{8} \frac{2-\nu}{1-\nu} \frac{\lambda}{d} \mu e^T - \frac{2\gamma_{TB}}{\lambda e^T} \tag{6.10}$$

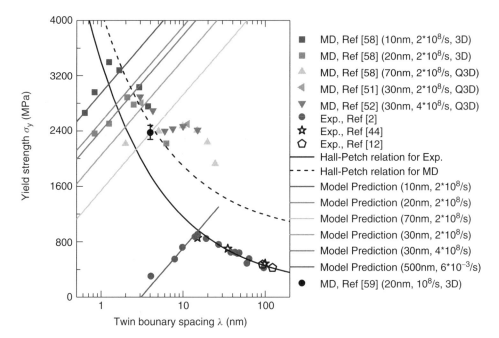

Figure 6.14 Summary of data from experimental and computational studies of nanotwinned bulk Cu [2,12,44,51,52,58,59]. Hall–Petch relations and predictions from the kinetic model of Equation (6.7) are plotted for comparison

where α is a constant between zero and one, μ is the shear modulus, ν is the Poisson ratio, and γ_{TB} is the TB energy. The second term on the right-hand side of Equation (6.10) represents the driving stress related to the elastic energy of nanoscale twins (equivalent as inclusions), and the third term corresponds to the change in interfacial energy. The predictions from Equation (6.10) are in good agreement with the experiment results.

Based on strain-gradient plasticity, Lu and coworkers [123] developed a mechanism-based model to describe the plastic deformation of nanotwinned metals. By introducing the concept of dislocation pile-up zones near TBs and GBs, they derived equations that govern the evolution of dislocation density, substituted the resultant expressions into the Taylor relationship, and obtained a constitutive relationship. In the dislocation pile-up zone near TBs, they considered shear strains from two types of dislocations: one from dislocations inclined to TBs (hard slip) and the other from dislocations on TBs (easy slip). This treatment distinguishes the contributions of two slip modes to plastic deformation. Coupling the failure and dislocation nucleation criteria, they implemented numerical simulations to investigate the twin-size effect on strength, ductility, and work hardening of nanotwinned metals. The results obtained are in excellent agreement with the experimental results, with predictions that the critical TB spacing for strengthening–softening transition should be linearly proportional to the average grain size, which is consistent with recent CMD simulations [58].

Deformation Mechanisms in Nanotwinned Nanowires and Nanopillars

Deng and Sansoz [66–71] conducted a series of atomistic simulations of nanotwinned Au nanowires under uniaxial tension, with TBs normal to the tensile direction. They observed that dislocations always nucleate from the intersections between TBs and free surfaces, and glide on the slip planes inclined to TBs. When dislocations approached TBs, the slip-transfer reaction occurred in the following steps [66]:

1. Shrinkage of leading $B\gamma$ and trailing γD partials: $B\gamma + \gamma D \to BD$.
2. Decomposition of BD into a transmission dislocation CD and a residual dislocation BC on TB: $BD \to BC + CD$.
3. Reaction of CD with the next TB, and its decomposition into a stair-rod dislocation $C\gamma$ on TB and a partial dislocation γD through TB: $CD \to C\gamma + \gamma D$.
4. Further dissociation of sessile stair-rod $C\gamma$ into full dislocation CB and partial dislocation $B\gamma$: $C\gamma \to CB + B\gamma$.

Based on these simulation results, Deng and Sansoz [66] concluded that the site-specific dislocation nucleation and the above slip-transfer reaction are controlling mechanisms in plastic deformation of nanotwinned nanowires. Considering site-specific surface dislocation emission and image forces due to the presence of TBs, they developed a model showing that the yield strength is inversely proportional to TB spacing [69], as described by Equation (6.2). This model is consistent with results from CMD simulations, as shown in Figure 6.5b.

Wu et al. [124] measured the mechanical properties of fivefold-twinned Ag nanowires by conducting three-point bending tests via lateral force atomic force microscopy. The diameter of the nanowires tested ranged from 16 to 35 nm, with fivefold twin structure along the entire wire length. Such nanowires exhibit brittle fracture without significant plastic deformation, which is attributed to fivefold TBs suppressing dislocation slip. Most recently, more comprehensive and quantitative measurements [125] of mechanical properties of fivefold-twinned Ag nanowires were reported by Zhu et al. They performed in-situ scanning electron microscopy tensile testing on Ag nanowires with diameter between 34 and 130 nm and found that the Young's modulus, yield strength, and ultimate tensile strength increase as the diameter of nanowires decreases. These remarkable size effects originate from the presence of the free surface and the internal fivefold TBs. Filleter et al. [126] conducted in-situ TEM tensile testing and MD simulations on fivefold twinned Ag nanowires, observing pronounced plastic deformation from dislocation nucleation at the intersections between fivefold TBs and the free surface.

Jang et al. [26] conducted experiments on uniaxial compression of nanotwinned Cu nanopillars with diameter 500 nm and TBs normal to the compressive axis. They found that the measured strength is much larger than that of twin-free single-crystalline pillars with the same diameter. Based on these observations, they concluded that TBs impeding dislocation slip are responsible for the strengthening of nanotwinned nanopillars. Recently, Brown and Ghoniem [75] performed atomistic simulations of nanotwinned Cu nanopillars of diameter 9 nm under uniaxial tension and compression, with TBs inclined to the loading axis. Their simulation results showed tension–compression asymmetry and a transition between reversible–irreversible plastic deformation: the plastic strain is found to be reversible when the resolved shear stress falls between 0.5 and 1.0 GPa and the axial strain is less than 3.3%; however, plastic irreversibility occurs at larger strains or stresses [75]. Such plastic reversible–irreversible transition has been attributed to competitions between partial dislocations gliding on TBs (for reversible plasticity) and partial nucleation inclined to the free surface (for irreversible plasticity) [75].

6.3.3 Fracture of Nanotwinned Copper

So far, there have been only a few studies on the effects of twin size on the fracture toughness of nanotwinned metals. Recent experiments [102–105] have demonstrated that the twin lamellar structures can significantly enhance the fracture toughness of nanotwinned Cu. Qin *et al.* [102] observed ductile fracture characterized by dimple-dominated fracture surfaces. It was found that the dimple structures are produced by voids nucleating ahead of a principal crack when the overall strain reaches a critical value [102]. In fracture of nanotwinned metals, a unique feature is that the typical dimples are coarser and deeper than those appearing on the fracture surfaces of nanocrystalline metals. These coarser/deeper dimples are created near the slender nanoscale twins, indicating that the highly anisotropic twin structures are effective in energy absorption and arresting crack propagation [102]. Shan *et al.* [127] conducted *in-situ* straining experiments on nanotwinned Cu and found that, when TB spacing is larger than 30 nm, the crack path is always along a specific crystallographic plane, implying a brittle-like behavior; however, when TB spacing is reduced to below 30 nm, the fracture path no longer follows the crystallographic plane. They concluded that such transition is attributed to full dislocation slip (for $\lambda > 30$ nm) switching to partial dislocation slip on TBs (for $\lambda < 30$ nm). Most recently, Kim *et al.* [128] performed *in-situ* tensile experiments on crack interactions with TBs in nanotwinned Cu films with a thickness of $\sim$100 nm. It was observed that a moving principal crack may jump over twin lamellar structures, leading to crack bridging by intact lamellae segments. TEM observations (see Figure 6.15) and complementary atomistic simulations showed that the

Figure 6.15 Crack jumping across a twin of thickness 80 nm in a Cu thin film 100 nm in thickness; the crack to the right of the first twin has grown through the middle thin twin and the matrix to the right and entered the third twin. "M" represents the matrix, "T" represents the twin, and "IC" represents the interfacial crack. Figure replotted from [128]

micro-crack bridging mechanism can be attributed to dislocation–TB interactions. During crack propagation, twin boundaries are impinged upon by numerous dislocations from the plastically deforming matrix; these dislocations react at the interface and evolve into substantially impenetrable dislocation walls that strongly confine crack nucleation and resist crack propagation [128].

Zhou and coworkers [129,130] performed Q3D CMD simulations of nanotwinned Ni with a pre-crack under uniaxial tension. They varied TB spacing in the simulated samples and found that the fracture toughness dramatically increases with decreasing TB spacing [129]. During simulations, four toughening mechanisms due to the presence of TBs were observed [130]:

1. crack blunting through dislocation accommodation along TBs;
2. crack deflection in intragranular propagation;
3. daughter crack formation along TBs; and
4. curved TB planes owing to an excessive pile-up of geometrically necessary dislocations.

6.4   Concluding Remarks

In this chapter we have presented a brief survey of recent advances in mechanics and mechanical behaviors of nanotwinned hierarchical metals. Experimental and computational studies have shown that nanotwinned metals exhibit a combination of ultrahigh strength, good tensile ductility, enhanced strain hardening, high strain-rate sensitivity, remarkable fracture toughness, and fatigue resistance, compared with their twin-free nanocrystalline or ultrafine-grained counterparts. We have addressed some of the microscopic deformation mechanisms responsible for the unusual mechanical properties of nanotwinned metals, with emphasis on the unique dislocation–TB interaction mechanisms, as well as the intriguing size effects associated with intrinsic length scales in these materials.

Numerous experimental and computational studies have shown that nanotwinned metals have an excellent combination of ultrahigh strength, good ductility, large fracture toughness, and high conductivities as a result of internal nanoscale twin structures. Because of these unique properties, nanotwinned metals have stimulated great interest as potential candidates for nanoelectronic and nanoelectromechanical devices. For example, nanotwinned bulk Cu can serve as high-strength interconnections in packaging technology, while fivefold twinned Ag nanowires may hold promising potential for applications in flexible/stretchable electronic devices.

In spite of the tremendous progresses made in recent years, our understanding of the mechanical and physical properties of nanotwinned hierarchical metals is still at an infant stage. There remain many open questions in the field, a few of which are listed below.

1. Nanotwinned metals have a higher strain-hardening rate, relative to ultrafine-grained and nanocrystalline metals, which leads to a good tensile ductility of nanotwinned metals. It is more amazing that the strain-hardening exponent increases with decreasing TB spacing in nanotwinned metals. This trend is opposite to that in twin-free ultrafine-grained and nanocrystalline metals, as illustrated in Figure 6.7. Such a size effect of

strain hardening is closely related to dislocation activities during plastic deformation. But what is the intrinsic mechanism/regime which determines the strain hardening exponent increasing with decreasing TB spacing in nanotwinned metals?

2. There have been various synthesis methods to introduce twin structures into pure materials. Currently, the embedded twins principally originate from growth, annealing, and deformation twinning. The fabrication and stability are two important issues all along in the studies of nanotwinned materials. Thus, two crucial questions concern what kinds of materials (metals or alloys) can be fabricated into a nanotwinned hierarchical material and whether there is a lower limit of the TB spacing for the stability of nanotwinned materials.

3. For nanotwinned nanowires and nanopillars, the yield strength depends on TB spacing and sample diameter, even the TB orientation. Atomistic simulations have revealed that, when the sample diameter is fixed and the TB is normal to loading axis, the yield strengths of nanotwinned nanowires and nanopillars are inversely proportional to TB spacing. This quantitative relationship needs to be verified by further experimental studies. A more important question is how the yield strengths of nanotwinned nanowires and nanopillars depend on a combination of sample diameter, TB spacing, and orientation.

4. It is known that material microstructures determine deformation behaviors of materials. Currently, there exist two distinct grain configurations in nanotwinned metals: equiaxed and columnar shapes. Thus, a question arising is how the grain configurations (e.g., equiaxed or columnar) of nanotwinned metals influence their mechanical behaviors, and whether distinct grain configurations induce different deformation mechanisms in grain interiors.

5. As a hierarchical material, nanotwinned metals have exhibited enhanced mechanical properties. In particular, nanotwinned metals have optimal strength corresponding to a certain grain size and TB spacing. The pursuit toward super-strong/strongest materials is always an untiring goal of studies in field of material science. Therefore, an urgent issue in nanotwinned metals is whether and how one could design the intrinsic structures of nanotwinned metals to control microscopic deformation mechanisms and/or the optimal strength.

Clearly, there are plenty of research opportunities to improve our understanding of nanotwinned hierarchical materials in the near future. There will also be urgent needs to perform further experimental and computational investigations on fabrication, thermal stability, mechanical testing, structural optimization, and applications of various forms of nanotwinned materials, including bulk materials, films, nanowires, and nanopillars. These efforts will be multidisciplinary and require the development of novel experimental and multiscale modeling techniques in the foreseeable future.

References

[1] Lu, L., Shen, Y., Chen, X. *et al.* (2004) Ultrahigh strength and high electrical conductivity in copper. *Science*, **304**, 422–426.
[2] Lu, L., Chen, X., Huang, X., and Lu, K. (2009) Revealing the maximum strength in nano-twinned copper. *Science*, **323**, 607–610.

[3] Anderoglu, O., Misra, A., Wang, H. *et al*. (2008) Epitaxial nanotwinned Cu films with high strength and high conductivity. *Applied Physics Letters*, **93**, 083108.

[4] Anderoglu, O., Misra, A., Ronning, F. *et al*. (2009) Significant enhancement of the strength-to-resistivity ratio by nanotwins in epitaxial Cu films. *Journal of Applied Physics*, **106**, 024313.

[5] Bufford, D., Wang, H., and Zhang, X. (2011) High strength, epitaxial nanotwinned Ag films. *Acta Materialia*, **59**, 93–101.

[6] Ma, E., Wang, Y.B., Lu, Q.H. *et al*. (2004) Strain hardening and large tensile elongation in ultrahigh-strength nano-twinned copper. *Applied Physics Letters*, **85**, 4932–4934.

[7] Wang, G., Li, G., Zhao, L. *et al*. (2010) The origin of the ultrahigh strength and good ductility in nanotwinned copper. *Materials Science and Engineering A*, **527**, 4270–4274.

[8] Idrissi, H., Wang, B., Colla, M.S., and Raskin, J.P. (2011) Ultrahigh strain hardening in thin Palladium films with nanoscale twins. *Advanced Materials*, **23**, 2119–2122.

[9] Algra, R.E., Verheijen, M.A., Borgstrom, M.T. *et al*. (2008) Twinning superlattices in indium phosphide nanowires. *Nature*, **456**, 369–372.

[10] Caroff, P., Dick, K.A., Johansson, J. *et al*. (2008) Controlled polytypic and twin-plane superlattices in III–V nanowires. *Nature Nanotechnology*, **4**, 50–55.

[11] Arbiol, J., Estrad, S., Prades, J.D. *et al*. (2009) Triple-twin domains in Mg doped GaN wurtzite nanowires: structural and electronic properties of this zinc-blende-like stacking. *Nanotechnology*, **20**, 145704.

[12] Spirkoska, D., Arbiol, J., Gustafsson, A. *et al*. (2009) Structural and optical properties of high quality zinc-blende/wurtzite GaAs nanowire heterostructures. *Physical Review B*, **80**, 245325.

[13] Lu, K., Lu, L., and Suresh, S. (2009) Strengthening materials by engineering coherent internal boundaries at the nanoscale. *Science*, **324**, 349–352.

[14] Kibey, S., Liu, J.B., Johnson, D.D., and Sehitoglu, H. (2007) Predicting twinning stress in fcc metals: linking twin-energy pathways to twin nucleation. *Acta Materialia*, **55**, 6843–6851.

[15] Xu, J., Lin, W., and Freeman, A.J. (1991) Twin-boundary and stacking-fault energies in Al and Pd. *Physical Review B*, **43**, 2018–2024.

[16] Kibey, S., Liu, J.B., Johnson, D.D., and Sehitoglu, H. (2006) Generalized planar fault energies and twinning in Cu–Al alloys. *Applied Physics Letters*, **89**, 191911.

[17] Fu, CL and Yoo, M.H. (1997) Interfacial energies in two-phase TiAl–Ti$_3$Al alloy. *Scripta Materialia*, **37**, 1453–1459.

[18] Zhang, W.J. and Appel, F. (2002) Weak-beam TEM study on planar fault energies of Al-lean TiAl-base alloys. *Materials Science and Engineering A*, **334**, 59–64.

[19] Merz, M.D. and Dahlgren, S.D. (1975) Tensile strength and work hardening of ultrafine-grained high-purity copper. *Journal of Applied Physics*, **46**, 3235–3237.

[20] Dahlgren, S.D., Nicholson, W.L., Merz, M.D. *et al*. (1977) Microstructural analysis and tensile properties of thick copper and nickel sputter deposits. *Thin Solid Films*, **40**, 345–353.

[21] Wang, B., Fei, G.T., Zhou, Y. *et al*. (2008) Controlled growth and phase transition of silver nanowires with dense lengthwise twins and stacking faults. *Crystal Growth and Design*, **8**, 3073–3076.

[22] Bernardi, M., Raja, S.N., and Lim, S.K. (2010) Nanotwinned gold nanowires obtained by chemical synthesis. *Nanotechnology*, **21**, 285607.

[23] Wang, C., Wei, Y., Jiang, H., and Sun, S. (2010) Bending nanowire growth in solution by mechanical disturbance. *Nano Letters*, **10**, 2121–2125.

[24] Zhong, S., Koch, Y., Wang, M. *et al*. (2009) Nanoscale twinned copper nanowire formation by direct electrodeposition. *Small*, **20**, 2265–2270.

[25] Burek, M.J. and Greer, J.R. (2010) Fabrication and microstructure control of nanoscale mechanical testing specimens via electron beam lithography and electroplating. *Nano Letters*, **10**, 69–76.

[26] Jang, D., Cai, C., and Greer, J.R. (2011) Influence of homogeneous interfaces on the strength of 500 nm diameter Cu nanopillars. *Nano Letters*, **11**, 1743–1746.

[27] Wang, Z.L., Dai, Z.R., Gao, R.P. *et al*. (2000) Side-by-side silicon carbide–silica biaxial nanowires: synthesis, structure, and mechanical properties. *Applied Physics Letters*, **77**, 3349–3351.

[28] Li, Q., Gong, X., Wang, C. *et al*. (2004) Size-dependent periodically twinned ZnSe naowires. *Advanced Materials*, **16**, 1436–1440.

[29] Hao, Y., Meng, G., Wang, Z.L. *et al*. (2006) Periodically twinned nanowires and polytypic nanobelts of ZnS: the role of mass diffusion in vapor–liquid–solid growth. *Nano Letters*, **6**, 1650–1655.

[30] Korgel, B.A. (2006) Twin cause kinks. *Nature Materials*, **5**, 521–522.

[31] Shim, H.W., Zhang, Y., and Huang, H. (2008) Twin formation during SiC nanowire synthesis. *Journal of Applied Physics*, **104**, 063511.

[32] Ross, F.M. (2009) Bringing order to twin-plane defects. *Nature Nanotechnology*, **4**, 17–18.

[33] Kim, Y.W. (1994) Ordered intermetallic alloys, part III: gamma titanium aluminides. *JOM*, **47**, 30–39.

[34] Appel, F. and Wagner, R. (1998) Microstructure and deformation of two-phase γ-titanium aluminides. *Materials Science and Engineering R*, **22**, 187–268.

[35] Kim, Y.W. (1998) Strength and ductility in TiAl alloys. *Intermetallics*, **6**, 623–628.

[36] Yao, K.F., Inui, H., Kishida, K., and Yamaguchi, M. (1995) Plastic deformation of V- and Zr-alloyed PST TiAl in tension and compression at room temperature. *Acta Metallurgica et Materialia*, **43**, 1075–1086.

[37] Umkoshi, Y., Yasuda, H.Y., and Nakano, T. (1996) Plastic anisotropy and fatigue of TiAl PST crystals: a review. *Intermetallics*, **4**, S65–S75.

[38] Umeda, H., Kishida, K., Inui, H., and Yamaguchi, M. (1997) Effects of Al-concentration and lamellar spacing on the room-temperature strength and ductility of PST crystals of TiAl. *Materials Science and Engineering A*, **239**, 336–343.

[39] Parthasarathy, T.A., Mendiratta, M.G., and Dimiduk, D.M. (1998) Strength flow behavior of PST and fully lamellar polycrysts of Ti–48Al in the microstrain regime. *Acta Materialia*, **46**, 4005–4016.

[40] Parthasarathy, T.A., Subramanian, P.R., Mendiratta, M.G., and Dimiduk, D.M. (2000) Phenomenological observations of lamellar orientation effects on the creep behavior of Ti–48at%Al PST crystals. *Acta Materialia*, **48**, 541–551.

[41] Qu, S., Zhang, P., Wu, S.D. *et al.* (2008) Twin boundaries: strong or weak? *Scripta Materialia*, **2008**, 1131–1134.

[42] Hong, S., Tao, N.R., Huang, X., and Lu, K. (2010) Nucleation and thickening of shear bands in nanoscale twin/matrix lamellae of a Cu–Al alloy processed by dynamic plastic deformation. *Acta Materialia*, **58**, 3103–3116.

[43] Ghoniem, N.M. and Han, X. (2005) Dislocation motion in anisotropic multilayer materials. *Philosophical Magazine*, **85**, 2809–2830.

[44] Dao, M., Lu, L., Shen, Y.F., and Suresh, S. (2006) Strength, strain-rate sensitivity and ductility of copper with nanoscale twins. *Acta Materialia*, **54**, 5421–5432.

[45] Jerusalem, A., Dao, M., Suresh, S., and Radovitzky, R. (2008) Three-dimensional model of strength and ductility of polycrystalline copper containing nanoscale twins. *Acta Materialia*, **56**, 4647–4657.

[46] Shen, F., Zhou, J., Liu, Y. *et al.* (2010) Deformation twinning mechanism and its effects on the mechanical behaviors of ultrafine grained and nanocrystalline copper. *Computational Materials Science*, **49**, 226–235.

[47] Wu, B., Zou, N., Tan, J. *et al.* (2011) Numerical simulation on size effect of copper with nano-scale twins. *The Transactions of Nonferrous Metals Society of China*, **21**, 364–370.

[48] Wolf, D., Yamakov, V., Phillpot, S.R. *et al.* (2005) Deformation of nanocrystalline materials by molecular-dynamics simulation: relationship to experiments? *Acta Materialia*, **53**, 1–40.

[49] Meyers, M.A., Mishra, A., and Benson, D.J. (2006) Mechanical properties of nanocrystalline materials. *Progress in Materials Science*, **51**, 427–556.

[50] Zhu, T. and Li, J. (2010) Ultra-strength materials. *Progress in Materials Science*, **55**, 710–757.

[51] Cao, A.J. and Wei, Y.G. (2007) Molecular dynamics simulation of plastic deformation of nanotwinned copper. *Journal of Applied Physics*, **102**, 083511.

[52] Shabib, I. and Miller, R.E. (2009) A molecular dynamics study of twin width, grain size and temperature effects on the toughness of 2D-columnar nanotwinned copper. *Modelling and Simulation in Materials Science and Engineering*, **17**, 055009.

[53] Shabib, I. and Miller, R.E. (2009) Deformation characteristics and stress–strain response of nanotwinned copper via molecular dynamics simulation. *Acta Materialia*, **57**, 4364–4373.

[54] Qu, S. and Zhou, H. (2011) Atomistic mechanisms of microstructure evolution in nanotwinned polycrystals. *Scripta Materialia*, **65**, 265–268.

[55] Froseth, A.G., Derlet, P.M., and Van Swygenhoven, H. (2004) Grown-in twin boundaries affecting deformation mechanisms in nc-metals. *Applied Physics Letters*, **85**, 5863–5866.

[56] Froseth, A.G., Van Swygenhoven, H., and Derlet, P.M. (2004) The influence of twins on the mechanical properties of nc-Al. *Acta Materialia*, **52**, 2259–2268.

[57] Li, L. and Ghoniem, N.M. (2009) Twin-size effects on the deformation of nanotwinned copper. *Physical Review B*, **79**, 075444.

[58] Li, X., Wei, Y., Lu, L. *et al.* (2010) Dislocation nucleation governed softening and maximum strength in nano-twinned metals. *Nature*, **464**, 877–880.

[59] Stukowski, A., Albe, K., and Farkas, D. (2010) Nanotwinned fcc metals: strengthening versus softening mechanisms. *Physical Review B*, **82**, 224103.

[60] Kulkarni, Y. and Asaro, R.J. (2009) Are some nano-twinned fcc metals optimal for strength, ductility and grain stability? *Acta Materialia*, **57**, 4835–4844.

[61] Kulkarni, Y., Asaro, R.J., and Farkas, D. (2009) Are nanotwinned structures in fcc metals optimal for strength, ductility and grain stability? *Scripta Materialia*, **60**, 532–535.

[62] Yue, L., Zhang, H., and Li, D.Y. (2010) Defect generation in nano-twinned, nano-grained and single crystal Cu systems caused by wear: a molecular dynamics study. *Scripta Materialia*, **63**, 1116–1119.

[63] Qu, S. and Zhou, H. (2010) Hardening by twin boundary during nanoindentation in nanocrystals. *Nanotechnology*, **21**, 335704.

[64] Cao, A.J., Wei, Y.G., and Mao, S. (2008) Strengthening mechanisms and dislocation dynamics in twinned metal nanowires. *JOM*, **60**, 85–88.

[65] Zhang, Y.F. and Huang, H.C. (2009) Do twin boundaries always strengthen metal nanowires? *Nanoscale Research Letters*, **4**, 34–38.

[66] Deng, C. and Sansoz, F. (2009) Enabling ultrahigh plastic flow and work hardening in twinned gold nanowires. *Nano Letters*, **9**, 1517–1522.

[67] Deng, C. and Sansoz, F. (2009) Fundamental differences in the plasticity of periodically twinned nanowires in Au, Ag, Al, Cu, Pb and Ni. *Acta Materialia*, **57**, 6090–6101.

[68] Deng, C. and Sansoz, F. (2009) Near-ideal strength in gold nanowires achieved through microstructural design. *ACS Nano*, **3**, 3001–3008.

[69] Deng, C. and Sansoz, F. (2009) Size-dependent yield stress in twinned gold nanowires mediated by site-specific surface dislocation emission. *Applied Physics Letters*, **95**, 091914.

[70] Deng, C. and Sansoz, F. (2010) Effects of twin and surface facet on strain-rate sensitivity of gold nanowires at different temperatures. *Physical Review B*, **81**, 155430.

[71] Deng, C. and Sansoz, F. (2010) Repulsive force of twin boundary on curved dislocations and its role on the yielding of twinned nanowires. *Scripta Materialia*, **63**, 50–53.

[72] Guo, X. and Xia, Y.Z. (2011) Repulsive force vs. source number: competing mechanisms in the yield of twinned gold nanowires of finite length. *Acta Materialia*, **59**, 2350–2357.

[73] Zhai, Y.T. and Gong, X.G. (2011) Understanding periodically twinned structure in nano-wires. *Physics Letters A*, **375**, 1889–1892.

[74] Afanasyev, K.A. and Sansoz, F. (2007) Strengthening in gold nanopillars with nanoscale twins. *Nano Letters*, **7**, 2056–2062.

[75] Brown, J.A. and Ghoniem, N.M. (2010) Reversible–irreversible plasticity transition in twinned copper nanopillars. *Acta Materialia*, **58**, 866–894.

[76] Zhu, T., Li, J., Samanta, A. *et al.* (2007) Interfacial plasticity governs strain rate sensitivity and ductility in nanostructured metals. *Proceedings of the National Academy of Sciences of the United States of America*, **104**, 3031–3036.

[77] Chen, Z., Jin, Z., and Gao, H. (2007) Repulsive force between screw dislocation and coherent twin boundary in aluminum and copper. *Physical Review B*, **75**, 212104.

[78] Wu, Z.X., Zhang, Y.W., and Srolovitz, D.J. (2009) Dislocation–twin interaction mechanisms for ultrahigh strength and ductility in nanotwinned metals. *Acta Materialia*, **57**, 4508–4518.

[79] Zheng, Y.G., Lu, J., Zhang, H.W., and Chen, Z. (2009) Strengthening and toughening by interface-mediated slip transfer reaction in nanotwinned copper. *Scripta Materialia*, **60**, 508–511.

[80] Hu, Q., Li, L., and Ghoniem, N.M. (2009) Stick-slip dynamics of coherent twin boundaries in copper. *Acta Materialia*, **57**, 4866–4873.

[81] Jin, Z.H., Gumbsch, P., Ma, E. *et al.* (2006) The interaction mechanism of screw dislocations with coherent twin boundaries in different face-centred cubic metals. *Scripta Materialia*, **54**, 1163–1168.

[82] Jin, Z.H., Gumbsch, P., Albe, K. *et al.* (2008) Interactions between non-screw lattice dislocations and coherent twin boundaries in face-centered cubic metals. *Acta Materialia*, **56**, 1126–1135.

[83] Ezaz, T., Sangid, M.D., and Sehitoglu, H. (2011) Energy barriers associated with slip–twin interactions. *Philosophical Magazine*, **1–25**, doi: 10.1080/14786435.2010.541166.

[84] Chassagne, M., Legros, M., and Rodeny, D. (2011) Atomic-scale simulation of screw dislocation/coherent twin boundary interaction in Al, Au, Cu and Ni. *Acta Materialia*, **59**, 1456–1463.

[85] Wang, J., Li, N., Anderoglu, O. *et al.* (2010) Detwinning mechanisms for growth twins in face-centered cubic metals. *Acta Materialia*, **58**, 2262–2270.

[86] Wang, W., Dai, Y., Li, J., and Liu, B. (2011) An atomic-level mechanism of annealing twinning in copper observed by molecular dynamics simulation. *Crystal Growth and Design*, **11**, 2928–2934.

[87] Barg, A.I., Rabkin, E., and Gust, W. (1995) Faceting transformation and energy of a Σ3 grain boundary in silver. *Acta Metallurgica et Materialia*, **43**, 4067–4074.

[88] Christian, J.W. and Mahajan, S. (1995) Deformation twinning. *Progress in Materials Science*, **39**, 1–157.

[89] Hofmeister, H. (1998) Forty years study of fivefold twinned structures in small particles and twin films. *Crystal Research and Technology*, **33**, 3–25.

[90] Huang, J.Y., Wu, Y.K., and Ye, H.Q. (1996) Allotropic transformation of cobalt induced by ball milling. *Acta Materialia*, **44**, 1201–1209.

[91] Liao, X.Z., Zhao, Y.H., Srinivasan, S.G. *et al.* (2004) Deformation twinning in nanocrystalline copper at room temperature and low strain rate. *Applied Physics Letters*, **84**, 592–594.

[92] Liao, X.Z., Zhao, Y.H., Zhu, Y.T. *et al.* (2004) Grain-size effect on the deformation mechanisms of nanostructured copper processed by high-pressure torsion. *Journal of Applied Physics*, **96**, 636–640.

[93] Zhu, Y.T., Liao, X.Z., and Valiev, R.Z. (2005) Formation mechanism of fivefold deformation twins in nanocrystalline face-centered-cubic metals. *Applied Physics Letters*, **86**, 103112.

[94] Huang, P., Dai, G.Q., Wang, D. *et al.* (2009) Fivefold annealing twin in nanocrystalline Cu. *Applied Physics Letters*, **95**, 203101.

[95] Cao, A.J. and Wei, Y.G. (2006) Formation of fivefold deformation twins in nanocrystalline face-centered-cubic copper based on molecular dynamics simulations. *Applied Physics Letters*, **89**, 041919.

[96] Bringa, E.M., Farkas, D., Caro, A. *et al.* (2008) Fivefold twin formation during annealing of nanocrystalline Cu. *Scripta Materialia*, **59**, 1267–1270.

[97] Koch, C.C. (2003) Optimization of strength and ductility in nanocrysatlline and ultrafine grained metals. *Scripta Materialia*, **49**, 657–662.

[98] Kumar, S., Haque, M.A., and Gao, H. (2009) Notch insensitive fracture in nanoscale thin films. *Applied Physics Letters*, **94**, 253104.

[99] Kumar, S., Li, X., Haque, M.A., and Gao, H. (2011) Is stress concentration relevant for nanocrystalline metals? *Nano Letters*, **11**, 2510–2516.

[100] Lu, L., Schwaiger, R., Shan, Z.W. *et al.* (2005) Nano-sized twins induce high rate sensitivity of flow stress in nano-sized twins induce high rate sensitivity of flow stress in pure copper. *Acta Materialia*, **53**, 2169–2179.

[101] Lu, L., Dao, M., Zhu, T., and Li, J. (2009) Size dependence of rate-controlling deformation mechanisms in nanotwinned copper. *Scripta Materialia*, **60**, 1062–1066.

[102] Qin, E.W., Lu, L., Tao, N.R., and Lu, K. (2009) Enhanced fracture toughness of bulk nanocrystalline Cu with embedded nanoscale twins. *Scripta Materialia*, **60**, 539–542.

[103] Qin, E.W., Lu, L., Tao, N.R. *et al.* (2009) Enhanced fracture toughness and strength in bulk nanocrystalline Cu with nanoscale twin bundles. *Acta Materialia*, **57**, 6215–6225.

[104] Singh, A., Tang, L., Dao, M. *et al.* (2011) Fracture toughness and fatigue crack growth characteristics of nanotwinned copper. *Acta Materialia*, **59**, 2437–2446.

[105] Shute, C.J., Myer, B.D., Xie, S. *et al.* (2011) Detwinning, damage and crack initiation during cyclic loading of Cu samples containing aligned nanotwins. *Acta Materialia*, **59**, 4569–4577.

[106] Mayadas, A.F. and Shatzkes, M. (1970) Electrical-resistivity model for polycrystalline films: the case of arbitrary reflection at external surfaces. *Physical Review B*, **1**, 1382–1389.

[107] Chen, K.C., Wu, W.W., Liao, C.N. *et al.* (2008) Observation of atomic diffusion at twin-modified grain boundaries in copper. *Science*, **321**, 1066–1069.

[108] Chen, K.C., Wu, W.W., Liao, C.N. *et al.* (2010) Stability of nanoscale twins in copper under electric current stressing. *Journal of Applied Physics*, **108**, 066103.

[109] Sleeswyk, A.W. and Verbraak, C.A. (1961) Incorporation of slip dislocations in mechanical twins – I. *Acta Metallurgica*, **9**, 917–927.

[110] Mahajan, S. and Chin, G.Y. (1973) Twin–slip, twin–twin and slip–twin interactions in Co–8wt.% Fe alloy single crystals. *Acta Metallurgica et Materialia*, **21**, 173–179.

[111] Hartley, C.S. and Blachon, DLA. (1978) Reactions of slip dislocations at coherent twin boundaries in face-centered-cubic metals. *Journal of Applied Physics*, **49**, 4788–4796.

[112] Remy, L. (1981) The interaction between slip and twinning systems and the influence of twinning on the mechanical behavior of fcc metals and alloys. *Metallurgical and Materials Transactions A*, **12**, 387–408.

[113] Zhu, Y.T., Li, X.Z., and Wu, X.L. (2012) Deformation twinning in nanocrystalline materials. *Progress in Materials Science*, **57**, 1–62.

[114] Wang, Y.B. and Sui, M.L. (2009) Atomic-scale *in situ* observation of lattice dislocations passing through twin boundaries. *Applied Physics Letters*, **94**, 021909.

[115] Zhu, Y.T., Wu, X.L., Liao, X.Z. *et al.* (2011) Dislocation–twin interactions in nanocrystalline fcc metals. *Acta Materialia*, **59**, 812–821.

[116] Wang, Y.B., Wu, B., and Sui, M.L. (2008) Dynamical dislocation emission processes from twin boundaries. *Applied Physics Letters*, **93**, 041906.

[117] Konopka, K., Mizera, J., and Wyrzykowski, J.W. (2000) The generation of dislocations from twin boundaries and its effect upon the flow stresses in FCC metals. *Journal of Materials Processing Technology*, **99**, 255–259.

[118] Sennour, M., Korinek, S.L., Champion, Y., and Hÿtch, M.J. (2007) HRTEM study of defects in twin boundaries of ultra-fine grained copper. *Philosophical Magazine*, **87**, 1465–1486.

[119] Schiøtz, J. and Jacobsen, K.W. (2003) A maximum in the strength of nanocrystalline copper. *Science*, **301**, 1357–1359.

[120] Xu, L., Xu, D., Tu, K.N. *et al.* (2008) Structure and migration of (112) step on (111) twin boundaries in nanocrystalline copper. *Journal of Applied Physics*, **104**, 113717.

[121] Wei, Y. (2011) The kinetics and energetics of dislocation mediated de-twinning in nano-twinned face-centered cubic metals. *Materials Science and Engineering A*, **528**, 1558–1566.

[122] Wei, Y. (2011) Scaling of maximum strength with grain size in nanotwinned fcc metals. *Physical Review B*, **83**, 132104.

[123] Zhu, L., Ruan, H., Dao, M. *et al.* (2011) Modeling grain size dependent optimal twin spacing for achieving ultimate high strength and related high ductility in nanotwinned metals. *Acta Materialia*, **59**, 5544–5557.

[124] Wu, B., Heidelberg, A., Boland, J.J. *et al.* (2006) Microstructure-hardened silver nanowires. *Nano Letters*, **6**, 468–472.

[125] Zhu, Y., Qin, Q., Xu, F. *et al.* (2012) Size effects on elasticity, yielding, and fracture of silver nanowires: *in situ* experiments. *Physical Review B*, **85**, 045443.

[126] Filleter, T., Ryu, S., Kang, K. *et al.* (2012) Nucleation-controlled distributed plasticity in penta-twinned silver nanowires. *Small*, in press.

[127] Shan, Z.W., Lu, L., Minor, A.M. *et al.* (2008) The effect of twin plane spacing on the deformation of copper containing a high density of growth twins. *JOM*, **60**, 71–74.

[128] Kim, S., Li, X., Gao, H., and Kumar, S. (2012) *In-situ* observations of crack arrest and bridging by nanoscale twins in copper thin films. *Acta Materialia*, **60**, 2959–2972.

[129] Zhou, H., Qu, S., and Yang, W. (2010) Toughening by nano-scaled twin boundaries in nanocrystals. *Modelling and Simulation in Materials Science and Engineering*, **18**, 065002.

[130] Zhou, H. and Qu, S. (2010) The effect of nanoscale twin boundaries on fracture toughness in nanocrystalline Ni. *Nanotechnology*, **21**, 035706.

7

Size-Dependent Strength in Single-Crystalline Metallic Nanostructures

Julia R. Greer

California Institute of Technology, USA

7.1 Introduction

The emergence of a substantial body of literature focusing on uniaxial compression experiments of micro- and nano-sized single-crystalline cylindrical papers has unambiguously demonstrated that, at these scales, the sample dimensions dramatically affect crystalline strength (for reviews, see [1–3]). In most of these experimental studies, cylindrical pillars with diameters ranging from ∼100 nm up to several micrometers were fabricated, largely by the use of the focused ion beam (FIB) method, with some non-FIB-based methods as well, and were subsequently compressed in a nanoindenter with a custom-made flat punch indenter tip. More recently, small-scale mechanical behavior has also been explored through uniaxial tensile experiments, usually performed inside of *in-situ* scanning electron microscopes (SEM)- or transmission electron microscopes (TEM)-with the custom-built mechanical deformation instruments by a small number of research groups [3–6]. Intriguingly, the results of all of these reports for single-crystalline metals with a variety of crystal structures – face-centered cubic (fcc), body-centered cubic (bcc), hexagonal close-packed (hcp), and tetragonal – show power-law dependence between flow stress and pillar diameter. Further, within the fcc family, the slopes of all metals tested converge on a unique value of approximately −0.6 [1–3,7] (see Figure 7.1), which is not the case for all other crystals.

This chapter is organized as follows: first, we present an overview of the state of the art on small-scale single-crystalline plasticity where specimen size is reduced to micro- or nano-scale in all but one dimension (i.e., nano-pillars, nano-wires, uniaxial tensile specimens, etc.). This is followed by the descriptions of sample fabrication techniques

Nano and Cell Mechanics: Fundamentals and Frontiers, First Edition. Edited by Horacio D. Espinosa and Gang Bao.
© 2013 John Wiley & Sons, Ltd. Published 2013 by John Wiley & Sons, Ltd.

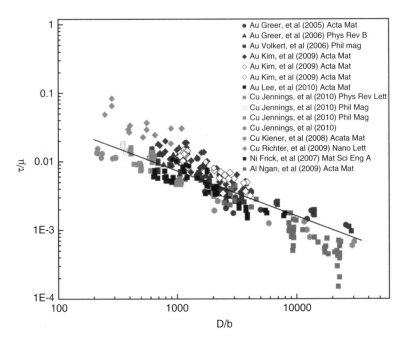

Figure 7.1 Shear strength normalized by shear modulus as a function of pillar diameter normalized by Burgers vector for a wide range of non-pristine fcc metals. Reprinted with permission from [1]. © 2011 Elsevier. Refer to Plate 5 for the colored version

and uniaxial deformation methodologies, outlining both the fabrication details and their effect on the resulting microstructure of the specimen, as well as the specific instrumental systems used in testing. We close with a discussion and outlook on the current state of the nanomechanical field in the context of current literature for three classes of single crystals: (1) face-centered and body-centered cubic structures, (2) hexagonal structures, and (3) tetragonal structures.

7.2 Background

7.2.1 Experimental Foundation

Uniaxial compression methodology to study small-scale mechanical deformation was first introduced by Uchic *et al.*, who reported higher compressive strengths attained by FIB-fabricated cylindrical Ni micro-pillars [5]. Greer and Nix extended this technique into the sub-micrometer regime, where single-crystalline Au nano-pillars with diameters below 1 μm sustained stresses of nearly 50 times higher than bulk [8]. Adapting this elegant testing methodology, several research groups explored the small-scale post-elastic compressive response of various materials, with the largest fraction focusing on fcc micro- and nano-pillars. In addition to these compression studies, understanding of plasticity in small structures has now been further enriched through uniaxial *tensile* experiments, usually performed inside scanning or transmission electron microscopes. These types of tensile experiments generally require a custom-built *in-situ* mechanical deformation instrument, as has been done by few research groups [3–6]. A comprehensive overview of quantitative *in-situ* micro- and nanomechanical testing is provided in Ref. [4]. To date, uniaxial compression and tension tests have been performed on a variety of cubic and hexagonal close-packed (hcp) single crystals, shape memory alloys, and tetragonal metals, as well as on boundary-containing pillars spanning from bi-crystalline and nanocrystalline to nanolaminates and metallic glass systems (this list is not meant to be exhaustive but rather representative of some key material systems). The pertinent details and key findings of these studies, as well as a critical discussion, are provided in the most recent review by Greer and De Hosson [1] and references therein, as well as in two previous reviews by Kraft *et al*. [2] and Uchic *et al*. [3]. Experimental reports on small-scale mechanical deformation of the following single and bi-crystalline metallic systems have been published to date:

- *fcc* – Ni and Ni-based superalloys [3,5–7], Au [8–11], Cu [12–20], Ag [21] and Al (as-fabricated [22,23] and intentionally passivated [24]);
- *bcc* – W, Nb, Ta, Mo [25–35] and V [36];
- *hcp* – Mg [37–39] and Ti [40,41];
- *tetragonal* – low-temperature metals, In [42,43] and Sn [44].

Most of these experimental reports revealed a remarkable dependence of the attained strength on sample diameter. While no definitive theory accounting for size-dependent plasticity exists, this field is now coming to maturity, in that many aspects of the size effect have been explained and are elaborated upon later in the chapter. In contrast, metallic systems where the intrinsic microstructural scale dominates mechanical behavior (e.g., nanocrystalline or nano-twinned metals, as well as nanolaminates), sample dimensions appear to play a less significant role. These concepts are illustrated in Figure 7.2, where flow stress normalized by shear modulus is plotted as a function of both the external diameter D and the critical microstructural length scale d, corresponding to grain size here, but, in general, can be any other microstructural intrinsic dimension, like interlayer nanolaminates spacing, precipitate size, and twin boundary spacing. While these combined effects of limiting internal and external sizes represent a fascinating discussion topic and are being actively pursued by the nanomechanics research community today, they are outside the scope of this chapter, which is concerned with the deformation taking place in micro- and nano-sized vertically oriented *single-crystalline* pillars spanning a variety of lattice structures. Intriguingly, with the exception of tetragonal metals like In and Sn, compressive strengths of all single-crystalline metals studied with diameters ranging between ~75 nm and several micrometers and containing initial dislocations (i.e., not pristine whiskers or nano-wires) increase significantly with reduced size; that is, samples with diameters ~150 nm have been shown to attain flow stresses more than 50 times greater than bulk. Figure 7.3 shows some representative stress–strain curves with a characteristic stochastic signature and post-mortem images revealing multiple shear slip lines on compressed single-crystalline Nb nano-pillars [35].

A notable distinction in the deformation of these small-scale material samples compared with their bulk counterparts is that, unlike Taylor hardening (whereby the flow strength scales linearly with the square root of the evolving density of mobile dislocations [45]), the rate of dislocation annihilation in these small crystals is often higher than

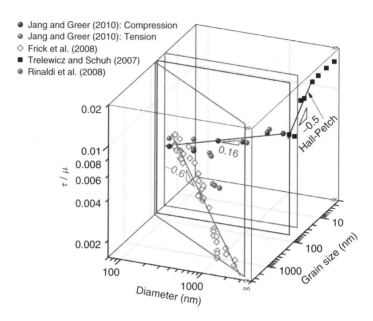

Figure 7.2   Strength versus internal (grain size) and external (pillar diameter) for nanocrystalline Ni. Reprinted with permission from [1]. © 2011 Elsevier

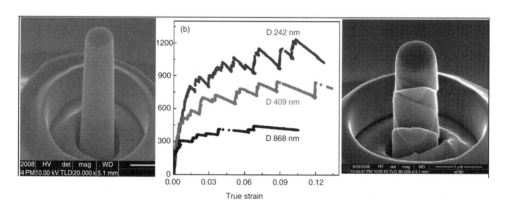

Figure 7.3   SEM images of a typical Nb nano-pillar before and after uniaxial compression. Typical stress–strain curves in the middle clearly exhibit the size effect (i.e., smaller is stronger) and stochastic, strain burst-ridden behaviour. Reprinted with permission from [1]. © 2011 Elsevier

that of dislocation storage through multiplicative processes, resulting in the overall net zero or even net loss of dislocations. The absence of dislocation multiplication via double cross-slip processes (i.e., via Frank–Read sources) and subsequent forest dislocation substructure formation, as is the case during bulk crystal deformation, is unique to the nanometer length scale and likely means that, in nano crystals, plasticity is governed by dislocation nucleation from alternative sources. This concept can be understood as follows. In bulk crystals, mobile glissile dislocations are in abundance and, hence, are always available to carry plastic strain, as they multiply in the course of deformation. In nano crystals, however, the supply of mobile dislocations is limited; in order to accommodate plasticity from the imposed load, therefore, the plastic strain is carried by the newly emitted dislocations from sources either in the bulk of the pillar or on the sample surface. The nature and operating mechanisms of the dislocation sources in nano- and micro-pillars is a topic of an intense ongoing debate, as both the so-called "truncated" sources (also known as single-arm sources) and the surface sources have been theorized, with the latter modeled analytically [46] and observed via molecular dynamics [47–51], and the former simulated through dislocation dynamics [3,52–60]. Recently, experimental evidence for the presence of both types of sources has also been reported [15,51,61].

7.2.2 Models

Several models attempting to explain the causality between either the declining or the marginally increasing dislocation density, dislocation source operation, and attained strengths have been proposed. For example, the dislocation starvation model hypothesizes that the mobile dislocations inside a small, deeply in the sub-micron range nano-pillar have a greater probability of annihilating at a free surface than of interacting with one another, thereby shifting plasticity into the nucleation-controlled regime [10,49,58,62–64]. These phenomena of dislocation annihilation and subsequent nucleation from the surface upon uniaxial loading have been predicted by analytical models [15,46,47,50,62] and molecular dynamics simulations [30,49,50,65,66] and have now been demonstrated by some *in-situ* TEM experiments [51,67] and strain-rate-controlled experiments [15]. *In-situ* TEM experiments on the compression of single-crystalline Ni, Cu, and Mo nano-pillars with diameters of 100–200 nm showed their initial dense dislocation population, consisting mainly of the FIB-induced small surface dislocation loops, which vanished immediately upon compressive loading – a process Minor and coworkers [17,67,68] coined as "mechanical annealing." In tension, 10 nm diameter Au nanowires were strained inside a high-resolution transmission electron microscope (HRTEM) and revealed the nucleation and emission of partial dislocations from the nanowire surface followed by their immediate annihilation at the opposite side [51].

Much more complex models, usually based on dislocation dynamics simulations, have been developed and many capture (at least qualitatively) the stochastic intermittent strain bursts, the strength increase, and the dislocation density evolution consistent with experimental results. Some of the more prominent of these models include "source exhaustion hardening" [56,60,69–71], "source truncation" [59,71,72], and "weakest link theory" [71,73]. The common underlying premise in these models involves representing dislocation emission from a particular single source or a randomly distributed array of sources, which operate in a discrete fashion, and then evaluating the effect of sample size on these sources' lengths and, therefore, on their operation strengths. These computational models capture the ubiquitously observed stochastic signature of the experimentally obtained compression results and show dislocation storage for micrometer-sized fcc samples [52,55,57,74] and virtually no dislocation storage for the smaller samples, consistent with the notion of dislocation starvation [51].

A thorough review and a critical discussion on the compression results and corresponding dislocation density evolution in single crystalline fcc micro-pillars was published by Uchic *et al.* [3] in 2009. Beyond a detailed assessment of the existing experimental and computational studies on micro- and nano-pillars with diameters from a couple of hundred nanometers up to tens of micrometers, with the aspect ratios (length/diameter) in the range between 2:1 and 5:1, the review in [3] includes details on some of the compression sample preparation methodologies via FIB, the various aspects of mechanical testing procedures, sample geometry, and the effects of ion-beam-based preparation on the stress–strain response. It clearly endorses the micro-compression experiments as a useful technique for gaining a fundamental understanding of dislocation-based deformation processes in small-scale crystals and calls for utilizing care when interpreting the results. Further, it deems the size effect as real, but that it cannot be explained through the known thin-film strengthening mechanisms like grain-size hardening [75], the confinement of dislocations within a thin film by the substrate [76], or to the presence of strong strain gradients [77,78]. A more recent review by Greer and De Hosson summarizes the compressive and tensile strength data published to date for a variety of single-crystalline metals (fcc, bcc, and hcp), as well as for shape memory alloys, nanolaminate-containing, bi-crystalline, and nanocrystalline pillars [1]. It appears that all single crystals exhibit size-dependent strengths in a power-law fashion:

$$\frac{\tau}{\mu} = A \left(\frac{D}{b} \right)^n$$

where μ is the shear modulus, τ is the resolved shear stress onto the appropriate slip system, D is the pillar diameter, and b is the magnitude of the Burgers vector. The exponent n of the power-law strengthening is not universal, but rather is a strong function of the material class; for example, nearly all fcc metals appear to have this exponent in a relatively tight range of -0.5 to -0.7; those for bcc metals were from -0.21 up to -0.98 and were found to correlate with the residual Peierls barrier, which in turn is a function of critical temperature; and hcp metals exhibited two distinct power-law strengthening exponents based on basal and prismatic slip versus twinning. The review by Kraft *et al.* [2] provides further context for the origins and possible mechanisms responsible for the size effect by also incorporating the results of thin-film experiments and simulations. These authors condensed the thin-film and pillar data for Cu and subdivided its strength versus size domain into three regions representative of the governing confined-dimensions plasticity mechanism: (i) dislocation multiplication in samples with critical dimensions above 1 μm; (ii) full dislocation nucleation in samples between 100 nm up to 1 μm; and (iii) partial dislocation nucleation in Cu samples below 100 nm; see Figure 7.4. Some of the more compelling summary and outlook aspects brought up by these reviews are:

1. The effect of size on strength of small-scale single crystals is a strong function of the sample's initial internal structure, lattice resistance to deformation, loading conditions, sample purity, surface roughness, and likely several others.

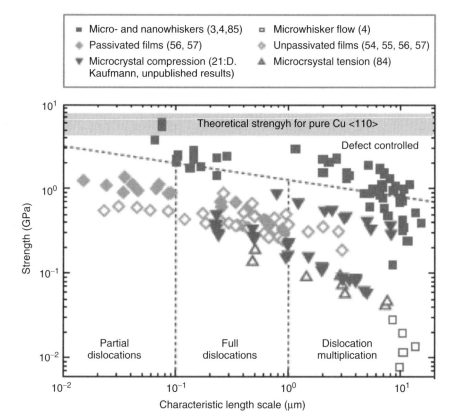

Figure 7.4 Compilation of strength data for Cu from experiments on thin films and small samples. Three regimes with different deformation mechanisms are identified: at smallest sizes, the governing mechanism is via nucleation of partial dislocations; at intermediate sizes, it is via nucleation of full dislocations and their subsequent mutual interactions; and at the largest sizes, it is via dislocation multiplication. Reprinted with permission from [2]. © 2010 Annual Reviews

2. While the size-dependent strength in, at least, all non-pristine fcc micro- and nano-sized metals tested to date is well captured by a unique power-law form, the mechanisms governing the deformation giving rise to the observed strengths are varied.
3. The length scale of the transition from homogenized defect-driven bulk behavior, classically described by the mean-field framework, into size-dependent deformation, where activity by a small numbers of defects is discernible, is unique for each material and is also a strong function not only of size, but also of the factors listed in (1).

All existing reviews enlist the merits of uniaxial micro- and nano-sized testing to elicit a number of fundamental scientific aspects of small-scale mechanical deformation, which, in turn, will bring the materials scientists closer towards solving the grand challenge of linking the atomic-level microstructural detail to the macro-scale mechanical response.

7.3 Sample Fabrication

While patterning via FIB is the most commonly utilized technique for creating micro-compression (and tension) samples, several non-FIB processes have been developed by a few research groups to avoid the adverse effects of Ga^+ ion bombardment like surface amorphization, injection of small dislocation loops, and ion implantation, as well as to broaden the accessible sample size range and initial dislocation content [79–82]. This section describes the sample fabrication approaches published to date and conveys their comparative attributes.

7.3.1 FIB Approach

Generally, most FIB instruments are equipped with a focused Ga^+ ion beam and a high-resolution field-emission scanning electron column. The FIB can locally etch the sample surface with sub-micrometer precision. Variables like the beam current, sample density, and sample atomic mass affect the sputtering rate of the sample, so that the FIB approach has to be custom-optimized for each application. For example, at a current of 20 nA (usually the highest current available in the FIB), the etch rate of single-crystal Au is approximately five times higher than that for Si. There are two main methodologies to create the free-standing vertically aligned micro-compression pillar specimens: the lathe method, which generally works for the micrometer-sized samples, and the top-down fabrication method, which has been demonstrated to be capable of creating nano-pillars down to $\sim$100 nm diameters. FIB-produced columns for compression testing generally have aspect ratios (height/diameter) between $\sim$2.5 : 1 to $\sim$6 : 1.

The *top-down approach* is a subtractive technique, whereby concentric ring geometries are iteratively applied to the sample in a top-down direction to etch away the material within the ring. Subsequent mechanical testing usually requires the pillar to be located in a recessed area to allow for sufficient clearance of the indenter tip. It is also helpful for the pillar of interest to be raised slightly above the "crater floor," on a pedestal, to avoid the possibility of the indenter coming in contact with other features if the sample is slightly misaligned with respect to the loading axis. Once inside the FIB vacuum chamber, the first step of the fabrication process involves locating the appropriate section on the sample for machining the specimen. After adjusting the imaging parameters under the electron beam, the sample and the stage are adjusted to the eucentric position (i.e., the position where both the ion beam and the electron beam are focused at the same working distance, usually close to 5 mm), while the sample is tilted at 52° with respect to the electron detector. At the eucentric position, the ion beam is coincident with the top view of the etch pattern, which makes the geometrical shape control more precise for the horizontal features than for vertical ones. Following the fabrication of the initial "crater" with a relatively large pedestal in the center, incrementally finer beam currents are then used to refine the structure by sequential etching of concentric rings, with smaller and smaller inner diameters, until the desired pillar size is reached. The intermediate currents for each concentric ring may vary from pillar to pillar, as the current choice depends strongly on the trade-off between the image quality, the etch rates, and the shape integrity. Several representative nano-pillars of different materials produced via this top-down technique are shown in Figure 7.5.

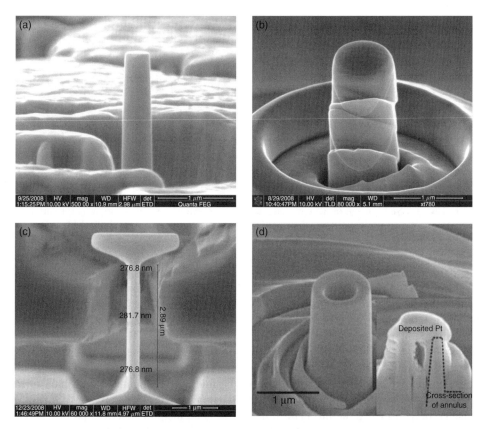

Figure 7.5   SEM images of FIB-fabricated samples: (a) 400 nm nanocrystalline Ni–4%W nano-pillar; (b) compressed 600 nm Nb pillar with significant slip offsets. Reproduced with permission from [35]; (c) typical dog-bone-shaped tensile Au sample. Reproduced with permission from [33]; (d) Mo anti-pillar with a hollow center. Reproduced with permission from [83].

Another frequently used methodology for FIB-based sample preparation is called *lathe milling*, which has generally been used to produce relatively large pillars, with diameters of 1 μm and greater. In this technique, the ion beam is positioned at an oblique angle to the bulk sample surface, while the specimen is continuously rotated by 360°, sequentially shaving off all sharp features on the pillar surface, and resulting in a very smooth final shape. An example of such a micro-pillar is shown in Figure 7.6. Samples fabricated by the lathe method generally require greater time and effort to produce, and have two distinct advantages: taper-free geometry and unambiguous pillar height determination (i.e., no crater around the pillar, as is the case with the top-down method). Lathe milling can also be advantageous when preparing multi-phase samples as it is less sensitive to the differences in the etch rate between dissimilar materials. It is, however, less effective when preparing samples with diameters smaller than 1 μm. Lathe method also requires having a stage rotation capability in the direction orthogonal to the sample stage.

Both of these FIB-based techniques have some key advantages over other fabrication methods; for example, precise control of the sample geometry, accurate marker or

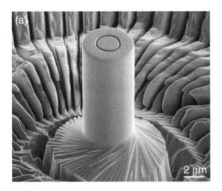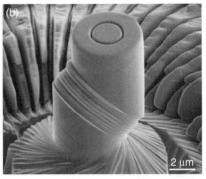

Figure 7.6   (a) SEM image of a 5 μm diameter microcrystal sample of Ni oriented for single slip. (b) SEM image of (a) after testing. Reprinted with permission from [3]. © 2009 Annual Reviews

boundary placement (if samples contain secondary phases), and material non-specificity (i.e. samples can be made from any electronically conductive material). They also have some disadvantages. Ion bombardment of the material has been shown to lead to multiple surface-damage processes, like injection of surface dislocation loops, creation of an amorphized outer layer, and ion implantation, as revealed by TEM [84], X-ray diffraction [18,85,86], and atom probe tomography [87]. In addition to these inevitable undesirable aspects associated with any ion beam machining, samples prepared via the top-down method often exhibit vertical tapering in the sub-micrometer regime, base trench formation, non-planarity of the pillar top, and surface damage due to Ga$^+$ ion bombardment, all of which may influence the mechanical data. The lathe method generally produces cylinders that are taper free and have well-defined heights, but is ill-suited for the fabrication of very small cylinders, in the sub-micrometer range. It also has slower throughput times compared with the top-down method and requires having a special rotating stage setup in the DualBeam instrument.

7.3.2 Directional Solidification and Etching

Perhaps the first non-FIB micrometer-sized mechanical sample preparation methodology was published by Bei *et al*. [34] who produced Mo-alloy square-cross-section micro-pillars by etching away the NiAl matrix from an NiAl–Mo composite, grown by directional solidification.[1] This composite initially consisted of a 55 at.% Ni–45 at.% Al matrix with embedded single-crystalline solid solution 86 at% Mo–10 at% Al–4 at.% Ni fibers oriented along the ⟨100⟩ crystallographic direction. The NiAl matrix was then chemically removed to reveal arrays of free-standing Mo-alloy fibers with side dimensions between ~360 nm and ~2.5 μm [91]. Unlike the FIB-produced samples, these columns had nearly

---

[1] The pristine whiskers in the classical studies of Brenner [88,89] and Fisher and Hollomon [90] would also fit this criterion, but their diameters were in the millimeter and hundreds of micrometers range.

square cross-sections, had relatively rough side and top surfaces, and (most importantly) contained no initial dislocations.

7.3.3 Templated Electroplating

To avoid the aforementioned adverse effects of the FIB and to enable the fabrication of nano-pillars with significantly smaller diameters and no vertical taper, Burek and Greer [92] developed a template-based method to manufacture many thousands of free-standing vertically oriented Cu and Au nano-pillars on a rigid substrate. This non-FIB method involves first spin-coating a thin layer of polymethylmethacrylate (PMMA) on Si wafer with a seed metal film, followed by exposing this resist layer according to a specifically designed pattern (in this case, an array of circles) by electron beam lithography. The pattern of through-holes to the underlying substrate is subsequently developed, thereby opening vertical cylindrical holes in the PMMA to the underlying metal seed layer, and the metal of choice is electroplated into the template. Full details on this type of fabrication process to create a variety of different metal samples can be found in Burek and Greer [92]. This fabrication process is capable of producing a wide variety of metallic microstructures: from single crystals to nano-twinned, bi-, poly-, and nano-crystalline mechanical testing specimens with diameters as large as $\sim 1\,\mu$m down to only 25 nm. An important distinction of this fabrication method from many other non-FIB methodologies is that the single-crystalline nano-pillars are not pristine but rather contain few in-grown dislocations, with dislocation densities on the order of 1×10^{14} m^{-2}, similar to those of the FIB-machined samples. SEM images of a representative single-crystalline Cu nano-pillar produced by electroplating and a TEM image revealing this seemingly high dislocation density are shown in Figure 7.7.

7.3.4 Nanoimprinting

Buzzi and Dietiker recently reported an embossing method to create Ag [21] and Au [9] nano- and micro-pillars by pushing a patterned Si template into a square platelet of

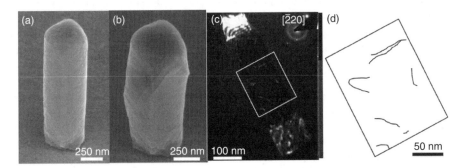

Figure 7.7 SEM images (a, b) of electroplated single-crystalline Cu nano-pillars before and after compression as evidenced by slip lines on the surface. (c, d) Weak-beam dark-field TEM image of a typical Cu nano-pillar showing initial dislocation density Reproduced with permission from [13]

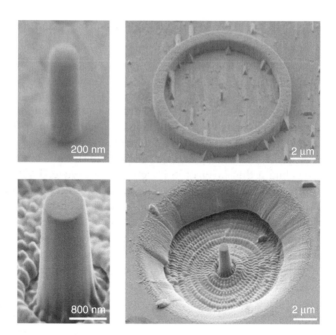

Figure 7.8   SEM images of nanoimprinted Au pillars: 210 nm diameter and 520 nm height in top two images and 1100 nm diameter and 2100 nm height in bottom two images. Bottom sample was produced by FIB milling using a two-step procedure. Reprinted with permission from [9]. © 2011 Elsevier

the desired material at elevated temperatures. The metal of interest was then formed into pillar shapes with diameters ranging from ~150 nm up to several micrometers. Pillars produced by this method were of different crystallographic orientations and not necessarily single crystalline, although the smallest samples with diameters less than 200 nm for Au and 500 nm for Ag were, in fact, single crystalline. Larger pillars were a mix of single crystalline and polycrystalline specimens. Dietiker *et al.* [9] showed no substantial difference between the Au nano-pillars produced by this embossing methodology and those machined by FIB. A representative set of pillars is shown in Figure 7.8. Compressions were performed in a "conventional" way; that is, by locating the pillar tops in the optical microscope on the nanoindenter and then performing the top-down compression in the nanoindenter.

7.3.5   Vapor–Liquid–Solid Growth

Richter *et al.* reported the fabrication of large arrays of pristine [111]-oriented Cu nanowires with diameters between 75 and 400 nm [16]. In contrast to pillars produced through all the aforementioned methods, the cross-sectional areas of these nanowires have equilibrium Wulff shapes, with atomically smooth side surfaces and virtually nonexistent initial dislocations. These samples, therefore, are quite similar to the millimeter-sized wires whose tensile deformation results were originally reported by Brenner [88,89]. A typical nanowire grown by these researchers is shown in Figure 7.9. Tensile testing was performed inside a DualBeam FIB by attaching the Omniprobe manipulator to the nanowire tips and pulling on them.

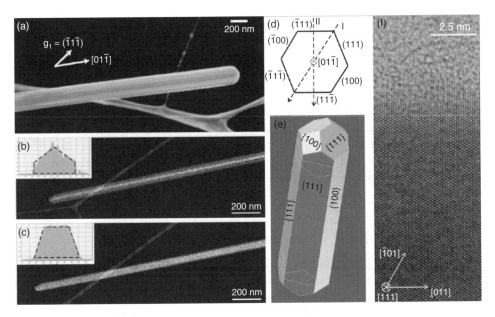

Figure 7.9 Morphology of Cu whisker. TEM images reveal pristine microstructure, growth direction of [011], and Wulff equilibrium side facet shapes. Reprinted with permission from [15]. © 2011 Elsevier

7.3.6 Nanowire Growth

Huang and coworkers prepared Au nanowires (NWs) by reducing AuCl(oleylamine) complex with 10 nm Ag nanoparticles (NPs) in hexane [51,93]. The Au nanowires were formed by mixing 10 mg AuCl and 0.28 ml oleylamine in 2.15 ml of hexane, heating to 60 °C, and then reacted with 0.05 ml of 10 nm Ag nanoparticles at 60 °C for 45 h. These nanowires had atomically smooth faceted surfaces and were generally oriented in $\langle 001 \rangle$ or $\langle 110 \rangle$ directions. The lateral surfaces in these <10 nm diameter nanowires contained multiple surface steps of two types enclosed by {111} planes: a single atomic-height step and a double atomic-height step.

7.4 Uniaxial Deformation Experiments

Most uniaxial deformation experiments on micro- and nano-pillar samples have been conducted in commercial nanoindenters with either custom-made or pre-ordered flat-punch diamond-indenter tips. An example of a custom-made flat-punch tip would be a commercially bought Berkovich or any other geometry tip, flattened at the apex either by etching it in the FIB or by mechanical polishing. The cross-sectional area of a tip made in such way can vary anywhere between several micrometers up to several tens of micrometers, depending on the sample sizes to be tested. Several groups, however, extended this type of nanomechanical testing to include the *in-situ* observing capability (i.e., in an electron microscope or in an X-ray setup), as well as expanding the loading type (i.e., tension and bending). This section provides an overview of the particular testing instrumental systems and procedures utilized in nanomechanical pillar testing.

7.4.1 Nanoindenter-Based Systems (Ex Situ)

A typical uniaxial compression procedure is performed by first locating the pillar top in an optical microscope and then positioning the flat-punch indenter tip above the sample. The compression is performed under either a constant loading rate or under a constant displacement rate. Generally, most nanoindenters are intrinsically load-controlled instruments and, therefore, require writing a special feedback loop software method in order to attain nominal displacement-rate control. The term "nominal" here is intended to signify that the stress–strain signature of the micro- and nano-scale-sized metals is highly stochastic, with intermittent strain bursts under both types of loading conditions, during which the strain rate can be one or two orders of magnitude higher than the prescribed deformation rate.

7.4.2 In-Situ Systems

In-situ systems have generally been built to monitor the evolution of sample morphology and/or microstructure – often concurrently with the mechanical data collection in the course of the deformation. Although several *in-situ* systems for assessing both global and local material mechanical behavior exist (like scanning probe microscopy (SPM) based, diffraction and spectroscopy based, microelectromechanical system (MEMS)-driven, etc.), in this chapter we focus on electron-microscope-based instrumentation, which has been predominantly utilized in studying uniaxially deforming micro- and nano-pillars. SEM- and TEM-based systems have long been utilized as effective instruments to glean evolving morphologies and deformation mechanisms in metallic systems (e.g. see Gane and Bowden [94]). Examples of such deformation characteristics are crystallographic slip, shear band formation, and crack propagation. Legros *et al.* [4] provide an overview of the current state of the art on *in-situ* tensile deformation inside electron microscopes, as well as other *in-situ* nanomechanical techniques.

SEM- and FIB-Based Systems

Although conducting nanomechanical testing using either SEM or FIB systems dates back to around the 1960s [94], it has recently garnered newly engaged scientific research interest. *In-situ* SEM approaches are versatile in the sense that they allow for testing a broad range of sample geometries and sizes (from nanometers up to several hundred micrometers), they do not require electron transparency, and they allow for multiple deformation modes mainly due to the relatively large chamber sizes. These types of instruments allow for conducting both uniaxial compression and tensile straining, as well as bending experiments – all while observing the deformation *in situ*. To date, most of these types of instruments are custom built, with the electron microscope component purchased as a stand-alone tool from the manufacturer and the nanomechanical module either constructed by a research group or through combined research group–vendor efforts. Currently, there are a few SEM-based *in-situ* mechanical deformation instruments being utilized for micro- and nano-tension and compression mechanical testing, and several new ones are being constructed. For example, one is located at the Wright–Patterson Air Force Base under the direction of M. Uchic, a pioneer in micro-compression testing, to conduct compression and tension experiments on micrometer-scale samples of superalloys and metals [3,6]. This is a custom-built system with a sensor-based loading stage inside a scanning electron microscope. Another *in-situ* SEM-based system is located in G. Dehm's group at the University of Leoben (Austria). This group focuses on investigating mechanical properties of small-scale metals and has published several important papers on micro-tensile and compressive results of such samples [20,95–97]. This setup has a microindenter (ASMEC UNAT) mounted inside a LEO Stereoscan 440 scanning electron microscope. The *in-situ* system in J. Michler's research group at EMPA (Switzerland) consists of a load cell fixed on a piezo-actuated positioning stage [98,99]. There is also a second, smaller stick–slip positioning stage fixed between load cell and specimen to allow full Cartesian positioning, with a travel range of several millimeters. Karlsruhe Institute of Technology (Germany) has a DualBeam (FEI Nova Nanolab 200) system with a micro-manipulator (Kleindiek Nanotechnik) [4,100,101]. This system has prototyping capability and uses an *in-situ* nanomechanical transducer (Hysitron, Inc.) rather than a dedicated nanomechanical testing module, requiring complex digital image correlation to compute stresses and strains. D. Gianola's group (University of Pennsylvania) has two home-built MEMS-based systems capable of performing nano-scale mechanical experiments inside of a variety of microscopes (FIB, SEM, optical, and atomic force microscope) [4,100,101]. J. Greer's group at Caltech has an *in-situ* nanomechanical instrument, SEMentor, which consists of a scanning electron microscope (FEI Quanta 200 FEG) and a nanomechanical arm similar to a nanoindenter (DCM module, Agilent) [12,14,102]. Owing to the increasing demand for their use, single-unit *in-situ* SEM systems are becoming commercially available.

TEM-Based Nanomechanical Instruments

Generally, the *in-situ* TEM setups are more complex than the SEM ones described in the previous section, mainly due to the very limited space between the pole piece and the sample holder within the chamber and to the necessity to build a custom-made, sophisticated sample holder capable of mechanical perturbation and measurements. The limited space leads to a restriction in the tilt angles attained, sometimes rendering positioning the electron beam in diffraction conditions impossible. Holders that are most commonly used for *in-situ* TEM nanomechanical experiments are room-temperature tensile-straining holders. These tensile holders are usually equipped with micro-actuators that can apply displacement rates between $10\,\mathrm{nm\,s^{-1}}$ and $1\,\mathrm{\mu m\,s^{-1}}$, but great care has to be taken to ensure that the displacement measurements are accurate. This is a particularly important point, since the elastic energy stored in the actuator–sample system can lead to a sudden rapid failure of the extremely thin specimens necessary for the TEM analysis [4,103]. Beyond these "conventional holders," several newer, nanoindenter-integrating systems are capable of performing uniaxial compression (e.g., see Refs. [39,68,104]), nanoindentation [105], tension [95,106,107], and even HRTEM tensile testing of nanowires cold welded to the tensile device [51,93]. These advancements in instrumental prowess have catalyzed a new wave in the *in-situ* TEM nanomechanical testing on a variety of metallic nano-pillars (Ni, Mg, Al, and Au to name a few), and the community is actively pursuing this type of *in-situ* TEM nanomechanical testing. To date, there are few *in-situ* TEM nanomechanical testing instruments, with prominent literature reports on small-scale plasticity published by

A. Minor's group (UC Berkeley), Z. Shan's group at Xi'an University (China), J. Huang's group at Sandia National Lab, M. Legros's group at CEMES–CNRS (France), and J. De Hossons's group at the University of Groningen (Netherlands).

7.5 Discussion and Outlook on Size-Dependent Strength in Single-Crystalline Metals

7.5.1 Cubic Crystals

Face-Centered Cubic Metals

It has been ubiquitously demonstrated that, at the micrometer and sub-micrometer dimensions, the sample size dramatically affects strength, as revealed by a great number of room-temperature uniaxial compression experiments on a variety of single-crystalline, *non-pristine* (i.e., containing defects) metallic nano-pillars (for reviews, see Refs. [1–3]). As described above, in most of these studies, cylindrical pillars with micrometer and sub-micrometer dimensions were fabricated mainly by the subtractive FIB-based techniques, as well as via some non-FIB, "bottom-up" synthesis. Noteworthy is that nearly all of these reports demonstrate that single-crystalline fcc metals exhibit an identical power-law dependence between strength and sample size, suggesting that this scaling for fcc crystals might be universal [108]. Figure 7.1 shows a compiled data plot of the resolved shear stresses normalized by the appropriate shear modulus as a function of pillar diameter normalized by the Burgers vector for nearly all currently published data on the non-pristine fcc nano-pillars subjected to uniaxial compression or tension.

Beyond compression, Dehm and coworkers reported a uniaxial *tensile* methodology inside a scanning electron microscope performed on single-crystalline micro-tensile Cu samples [95–97,109]. These samples had diameters between 500 nm and 8 μm with the aspect ratios between 1 : 1 and 13.5 : 1 and were fabricated by the subtractive FIB technique. These authors report size-dependent strengths, although in these tension experiments the strength appears to be less sensitive to the diameter while being strongly dependent on the sample aspect ratio. This team of researchers has also done extensive work on quantifying the effects of experimental constraints, due both to sample geometry and to the loading mechanism [97,109]. Concurrently, tensile experiments were conducted by several other groups; for example, Jennings and coworkers strained single-crystalline ⟨111⟩-oriented electroplated Cu nano-pillars inside a scanning electron microscope and reported neck formation and a nearly identical size effect as with compression [12,13]. These samples had diameters between 75 and 165 nm, and their tensile signature was characterized by limited homogeneous deformation prior to reaching the ultimate tensile strength, followed by an extensive strain burst, correlated with the formation of a ∼20 nm diameter neck. The authors reported a nearly identical power law for the strength as a function of sample size as in both electroplated and FIB-machined fcc pillars and attributed the lack of homogeneous ductility to the annihilation of dislocations at the free surfaces (i.e., the notion of "dislocation starvation"). Richter *et al.* [16] applied tensile loading by attaching a micromanipulator in a dual-beam FIB system to pristine (i.e., defect-free) Cu nano-whiskers with diameters between 75 and 300 nm. Unlike nearly all other such studies, these Cu nanowires did not exhibit a size effect, but rather deformed in a brittle fashion (i.e., no ductility) at nearly theoretical strengths – akin to the classical reports by Brenner [88,89], who demonstrated the attainment of theoretical strengths in Cu, Fe, and Ag micro-wires under uniaxial tension. Importantly, all nanowires tested fractured at stresses between 1 and 7 GPa, largely orthogonal to the loading axes, lacking any plasticity, all of which suggests that the fracture was brittle, probably due to the absence of dislocations. This lack of size effects in pristine single crystals is also consistent with the work of Bei and coworkers [29,34,110] on Mo (bcc) alloy micro-pillars, as described in further detail in Section 7.5.1. When considering the mechanistic origins of the size-dependent strength (or lack thereof), it becomes clear that conventional things are reversed in the nano-sized crystals, as pre-straining actually weakens the crystal [10,27,29]. These phenomena are antipodal to classical plasticity, where dislocations multiply in the course of deformation, forming numerous pinned segments and a densely populated substructure (i.e., a forest of dislocations), whereby this so-called "forest hardening" is the dominant mechanism responsible for work hardening, or requiring higher applied stresses at greater strains. The community has now unambiguously demonstrated in small single-crystalline samples that the initial dislocation density has an enormous influence on the presence of the size effects [49,52,56,57,71,72,74]. An example of this concept is in the report by Jennings *et al.* [13], which revealed that single-crystalline electroplated Cu nano-pillars with similar initial dislocation densities as FIB-produced nano-pillars exhibit an identical size effect as the ubiquitous power law for fcc metals with an exponent of approximately −0.6.

To further elucidate the role of dislocation nucleation-driven plasticity, some research groups investigated the compressive behavior of passivated, or coated, micro- and nano-pillars to intentionally prevent dislocations from escaping at the free surfaces. For example, Ng and Ngan [24] trapped mobile dislocations inside Al micro-pillars with diameters from ∼1.2 to 6 μm, either surface coated or center filled with a tungsten-based compound deposited in the FIB. Consistent with the dislocation dynamics simulations (with the most recent one by El-Awady *et al.* [111]), the passivated samples exhibited a markedly higher strain-hardening rate and a much smoother stress–strain relationship, implying the suppression of dislocation avalanches and the lack of nucleation-controlled plasticity. Experiments on nano-sized Cu pillars with conformal atomic-layer-deposited coatings of Al_2O_3 and TiO_2 also corroborate these conclusions, as their stress–strain curves undergo hardening (unlike their uncoated counterparts), hysteresis loops were observed upon interim unloading–reloading cycles, and there was some suppression of dislocation avalanches [112]. These features have indeed been reported before for thin films, both experimentally and through dislocation dynamics computations [113,114]. TEM analysis from both of the aforementioned compressions of coated pillars revealed much greater dislocation densities and even dislocation cell formation in the post-deformed specimens [24,112], proving that the dislocations are indeed trapped inside the deforming coated pillars rather than annihilating at the free surface, like in the as-fabricated small-scale specimens.

What emerges, as suggested by Kraft *et al.* [2], is that the "phase diagram" for size-dependent plasticity in fcc metals may be envisioned as having three distinct areas, which are defined by size and by initial dislocation density, as shown in Figure 7.4. In the *smallest regime*, sample sizes are below ∼100 nm, and plasticity is governed by surface dislocation nucleation. In the *intermediate regime* with sample dimensions between 100 nm and 1 μm, the strength is controlled by the activation of a small number of single-arm sources. The operation nature of these single-arm sources can be quite complex, as well as is statistically determined because the sources can become exhausted and cease operation in the course of deformation [69,70,115]. These sources have also been shown to not be immortal, in the sense that they may become unpinned and destabilized [56,58]. Finally, in the *large regime*, with the length scale of greater than 1 μm, strength is defined by dislocation multiplication in a fashion similar to bulk, where dislocation interactions and subsequent network formation are well-known strength-determining mechanisms.

Body-Centered Cubic Structures

Most small-scale plasticity studies have focused on fcc metals; however, few research groups ventured out beyond those, with bcc structures occupying probably the next largest niche. Plasticity in bcc metals is significantly more complex than in fcc metals, as there is usually a strong thermal dependence, which means that factors like temperature, strain rate, and orientation can significantly affect the course of deformation. Another peculiarity is that bcc metals exhibit tension–compression asymmetry due to their unconventional slip behavior (i.e., not according to the well-known Schmid law). Non-Schmid behavior is manifested by crystallographic slip occurring on atomic planes other than the one with the maximum resolved shear stress [116,117]. These phenomena can be explained through the understanding of screw dislocation structure and behavior. Since the closest packed {110} atomic planes in bcc structures are not fully close-packed, the cores of screw dislocations are nonplanar, resulting in their motion in response to both glide and non-glide components of the applied shear stress tensor. This nonplanarity is also the key reason behind the much lower mobility of screw dislocations compared with their edge counterparts; therefore, their motion represents the overall strength-controlling element [118]. Further, the motion of screw dislocations in bcc crystals is not restricted to any particular set of crystallographic planes, as is the case for fcc metals, but rather can easily cross-slip among mutually intersecting planes.

To date, most of the reports on small-scale bcc deformation have described compressions on single crystalline Mo. For example, Bei and coworkers [29,34,79] conducted uniaxial compression experiments on directionally solidified Mo-alloy pillars with diameters between 360 nm and 1.5 μm, which contained either no initial dislocations or were intentionally pre-strained to introduce dislocations into the initial structure (Figure 7.10). Unlike all reports on the non-pristine structures (i.e., not nanowires), these authors found both the yield and the flow strengths to be independent of sample size when they are initially pristine (0% pre-strain) and with significant, 11% pre-straining, where even the smallest pillars deform similarly to bulk. Size effects were observed by these authors at the intermediate pre-strains of 4–8%. The authors considered the attainment of theoretical strengths by pillars of all sizes and the lack of size effects in the defect-free pillars not surprising. They attributed the lower stresses reported by others to the presence of microstructural debris like small dislocation loops and vacancies introduced into the material surfaces via FIB milling (see Section 7.3.1). Since the FIB has been an instrumental micro-fabrication instrument to create these small mechanical specimens, several attempts to quantify the adverse effects of using this technique have been published (e.g., see [79,81,84,85,119–121]). Most importantly, etching by ion

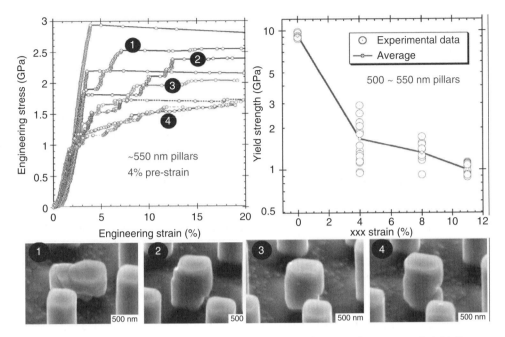

Figure 7.10 Compressive stress–strain curves and SEM images of compressed, initially pre-strained Mo nano-pillars fabricated by selective etching technique. Reprinted with permission from [29]. © 2008 Elsevier

bombardment alters the initial dislocation density in single-crystalline samples, which (as discussed in the previous section) is what determines the size effect.

Nano- and micrometer-sized bcc metals, which contained initial dislocations, appeared to behave quite differently from fcc nano-crystals. In comparing uniaxial compression results of Mo (bcc) and Au (fcc) nano-pillars, Brinckmann *et al.* [122] found that while both metals exhibited size-dependent strengths and stochastic discrete strain bursts, the *amount* of size-induced strengthening and the relative fraction of the flow stress to theoretical strength (~40% for Au and only ~7% for Mo) were notably different. These discrepancies led to the conclusion that fundamentally different plasticity mechanisms operate in fcc versus bcc metals, as revealed by the atomistic simulations [30,49]. According to these molecular dynamics simulations, the mobile dislocations initially residing in fcc nano-pillars quickly escape the crystal at the free surfaces upon compression, as consistent with the "dislocation starvation" notion. In bcc metals, however, a single pre-existing dislocation loop does not immediately glide to the surface when compressed; rather, it forms a kink, whose "arms" propagate through the crystal on different crystallographic planes and in the course of travel generate other dislocation segments and debris.

To date, five different bcc metals with pre-existing dislocations have been studied at the mirometer and sub-micrometer dimensions: W, Ta, Mo, V, and Nb [25–27,31,32,35,36]. These metals cover a wide range of critical temperatures and Peierls barriers, and it has been found that their mechanical response correlates with both of these physical properties. For example, Schneider *et al.* [25] observed that the power-law exponents in size-dependent strength of W, Ta, Nb, and Mo inversely correlate with the critical temperature for each metal. They attributed this to the low mobility of screw dislocations below T_c, requiring higher stresses to overcome the Peierls barrier in the metals with high T_c. As a corollary, at temperature close to T_c, as is the case for Nb, stress–strain relationships resembled those of fcc metals. These researchers also find that the strength difference among these metals is minimized at the smallest diameters of 20 nm, implying that the Peierls mechanism becomes less dominant as the pillar size decreases. The authors propose two possibilities for this observation: (1) relative ease of kink-pair nucleation at the sample surface and (2) reduction in dislocation back-stresses due to the smaller screw dislocation pile-ups near dislocation sources. On the other hand, Kim *et al*. [32] report a quantitative comparison of tension and compression on (001)-oriented single-crystalline W, Mo, Ta, and Nb and, while also observing strong size effects, find the power-law slope of −0.93 for Nb to be much higher than those for the remaining metals: −0.44 for W, −0.44 for Mo, and −0.43 for Ta. These findings are attributed to the twinning versus anti-twinning slip common in bcc metals, the nonplanar character of screw dislocation cores, and the differences in lattice resistance to dislocation motion at room temperature inferred from the Peierls stress and critical temperature. Notably, no systematic correlation between size-dependent strength and critical temperature was found in this work.

Statistical evaluation of stress–strain curves on compressed Mo nano-pillars led to the conclusion that the presence of a Peierls barrier does not suppress avalanche-like deformation, in the sense that these curves are still ridden with a large number of intermittent strain bursts [28]. The experiments on tension and compression of nano- and micro-pillars made from these metals via FIB methodology revealed strong size effects, but (unlike in fcc metals) each bcc metal had a unique power-law exponent, which was also found to be a function of the loading direction; that is, a difference between compression and tension for the same material [1]. The difference in strength as a function of loading direction naturally leads to tension–compression asymmetry, which is another characteristic typical of bcc transition metals. Interestingly, this tension–compression asymmetry appears to remain at the nano-scale level, but is in itself also a function of size for the small pillars, with diameters of 800 nm and smaller, whereas the strength differential between tension and compression for larger nano-pillars approaches size-independent bulk values [32].

Bcc metals are also quite sensitive to strain rate, as the thermal fluctuations contribute greatly to the screw dislocation mobility through the potential energy landscape. Some experiments studying the effects of strain rate on the deformation of bcc nano-pillars have been done, and activation volumes were determined from the yield strength versus strain-rate plots [26,31,33]. The activation volumes determined on the compression of Mo nano-pillars were between $1b^3$ and $5b^3$, values only marginally smaller than those for bulk. This *independence* between the activation volume and sample size suggests that nano-scale bcc deformation occurs via thermally activated kink pair nucleation, a typical screw dislocation propagation mechanism in bulk bcc metals.

Consistent with the overall complexity of bcc nano-scale plasticity, crystallographic orientation also plays a role in the strengths attained. As an example of this, Kim *et al*. [31] measured uniaxial tensile and compressive strengths of (001)- versus (110)-oriented Mo nano-pillars and found that compressive flow stresses are higher than tensile ones in the [001] orientation and vice versa in the [011] orientation. Also, it was demonstrated that, unlike in bulk, the difference between flow stresses attained in tension versus compression of [001]- and [011]-oriented single-crystalline Mo nano-pillars depends on the pillar size, resulting in nontrivial tension–compression asymmetry. This asymmetry was attributed to the differences in the Peierls stress in twinning versus anti-twinning direction and to a strong dependence of the critical resolved shear stress on the non-glide applied stress tensor components.

Statistical Evaluation of Nano-Scale Cubic Single-Crystal Deformation

The stress–strain curves generated by uniaxial testing of micro- and nano-sized single crystals consist of numerous, highly stochastic, intermittent strain bursts. Naturally, statistical data treatments were performed in order to understand how these slip events may be related to the energy dissipation of dislocation avalanches [63,123–127]. Dimiduk *et al.* [127] characterized strain bursts during compressions of Ni micro-pillars by utilizing the framework of self-organized criticality (SOC). SOC is a common method of describing nonequilibrium dynamic systems which organize themselves to a critical point, with one of its important characteristics being that it is scale free. These authors found that the cumulative probability of discrete strain bursts as a function of burst extent obeyed a scale-free distribution of the form $P(\Delta l) \propto \Delta l^k$, where $k \approx 1.5$, implying that dislocation motion in micro-scale single crystals is closely related to other scale-free SOC phenomena like magnetic-domain dynamics and earthquakes. Several other statistical treatments of the strain burst data during compression of small-scale crystals also revealed their scale-free nature, likening the discrete nature of these deformations to crackling noise, acoustic emissions, and earthquakes [123,128]. Ng and Ngan proposed a theory based on statistical mechanics and Monte Carlo simulations proposing the presence of a transition in the Al pillar deformation behavior from deterministic to stochastic [63,70]. The premise of this theory is that if either the number of dislocation sources or the emission rate increases with burst order, the result is that the large-strain deformation is deterministic. Analogously, when the sources become depleted, the deformation shifts towards being stochastic. For samples made from the same material with identical initial dislocation densities, this deterministic-to-stochastic transition will occur once the size becomes sufficiently small. Friedman *et al.* [124] statistically analyzed the deformation of nano- rather than micro-pillars and reported a notable difference between the two size regimes: in the micro-sized samples, which are usually loaded at relatively fast rates, the temporal overlap of dislocation avalanches leads to a power-law exponent of 1.5, while the stress-integrated exponent of 2, observed for the slower-deformed nano-pillars, is consistent with the mean field theory (MFT) [126]. The results of this work suggest that MFT may be applicable to single-crystalline metals spanning four size decades (from fractions of millimeters to only tens of nanometers), implying that the same physical mechanisms are controlling the statistics of dislocation avalanches in metallic single crystals, regardless of size.

7.5.2 Non-Cubic Single Crystals

Hexagonal Close-Packed (HCP)

Deformation of small-scale hcp metals has generated some, albeit much less, attention compared with the cubic crystals; hence, understanding of their nano-scale plastic response is only at a preliminary stage. So far, Mg and Ti have been studied, and size effects were found to occur in both [37–41]. Interestingly, when compressed along (0001) orientation, Mg did not twin; rather, the crystals deformed via multiple slip on pyramidal planes, causing pronounced hardening. The size effect in Mg was attributed to two distinct mechanisms: basal plane sliding and extension twinning [39]. *In-situ* TEM analysis confirmed the dependence of twin nucleation stress on pillar size, and regarded the availability of pre-existing dislocations to be the enabling factor for twin formation. Another example of small-scale hcp deformation is single-crystalline Ti alloy [41]. Here, a transition in the strength versus size trend is observed: from size dependent in the micro-scale regime, where the twinning stress strongly increases with decreasing sample size, to size *independent*, close-to-theoretical strength in the sub-micrometer diameter pillars, which deformed via crystallographic slip rather than twinning. As in the case of Mg compression, deformation twinning in Ti appears to be a strong function of the external sample dimensions [41]. An unambiguous size effect with the power law slope of -0.5 was also found in the deformation of pure single-crystalline Ti oriented for double prismatic slip [40]. It was found that the critical resolved shear stress scaled inversely with sample size, yet again strongly suggesting that plasticity here is dominated by dislocation-source nucleation.

Tetragonal Metals

An interesting group of metals that has been only scarcely studied is low-melting-temperature materials like Sn, Bi, and In [42–44]. Unlike all the metals described above, small-scale behavior of these metals appears to be size-independent even with some pre-existing dislocations, albeit with strengths still much higher than bulk. Lee *et al.* report that, unlike all other single crystalline metals, which deform by crystallographic slip, most of the deformed In pillars flow in a viscous fashion, forming surface wrinkles and folds [42]. The yield strength was also shown to depend greatly on strain rate and nearly negligibly on sample size. This suppression of size dependence was attributed to the ability of dislocations not only to glide, but also to climb at room temperature, thereby precluding dislocation starvation and nucleation-controlled plasticity, which are thought to be dominating in all other nano-sized single crystals.

7.6   Conclusions and Outlook

The field of small-scale plasticity is vibrant and active, yet it is nearing the first stages of maturity with the nascent understanding of size-dependent mechanical properties and nano- and micrometer-scale-specific deformation mechanisms. What emerges is that the strength of crystalline materials with micrometer and sub-micrometer dimensions is indeed size dependent, and for metals deforming by crystallographic slip can be represented by a power law of the form $\tau/\mu = (D/b)^n$, where τ is the shear strength, μ is the shear modulus, D is the pillar diameter, and b is the magnitude of the Burgers vector. The exponent n varies in accordance with material, initial dislocation content, loading direction, and Peierls barrier, albeit that it is universally found to be approximately -0.6 for non-pristine fcc metals.

Our understanding of the size-dependent mechanical response of fcc metals is converging with the advent of cutting-edge experimental and computational tools and the ever-approaching size and temporal scales between the two. Through these investigations, it is now known that in the micrometer-sized fcc samples with $\sim 10^{12} - 10^{14}\,\mathrm{m}^{-2}$ initial dislocation densities, plasticity is governed via the operation of single-arm dislocation sources, whose lengths and, therefore, operational strengths are strongly coupled with the pillar diameter. Size-induced strengthening in these pillars arises from the source truncation and source exhaustion. In contrast, deeply in the sub-micrometer and nanoscale regime, plasticity is carried by dislocations nucleating from surface sources, and the higher strengths at smaller sizes are attained due to the statistical nature of the shortage of weak sources in smaller structures. In both size regimes, plasticity is shifted into nucleation rather than propagation control, and the stress–strain behavior is highly stochastic, with multiple strain bursts of various extents dominating the deformation.

More complex single-crystalline nanostructures (like bcc, hcp, tetragonal, etc.) are characterized by much more sophisticated processes occurring during their deformation. Unlike fcc crystals, these materials are susceptible to the intrinsic lattice resistance, less restricted dislocation motion, and thermal effects and, therefore, are not easily explained through the mechanisms outlined above.

It is encouraging that the number of material classes being explored in the context of size-dependent mechanical properties is growing, with some of the more recent additions containing internal boundaries: nanocrystalline metals, multilayered composites, shape memory alloys, and bi-crystalline metals. It is critical that experimentalists practice care while conducting these uniaxial nanomechanical experiments, as the results obtained vary greatly with experimental sources of error: indenter–sample misalignment, vertical pillar taper, surface damage due to FIB etching, sample aspect ratio, and nonplanarity of pillar tops. Even more importantly, the effect of nanoindenter shaft stiffness in compression and tension is an important consideration that needs to be understood.

The substantial body of work on this topic that has emerged in the last decade, as well as classic metallurgy, dictates that size-dependent yield and flow stresses can arise due to both intrinsic (i.e., grain size, boundary spacing, and precipitate size) and extrinsic (i.e., sample dimensions) characteristic length scales, although very few studies address the combined effects of these dimensions. The next task for us as a community is to populate the strength–d–D space (where d is the internal characteristic length scale and D is the external one) with data points beyond single crystals. When such a full layout of size effects is obtained and explained, "size" will be a critical "knob" to be used as a design parameter in future technological applications.

References

[1] Greer, J.R. and De Hosson, J.T.M. (2011) Plasticity in small-sized metallic systems: Intrinsic versus extrinsic size effect. *Progress in Materials Science*, **56**, 654–724.
[2] Kraft, O., Gruber, P.A., Mönig, R., and Weygand, D. (2010) *Annual Review of Materials Research*, **40**, 293–317.
[3] Uchic, M.D., Shade, P.A., and Dimiduk, D. (2009) *Annual Review of Materials Research*, **39**, 361–386.

[4] Legros, M., Gianola, D.S., and Motz, C. (2010) Quantitative *in situ* mechanical testing in electron microscopes. *MRS Bulletin*, **35**, 354–360.

[5] Uchic, M.D., Dimiduk, D.M., Florando, J.N., and Nix, W.D. (2004) Sample dimensions influence strength and crystal plasticity. *Science*, **305**, 986–989.

[6] Wheeler, R., Shade, P., Uchic, M.D. *et al*. (2008) Advances in instrumentation and techniques – FIB-based applications and instrumentation advances for the physical and biological sciences. *Microscopy and Microanalysis*, **14**, 100–101.

[7] Frick, C.P., Clark, B.G., Orso, S. *et al*. (2008) Size effect on strength and strain hardening of [111] nickel sub-micron compression pillars. *Materials Science and Engineering A*, **489**, 319–329.

[8] Greer, J.R., Oliver, W.C., and Nix, W.D. (2005) Size effects in mechanical properties of gold at the micron scale in the absence of strain gradients. *Acta Materialia*, **53**, 1821–1830.

[9] Dietiker, M., Buzzi, S., Pigozzi, G. *et al*. (2011) Deformation behavior of gold nano-pillars prepared by nanoimprinting and focused ion-beam milling. *Acta Materialia*, **59**, 2180–2192.

[10] Lee, S.-W., Han, S.M., and Nix, W.D. (2009) Uniaxial compression of fcc Au nanopillars on an MgO substrate: the effects of prestraining and annealing. *Acta Materialia*, **57**, 4404–4415.

[11] Volkert, C.A. and Lilleodden, E.T. (2006) Size effects in the deformation of sub-micron Au columns. *Philosophical Magazine*, **86**, 5567–5579.

[12] Jennings, A.T. and Greer, J.R. (2010) Tensile deformation of electroplated copper nanopillars. *Philosophical Magazine*, **91**, 1108–1120.

[13] Jennings, A.T., Burek, M.J., and Greer, J.R. (2010) Microstructure versus size: mechanical properties of electroplated single crystalline Cu nanopillars. *Physical Review Letters*, **104**, 135503.

[14] Greer, J.R., Kim, J.-Y., and Burek, M.J. (2009) *In-situ* mechanical testing of nano-scale single crystalline nano-pillars. *Journal of Materials*, **61**, 19–25.

[15] Jennings, A.T., Li, J., and Greer, J.R. (2011) Emergence of strain rate sensitivity in Cu nano-pillars: transition from dislocation multiplication to dislocation nucleation. *Acta Materialia*, **59**, 5627–5637.

[16] Richter, G., Hillerich, K., Gianola, D.S. *et al*. (2009) Ultrahigh strength single crystalline nanowhiskers grown by physical vapor deposition. *Nano Letters*, **9**, 3048–3052.

[17] Kiener, D. and Minor, A.M. (2011) Source controlled yield and hardening of Cu (100) studied by *in situ* TEM. *Acta Materialia*, **59**, 1328–1337.

[18] Maaß, R., Van Petegem, S., Grolimund, D. *et al*. (2008) Crystal rotation in Cu single crystal micropillars: *in situ* Laue and electron backscatter diffraction. *Applied Physics Letters*, **92**, 071905.

[19] Kiener, D., Motz, C., Schöberl, T. *et al*. (2006) Determination of mechanical properties of copper at the micron scale. *Advanced Engineering Materials*, **8**, 1119–1125.

[20] Kiener, D., Motz, C., and Dehm, G. (2009) Micro-compression testing: a critical discussion of experimental constraints. *Materials Science and Engineering A*, **505**, 79–87.

[21] Buzzi, S., Dietiker, M., Kunze, K. *et al*. (2009) Deformation behavior of silver submicrometer-pillars prepared by nanoimprinting. *Philosophical Magazine*, **89**, 869–884.

[22] Ng, K.S. and Ngan, A.H.W. (2009) Deformation of micron-sized aluminium bi-crystal pillars. *Philosophical Magazine*, **89**, 3013–3026.

[23] Kunz, A., Pathak, S., and Greer, J.R. (2011) Compressive properties of bi-crystalline Al nano-pillars. *Acta Materialia*, **59**, 4416–4424.

[24] Ng, KS. and Ngan, A.H.W. (2009) Effects of trapping dislocations within small crystals on their deformation behavior. *Acta Materialia*, **57**, 4902–4910.

[25] Schneider, A.S., Kaufmann, D., Clark, B. *et al*. (2009) Correlation between critical temperature and strength of small-scale bcc pillars. *Physical Review Letters*, **103**, 105501.

[26] Schneider, A.S., Clark, B.G., Frick, C.P. *et al*. (2009) Effect of orientation and loading rate on compression behavior of small-scale Mo pillars. *Materials Science and Engineering A*, **508**, 241–246.

[27] Schneider, A.S., Clark, B.G., Frick, C.P. *et al*. (2010) Effect of pre-straining on the size effect in molybdenum pillars. *Philosophical Magazine Letters*, **90**, 841–849.

[28] Zaiser, M., Schwerdtfeger, J., Schneider, A.S. *et al*. (2008) Strain bursts in plastically deforming molybdenum micro- and nanopillars. *Philosophical Magazine*, **88**, 3861–3874.

[29] Bei, H., Shim, S., Pharr, G., and George, E. (2008) Effects of pre-strain on the compressive stress–strain response of Mo-alloy single-crystal micropillars. *Acta Materialia*, **56**, 4762–4770.

[30] Greer, J.R., Weinberger, C., and Cai, W. (2008) Comparing strengths of FCC and BCC sub-micrometer pillars: compression experiments and dislocation dynamics simulations. *Materials Science and Engineering A*, **493**, 21–25.

[31] Kim, J.-Y., Jang, D., Greer, J.R. (2011) Crystallographic orientation and size dependence of tension-compression asymmetry in molybdenum nano-pillars. *International Journal of Plasticity*, **28**, 46–52.

[32] Kim, J.-Y., Jang, D., and Greer, J.R. (2010) Tensile and compressive behavior of tungsten, molybdenum, tantalum, and niobium at the nanoscale. *Acta Materialia*, **58**, 2355–2363.

[33] Kim, J.-Y. and Greer, J.R. (2009) Tensile and compressive behavior of gold and molybdenum single crystals at nano-scale. *Acta Materialia*, **57**, 5245–5253.

[34] Bei, H., Shim, S., George, E. *et al.* (2007) Compressive strengths of molybdenum alloy micro-pillars prepared using a new technique. *Scripta Materialia*, **57**, 397–400.

[35] Kim, J.-Y., Jang, D., and Greer, J.R. (2009) Insight into the deformation behavior of niobium single crystals under uniaxial compression and tension at the nanoscale. *Scripta Materialia*, **61**, 300–303.

[36] Han, S.M., Bozorg-Grayeli T., Groves, J.R., and Nix, W.D. (2010) Size effects on strength and plasticity of vanadium nanopillars. *Scripta Materialia*, **63**, 1153–1156.

[37] Byer, C.M., Li, B., Cao, B., and Ramesh, K. (2010) Microcompression of single-crystal magnesium. *Scripta Materialia*, **62**, 536–539.

[38] Lilleodden, E. (2010) Microcompression study of Mg (0001) single crystal. *Scripta Materialia*, **62**, 532–535.

[39] Ye, J., Mishra, R.K., Sachdev, A.K., and Minor, A.M. (2010) *In situ* TEM compression testing of Mg and Mg–0.2 wt.% Ce single crystals *Scripta Materialia*, **64**, 292–295.

[40] Sun, Q., Guo, Q., Yao, X. *et al.* (2011) Size effects in strength and plasticity of single-crystalline titanium micropillars with prismatic slip orientation. *Scripta Materialia*, **65**, 473–476.

[41] Yu, Q., Shan, Z.-W., Li, J. *et al.* (2010) Strong crystal size effect on deformation twinning. *Nature*, **463**, 335–338.

[42] Lee, G., Kim, J.-Y., Budiman, A. *et al.* (2010) Fabrication, structure, and mechanical properties of indium nanopillars. *Acta Materialia*, **58**, 1361–1368.

[43] Lee, G., Kim, J.-Y., Burek, M.J. *et al.* (2011) Plastic deformation of indium nanostructures. *Materials Science and Engineering A*, **528**, 6112–6120.

[44] Burek M.J., Budiman A.S., Jahed Z., Tamura N., Kunz M., Jin S., Han S.M.J., Lee G., Zamecnik C., and Tsui T.Y. (2011) Fabrication, microstructure, and mechanical properties of tin nanostructures. *Materials Science and Engineering A*, **528**, 5822–5832.

[45] Taylor GI. (1938) Plastic strain in metals. *Journal of the Institute of Metals*, **62**, 307–324.

[46] Nix, W.D. and Lee, S.-W. (2011) Micro-pillar plasticity controlled by dislocation nucleation at surfaces. *Philosophical Magazine*, **91**, 1084–1096.

[47] Zhu, T., Li, J., Samanta, A. *et al.* (2008) Temperature and strain-rate dependence of surface dislocation nucleation. *Physical Review Letters*, **100**, 025502.

[48] Li, J. (2007) The mechanics and physics of defect nucleation. *MRS Bulletin*, **32**, 151–159.

[49] Weinberger, C.R. and Cai, W. (2008) Surface-controlled dislocation multiplication in metal micropillars. *Proceedings of the National Academy of Sciences of the United States of America*, **105**, 14304–14307.

[50] Weinberger, C.R., Jennings, A.T., Kang, K., and Greer, J.R. (2012) Atomistic simulations and continuum modeling of dislocation nucleation and strength in gold nanowires. *Journal of the Mechanics and Physics of Solids*, **60**, 84–103.

[51] Zheng, H., Cao, A., Weinberger, C.R. *et al.* (2010) Discrete plasticity in sub-10-nm-sized gold crystals. *Nature Communications*, **1**, 144, doi: 10.1038/ncomms1149.

[52] Weygand, D., Poignant, M., Gumbsch, P., and Kraft, O. (2008) Three-dimensional dislocation dynamics simulation of the influence of sample size on the stress–strain behavior of fcc single-crystalline pillars. *Materials Science and Engineering A*, **483–484**, 188–190.

[53] Lee, S.-W. and Nix, W.D. (2010) Geometrical analysis of 3D dislocation dynamics simulations of FCC micro-pillar plasticity. *Materials Science and Engineering A*, **527**, 1903–1910.

[54] Han, C.-S., Hartmaiera, A., Gao, H., and Huang, Y. (2006) Discrete dislocation dynamics simulations of surface induced size effects in plasticity. *Materials Science and Engineering A*, **415**, 225–233.

[55] Motz, C., Weygand, D., Senger, J., and Gumbsch, P. (2009) Initial dislocation structures in 3-D discrete dislocation dynamics and their influence on microscale plasticity. *Acta Materialia*, **57**, 1744–1754.

[56] Tang, H., Schwarz, K., and Espinosa, H. (2007) Dislocation escape-related size effects in single-crystal micropillars under uniaxial compression. *Acta Materialia*, **55**, 1607–1616.

[57] Guruprasad, P. and Benzerga, A. (2008) Size effects under homogeneous deformation of single crystals: a discrete dislocation analysis. *Journal of the Mechanics and Physics of Solids*, **56**, 132–156.

[58] Tang, H., Schwarz, K., and Espinosa, H. (2008) Dislocation-source shutdown and the plastic behavior of single-crystal micropillars. *Physical Review Letters*, **100**, 185503.

[59] Rao, S.I., Dimiduk, D., Tang, M. *et al.* (2007) Estimating the strength of single-ended dislocation sources in micron-sized single crystals. *Philosophical Magazine*, **87**, 4777–4794.

[60] Liu, Z., Liu, X., Zhuang, Z., and You, X. (2009) Atypical three-stage-hardening mechanical behavior of Cu single-crystal micropillars. *Scripta Materialia*, **60**, 594–597.

[61] Oh, S.H., Legros, M., Kiener, D., and Dehm, G. (2009) *In situ* observation of dislocation nucleation and escape in a submicrometre aluminium single crystal. *Nature Materials*, **8**, 95–100.

[62] Greer, J.R. and Nix, W.D. (2006) Nanoscale gold pillars strengthened through dislocation starvation. *Physical Review B*, **73**, 245410.

[63] Ng, K. and Ngan, A.H.W. (2008) Stochastic nature of plasticity of aluminum micro-pillars. *Acta Materialia*, **56**, 1712–1720.

[64] Zhu, T. and Li, J. (2010) Ultra-strength materials. *Progress in Materials Science*, **55**, 710–757.

[65] Rabkin, E., Nam, H.-S., and Srolovitz, D.J. (2007) Atomistic simulations of the deformation of gold nanopillars. *Acta Materialia*, **55**, 2085–2099.

[66] Rabkin, E. and Srolovitz, D.J. (2007) Onset of plasticity in gold nanopillar compression. *Nano Letters*, **7**, 101–107.

[67] Shan, Z.W., Mishra, R.K., Asif, S.A.S. *et al.* (2008) Mechanical annealing and source-limited deformation in submicrometre-diameter Ni crystals. *Nature Materials*, **7**, 115–119.

[68] Lowry, M.B., Kiener, D., LeBlanc, M.M. *et al.* (2010) Achieving the ideal strength in annealed molybdenum nanopillars. *Acta Materialia*, **58**, 5160–5167.

[69] Espinosa, H., Berbenni, S., Panico, M., and Schwarz, K.W. (2005) An interpretation of size-scale plasticity in geometrically confined systems. *Proceedings of the National Academy of Sciences of the United States of America*, **102**, 16933–16938.

[70] Ngan, A.H.W. and Ng, K.S. (2010) Transition from deterministic to stochastic deformation. *Philosophical Magazine*, **90**, 1937–1954.

[71] Norfleet, D.M., Dimiduk, D.M., Polasik, S.J. *et al.* (2008) Dislocation structures and their relationship to strength in deformed nickel microcrystals. *Acta Materialia*, **56**, 2988–3001.

[72] Parthasarathy, T.A., Rao, S.I., Dimiduk, D.M. *et al.* (2007) Contribution to size effect of yield strength from the stochastics of dislocation source lengths in finite samples. *Scripta Materialia*, **56**, 313–316.

[73] El-Awady, J.A., Wen, M., and Ghoniem, N.M. (2009) The role of the weakest link mechanism in controlling the plasticity of micropillars. *Journal of the Mechanics and Physics of Solids*, **57**, 32–50.

[74] Senger, J., Weygand, D., Gumbsch, P., and Kraft, O. (2008) Discrete dislocation simulations of the plasticity of micro-pillars under uniaxial loading. *Scripta Materialia*, **58**, 587–590.

[75] Thompson, C.V. (1993) The yield stress of polycrystalline thin films. *Journal of Materials Research*, **8**, 237–238.

[76] Von Blanckenhagen, B., Gumbsch, P., and Arzt, E. (2003) Dislocation sources and the flow stress of polycrystalline thin metal films. *Philosophical Magazine Letters*, **83**, 1–8.

[77] Fleck, N.A., Muller, G.M., Ashby, M.F., and Hutchinson, J.W. (1994) Strain gradient plasticity: theory and experiment. *Acta Materialia*, **42**, 475–487.

[78] Fleck, N.A. and Hutchinson, J.W. (2001) A reformulation of strain gradient plasticity. *Journal of the Mechanics and Physics of Solids*, **49**, 2245–2271.

[79] Shim, S., Bei, H., Miller, M.K. *et al.* (2009) Effects of focused ion beam milling on the compressive behavior of directionally solidified micropillars and the nanoindentation response of an electropolished surface. *Acta Materialia*, **57**, 503–510.

[80] Kempshall, B.W., Schwarz, S.M., Prenitzer, B.I. *et al.* (2001) Ion channeling effects on the focused ion beam milling of Cu. *Journal of Vacuum Science and Technology B: Microelectronics and Nanometer Structures*, **19**, 749–754.

[81] MoberlyChan, W.J., Adams, D.P., Aziz, M.J. *et al.* (2007) Fundamentals of focused ion beam nanostructural processing: below, at, and above the surface. *MRS Bulletin*, **32**, 424–433.

[82] MoberlyChan, W.J., Sanchez, E.J., Stark, P.R.H., and Krug, J.T. (2003) Applications of focused ion beam using FEI DualBeam DB235: how deep is how small a hole? & How to Drill it deeper & smaller? *Microscopy and Microanalysis*, **9**, 886–887.

[83] Kim, J.-Y., Greer, J.R., (2008) Size-Dependent Mechanical Properties of Molybdenum Nanopillars *Applied Phys. Lett.* **93** (10), 101916

[84] McCaffrey, J.P., Phaneuf, M.W., and Madsen, L.D. (2001) Surface damage formation during ion-beam thinning of samples for transmission electron microscopy. *Ultramicroscopy*, **87**, 97–104.

[85] Maaβ, R., Grolimund, D., Van Petegem, S. *et al.* (2006) Defect structure in micropillars using X-ray microdiffraction. *Applied Physics Letters*, **89**, 151905.

[86] Kiener, D., Motz, C., and Dehm, G. (2008) Dislocation-induced crystal rotations in micro-compressed single crystal copper columns. *Journal of Materials Science*, **43**, 2503–2506.

[87] Thompson, K., Lawrence, D., Larson, D.J. *et al.* (2007) *In situ* site-specific specimen preparation for atom probe tomography. *Ultramicroscopy*, **107**, 131–139.

[88] Brenner, S.S. (1958) Growth and properties of "whiskers". *Science*, **128**, 568–575.

[89] Brenner, S.S. (1956) Tensile strength of whiskers. *Journal of Applied Physics*, **27**, 1484–1491.

[90] Fisher, J.C. and Hollomon, J.H. (1947) A statistical theory of fracture, Technical Publication 2218, American Institute of Mining and Metallurgical Engineers.

[91] Bei, H. and George, E.P. (2005) Microstructures and mechanical properties of a directionally solidified NiAl–Mo eutectic alloy. *Acta Materialia*, **53**, 69–77.

[92] Burek, M.J. and Greer, J.R. (2010) Fabrication and microstructure control of nanoscale mechanical testing specimens via electron beam lithography and electroplating. *Nano Letters*, **10**, 69–76.

[93] Lu, Y., Huang, J.Y., Wang, C. *et al.* (2010) Cold welding of ultrathin gold nanowires. *Nature Nanotechnology*, **5**, 218–224.

[94] Gane, N. and Bowden, F.P. (1968) Microdeformation of solids. *Journal of Applied Physics*, **39**, 1432–1435.

[95] Dehm, G. (2009) Miniaturized single-crystalline fcc metals deformed in tension: new insights in size-dependent plasticity. *Progress in Materials Science*, **54**, 664–688.

[96] Kiener, D., Grosinger, W., and Dehm, G. (2009) On the importance of sample compliance in uniaxial microtesting. *Scripta Materialia*, **60**, 148–151.

[97] Kiener, D., Grosinger, W., Dehm, G., and Pippan, R. (2008) A further step towards an understanding of size-dependent crystal plasticity: *in situ* tension experiments of miniaturized single-crystal copper samples. *Acta Materialia*, **56**, 580–592.

[98] Moser, B., Wasmer, K., Barbieri, L., and Michler, J. (2007) Strength and fracture of Si micropillars: a new scanning electron microscopy-based micro-compression test. *Journal of Materials Research*, **22**, 1004–1011.

[99] Östlund, F., Rzepiejewska-Malyska, K., Leifer, K. *et al.* (2009) Brittle-to-ductile transition in uniaxial compression of silicon pillars at room temperature. *Advanced Functional Materials*, **19**, 2439–2444.

[100] Gianola, D.S., Sedlmayr, A., Mönig, R. *et al.* (2011) *In situ* nanomechanical testing in focused ion beam and scanning electron microscopes. *Review of Scientific Instruments*, **82**, 063901.

[101] Gianola D.S. and Eberl C. (2009) *Journal of Materials*, **61**, 24–35.

[102] Greer, J.R., Jang, D., Kim, J.-Y., and Burek, M.J. (2009) Emergence of the mechanical functionality in materials via size reduction. *Advanced Functional Materials*, **19**, 2880–2886.

[103] Dehm, G., Legros, M., and Heiland, B. (2006) Micro- and nanoscale tensile testing of materials. *Journal of Materials Science*, **41**, 4484–4489.

[104] Chen, C.Q., Pei, Y.T., and De Hosson, J.T.M. (2010) Effects of size on the mechanical response of metallic glasses investigated through *in situ* TEM bending and compression experiments. *Acta Materialia*, **58**, 189–200.

[105] Minor, A.M. Asif, S.A.S., Shan, Z. *et al.* (2006) A new view of the onset of plasticity during the nanoindentation of aluminium. *Nature Materials*, **5**, 697–702.

[106] Legros, M., Gianola, D.S., and Hemker, K.J. (2008) *In situ* TEM observations of fast grain boundary motion in stressed nanocrystalline aluminum films. *Acta Materialia*, **56**, 3380–3393.

[107] Haque, M.A. and Saif, M.T.A. (2005) *In situ* tensile testing of nanoscale freestanding thin films inside a transmission electron microscope. *Journal of Materials Research*, **20**, 1769–1777.

[108] Dou, R. and Derby, B. (2009) A universal scaling law for the strength of metal micropillars and nanowires. *Scripta Materialia*, **61**, 524–527.

[109] Kiener, D., Motz, C., Dehm, G., and Pippan, R. (2009) Overview on established and novel FIB based miniaturized mechanical testing using *in-situ* SEM. *International Journal of Materials Research*, **100**, 1074–1087.

[110] Bei, H., Shim, S., Miller, M.K. *et al.* (2007) Effects of focused ion beam milling on the nanomechanical behavior of a molybdenum-alloy single crystal. *Applied Physics Letters*, **91**, 111915.

[111] El-Awady, J.A., Rao, S.I., Woodward, C. *et al.* (2010) Trapping and escape of dislocations in micro-crystals with external and internal barriers. *International Journal of Plasticity*, **27**, 372–387

[112] Jennings, A.T., Gross, C., Greer, F., Aitken, Z.H., Lee, S.-W., and Greer, J.R. *et al.* (2012) Higher compressive strengths and Bauschinger effect in conformally-passivated copper nanopillars. *Acta Materialia*, **60**, 3444–3455.

[113] Nicola, L., Xiang, Y., Vlassak, J. *et al.* (2006) Plastic deformation of freestanding thin films: Experiments and modeling. *Journal of the Mechanics and Physics of Solids*, **54**, 2089–2110.

[114] Espinosa, H. and Prorok, B.C. (2002) Effects of nanometer thick passivation layers on the mechanical response of thin gold films. *Journal of Nanoscience and Nanotechnology*, **2**, 427–433.

[115] Benzerga, A.A. (2008) An analysis of exhaustion hardening in micron-scale plasticity. *International Journal of Plasticity*, **24**, 1128–1157.

[116] Duesbery, M.S. and Vitek, V. (1998) Plastic anisotropy in b.c.c. transition metals. *Acta Metallurgica*, **46**, 1481–1492.

[117] Gröger, R., Bailey, A., and Vitek, V. (2008) Multiscale modeling of plastic deformation of molybdenum and tungsten: I. Atomistic studies of the core structure and glide of 1/2 ⟨111⟩ screw dislocations at 0 K. *Acta Materialia*, **56**, 5401–5411.

[118] Argon, A.S. (2008) *Strengthening Mechanisms in Crystal Plasticity*, Oxford University Press, Oxford.

[119] Kiener, D., Motz, C., Rester, M. *et al.* (2006) FIB damage of Cu and possible consequences for miniaturized mechanical tests. *Materials Science and Engineering A*, **459**, 262–272.

[120] El-Awady, J.A., Woodward, C., Dimiduk, D., and Ghoniem, N.M. (2009) Effects of focused ion beam induced damages on the plasticity of micropillars. *Physical Review B*, **80**, 104104.

[121] Cantoni M, Bobard F. (2004) TEM lamellae preparation FIB at CMI EPFL focussed ion beam. Why use ions instead of electrons? pp. 1–17.

[122] Brinckmann, S., Kim, J.-Y., and Greer, J.R. (2008) Fundamental differences in mechanical behavior between two types of crystals at nano-scale. *Physical Review Letters*, **100**, 155502.

[123] Csikor, F.C., Motz, C., Weygand, D. *et al.* (2007) Dislocation avalanches, strain bursts, and the problem of plastic forming at the micrometer scale. *Science*, **318**, 251–254.

[124] Friedman, N., Jennings, A.T., Tsekenis, G. *et al.* *Physical Review Letters* 109 (9), 095507 (2012)

[125] Zaiser, M. and Nikitas, N. (2007) Slip avalanches in crystal plasticity: scaling of the avalanche cut-off. *Journal of Statistical Mechanics: Theory and Experiment*, **2007**, P04013.

[126] Dahmen, K., Ben-Zion, Y., and Uhl, J. (2009) A micromechanical model for deformation in solids with universal predictions for stress–strain curves and slip avalanches. *Physical Review Letters*, **102**, 175501.

[127] Dimiduk, D., Woodward, C., Lesar, R., and Uchic, M.D. (2006) Scale-free intermittent flow in crystal plasticity. *Science*, **312**, 1188–1190.

[128] Sethna, J.P., Dahmen, K.A., and Myers, C.R. (2001) Crackling noise. *Nature*, **410**, 242–250.

Part Three

Experimentation

8

In-Situ TEM Electromechanical Testing of Nanowires and Nanotubes[1]

Horacio D. Espinosa[1], Rodrigo A. Bernal[1], and Tobin Filleter[2]

[1]*Northwestern University, USA*
[2]*Univeristy of Toronto, Canada*

8.1 Introduction

One-dimensional nanostructures, namely nanowires and nanotubes, have emerged as feasible building blocks for the next generation of devices and materials, with applications envisioned in electronics, sensors, and energy technologies. Nanotubes, long used as test beds of fundamental phenomena, are now viable alternatives for electronic systems [1] and are used as the fundamental constituent of high-performance fibers [2], given their outstanding mechanical [3] and transport properties [4]. On the other hand, semiconducting and metallic nanowires are envisioned as interconnects in next-generation electronics [5], building blocks of nanophotonic systems [6], and energy conversion elements for self-powered nanodevices [7].

The wide technological applicability of nanowires and nanotubes has elicited demand for their extensive characterization. Such characterization is critical in order to quantify their mechanical, electrical, thermal, chemical, and physical properties and generate data that can be used at the design and manufacturing stage to optimize synthesis, device architecture, and performance.

From the different types of characterization that can be carried out, measurement of mechanical and electromechanical properties are of high relevance to most applications. These measurements provide parameters for operation of devices that require mechanical movement to achieve functionality and, more importantly, offer insight into failure

---

[1] This chapter has been modified to feature as a Review: H. D. Espinosa, R. A. Bernal, T. Filleter, *Small*, 2012, **8**, 10.1002/smll.201200522, published online 25 Jul 2012.

limits and associated mechanisms in nanostructures. In turn, this provides much needed data towards reliable and robust designs, which are critical if nanostructures are to find widespread usage in future consumer applications [8].

The characterization of mechanical and electromechanical properties of one-dimensional nanostructures is becoming a mature field in which numerous researchers pursue characterization using a vast array of techniques. However, this undertaking continues to prove challenging for two main reasons. The first is the characteristic size of the specimen under study (e.g., diameter), which ranges from a few nanometers to up to several hundred nanometers. This renders specimen manipulation and preparation for testing very difficult, and it imposes highly demanding requirements in metrology and instrumentation. The second is the diverse variety of one-dimensional nanostructures and their marked structural dependence on synthesis methods. For instance, for carbon nanotubes (CNTs), the chirality and diameter, which control many properties, have not yet been controlled during growth, and only post-growth sorting methods allow the obtention of monodisperse solutions of nanotubes [9]. For nanowires, although diameter and structural control are fairly good [10], dopant concentrations may be nonuniform [11], systematic defects such as stacking faults are possible [12], and incorporated impurities during growth can be present [13], among other complications. As a result, it is difficult to propose a universal technique or testing method that is suitable for every nanostructure, as sample preparation and testing will require tailoring to the specific nanomaterial to varying degrees. More importantly, the possible presence of nonuniform defects or impurities, even in nanostructures of the same material, imposes the need to thoroughly characterize the very same nanostructure that is mechanically or electromechanically tested in order to establish unambiguous synthesis–structure–property relations for a given nanomaterial.

In this chapter, we review the mechanical and electromechanical testing of one-dimensional nanostructures carried out by *in-situ* transmission electron microscopy (TEM). It will be argued that, given the aforementioned constraints for establishing structure–property–relations for these specimens, *in-situ* TEM testing is a technique that excels at enhancing our understanding of one-dimensional nanostructures. The chapter presents a critical review of the available experimental techniques and several examples that showcase the capabilities of *in-situ* TEM to uncover new phenomena in several nanomaterials. Throughout this review, it will be argued that, given the challenges in mechanical and electromechanical characterization of nanostructures, the extremely high atomic resolution achieved in the TEM in addition to its analytical capabilities make *in-situ* TEM one of the most suitable techniques for carrying out such measurements.[2]

8.1.1 Relevance of Mechanical and Electromechanical Testing for One-Dimensional Nanostructures

As mentioned above, one-dimensional nanostructures have emerged as viable alternatives for new materials and devices. For instance, CNTs are being used to develop novel

---

[2] A technical account of the operation of the TEM or its application to a specific material system is beyond the scope of this review. Hence, the reader should refer to several specialized books about TEM (e.g., [14]) and TEM of nanostructures (e.g., [15,16]).

yarns [2] with specific energy to failure similar to that established for spider silk but with higher specific strengths [2]. Still, yarn strength is much lower than that of pristine CNTs (1 TPa modulus and 100 GPa strength [3]). Hence, mechanical testing of CNTs, CNT bundles, and their interaction, with atomic resolution, presents a unique opportunity for multiscale design of materials with properties scaled to the macroscale [17]. In the case of other nanomaterials, such as semiconducting and metallic nanowires, with applications in electronics, plasmonics, and photonics, the mechanical and electromechanical properties are of relevance, although a direct link sometimes may not be explicit. However, there are several examples in which these properties are at the core of establishing the behavior of nanoscale materials and devices. Take, for example, the case of nanogenerators engineered with semiconducting, piezoelectric nanowires [18] and strips or fibers of piezoelectric materials such as lead zirconate titanate (PZT) [19]. These devices produce electrical energy, which is envisioned to power personal electronic devices, by harvesting mechanical energy coming from, for example, body motions or vibrations and converting it to electricity via the piezoelectric effect. In the case of semiconducting nanowires, mechanical and piezoelectric properties of ZnO nanowires were necessary to establish the amount of voltage produced by the nanogenerating scheme [20]. Charges were estimated by using a mathematical model that linked the elastic, piezoelectric, and dielectric properties of the material, as well as its carrier concentration. In the case of PZT nanoribbons, the mechanical properties were necessary for designing a device that can buckle and stretch the fibers to generate electricity without undergoing fracture [21].

Another example where the knowledge of mechanical and electromechanical properties has proved important is in the investigation of piezoresistance in silicon nanowires. In order to extract the piezoresistive coefficient and decouple the changes in resistance from dimensional changes, knowledge of stresses and strains in the nanowire is necessary. In a highly cited paper where giant piezoresistance in silicon nanowires was demonstrated, calculations were performed using the values of bulk silicon. However, it has been found that, depending on the orientation, the modulus of silicon nanowires changes as their dimensions decrease [22]. This points to the need for measuring all nanowire properties during electromechanical tests. Given that size-dependent elastic behavior [23] and surface states that induce electron and hole trapping [24] were not considered in the initial report, giant piezoresistance continues to be a subject of debate.

As a final example, one can cite the usage of silicon nanowires for lithium batteries, where knowledge of their mechanical properties and theoretical modeling of the deformation processes during lithiation are fundamental to understand why the nanowires withstand the extreme volume changes that make them good electrode materials [25].

The previous examples showcase that knowledge of mechanical and electromechanical properties is paramount to understand and optimize the behavior of nanostructured materials and nanoscale devices. Hence, characterization of mechanical properties in one-dimensional nanomaterials has played, and will continue to play, an important role in addressing challenges of reliability, robustness, and functionality of systems with nanoscale architectures. Having established that mechanical property characterization is an important field of study, we next discuss one of the most accurate and powerful techniques for doing so.

8.1.2 Mechanical and Electromechanical Characterization of Nanostructures: The Need for In-Situ TEM

There are several requirements for carrying out accurate mechanical and electromechanical characterization at the nanoscale. These are:

1. Precise measurements of forces and displacements/strains. We highlight the fact that it is much preferred for both of them to be measured, as otherwise an assumption of constitutive behavior (which is not always available for nanoscale materials) has to be made.
2. Precise measurement of the specimen cross-sectional area. This is important for calculating stress from the measurement of force [26]. Given the dimension of nanoscale specimens, resolution is important and errors in measurement of the cross-sectional area become more important as size decreases.
3. Knowledge of preexisting defects and crystalline and surface quality of the specimen.
4. In the case of electromechanical measurements, adequate resolution of the electrical measures (current and voltage).
5. Real-time or near-real-time observation of the experiment in order to establish cause–effect relations, metastable states, and mechanisms of fracture or failure [27,28].

The measurements of force and electrical quantities are typically more dependent on the particular setup used to deform the nanostructure. Depending on the sensor employed, the measurement of force may not depend on spatial input but rather on other electrical variables, such as electronic noise. In the same vein, the measurement of electrical quantities depends primarily on the instrumentation available, and in the establishment of proper electrical contacts to the nanostructure (ohmic or Schottky, depending on the desired measurement).

On the other hand, the measurement of strains, displacements and cross-sectional areas, and the identification of defects depend directly on the temporal and spatial resolution of the instrument where the mechanical deformation is visualized. Given the dimensions of nanostructures, optical microscopes are mostly inadequate, narrowing the possible choices to electron microscopy and scanning-probe microscopy (SPM) – the latter with some limitations, as explained below. Within electron microscopy, conventional TEM and aberration-corrected TEM have nanometer and sub-ångström resolution [29], respectively, which is much superior to conventional scanning-electron microscopy (SEM). Additionally, a number of analytical capabilities (with a probe size on the nanometer scale) complement resolution, providing information about crystalline quality, preexisting defects, and chemical composition [14,30]. Temporal resolution is another advantage of TEM, as it usually can capture images at TV rate (∼30 frames per second) [31]. This makes it superior for real-time *in-situ* testing over similar-resolution techniques such as scanning tunneling microscopy (STM) or other scanning-probe techniques (atomic-force microscopy (AFM)), where slow scanning is often required to achieve ultimate resolution. Thus, TEM is able to resolve a given area with atomic resolution in a shorter time span than would be possible with scanning-probe techniques. Recently, efforts have been reported to extend the time resolution of TEM up to 15 ns [32]; however, high spatial

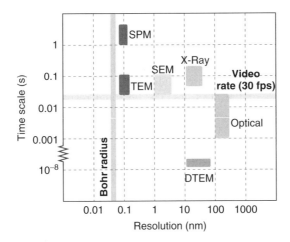

Figure 8.1 Temporal and spatial resolution of several microscopy techniques. TEM possesses high resolution in displacement without compromising time resolution. X-ray data from [34]. DTEM data from [32,33]

resolution is compromised [33]. A summary of the advantages of TEM in temporal and spatial resolution are illustrated in Figure 8.1 [30].

The identification of preexisting defects in one-dimensional nanostructures and their relevance to mechanical and electromechanical properties deserves special attention. Crystalline defects, such as stacking faults and dislocations, are known to influence the mechanical response. However, their influence on electromechanical properties is often overlooked. Inversion domain boundaries and dislocations are known to have an effect on piezoelectricity [35]. On the other hand, the role of stacking faults and surface defects is not well established , although stacking faults affect locally the band structure [35]. In the case of dislocations, given that they alter the local strain fields, a local piezoelectric polarization will be created [35]. For the case of inversion domain boundaries (i.e., the coexistence of two inverted regions of the crystal separated by a single-atom boundary [35]), the overall piezoelectric response may be cancelled. All of the aforementioned defects have been identified in nanostructures. More importantly, their identification was achieved exclusively using TEM [12,36–39], which attests to its suitability to evaluate crystalline structure in one-dimensional nanostructures.

TEM clearly possesses superior capabilities to thoroughly and unambiguously characterize the mechanical and electromechanical response of one-dimensional nanostructures. In the following sections we present a survey of the most impactful experimental setups and scientific discoveries that demonstrate how TEM has and will continue to provide unprecedented insight into the fundamental synthesis-structure-properties relations for nanostructures.

8.2 *In-Situ* TEM Experimental Methods

In Section 8.1.2, it was established that *in-situ* TEM is the technique of choice to carry out mechanical and electromechanical characterization of nanoscale specimens. Here, we will provide a short summary of the basic principles of TEM, followed by a review of the different experimental setups that have been developed to carry out testing inside the microscope, as well as the specimen preparation techniques used for positioning one-dimensional nanostructures in the testing setups.

The TEM operates by passing a coherent beam of electrons through a thin specimen. Given the wavelength of the electrons, the potential resolution is well below the size of an atom [14]. Indeed, recent advances in aberration-corrected TEMs have allowed resolutions of 80 pm [29]. The electron beam is focused through a series of electron lenses after passing through the sample, and finally impinges an electroluminescent screen or charge-coupled device (CCD) detector that provides a visible output directly correlated to the image created by the electrons passing through the sample. Movies may be recorded by continuously acquiring images from the CCD. A simplified schematic of this concept is shown in Figure 8.2a.

Given that the high-energy electrons ($\sim$50–200 keV in conventional TEMs) are ionizing radiation, some electrons from the sample can interact with the beam, causing scattering of electrons, which in turn provides information about the specimen structure and composition [14]. Many of the analytical techniques in TEM that allow elemental or other types of characterization are based on these scattered electrons. On the other hand, electrons that pass unscattered through the sample (the direct beam) are used to form the so-called bright-field image, which represents the projection of the sample along the direction of the beam. A schematic of this is shown in Figure 8.2b.

In addition, as with any wave, the electrons are diffracted by the specimen. These diffracted beams (Figure 8.2c) are directly correlated to the crystalline structure and periodicity of the sample. Thus, diffraction in the microscope can be used to identify

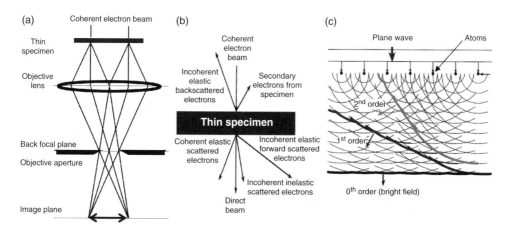

Figure 8.2 Basic concepts of TEM. (a) Simplified schematic of the electron path inside a transmission electron microscope column. (b) Schematic of the different interactions of the incident electrons with the electrons from the specimen. These interactions are useful in several analytical techniques in TEM. (c) Schematic of the formation of diffraction spots. The structure and periodicity of the diffraction pattern is directly related to the structure and periodicity of the atoms in the specimen. Adapted with permission from [14]. © 2009 Springer Science and Business Media

crystalline structure and to obtain interplanar spacing, which is useful for measuring strains in the specimen.

Of particular relevance to the investigation of one-dimensional nanostructures is the transmission electron microscope's capability of performing analytical studies, diffraction, and imaging on the scale of a few nanometers or even atomic scale, which enables thorough characterization of the nanospecimen and may uncover localized phenomena, such as amorphization, created by large inhomogeneous strains [40].

8.2.1 Overview of TEM Specimen Holders

One of the challenges of carrying out *in-situ* testing in TEM arises because the functional part of the experimental setup must fit within the limited volume available in the microscope for the specimen holder. The thickness, which is controlled by the gap between the pole pieces of the lenses, is one of the most stringent dimensional constraints if complex microsystems are used in the experimental setup. This constraint becomes more prominent as resolution increases in view of the fact that higher resolution usually means a smaller gap, the exception being some specialized aberration-corrected instruments where the gap can be up to 20 mm [27]. An illustration of the geometrical characteristics of typical TEM specimen holders is given in Figure 8.3.

For carrying out mechanical and electromechanical testing, researchers have engineered several types of holders and experimental techniques to work around these constraints. Although this may sometimes result in compromises in functionality (e.g., only having primary tilt capabilities [31]), technical advances continue to push the boundaries of what can be done inside the microscope; for example, micro-electromechanical systems (MEMS) devices that miniaturize the setup so that the size of movable parts is no longer a major concern. In Section 8.2.2, we discuss several methods and setups for mechanical and electromechanical testing and their application to the testing of one-dimensional nanostructures. In each method we present the physical principles that govern the particular measurement technique and the essential technical details of how the method is implemented. We also provide representative results obtained with each technique to

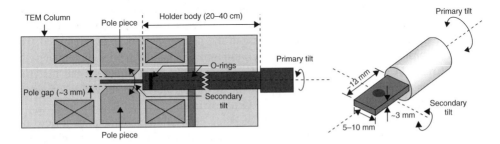

Figure 8.3 Schematic of a typical TEM specimen holder. *Left*: cross-section of a TEM column and the position of the holder. *Right*: perspective view of the holder-tip. The specimen typically has to be within a circle of 3 mm diameter (dark gray area). These dimensional constraints result in a limited size for the experimental setup

illustrate the capabilities of *in-situ* TEM testing to the mechanical and electromechanical characterization of one-dimensional nanostructures.

8.2.2   Methods for Mechanical and Electromechanical Testing of Nanowires and Nanotubes

Resonance

Resonance methods deserve a special mention, given their historical importance and simplicity. The first reports on the experimental measurement of the modulus of CNTs were published using this method [41,42]. Given the success of measuring CNT properties and its relative simplicity, the technique gained quick acceptance and has been used to characterize the properties of other materials [43,44].

In the resonance characterization method, the modes of vibration of a freestanding nanostructure are characterized and the mechanical properties are extracted based on continuum and statistical mechanical models. The nanostructure is cantilevered on a substrate and is free to vibrate at the tip. In order to observe the vibration, the nanostructure axis must be perpendicular to the axis of the electron beam and any misalignment must be prevented, using the microscope focus as a guide [41].

The resonance of the structure can either be thermal or electrostatic. Thermal resonance is observed as blurriness of the tip of the nanostructure and is intrinsic to any system. Electrostatic resonance is induced by the application of an oscillating electric field to the nanostructure.

Thermal resonance has been used to characterize the elastic moduli of carbon [41] and boron nitride nanotubes [45]. A continuum mechanics model of a cylinder, excited by thermal vibrations, is used to relate the geometry, temperature, and modulus to the measured vibration amplitude. The vibration amplitude at the nanostructure tip as a function of temperature [41], or as a function of position (keeping the temperature fixed) [45], can be used to extract the elastic modulus.

On the other hand, electrostatic resonance has been used to characterize the elastic moduli of CNTs [42], tungsten oxide nanowires [43], and gallium nitride nanowires [44]. In this method, an oscillating voltage is applied between the specimen and a counterelectrode positioned nearby. This creates an oscillating electric field that, in turn, induces oscillations in the nanostructure. By sweeping the frequency of the voltage, the resonance frequencies (fundamental or harmonics) of the nanostructure can be identified. The mechanical properties are extracted via a continuum mechanics model that relates the measured resonance frequency, geometry, density, and elastic modulus of the material. Note that this method has also been applied in SEM for the characterization of elastic modulus of other nanostructures, such as ZnO nanowires [46].

The resonance method has the advantage of being easily implementable, as it does not require a very sophisticated TEM holder (see Figure 8.4a). In addition, sample preparation is relatively straightforward (see "Random sample preparation" in Section 8.2.3). However, only the measurement of elastic modulus is possible, while measurements of stress–strain response and fracture strength and strain are not. The difficulty of precisely identifying the amplitude of thermal vibration has also been pointed out to lead to experimental uncertainties [42].

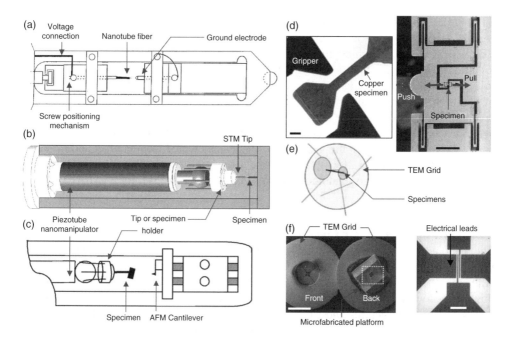

Figure 8.4 Schematics of some of the different setups available for TEM electromechanical characterization; (a)–(c) show the particular TEM holder tip. (a) Holder tip for the electrostatic resonance method. Note the two terminals used to introduce an AC electrostatic field to the specimen. Reprinted with permission from [106]. © 2000 International Union of Pure and Applied Chemistry. (b) STM–TEM system where an STM head is used to deform the sample and electrical measurements are possible. Reprinted with permission from [64]. © 2003 American institute of Physics. (c) An evolution of STM–TEM is the AFM–TEM, where a cantilever can be used to measure force and, in selected cases, also apply electrical signals. Reprinted with permission from [80]. © 2008 IEEE. (d) Nanoindenter-based extensions for one-dimensional nanostructure testing. On the left, a gripper tensions an FIB-fabricated copper specimen. Scale bar: 200 nm. Reprinted with permission of [60,61]. © 2011 American Chemical Society. On the right, a microfabricated flexure converts the compressing motion of the indenter to tension. Scale bar: 100 μm. (e) TEM grid covered with a collodion thin film that stretches under e-beam irradiation, thus straining the specimens. Reprinted with permission from [45]. © 2007 American Chemical Society. (f) TEM grid with a microfabricated platform allowing four-point measurements. The image on the right shows a magnified view of the dashed square on the left. Scale bars: 1 mm, 50 μm. Reprinted with permission from [104]. © 2006 Institute of Physics Publ. Ltd

TEM–SPM and Nanoindenter-Based Compression, Tension and Bending

Some of the resonance setups that were proposed in the early years had, as one of the two electrodes, a small movable tip actuated by a piezoelectric tube scanner, similar to the early setups of STM. This initial idea of having a nanometer-precision movable tip opposing another electrode, all embedded in a TEM specimen holder, has evolved into three main setup types. When these are combined, they account for the majority of reported results of mechanical, electrical, and electromechanical tests of nanostructures in *in-situ* TEM. These classes are respectively, the nanoindenter setup, the so-called

STM–TEM (STM–Scanning Tunneling Microscope), and the so-called AFM–TEM (AFM–Atomic Force Microscope). The latter two are often grouped under the name TEM–SPM (SPM–Scanning Probe Microscopy). The common idea in these setups is to utilize the moving tip to deform the sample in some fashion while visualizing it in *in-situ* TEM. Depending on the setup, the electrical and mechanical consequences of said deformations can be measured and correlated. Common to all the setups is a predominance of compressive deformations, either homogeneous in the case of pillars or heterogeneous when slender specimens are compressed and later buckled or bent.

The exception to this are the atom-sized specimens created from contact and retraction of the two electrodes (e.g., [53]). These three setups will be briefly explained below; for a comprehensive review of nanoindentation setups, see Stach [54]; for STM/AFM–TEM, see Nafari *et al*. [31].

Nanoindentation and Pillar Compression

The nanoindenter setup was initially implemented using a piezo-tube scanner pushing against the specimen to impose deformation through a diamond tip or punch [55]. Although indentation as a technique to impose deformation is outside the scope of this review, we mention it because this setup has evolved towards the testing of one-dimensional nanostructures and can be considered a precursor of the TEM–STM setup.

Nowadays, commercial nanoindenter setups exist and are widely used. They consist of an electrostatic actuator that employs capacitive sensing for the measurement of force [56]. These setups are typically used to impose forces in the order of micronewtons to pillar samples fabricated by focused-ion-beam (FIB) milling. Until very recently, the pillars were in the micrometer regime and, therefore, are not considered one-dimensional nanostructures. Nevertheless, there exists an extensive body of literature on metallic nanopillars, and important phenomena such as size-scale plasticity were investigated using these setups; see Greer and De Hosson [57] and Agrawal and Espinosa [58] for recent reviews. Nowadays, some limited reports exist for submicrometer experiments in *in-situ* TEM (e.g., [59]) and SEM. One example where this class of setups has been used for nanostructures is the work by Huang *et al*. [60], where submicrometer GaN nanowires with very low aspect ratio (in the guise of pillars) were compressed to failure (see "*In-situ* HRTEM of crystalline nanospecimens" in Section 8.3.1).

An important new development that will likely extend the range of this technique towards submicrometer one-dimensional nanostructures is the development of modified nanoindenter tips and microfabricated stages for tensile testing. In the first case, dog-bone specimens with dimensions in the 100–200 nm diameter regime, fabricated by FIB, are pulled by a nanoindenter tip in the shape of a gripper. This setup was employed in copper specimens to prove that, in the submicrometer regimes, dislocation sourcing and exhaustion influence simultaneously the hardening behavior in metallic specimens [50]. In the second case, a microfabricated flexure was used to convert the compressing motion of the nanoindenter into tension. With this implementation, a 270 nm vanadium dioxide (VO_2) nanowire was strained to achieve phase transformation. The Young's moduli of both M1 and M2 phases in VO_2 were measured (128 ± 10 and 156 ± 10 GPa) [51]. It must be pointed out that, given the resolution of the nanoindenter setup (sub-micronewton [56]), tests of smaller diameter samples, as achieved by other techniques, will require the development of load sensors with greater resolution.

TEM–STM

Using TEM–STM, several interesting phenomena have been probed, such as superplasticity in CNTs [61], wall-by-wall current-induced breakdown of multi-walled CNTs (MWCNTs) [62], and the first report on quantized conductance through rows of individual atoms of gold [53]. These are very relevant examples of the employment of high-resolution TEM (HRTEM) to discover new nanoscale phenomena.

The TEM–STM setup consists of two opposing electrodes in the frame of the TEM specimen holder. One of them is fixed and the other is moved by a piezoelectric setup that can operate in fine and, in some cases, coarse motion-regimes, the latter enabled by an inertial mechanism [48]. Depending on the type of experiment, the specimen can be positioned in either electrode and the deformation is imposed by the movement of the movable electrode. In most cases, a tip is positioned in the movable electrode in order to be able to target individual nanostructures. An example of the experimental setup is shown in Figure 8.4b.

The initial motivation for the development of a TEM–STM setup was to observe the nature of the contact between tip and surface in an STM experiment [63]. This continued to be an active area of research, with recent demonstrations of atomic-resolution scans performed in this type of instrument [48]. However, nowadays this technique is rarely used in the traditional sense of an STM experiment [31], where the tip performs a raster scan. Rather, the movement of the tip is used to introduce deformations in the nanostructure under study while the electrical properties (current versus voltage relation) are measured using the electrical capabilities provided by the STM tip. Incidentally, this type of setup has also been used for studying friction in surfaces *in-situ* in TEM [64].

One of the limitations of this setup is the lack of measurement of force. As a result, the experiments are limited to correlating the observed deformations with changes in electrical properties, but knowledge of the forces is nonexistent or estimated from the strain state. The most usual type of test is the compression and posterior bending, buckling, and fracture of a nanowire or nanotube and the high-resolution observation of the process; for example, see Golberg and coworkers [65,66]. Nevertheless, given its relative simplicity, STM–TEM continues to be used, accounting for a large portion of the literature on *in-situ* TEM testing of nanostructures.

TEM–AFM

Following the developments of STM–TEM and nanoindenters in *in-situ* TEM, more sophisticated setups have been proposed. In particular, instead of just a simple electrode opposing the nanomanipulator, one can position an AFM cantilever, which serves as a force sensor. In this way, mechanical measurements with nanonewton force resolution are possible. Additionally, if the AFM cantilever is conductive and connected to an electrode, coupled electrical and mechanical measurements are possible. An example of the experimental setup is shown in Figure 8.4c.

With this implementation, kinking of CNTs was observed, where a yield strength of 1.7 GPa was measured [67]. Furthermore, atom-sized specimens, obtained by pulling apart a substrate and an AFM tip coated with the same material, have been created. In particular, Au, Cu, and Pd point contacts [68–70] and silicon nanowires [71,72] have been tested. Additional mechanical measurements have been carried out, including filled CNTs [66,73–75] and ZnO nanowires [76]. In the first studies, the difference in mechanical properties between filled and empty nanotubes was established, as well as several mechanisms of failure, such as kinking. In the latter study, an elasticity size effect in nanowires previously reported by Agrawal *et al.* [77] was observed; additionally, amorphization of the nanowires under repeated bending loads was demonstrated using high-resolution imaging and diffraction (see Section 8.3).

There are three main sub-classes of this type of holder, differentiated by the method utilized to measure the force; that is, the deflection of the AFM cantilever. In the first case, the deflection is simply measured from images obtained in the bright-field mode. In the second case, the laser optical lever system used in traditional AFMs is implemented inside the transmission electron microscope. In the third case, a piezoresistive cantilever is used to electronically sense the deformation.

Direct imaging of the cantilever has the advantage of simplicity, but it can compromise resolution and accuracy because relatively low magnifications are required if the whole cantilever is imaged. One can use high resolution and image just the contact zone and measure the cantilever deflection based on the displacement of the AFM tip, but this assumes that the field of view of the transmission electron microscope does not drift, which is not always the case. Furthermore, time resolution is limited to the rate of images captured, thus presenting a compromise: the faster the images are captured, the better the time resolution is, but the image quality decreases. This is a problem if atomic resolution is needed.

On the other hand, implementing an optical laser system to detect the cantilever deflection is very challenging, but it provides considerable advantages in terms of force and time resolution. This setup was implemented by Kizuka *et al.* [68], who used one of the goniometer ports of the transmission electron microscope to implement the optical setup necessary for detecting the motion of the AFM cantilever. The advantage of this system compared with the previous one is the possibility of obtaining real-time measurements of force with sub-nanonewton resolution. A remarkable example of the advantages of real-time, high-resolution measurements of force is the synthesis of stable, high aspect ratio, atomic-scale-width silicon nanowires by withdrawing a point contact, which required continued monitoring and adjusting of the pulling force in order to obtain a stable structure [72]. Coupled mechanical and electrical measurements and atomic-resolution imaging of this specimen were performed, establishing that these atom-sized nanowires can withstand $10^9 - 10^{11}$ A m^{-2} of current density.

Finally, implementing a piezoresistive sensing of the cantilever deflection provides a compromise between the two aforementioned approaches: force resolution is lower, on the order of 15 nN [49], but real-time measurement of force is possible. One possible caveat is the drift induced in the sensors by irradiation of the TEM beam [49]. Additionally, custom microfabrication is required to deposit a piezoresistive film over the cantilever and interface it with the setup.

MEMS-Based Testing

MEMS have emerged as the most advanced alternative to create test setups for the mechanical and electrical characterization of nanostructures and other specimens. The use of traditional microfabrication techniques allows the creation of diverse geometries for mechanical testing in a very precise way, which is beneficial for creating tests with carefully controlled boundary conditions. By combining electronically controlled actuation and sensing, these devices have significant advantages, as they apply and measure load and strain independently while simultaneously allowing for uninterrupted observation of the specimen in HRTEM [78].

Several nanomaterials have been tested with these devices. In particular, size effects in the elastic and fracture behavior of ZnO and GaN nanowires [26,77,79], irradiation improvements in the mechanical properties of CNTs and CNT-based materials [3,80], and plasticity in penta-twinned silver nanowires [81] have been investigated.

The precursor of the aforementioned MEMS for mechanical testing was developed by Haque and Saif [82]. These devices were not strictly MEMS, since electronic sensing and actuation were not implemented. Instead, microfabricated flexures were adapted to be actuated by traditional TEM straining holders [83]. These devices allowed measurement of strain by TEM observation, but the measurement of force required tracking the movement of a structure with known stiffness. This tracking was achieved by TEM observation as well, which implies that the beam needed to be shifted from the specimen. Moreover, these devices have been mostly used to test thin films or micrometer-sized samples [84,85].

A MEMS device for nanostructure testing, combining electronic sensing and actuation, was developed by Espinosa and coworkers [78,85–88]. The complete setup for *in-situ* TEM testing using this MEMS device is shown in Figure 8.5. A custom TEM holder (Figure 8.5a–c) allows positioning of the MEMS chip inside the microscope and electrical addressing of the electronics for sensing and actuation [85,89]. Figure 8.5a shows a

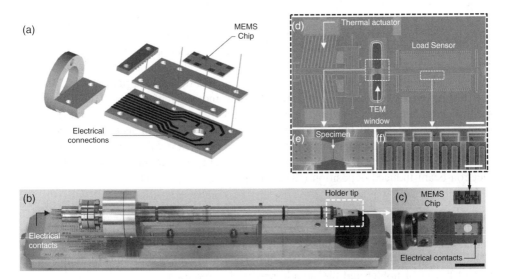

Figure 8.5 MEMS devices for *in-situ* TEM testing. (a) TEM holder tip showing the interfacing of electrical connections with the terminals of the MEMS chip. Reprinted with permission from [92]. © 2005 Cambridge University Press (b, c) TEM holder for the MEMS chip. Reprinted with permission from [88]. © 2010 Cambridge University Press. Scale bar in (c) is 10 mm. (d) SEM micrograph of the MEMS device developed by Espinosa and coworkers. Note the thermal actuator, load sensor, and the window allowing observation of the sample in TEM. Scale bar: 100 μm. (e) Magnified view of the shuttles where the specimen is positioned. Scale bar: 40 μm. (f) Magnified view of the differential capacitors for detecting displacement of the load sensor. Scale bar: 15 μm

detailed view of the holder tip, where it can be seen how the device chip fits in the holder and the way electrical connections are achieved. These connections are then routed outside the TEM by contacts running inside the holder and through a flange (Figure 8.5b).

The MEMS devices have a thermal actuator and a capacitive load sensor (Figure 8.5d). The specimen is located in between these two (Figure 8.5e) and is positioned in the device using a nanomanipulator. Determination of the specimen strain is achieved by direct observation in TEM, either in bright-field or in diffraction mode (see Section 8.3.2). For force measurement, the displacement of a previously calibrated load sensor is measured using a set of differential capacitors (Figure 8.5f).

Although this technique allows carefully controlled testing, the sample preparation and operation of the device is challenging (see "Nanomanipulation" in Section 8.2.3), lowering the overall throughput of testing. Advances in *in-situ* growth of nanostructures and directed growth of nanostructures should contribute to increase the number of tests that can be performed (see Section 8.2.3). This becomes relevant when strength and failure of nanostructures are studied, as these phenomena are inherently stochastic, requiring performing several tests to achieve reliable statistics [79].

It should be noted that, recently, many other groups have developed similar MEMS for similar applications, although *in-situ* TEM operation has not been demonstrated [90–93]. Nanofibers [94] and biological specimens such as collagen fibrils [95] have been tested either in scanning electron or optical microscopes.

Important new developments in this type of *in-situ* TEM testing are the introduction of devices with the capability to perform mechanical and electrical measurements either separately [96] or simultaneously [85,92,93]. This should allow the thorough probing of properties such as piezoelectricity and piezoresistance [93] in the near future.

In particular, Espinosa and coworkers have recently developed a four-point microelectromechanical testing system that combines the aforementioned advantages in mechanical testing with true resistance measurements [97]. In this device, the platform for mechanical testing was extended to accommodate four independent electrical connections to the nanostructure. As shown in Figure 8.6a and b, four traces from the outside electronics are connected to the specimen (Figure 8.6b). These connections are fabricated on top of insulating freestanding silicon nitride layers [85] to ensure that all electrical signals are independent from each other. The specimen is contacted by ion- (IBID) or electron-beam-induced deposition (EBID) of platinum (Figure 8.6c).

Other Techniques

Other, less-reported techniques exist for mechanical and electrical testing in *in-situ* TEM. They incorporate functionality in a standard TEM grid which simplifies the implementation of the TEM holder and the sample preparation required.

Han and coworkers [40,98] used TEM copper grids covered with a collodion thin film (Figure 8.4e) that bends under electron-beam irradiation as it undergoes polymerization, presumably the result of electron-beam-induced heating. Silicon or silicon carbide (SiC) nanowires are deposited randomly on the grids and get deformed either in tension or

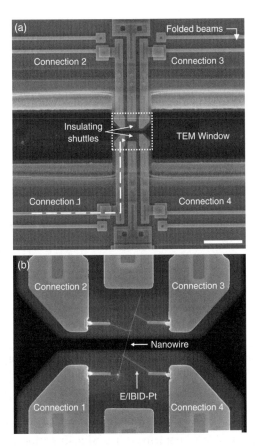

Figure 8.6 MEMS device for four-point electromechanical measurements. (a) Using the same platform shown in Figure 8.5d, the shuttles of the MEMS device have been modified to perform four-point measurements. Four electrical connections extend on top of insulating silicon nitride shuttles and come close to the specimen. The path of one of the connections is indicated by a dashed line. Scale bar: 40 μm. A detail of the dashed square is provided in (b) where a nanowire is connected for four-point measurements using electron and ion beam deposition of platinum. Scale bar 5 μm [97]

bending as the supporting film is deformed under the electron beam. One important aspect is that the collodion film is able to shrink up to 4–5%, which induces large strains in the nanowires, up to 125% [98]. One disadvantage of the method is the lack of measurement of force.

On the other hand, Xu *et al.* [52] developed a microfabricated platform for four-point electrical measurements of nanowires (Figure 8.4f) that can be bonded in a standard TEM single-hole aperture grid. The four probes have a separation of 1–2 μm, which allows

Table 8.1 Comparative summary of the loading mode, capabilities, resolution, and throughput of several techniques for *in-situ* TEM mechanical and electromechanical testing of one-dimensional nanostructures

| Method | Loading mode | Measurement capabilities[a] | | Electronic readout *in-situ* TEM | Force resolution | Testing throughput |
		Mechanical	Electrical			
Resonance TEM–SPM	Bending	E	N/A	No	N/A	High
Nanoindenter based	Compression/ tension	E, σ, ε, σ_f, ε_f	N/A	Yes	Sub μN	Low
TEM–STM	Bending/ tension	E, ε, ε_f	R	Yes	N/A	High
TEM–AFM	Bending/ tension	E, σ, ε, σ_f, ε_f	R	In some setups	15 nN [49]	Medium
MEMS	Compression/ tension	E, σ, ε, σ_f, ε_f	R, ρ	Yes	12 nN [78]	Low
Straining grid	Tension (not well controlled)	ε, ε_f	N/A	No	N/A	High
Four-point grid	N/A	N/A	R, ρ	No	N/A	Low

[a] E: elastic modulus; σ: stress; ε: Strain; σ_f: failure strength; ε_f: failure strain; R: electrical resistance of specimen and contacts; ρ: resistivity of specimen.

testing of the electrical properties of nanowires. With this setup, post-mortem inspection of pretested nanowires can be carried out in TEM. Even when this is not an *in-situ* technique, it could be possible to implement it *in situ* if the four contacts are addressed with a custom TEM holder.

As a conclusion to this section, we summarize the loading modes, capabilities, resolution, and throughput of the techniques that were discussed in Table 8.1.

8.2.3 Sample Preparation for TEM of One-Dimensional Nanostructures

As the reader may appreciate, all of the aforementioned techniques for nanomechanical testing require isolating an individual nanostructure in order to carry out the experiment. Depending on the technique, the sample preparation varies in method and level of difficulty. Here, we survey the different methods of sample preparation. Nafari *et al.* [31] and Costa and coworkers [66] have also recently reviewed some of the methods described herein. Note also that we review here some techniques that have been applied in other contexts aside from *in-situ* techniques but that are suitable for TEM sample preparation for mechanical and electromechanical testing.

Random Sample Preparation

For techniques that use a nanomanipulator, tip, or electrode to compress, stretch, or resonate nanostructures, sample preparation relies on sheer statistics. That is, a large number of specimens are prepared at the same time and a suitable sample is located *in situ* for the experiment. For CNTs, a CNT-rich fiber, bundle, or buckypaper is directly attached to a metallic wire [41,42] or using silver paint to ensure electrical contact [47,99,100]. A similar method consists of lightly rubbing a wire, previously dipped in silver paste, against CNT powder [101–103]. This last method has also been used for boron nitride nanotubes [65,104], tungsten [105], and ZnO nanowires [76]. Alternatively, for the testing of GaN nanowires and InAs nanowhiskers a piece of the growth substrate was attached directly to one of the electrodes using silver paste [44,106]. A caveat of this technique is that only nanostructures on the edges of the substrate are accessible because of unavoidable small misalignments in the mounting. This means that, in order to preserve the nanostructures, proper care needs to be taken in the cutting of the substrate (cleaving is preferred) [31].

Dielectrophoresis has also been used in order to obtain many specimens in a tip that is later attached to the sample holder. Here, an AC voltage is applied between two electrodes while they are immersed in a solution containing the nanostructures, resulting in many of them sticking out of an electrode once the solution dries up. This method has been used to prepare CNTs [101] and ZnO nanowires [107].

Nanomanipulation

We refer to nanomanipulation as the method where an individual nanowire or nanotube is inspected and selected for testing and later positioned on the testing stage. This technique was originally reported for positioning of nanowires in a MEMS stage for mechanical testing [78]. Here, we review the methods used to position individual nanostructures in the test stages using nanomanipulators either *in situ* or *ex situ* the transmission electron microscope. We will digress into sample preparation methods that do not necessarily target the transmission electron microscope as a testing instrument because the techniques used to mount nanostructures on stages (even if they are not intended for TEM) can be applicable to TEM if the stage itself can be incorporated in the TEM holder (see "MEMS-based testing" in Section 8.2.2).

Two main manipulation targets (meaning the final destination of the nanostructure) can be identified in the literature: microfabricated stages or MEMS devices (which may or may not be used in the transmission electron microscope *in situ*) [3,26,77,79,80,91,93] or *in-situ* TEM holders which have a tip holder [108].

Manipulation with a testing stage as target proves to be difficult, since the nanostructure has to be attached to two surfaces. Three stages are necessary: first, a suitable nanostructure source is prepared; second, a nanostructure is selected and detached from the source with a nanomanipulator; and third, the manipulator is used to place the nanostructure on the target setup (see Figure 8.7a)

For the first stage, most successful results are obtained when the nanostructures are deposited on TEM grids [3,8,26,77,79,80]. They provide a suitable support structure, while some portion of the nanostructure sticks out, allowing for manipulation. Nanostructures can be mass-placed in the TEM grid either by sonication followed by drop-casting [77], or by mechanical exfoliation, achieved by sliding half-cut TEM grids over the substrate or growth source [2,8,110].

After the nanostructures are in the grid, detachment of them from the grid is performed. Typically, a very sharp tungsten tip is attached to a manipulator and then is moved toward

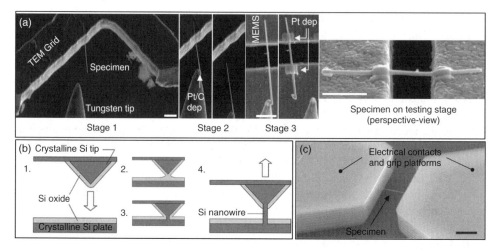

Figure 8.7 Examples of current and potential sample preparation methods for *in-situ* TEM testing. (a) The sequence of nanomanipulation, where a specimen is transported from a TEM grid to a testing stage with the aid of a nanomanipulator and metal deposition in *in-situ* SEM. See text for further details. All scale bars: 2 μm. (b) *In-situ* TEM specimen preparation, in this case of silicon nanowires. A tip and surface of the same material are brought together and then separated, forming a nanoscale specimen. Reprinted with permission from [73]. © 2007 The Japan Society of Applied Physics. (c) Illustrates a promising approach for improving testing throughput as yet not implemented for *in-situ* TEM studies. The specimen is grown directly in the testing platforms, which ensure good electrical contact and well-defined boundary conditions. Scale bar: 2 μm. Reprinted with permission from [22] © 2006 Nature Publishing Group

the nanostructure. Once contact is established, some form of attachment needs to be enforced in order to detach or break off the nanostructure from the grid. It is desirable that the adhesion of the specimen to the tip is greater than what it is to the grid [78,91]. This may happen spontaneously, depending on the materials and conditions of the sample and manipulator tip, but more often EBID of an additional material is required [78]. This deposition can be either of residual hydrocarbons in the SEM chamber, resulting in an amorphous carbon deposit [80], or of other materials such as platinum [78], copper, or tungsten [111]. After attachment of the specimen to the tip is achieved, detachment from the grid is attempted. This requires a careful retraction, or a controlled way of cutting the specimen; for example, by electron-beam etching [2]. Although FIB may be used to perform controlled cutting, its effect on the specimen may be detrimental and should be avoided.

Once the nanostructure is attached to the tip, the manipulator is used to move the tip close to the testing setup. A careful approach to the target is required, typically using the focus of the SEM to judge depth. By careful approach, one can assess contact when electrostatic attraction of the sample to the surface of the testing setup occurs. Once contact is established, the nanostructure may be detached from the nanomanipulator if the adhesion to the surface is greater than the adhesion to the tip–nanostructure weld [78,91]. If this is not possible, EBID is used again followed by detachment of the tip. One of the most challenging aspects of this stage of nanomanipulation is judging depth and perspective based on the two-dimensional image of the SEM. A strategy to overcome this difficulty is the patterning of guiding structures in the target [91].

The nanomanipulation technique may also be applied to mechanical testing of nanostructures in the SPM–TEM setups. Most of these TEM holders that integrate a nanomanipulator have the advantage of possessing a detachable tip that can be used for sample preparation *ex situ*. Indeed, Asthana and coworkers [108,112] used a nanomanipulator inside an FIB system to pick a nanowire from the growth substrate and attach it to a tungsten probe, using IBID of tungsten. After this, the tip is mounted on an STM–TEM-type *in-situ* TEM holder. Under this methodology of preparation, ZnO nanobelts and titanium dioxide nanotubes [112] were tested.

Other Techniques

Other techniques have been used for TEM sample preparation. They include *in-situ* sample preparation and directed growth or cofabrication of the nanostructure in the testing setup.

For the first method, samples are *in-situ* prepared for testing using a combination of mechanical forces and adhesion. Kizuka *et al.* [72], Luo *et al.* [113], and Moore *et al.* [114] used a nanomanipulator tip in *in-situ* TEM in order to press a surface of the material of interest. Retraction of the tip causes pulling of the material in a nanowire shape that, although not very well defined, is narrow, long, and with very small dimensions (see Figure 8.7a). Silicon nanowires with extremely high fracture strains (30%) [72], superelongation (200%) of metallic glasses at room temperature [113], and superplastic (elongation 280%) salt nanowires [114] were demonstrated using this approach. The point contacts mentioned in "TEM–AFM" in Section 8.2.2 are fabricated using this method.

Using a combination of the random dispersion method and *in-situ* sample preparation, Lu *et al.* demonstrated cold welding of gold nanowires [115]. They started with ultrathin nanowires prepared by a chemical method, attached them to a nanomanipulator tip and to an opposing wire using methods similar to those outlined in "Random sample preparation" in Section 8.2.3. When two of these ultrathin nanowires are brought into contact they cold weld together by applying pressures of less than 4.7 MPa, which is much lower than typical cold welding of bulk metals. The welded nanowire does not have grain boundaries or additional defects, and this was demonstrated by electrical and mechanical measurements where the current conduction was not affected by the welding and the nanowires fractured in a different location far from the weld. Another type of attachment mechanism is the amorphous carbon produced by concentrating the TEM in a very small area [116,117].

In general, *in-situ* specimen preparation has been demonstrated to be useful in order to test small-sized samples (<10 nm) which are typically employed in testing fundamental phenomena, such as quantized conduction [53]. However, the technique is not amenable for testing batch-produced nanotubes or nanowires (used in the majority of demonstrations of new electronic devices) and, therefore, its applicability is limited.

Directed growth is an alternative for sample preparation where the nanostructure is cofabricated with the specimen or grown directly in the testing setup. It has advantages because it may lead to more robust boundary conditions of the specimen. Moreover, electron-beam-deposition processing steps are avoided, lessening the potential for sample contamination [43]. In addition, there is the potential to improve throughput. Liu *et al.* [43]

were able to grow tungsten oxide nanowires in tungsten tips. These tips were later used inside the transmission electron microscope as one of the electrodes in the resonance technique. This technique was remarkable, as it was possible to test nanowires as small as 16 nm, obtaining atomic resolution and characterization of the specimens under investigation. They discovered that the modulus of tungsten oxide nanowires increases as much as 300% for small samples. In a similar approach, but not *in-situ* TEM, He and Yang [109] (see Figure 8.7c) grew sub-100 nm silicon nanowires between two separate platforms that were used to later uniaxially stretch the nanowires while measuring their current–voltage response, providing the first evidence of size-dependent piezoresistance of silicon nanowires. In both cases, TEM inspection of the samples was performed, confirming the suitability of the electrical (ohmic) and mechanical boundary conditions (clamped). This setup was later used to carry out mechanical bending experiments using AFM [118]. Overall, although directed growth is an attractive approach, many challenges remain in order to be able to grow and test the full spectrum of materials (metallic, semiconducting, CNTs) in this fashion. The particular synthesis and processing conditions will need to be investigated for each material and may conflict with other parts of the microfabrication of the testing setup.

In terms of cofabricated samples, strictly speaking, one-dimensional nanostructures have not been tested in *in-situ* TEM. Instead, only very thin films of aluminum [82,84] and silicon [119] have been studied. However, advancements in the minimal sample dimension that is obtainable in this microfabrication top-down approach [120] may allow the testing of true one-dimensional nanostructures in the near future.

8.3   Capabilities of *In-Situ* TEM Applied to One-Dimensional Nanostructures

It was mentioned in Section 8.1 that TEM is the technique of choice in order to get appropriate measurements of the mechanical properties of nanowires and nanotubes. Up to this point we have given an overview of the experimental techniques and sample preparation methods that allow the testing of nanostructures in *in-situ* TEM. In this section, we aim to illustrate with specific examples why TEM gives superior measurements of mechanical and electromechanical properties. As mentioned before, atomic-scale resolution is the most obvious advantage, since preexisting defects and atomic structure can be identified and correlated to the mechanical properties. However, TEM allows for a number of other analytical techniques, which when coupled to the high resolution allow the thorough characterization of one-dimensional nanostructures. In this section, we aim to illustrate examples where these extra capabilities have played a role in determining the mechanical and electromechanical properties of one-dimensional nanostructures. Clearly, nanostructure characterization in TEM has been carried out in several other cases, but we limit ourselves here to cases where it has been used in conjunction with mechanical or electromechanical testing.

8.3.1   HRTEM

High resolution is, perhaps, the most obvious advantage of TEM. As such, we aim in this section to provide a few case studies where the use of HRTEM was critical to obtain conclusions related to mechanical or electromechanical behavior. In particular, we aim to highlight a few examples where the high resolution of TEM was directly linked to measurements leading to a structure–properties correlation. We first present examples on mechanical and electromechanical characterization of nanotubes, followed by studies on nanowires. The most prominent advantages of HRTEM in mechanical and electromechanical testing are the identification of preexisting defects, the precise quantification of load and the current or load-bearing area, and the imaging of the failure mechanisms with atomic resolution.

In-Situ HRTEM Testing of Nanotubes

The majority of *in-situ* electromechanical TEM studies at high resolution have focused on CNTs. Fracture surfaces, kinks, and their atomic structure under deformation have been reported. Here, we highlight studies in which a direct correlation was demonstrated between measured properties and high-resolution imaging.

In terms of mechanical properties, measurements of modulus, fracture strength, and elucidation of the fracture mechanisms were achieved. One of the earliest examples was given by Poncharal *et al*. [42]. They measured the elastic modulus of CNTs for several diameters using the electromechanical resonance method (see "Resonance" in Section 8.2.2 for a review of the method). It was reported that the elastic modulus for very thin CNTs agreed with theoretical estimates of 1 TPa, while for diameters greater than 12 nm there was a sharp transition and the measured modulus dramatically decreased to around 100 GPa for larger diameters. However, HRTEM revealed that CNTs of larger diameter developed a waving/rippling of the shells when they undergo large bending deformation (see Figure 8.8a). This provided an explanation for the decrease in modulus for large diameters, as a result not of material-property degradation with size, but of a change in the deformation mode.

More recently, Peng *et al*. [3] measured the fracture strength of multi-walled CNTs (MWCNTs) in *in-situ* TEM using a MEMS-based tensile testing device (see "MEMS-based testing" in Section 8.2.2). Here, observation of the number of fractured shells in the nanotube was possible, providing an unambiguous and precise measurement of the load-bearing cross-sectional area (see Figure 8.8b). Sword-in-sheath failure was observed, in which fracture of only one (or in some cases a few) of the load-bearing shells of the MWCNT occurred. The number of fractured shells was found to increase when the MWC-NTs were subjected to increasing doses of electron irradiation, which introduced covalent cross-linking defects between shells. These findings proved the benefits of irradiation-induced cross-linking on the load-bearing capabilities of MWCNTs. In this case, HRTEM observation of the number of fractured shells was a critical requirement for the accurate interpretation of the experimental data.

A similar study, this time carried out with an STM–TEM setup (see "TEM–STM" in Section 8.2.2), demonstrated a sword-in-sheath fracture mechanisms for tungsten disulfide (WS_2) nanotubes [121]. In this work, the authors performed some of the testing by *in-situ* SEM, which does not allow for the direct and unambiguous identification of the failure mechanisms, thus exemplifying the contrast between TEM and SEM in terms of resolution and suitability of the techniques for mechanical testing of nanostructures.

Recently, Filleter *et al*. [80] carried out an *in-situ* TEM study on double-walled CNT (DWCNT) bundles. Again, the direct observation of failure mechanisms of the bundles

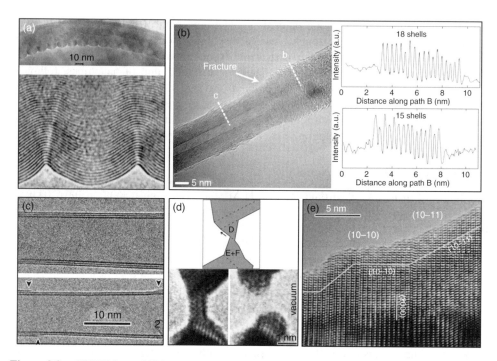

Figure 8.8 HRTEM capabilities applied to several electromechanical characterizations of nanostructures. (a) Observation of bending-induced kinking in MWCNTs. Reprinted with permission from [47]. © 1999 The American Association for the Advancement of Science. (b) Identification of number of shells fractured in a tensile test of an MWCNT. Reprinted with permission from [3] © 2008 Nature Publishing Group. (c) Shell-by-shell current-induced failure of MWCNTs. Reprinted with permission from [63] © 2005 The American Physical Society. (d) Observation of slip (indicated by an arrow) occurring as a precursor to failure in palladium nanocontacts. Reprinted with permission from [72]. © 2009 The Japan Society of Applied Physics. (e) Identification of fracture planes in GaN nanowires subjected to uniaxial compression. Reprinted with permission from [59]. © 2011 American Chemical Society

was critical in understanding the effect of electron irradiation dose. Here, by directly measuring the number of CNTs across the diameter of the bundles, the effective modulus was accurately determined and was found to increase by up to one order of magnitude (30–60 GPa to 693 GPa) at an optimal dose of $(8.9 \pm 0.3) \times 10^{20}$ e^- cm^{-2}. Moreover, a transition between sword-in-sheath and full cross-section failure was observed, demonstrating the benefits of cross-linking in promoting the full utilization of the CNTs at the bundle level. Likewise, the effective strength increased from 2–3 GPa to 17.1 GPa at a dose of $(11.3 \pm 0.3) \times 10^{20}$ e^- cm^{-2}. Such a significant improvement in mechanical performance at higher levels of hierarchy in CNT-based materials suggests promise in developing macroscopic materials that approach the exceptional properties of individual nanostructures.

In the context of electrical properties, a very compelling example of the power of HRTEM was provided by Huang *et al.* [62]. They imaged the wall-by-wall breakdown of an MWCNT when a critical current density was applied to it (see Figure 8.8c). Discrete jumps in the current through the MWCNT were directly correlated with one-by-one shell failure. Interestingly, they were able to establish that failure does not necessarily progress from the outer to the inner shells, but rather in an alternate way in which outer-to-inner and inner-to-outer breakdown sequences are possible. These findings, which have only become possible by the direct visualization, provided by *in-situ* TEM, may have significant implications on the use of CNTs in electronic devices.

In-Situ HRTEM of Crystalline Nanospecimens

Several electromechanical studies of crystalline nanospecimens have been carried out by *in-situ* TEM, including nanowires of metallic and semiconducting materials and point junctions in metallic specimens. Perhaps the most striking example of the use of HRTEM in electromechanical experiments is given by experiments using break junctions where specimens of atomic width have been tested. These types of specimens are fabricated by putting a tip in contact with an opposing surface of the same material and carefully retracting the tip. A review of this work is given by Agraït *et al.* [122]; here, we present some relevant examples.

In a classic paper, Ohnishi *et al.* [53] fabricated break junctions of gold and were able to stretch the junctions so that the cross-section was reduced by one row of atoms at a time. They simultaneously measured the conductance of the junction, demonstrating that it is quantized by the unit conductance $G_0 = 2e^2/h$, where e is the electron charge and h is Planck's constant. As the cross-section reduced by one row of atoms, the conductance reduced by one unit of G_0. Here, HRTEM was critical to establish the number of atomic rows present in the specimen. The width of the specimen was established by direct imaging, because the crystalline structure allowed counting atomic rows. The depth of the specimen was estimated from the gray intensity present in the images.

The majority of these experiments were carried out in an STM–TEM configuration (see "TEM–SPM and nanoindenter-based compression, tension and bending" in Section 8.2.2), meaning that force was not measured; however, recently Kizuka and coworkers developed a TEM–AFM with real-time measurement capabilities (see "TEM–SPM and nanoindenter-based compression, tension and bending" in Section 8.2.2) [68], which allows force measurements. With this setup they were able to test metallic and semiconducting specimens, specifically silicon wires of nanometer width [71,72], and copper and palladium nanocontacts [69,70]. In terms of HRTEM, their last work is very insightful, in the sense that several domains (grains), a few atoms wide, were directly imaged and their evolution, as a function of strain, was imaged. In particular, it was established that grain evolution was dominated by slip events (see Figure 8.8d). The discrete nature of the events, where the slip distance is a multiple of the lattice constant, was captured. The final stage consisted of a single crystal pillar that failed in shear. This allowed the accurate measurement of the critically resolved shear stress for palladium (0.3–0.4 GPa).

On the other hand, for nanospecimens synthesized by chemical methods, such as nanowires, HRTEM has been used for establishing the cross-section of the specimens and for imaging the lattice distortions caused by mechanical deformation [77]. TEM also enables accurate determination of the cross-section of the specimen [44], because it allows the differentiation of the specimen's atomic structure from surface contaminants, which in other imaging techniques may appear to be part of the specimen. Additionally, combined with diffraction and a knowledge of the crystalline structure of the specimen material, the exact orientation of the specimen can be established [26].

In terms of the lattice distortion caused by applied strain, examples have been reported for SiC [40], Si [98], and GaN nanowires [60]. In all of these cases, imaging of individual dislocations and measurement of the Burgers vector and circuit was possible. Furthermore, the evolution from pristine structure to nanowires containing dislocations and leading to amorphization or fracture (see Figure 8.8e) was clearly observed.

8.3.2 Diffraction

The different modes of diffraction in TEM give useful information about crystal structure, lattice spacing and strain, and presence or nucleation of defects in the sample. In addition, TEM allows probing very small volumes, enabling local measurements within a nanostructure. In particular, selected-area electron diffraction (SAED) has a strain resolution of up to 0.1% [123], nano-beam electron diffraction can probe an even smaller (few nanometers) region with 0.06% strain resolution [123], and convergent-beam electron diffraction (CBED) provides 0.02% strain resolution. Diffraction also has advantages in terms of structural characterization and defect identification. A diffraction pattern can reveal twinning, phase changes, and amorphization. Subtle defects that are difficult to locate in bright-field imaging (even at high resolution) like inversion domain boundaries (important for correlation to piezoelectric response [35]) can be identified by comparing simulated and experimental results from CBED patterns [124]. Despite the various capabilities of diffraction mode in TEM, it has to date been used mostly for strain measurements and crystalline characterization in the context of mechanical characterization of nanowires; however, it is a very promising *in-situ* approach to reveal novel electromechanical phenomena in future studies.

For the determination of strain, SAED has been used in the mechanical testing of nanowires and nanobelts. Agrawal *et al.* [77] acquired diffraction patterns of ZnO nanowires while they were uniaxially tensioned. The local strain obtained from diffraction was compared with the overall strain obtained from bright-field measurements of the change in length of the gage region. The difference between the two strains was comparable to the experimental error, proving that the experimental setup imposed uniaxial tensile boundary conditions and that there was no slippage between the nanowire and the shuttles in the microsystem (see Figure 8.9a–c). Similarly, Vaughn and Kordesch [125] used a TEM holder with an embedded manipulator to deform gallium oxide nanobelts and observed diffraction patterns as a function of deformation. They were able to measure the change in the a and c spacings of the monoclinic lattice with ångström-scale precision.

On the other hand, Han *et al.* [40] performed SAED on the highly deformed parts of silicon carbide (SiC) nanowires, proving the amorphization of the nanowires, which took place at large strains. Halo-ring segments on the patterns revealed the development of amorphous regions in the sample and dark-field imaging allowed their visualization. In a similar fashion, Asthana *et al.* [76] observed the amorphization of ZnO nanowires under repeated bending loads. HRTEM images showed the deformation of the lattice and SAED patterns allowed the confirmation of this amorphization, as well as the nucleation of defects, evidenced by streaks and arc-like diffraction spots.

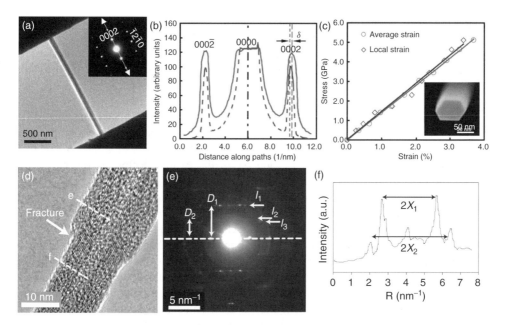

Figure 8.9 Use of diffraction capabilities in the TEM to perform mechanical characterization. (a)–(c) Uniaxial testing of ZnO nanowires. (a) The TEM image of the nanowire being pulled at two ends and its diffraction pattern. (b) An intensity profile along the path indicated. The peaks indicate the position of the diffraction spot, which shifts by a reciprocal distance δ as a result of strain. Local strain can be computed from measurement of δ. Comparison of this local strain with the average strain (c) allows discarding slippage effects in the grips of the tensile-testing device. Reprinted with permission from [79]. © 2008 American Chemical Society. (d)–(f) Uniaxial testing of MWCNTs. (d) HRTEM image of the fractured nanotube. (e) The corresponding diffraction pattern. The intensity profile (f) along I_1 can be fitted to Bessel functions and the chirality of the nanotube can be determined. Reprinted with permission from [3] © 2008 Nature Publishing Group

For CNTs, the chirality of the specimens under test can be determined using SAED [3] (see Figure 8.9d–f). Intensity profiles of principal layer lines (I_1, I_2, I_3 in Figure 8.9e), common in diffraction patterns of CNTs, are fitted to Bessel functions, the order of which can then be related to the chiral index [126].

8.3.3 Analytical Techniques

The interaction of impinging electrons with atomic structures produces several sub-products and physical phenomena that are used for analytical purposes. Taking advantage of this, TEM can be used to carry out spectroscopy and other types of analytical measurements. Using X-ray electron dispersive spectroscopy (EDS), elemental analysis of the specimen can be carried out with a resolution of a few nanometers, depending on the thickness of the sample [14]. Furthermore, electron-energy-loss spectroscopy (EELS) gives elemental information, as well as information on the bonding and thickness of the sample [14]. The resolution of these techniques in TEM is of particular relevance when nanostructures are tested, as local effects can be probed accurately.

In the framework of mechanical characterization, analytical TEM has been used to characterize the change in chemical bonding as a function of deformation, the diffusion of material within the specimen, as it is strained, and to characterize the shape and cross-sectional areas used to calculate stresses from measured forces.

In the context of chemical bonding studies, Aslam *et al.* [102] analyzed the EELS spectra of single-walled CNT bundles as they were bent and buckled under compressive load. In particular, they investigated the dependence of the ratios of π and σ covalent bonding (represented by the peaks π^* and σ^* in the EELS spectra) as reversible and irreversible deformation was imposed on the bundles. The ratio $\pi^*/(\pi^*+\sigma^*)$ increased in reversible deformation, showing that the overlap of the σ bonds decreased, created by the nonplanarity induced by the applied strain. When irreversible deformation was imposed and permanent defects were introduced in the bundle, the ratio $\pi^*/(\pi^*+\sigma^*)$ decreased, revealing a decrease of the π bonding, resulting from non-hexagonal defects being introduced in the structure of the CNT.

In the same vein, Han *et al.* [40] analyzed EELS spectra of SiC nanowires in order to establish the appearance of an amorphous phase under large-strain plasticity. Comparing with EELS spectra of crystalline SiC, they demonstrated that the broadening of some peaks corresponds to the manifestation of an amorphous phase in the highly strained region of the nanowire. Further demonstration of this amorphous phase was established by diffraction studies (see Section 8.3.2).

The same group also applied the EELS and EDS techniques to analyze the large-strain plasticity of silicon nanowires and demonstrated that oxygen diffusion did not drive this process. While straining the nanowire, several EELS spectra were taken across the diameter of the specimen. The results showed that a silicon oxide layer (which natively covers the silicon nanowires) does not migrate to the center of the wire while it is strained. Complementary elemental EDS spectra showed similar results. This helped establish that large-strain plasticity was in fact driven by dislocation nucleation, which led to a disordered atomic structure, and not by oxygen diffusion.

Filleter *et al.* [80], Bernal *et al.* [26], and Richter *et al.* [127] used EELS thickness maps in order to obtain information about the cross-section of the specimens under *in-situ* testing. Filleter *et al.* [80] applied this technique in the tensile testing of DWCNT bundles in order to establish their circular cross-section, therefore supporting the model applied for calculating stresses in the specimen. In a similar fashion, Bernal *et al.* [26] confirmed the geometry used for calculating stresses in GaN nanowires by comparing an EELS thickness map of the specimen with the expected thickness resulting from a polygonal cross-section specimen. Richter *et al.* [127] used the same technique to establish the octagonal cross-section of copper nanowhiskers, although the tensile tests were not carried out by *in-situ* TEM.

8.3.4 In-Situ *Specimen Modification*

Another advantage of *in-situ* TEM experiments is that in some cases the high-energy electron beam can also be used to advantageously modify the specimen under investigation in *in-situ* TEM, followed by mechanical characterization. Owing to the high resolution

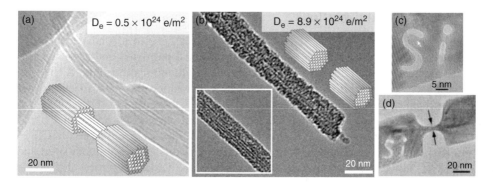

Figure 8.10 Examples of *in-situ* TEM specimen modification of samples and its potential application in mechanical problems. (a) and (b) The *in-situ* irradiation of CNT bundles. With a low irradiation dose (a) the fracture occurs at low effective stress levels and in a sword-in-sheath mode. By increasing the electron irradiation (b) the load-carrying capacity of all the shells in the nanotubes is utilized, resulting in higher effective fracture strengths and a brittle-like failure. Adapted from [80]. Electron irradiation may also be used to selectively remove material in the specimen. (c) and (d) A silicon nanowire where letters and a dog-bone shape have been patterned with the high-intensity electron beam in the microscope. These modifications could potentially be used in tensile testing and determination of stress-intensity factors at the nanoscale. Adapted from [116]

of TEM and the precise control of atomic modifications, new avenues are opened for the testing of mechanical systems based on nanostructures.

One of the most relevant examples is the use of electron irradiation to achieve covalent cross-linking of CNTs and shells inside MWCNTs [3] and CNT bundles [80,128] (Figure 8.10a and b). The high-energy electron beam induces knock-off of atoms in the nanotube shells, creating structural and interstitial defects that link adjacent shells (in the case of a multiwalled tube) and adjacent nanotubes (in the case of nanotube bundles). As alluded to before, this *in-situ* modification has been shown to significantly increase the effective strength and stiffness of these carbon materials, opening an avenue for their most effective use in nanocomposites and macroscopic fibers [17]. Although many reports exist on the modification of carbon-based materials using electron irradiation [129], the modification of crystalline materials and the mechanical effects of irradiation on nanowires remains relatively unexplored.

Although mechanical tests were not carried out, an interesting example of nanowire modification is given by Xu *et al.* [116]. By focusing a 200 keV TEM beam in a spot of a few nanometers they achieved current densities ranging from 10^3 to 10^6 A cm^{-2} passing across the specimen. This was used to create a variety of very controlled shapes, such as holes, letters, dog-bone specimens, and to generate welds between nanowires of several materials, namely silicon, gold, copper, silver, and tin (Figure 8.10c and d) . This type of technique may be useful to expand the possibilities of testing beyond the usual tension, compression, and bending experiments. One can envision, for example, the creation of pre-cracked specimens in order to compute stress-intensity factors at the nanoscale or to trigger failure at a certain location in the specimen while simultaneously performing high-resolution imaging.

8.4 Summary and Outlook

In this review we have illustrated how *in-situ* TEM has played a fundamental role in the characterization of mechanical and electromechanical properties and associated phenomena in one-dimensional nanostructures. The unique atomic-scale resolution, analytical, and spectroscopic capabilities of TEM, combined with a variety of *in-situ* mechanical and electromechanical experimental setups has allowed the achievement of several milestones in the study of one-dimensional nanostructures. Among the most relevant and impactful are the measurement of the elastic modulus [42], strength [3], and superplasticity of CNTs [61], the experimental observation of quantized conductance in atom-sized specimens [53], the identification of size-scale effects in the elastic properties of nanowires [26,77], and the measurements of room-temperature large-strain plasticity in silicon and SiC nanowires [40,98].

We have shown that the experimental setups have evolved in complexity, starting from relatively simple two-electrode systems to produce resonance in nanostructures (see "Resonance" in Section 8.2.2), to more sophisticated systems with *in-situ* TEM manipulators to achieve nanostructure deformation, force measurements, and electrical addressing (see "TEM–SPM and nanoindenter-based compression, tension and bending" in Section 8.2.2), ending in complex lab-on-a-chip MEMS approaches, where nanoscale tensile and compressive devices are used to impose and measure forces acting on the nanostructures (see "MEMS-based testing" in Section 8.2.2). One pattern emerging is the compromise that exists between experimental complexity and testing throughput. Although MEMS-based testing offers the best control of boundary conditions in a mechanical or electromechanical test, sample preparation is more challenging and setup complexity is greater, leading to a lower overall testing throughput. This explains the greater number of reports on nanostructure testing using the TEM–SPM approaches, which, at the expense of sometimes compromising homogeneity of applied deformation (by inducing buckling or bending), allow the realization of more tests in less time. In the medium term, the appearance of more groups working on MEMS-based characterization and the development of new specimen preparation techniques will most likely reverse this trend. Nevertheless, TEM–SPM will continue to be an important technique given the wide availability of commercial implementations, which lowers the entry barrier for new researchers.

A common challenge in all methods will continue to be the reduction of constraints, geometrical or of other kinds, so that the full capabilities of TEM can be employed in the testing of nanostructures. One notable omission in the literature is the lack of reports using double-tilting, which is illustrative of the still-untapped potential of TEM to discover more new phenomena in nanostructures. Incidentally, TEM–SPM double-tilt holders have been very recently made commercially available [130]. Similarly, the development of new MEMS and TEM–SPM approaches which allow coupled mechanical and electrical testing [85,93] will likely shift the focus of this field from purely mechanical tests to studies which probe electromechanical phenomena, such as piezoelectricity and piezoresistivity of nanostructures.

Lastly, merging of state-of-the-art TEM techniques, such as high-speed TEM [32] (dynamic TEM), with *in-situ* testing will likely result in the discovery of previously unobserved phenomena [17,58]. High-strain-rate experiments in nanostructures remain an unexplored area of research, even when presumably it is of high relevance, given that some of the mechanical and electromechanical applications for nanostructures will impose megahertz or gigahertz cycling [1]. In this regard, it should be noted that the current MEMS-based methods discussed herein can, in principle, approach testing rates at least in the kilohertz regime. Furthermore, although bridging of spatial scales between nanostructure computational simulations and experiments has been achieved to some extent [77,81], the bridging of strain rate will likely provide many insights into the suitability of the currently used atomistic approaches for nanomaterial modeling, such as molecular dynamics.

Acknowledgments

HDE acknowledges the support of the NSF through award no. DMR-0907196, ONR through award no. N00014-08-1-0108, and ARO through MURI award no. W911NF-09-1-0541. We thank Dr. Allison Beese for a critical review of the manuscript and helpful discussions.

References

[1] Rutherglen, C., Jain, D., and Burke, P. (2009) Nanotube electronics for radiofrequency applications. *Nature Nanotechnology*, **4** (12), 811–819.

[2] Naraghi, M., Filleter, T., Moravsky, A. *et al*. (2010) A multiscale study of high performance double-walled nanotube-polymer fibers. *ACS Nano*, **4** (11), 6463–6476.

[3] Peng, B., Locascio, M., Zapol, P. *et al*. (2008) Measurements of near-ultimate strength for multiwalled carbon nanotubes and irradiation-induced crosslinking improvements. *Nature Nanotechnology*, **3** (10), 626–631.

[4] Charlier, J.-C., Blase, X., and Roche, S. (2007) Electronic and transport properties of nanotubes. *Reviews of Modern Physics*, **79** (2), 677–732.

[5] Sun, Y., Gates, B., Mayers, B., and Xia, Y. (2002) Crystalline silver nanowires by soft solution processing. *Nano Letters*, **2** (2), 165–168.

[6] Yang, P., Yan, R., and Fardy, M. (2010) Semiconductor nanowire: what's next? *Nano Letters*, **10** (5), 1529–1536.

[7] Xu, S., Qin, Y., Xu, C. *et al*. (2010) Self-powered nanowire devices. *Nature Nanotechnology*, **5** (5), 366–373.

[8] Loh, O., Wei, X., Ke, C. *et al*. (2010) Robust carbon-nanotube-based nano-electromechanical devices: understanding and eliminating prevalent failure modes using alternative electrode materials. *Small*, **7** (1), 79–86.

[9] Arnold, M.S., Green, A.A., Hulvat, J.F. *et al*. (2006) Sorting carbon nanotubes by electronic structure using density differentiation. *Nature Nanotechnology*, **1** (1), 60–65.

[10] Wei, L. and Charles, M.L. (2006) Semiconductor nanowires. *Journal of Physics D: Applied Physics*, **39** (21), R387.

[11] Koren, E., Hyun, J.K., Givan, U. *et al*. (2011) Obtaining uniform dopant distributions in VLS-grown Si nanowires. *Nano Letters*, **11** (1), 183–187.

[12] Tham, D., Nam, C.Y., and Fischer, J.E. (2006) Defects in GaN nanowires. *Advanced Functional Materials*, **16**, 1197–1202.

[13] Allen, J.E., Hemesath, E.R., Perea, D.E. *et al*. (2008) High-resolution detection of Au catalyst atoms in Si nanowires. *Nature Nanotechnology*, **3** (3), 168–173.

[14] Williams, D.B. and Carter, C.B. (2009) *Transmission Electron Microscopy: A Textbook for Materials Science*, 2nd edn., Springer, New York, NY.

[15] Wang, Z.L. and Hui, C. (eds.) 2003 *Electron Microscopy of Nanotubes*, Springer.

[16] Yao, N. and Wang, Z.L. (eds.) 2005 *Handbook of Microscopy for Nanotechnology*, Springer.

[17] Espinosa, H.D., Filleter, T., and Naraghi, M. (2012) Multiscale experimental mechanics of hierarchical carbon-based materials. *Advanced Materials*, **24**, 2805–2823.

[18] Wang, Z.L. and Song, J. (2006) Piezoelectric nanogenerators based on zinc oxide nanowire arrays. *Science*, **312** (5771), 242–246.

[19] Qi, Y., Jafferis, N.T., Lyons, K. *et al*. (2010) Piezoelectric ribbons printed onto rubber for flexible energy conversion. *Nano Letters*, **10** (2), 524–528.

[20] Gao, Y. and Wang, Z.L. (2009) Equilibrium potential of free charge carriers in a bent piezoelectric semiconductive nanowire. *Nano Letters*, **9** (3), 1103–1110.

[21] Qi, Y., Kim, J., Nguyen, T.D. *et al*. (2011) Enhanced piezoelectricity and stretchability in energy harvesting devices fabricated from buckled PZT ribbons. *Nano Letters*, **11** (3), 1331–1336.

[22] Zhu, Y., Xu, F., Qin, Q. *et al*. (2009) Mechanical properties of vapor–liquid–solid synthesized silicon nanowires. *Nano Letters*, **9** (11), 3934–3939.

[23] Zhang, J.-H., Huang, Q.-A., Yu, H. *et al*. (2009) Effect of temperature and elastic constant on the piezoresistivity of silicon nanobeams. *Journal of Applied Physics*, **105** (8), 086102.

[24] Milne, J.S., Rowe, A.C.H., Arscott, S., and Renner, C. (2010) Giant piezoresistance effects in silicon nanowires and microwires. *Physical Review Letters*, **105** (22), 226802.

[25] Ryu, I., Choi, J.W., Cui, Y., and Nix, W.D. (2011) Size-dependent fracture of Si nanowire battery anodes. *Journal of the Mechanics and Physics of Solids*, **59**, 1717–1730.

[26] Bernal, R.A., Agrawal, R., Peng, B. *et al*. (2011) Effect of growth orientation and diameter on the elasticity of GaN nanowires. *A combined in situ TEM and atomistic modeling investigation. Nano Letters*, **11** (2), 548–555.

[27] Ferreira, P.J., Mitsuishi, K., and Stach, E.A. (2008) *In situ* transmission electron microscopy. *MRS Bulletin*, **33**, 83–85.

[28] Cummings, J., Olsson, E., Petford-Long, A.K., and Zhu, Y. (2008) Electric and magnetic phenomena studied by *in situ* transmission electron microscopy. *MRS Bulletin*, **33**, 101–106.

[29] Urban, K.W. (2008) Studying atomic structures by aberration-corrected transmission electron microscopy. *Science*, **321** (5888), 506–510.

[30] Legros, M., Gianola, D.S., and Motz, C. (2010) Quantitative *in situ* mechanical testing in electron microscopes. *MRS Bulletin*, **35**, 354–360.

[31] Nafari, A., Angenete, J., Svensson, K. *et al*. (2011) Combining scanning probe microscopy and transmission electron microscopy, in *Scanning Probe Microscopy in Nanoscience and Nanotechnology 2* (ed. B. Bhushan), Springer, pp. 59–100.

[32] Kim, J.S., LaGrange, T., Reed, B.W. *et al*. (2008) Imaging of transient structures using nanosecond *in situ* TEM. *Science*, **321** (5895), 1472–1475.

[33] Armstrong, M.R., Boyden, K., Browning, N.D. *et al*. (2007) Practical considerations for high spatial and temporal resolution dynamic transmission electron microscopy. *Ultramicroscopy*, **107** (4–5), 356–367.

[34] Thibault, P., Dierolf, M., Menzel, A. *et al*. (2008) High-resolution scanning X-ray diffraction microscopy. *Science*, **321** (5887), 379–382.

[35] Morkoç, H. (2008) *Handbook of Nitride Semiconductors and Devices*, vol. **1**, Wiley–VCH.

[36] Levin, I., Davydov, A., Nikoobakht, B. *et al*. (2005) Growth habits and defects in ZnO nanowires grown on GaN/sapphire substrates. *Applied Physics Letters*, **87** (10), 103110.

[37] Ding, Y. and Wang, Z.L. (2009) Structures of planar defects in ZnO nanobelts and nanowires. *Micron*, **40** (3), 335–342.

[38] Lee, K.H., Lee, J.Y., Kwon, Y.H. *et al*. (2009) Effects of defects on the morphologies of GaN nanorods grown on Si (111) substrated. *Journal of Materials Research*, **24** (10), 3032–3037.

[39] Jacobs, B.W., Crimp, M.A., McElroy, K., and Ayres, V.M. (2008) Nanopipes in gallium nitride nanowires and rods. *Nano Letters*, **8** (12), 4353–4358.

[40] Han, X.D., Zhang, Y.F., Zheng, K. *et al*. (2007) Low-temperature *in situ* large strain plasticity of ceramic SiC nanowires and its atomic-scale mechanism. *Nano Letters*, **7** (2), 452–457.

[41] Treacy, M.M.J., Ebbesen, T.W., and Gibson, J.M. (1996) Exceptionally high Young's modulus observed for individual carbon nanotubes. *Nature*, **381** (6584), 678–680.

[42] Poncharal, P., Wang, Z.L., Ugarte, D., and de Heer, W.A. (1999) Electrostatic deflections and electromechanical resonances of carbon nanotubes. *Science*, **283** (5407), 1513–1516.

[43] Liu, K.H., Wang, W.L., Xu, Z. *et al*. (2006) *In situ* probing mechanical properties of individual tungsten oxide nanowires directly grown on tungsten tips inside transmission electron microscope. *Applied Physics Letters*, **89** (22), 221908.

[44] Nam, C.-Y., Jaroenapibal, P., Tham, D. *et al*. (2006) Diameter-dependent electromechanical properties of GaN nanowires. *Nano Letters*, **6** (2), 153–158.

[45] Chopra, N.G. and Zettl, A. (1998) Measurement of the elastic modulus of a multi-wall boron nitride nanotube. *Solid State Communications*, **105** (5), 297–300.

[46] Chen, C.Q., Shi, Y., Zhang, Y.S. *et al.* (2006) Size dependence of Young's modulus in ZnO nanowires. *Physical Review Letters*, **96**, 075505.

[47] Wang, Z.L., Poncharal, P., and de Heer, W.A. (2000) Nanomeasurements of individual carbon nanotubes by *in situ* TEM. *Pure and Applied Chemistry*, **72** (1–2), 209–219.

[48] Svensson, K., Jompol, Y., Olin, H., and Olsson, E. (2003) Compact design of a transmission electron microscope-scanning tunneling microscope holder with three-dimensional coarse motion. *Review of Scientific Instruments*, **74** (11), 4945–4947.

[49] Nafari, A., Karlen, D., Rusu, C. *et al.* (2008) MEMS sensor for *in situ* TEM atomic force microscopy. *Journal of Microelectromechanical Systems*, **17** (2), 328–333.

[50] Kiener, D. and Minor, A.M. (2011) Source truncation and exhaustion: insights from quantitative *in situ* TEM tensile testing. *Nano Letters*, **11** (9), 3816–3820.

[51] Guo, H., Chen, K., Oh, Y. *et al.* (2011) Mechanics and dynamics of the strain-induced M1–M2 structural phase transition in individual VO$_2$ nanowires. *Nano Letters*, **11** (8), 3207–3213.

[52] Xu, S.Y., Xu, J., and Tian, M.L. (2006) A low cost platform for linking transport properties to the structure of nanomaterials. *Nanotechnology*, **17** (5), 1470–1475.

[53] Ohnishi, H., Kondo, Y., and Takayanagi, K. (1998) Quantized conductance through individual rows of suspended gold atoms. *Nature*, **395** (6704), 780–783.

[54] Stach, E.A. (2008) Real-time observations with electron microscopy. *Materials Today*, **11** (Suppl 1), 50–58.

[55] Stach, E.A., Freeman, T., Minor, A.M. *et al.* (2001) Development of a nanoindenter for *in situ* transmission electron microscopy. *Microscopy and Microanalysis*, **7**, 507–517.

[56] Hysitron, PI 95 TEM Picoindenter (2011) http://www.hysitron.com/LinkClick.aspx?fileticket= 4kfuCAsVUNA%3d&tabid=178 (accessed July 2012).

[57] Greer, J.R. and De Hosson, J.T.M. (2011) Plasticity in small-sized metallic systems: intrinsic versus extrinsic size effect. *Progress in Materials Science*, **56** (6), 654–724.

[58] Agrawal, R. and Espinosa, H.D. (2009) Multiscale experiments: state of the art and remaining challenges. *Journal of Engineering Materials and Technology: Transactions of the ASME*, **131** (4), 0412081.

[59] Shan, Z.W., Mishra, R.K., Asif, S.A.S. *et al.* (2008) Mechanical annealing and source-limited deformation in submicrometre-diameter Ni crystals. *Nature Materials*, **7** (2), 115–119.

[60] Huang, J.Y., Zheng, H., Mao, S.X. *et al.* (2011) *In situ* nanomechanics of GaN nanowires. *Nano Letters*, **11** (4), 1618–1622.

[61] Huang, J.Y., Chen, S., Wang, Z.Q. *et al.* (2006) Superplastic carbon nanotubes. *Nature*, **439** (7074), 281–281.

[62] Huang, J.Y., Chen, S., Jo, S.H. *et al.* (2005) Atomic-scale imaging of wall-by-wall breakdown and concurrent transport measurements in multiwall carbon nanotubes. *Physical Review Letters*, **94** (23), 236802.

[63] Naitoh, Y., Takayanagi, K., and Tomitori, M. (1996) Visualization of tip-surface geometry at atomic distance by TEM–STM holder. *Surface Science*, **357–358**, 208–212.

[64] Liao, Y., EswaraMoorthy, S.K., and Marks, L.D. (2010) Direct observation of tribological recrystallization. *Philosophical Magazine Letters*, **90** (3), 219–223.

[65] Golberg, D., Bai, X.D., Mitome, M. *et al.* (2007) Structural peculiarities of *in situ* deformation of a multi-walled BN nanotube inside a high-resolution analytical transmission electron microscope. *Acta Materialia*, **55** (4), 1293–1298.

[66] Golberg, D., Costa, P.M.F.J., Mitome, M., and Bando, Y. (2009) Properties and engineering of individual inorganic nanotubes in a transmission electron microscope. *Journal of Materials Chemistry*, **19** (7), 909–920.

[67] Jensen, K., Mickelson, W., Kis, A., and Zettl, A. (2007) Buckling and kinking force measurements on individual multiwalled carbon nanotubes. *Physical Review B*, **76** (19), 195436.

[68] Kizuka, T., Ohmi, H., Sumi, T. *et al.* (2001) Simultaneous observation of millisecond dynamics in atomistic structure, force and conductance on the basis of transmission electron microscopy. *Japanese Journal of Applied Physics*, **40**, L170–L173.

[69] Fujisawa, S., Kikkawa, T., and Kizuka, T. (2003) Direct observation of electromigration and induced stress in Cu nanowire. *Japanese Journal of Applied Physics*, **42**, L1433–L1435.

[70] Matsuda, T. and Kizuka, T. (2009) Slip sequences during tensile deformation of palladium nanocontacts. *Japanese Journal of Applied Physics*, **48**, 115003.

[71] Kizuka, T. and Takatani, Y. (2007) Growth of silicon nanowires by nanometer-sized tip manipulation. *Japanese Journal of Applied Physics*, **46**, 5706–5710.

[72] Kizuka, T., Takatani, Y., Asaka, K., and Yoshizaki, R. (2005) Measurements of the atomistic mechanics of single crystalline silicon wires of nanometer width. *Physical Review B*, **72** (3), 035333.

[73] Costa, P.M.F.J., Cachim, P.B., Gautam, U.K. *et al.* (2009) The mechanical response of turbostratic carbon nanotubes filled with Ga-doped ZnS: II. Slenderness ratio and crystalline filling effects. *Nanotechnology*, **20** (40), 405707.

[74] Costa, P.M.F.J., Cachim, P.B., Gautam, U.K. *et al.* (2009) The mechanical response of turbostratic carbon nanotubes filled with Ga-doped ZnS: I. Data processing for the extraction of the elastic modulus. *Nanotechnology*, **20** (40), 405706.

[75] Costa, P.M.F.J., Gautam, U.K., Wang, M. *et al.* (2009) Effect of crystalline filling on the mechanical response of carbon nanotubes. *Carbon*, **47** (2), 541–544.

[76] Asthana, A., Momeni, K., Prasad, A. *et al.* (2011) *In situ* observation of size-scale effects on the mechanical properties of ZnO nanowires. *Nanotechnology*, **22** (26), 265712.

[77] Agrawal, R., Peng, B., Gdoutos, E.E., and Espinosa, H.D. (2008) Elasticity size effects in ZnO nanowires – a combined experimental–computational approach. *Nano Letters*, **8** (11), 3668–3674.

[78] Zhu, Y. and Espinosa, H.D. (2005) An electromechanical material testing system for *in situ* electron microscopy and applications. *Proceedings of the National Academy of Sciences of the United States of America*, **102** (41), 14503–14508.

[79] Agrawal, R., Peng, B., and Espinosa, H.D. (2009) Experimental–computational investigation of ZnO nanowires strength and fracture. *Nano Letters*, **9** (12), 4177–4183.

[80] Filleter, T., Bernal, R., Li, S., and Espinosa, H.D. (2011) Ultrahigh strength and stiffness in cross-linked hierarchical carbon nanotube bundles. *Advanced Materials*, **23**, 2855–2860.

[81] Filleter, T., Ryu, S., Kang, K. *et al.* (2012) Nucleation-controlled distributed plasticity in penta-twinned silver nanowires. *Small*, doi: 10.1002/smll.201200522.

[82] Haque, M.A. and Saif, M.T.A. (2002) Application of MEMS force sensors for *in situ* mechanical characterization of nano-scale thin films in SEM and TEM. *Sensors and Actuators A: Physical*, **97–98**, 239–245.

[83] Haque, M. and Saif, M. (2003) A review of MEMS-based microscale and nanoscale tensile and bending testing. *Experimental Mechanics*, **43** (3), 248–255.

[84] Haque, M.A. and Saif, M.T.A. (2005) *In situ* tensile testing of nanoscale freestanding thin films inside a transmission electron microscope. *Journal of Materials Research*, **20** (7), 1769–1777.

[85] Haque, M.A., Espinosa, H.D., and Lee, H.J. (2010) MEMS for in situ testing – handling, actuation, loading, displacement measurement. *MRS Bulletin*, **35**, 375–381.

[86] Espinosa, H.D., Zhu, Y., and Moldovan, N. (2007) Design and operation of a MEMS-based material testing system for nanomechanical characterization. *Journal of Microelectromechanical Systems*, **16** (5), 1219–1231.

[87] Zhu, Y., Corigliano, A., and Espinosa, H.D. (2006) A thermal actuator for nanoscale *in situ* microscopy testing: design and characterization. *Journal of Micromechanics and Microengineering*, **16** (2), 242–253.

[88] Zhu, Y., Moldovan, N., and Espinosa, H.D. (2005) A microelectromechanical load sensor for in situ electron and X-ray microscopy tensile testing of nanostructures. *Applied Physics Letters*, **86** (1), 013506.

[89] Zhang, M., Olson, E.A., Twesten, R.D. *et al.* (2005) *In situ* transmission electron microscopy studies enabled by microelectromechanical system technology. *Journal of Materials Research*, **20** (7), 1802–1807.

[90] Zhang, D., Breguet, J.M., Clavel, R. *et al.* (2010) *In situ* electron microscopy mechanical testing of silicon nanowires using electrostatically actuated tensile stages. *Journal of Microelectromechanical Systems*, **19** (3), 663–674.

[91] Zhang, D., Breguet, J.-M., Clavel, R. *et al.* (2009) *In situ* tensile testing of individual Co nanowires inside a scanning electron microscope. *Nanotechnology*, **20** (36), 365706.

[92] Zhang, D., Drissen, W., Breguet, J.-M. *et al.* (2009) A high-sensitivity and quasi-linear capacitive sensor for nanomechanical testing applications. *Journal of Micromechanics and Microengineering*, **19** (7), 075003.

[93] Zhang, Y., Liu, X., Ru, C. *et al.* (2011) Piezoresistivity characterization of synthetic silicon nanowires using a MEMS device. *Journal of Microelectromechanical Systems*, **20** (4), 959–967.

[94] Naraghi, M., Ozkan, T., Chasiotis, I. *et al*. (2010) MEMS platform for on-chip nanomechanical experiments with strong and highly ductile nanofibers. *Journal of Micromechanics and Microengineering*, **20** (12), 125022.

[95] Eppell, S.J., Smith, B.N., Kahn, H., and Ballarini, R. (2006) Nano measurements with micro-devices: mechanical properties of hydrated collagen fibrils. *Journal of The Royal Society Interface*, **3** (6), 117–121.

[96] Han, J.H. and Saif, M.T.A. (2006) *In situ* microtensile stage for electromechanical characterization of nanoscale freestanding films. *Review of Scientific Instruments*, **77** (4), 045102.

[97] Espinosa, H.D. and Bernal, R.A. International Patent Application WO 2011/053346 A1.

[98] Han, X.D., Zheng, K., Zhang, Y.F. *et al*. (2007) Low-temperature *in-situ* large-strain plasticity of silicon nanowires. *Advanced Materials*, **19** (16), 2112–2118.

[99] Cumings, J. and Zettl, A. (2000) Low-friction nanoscale linear bearing realized from multiwall carbon nanotubes. *Science*, **289** (5479), 602–604.

[100] Jaroenapibal, P., Luzzi, D.E., Evoy, S., and Arepalli, S. (2004) Transmission-electron-microscopic studies of mechanical properties of single-walled carbon nanotube bundles. *Applied Physics Letters*, **85** (19), 4328–4330.

[101] Wang, M.S., Wang, J.Y., Chen, Q., and Peng, L.M. (2005) Fabrication and electrical and mechanical properties of carbon nanotube interconnections. *Advanced Functional Materials*, **15** (11), 1825–1831.

[102] Aslam, Z., Abraham, M., Brown, A. *et al*. (2008) Electronic property investigations of single-walled carbon nanotube bundles *in situ* within a transmission electron microscope: an evaluation. *Journal of Microscopy*, **231** (1), 144–155.

[103] Golberg, D., Costa, P., Mitome, M., and Bando, Y. (2008) Nanotubes in a gradient electric field as revealed by STM TEM technique. *Nano Research*, **1** (2), 166–175.

[104] Bai, X., Golberg, D., Bando, Y. *et al*. (2007) Deformation-driven electrical transport of individual boron nitride nanotubes. *Nano Letters*, **7** (3), 632–637.

[105] Costa, P.M.F.J., Fang, X., Wang, S. *et al*. (2009) Two-probe electrical measurements in transmission electron microscopes – behavioral control of tungsten microwires. *Microscopy Research and Technique*, **72** (2), 93–100.

[106] Larsson, M.W., Wallenberg, L.R., Persson, A.I., and Samuelson, L. (2004) Probing of individual semiconductor nanowhiskers by TEM–STM. *Microscopy and Microanalysis*, **10** (1), 41–46.

[107] Liu, K.H., Gao, P., Xu, Z. *et al*. (2008) *In situ* probing electrical response on bending of ZnO nanowires inside transmission electron microscope. *Applied Physics Letters*, **92** (21), 213105.

[108] Asthana, A., Momeni, K., Prasad, A. *et al*. (2009) *In situ* probing of electromechanical properties of an individual ZnO nanobelt. *Applied Physics Letters*, **95** (17), 172106.

[109] He, R.R. and Yang, P. (2006) Giant piezoresistance effect in silicon nanowires. *Nature Nanotechnology*, **1**, 42–46.

[110] Desai, A.V. and Haque, M.A. (2007) Sliding of zinc oxide nanowires on silicon substrate. *Applied Physics Letters*, **90**, 033102.

[111] Zhang, D., Breguet, J.M., Clavel, R. *et al*. (2009) *In situ* tensile testing of individual Co nanowires inside a scanning electron microscope. *Nanotechnology*, **20** (36), 365706.

[112] Asthana, A., Shokuhfar, T., Gao, Q. *et al*. (2010) A study on the modulation of the electrical transport by mechanical straining of individual titanium dioxide nanotube. *Applied Physics Letters*, **97** (7), 072107.

[113] Luo, J.H., Wu, F.F., Huang, J.Y. *et al*. (2010) Superelongation and atomic chain formation in nanosized metallic glass. *Physical Review Letters*, **104** (21), 215503.

[114] Moore, N.W., Luo, J., Huang, J.Y. *et al*. (2009) Superplastic nanowires pulled from the surface of common salt. *Nano Letters*, **9** (6), 2295–2299.

[115] Lu, Y., Huang, J.Y., Wang, C. *et al*. (2010) Cold welding of ultrathin gold nanowires. *Nature Nanotechnology*, **5** (3), 218–224.

[116] Xu, S., Tian, M., Wang, J. *et al*. (2005) Nanometer-scale modification and welding of silicon and metallic nanowires with a high-intensity electron beam. *Small*, **1** (12), 1221–1229.

[117] Wang, M.S., Wang, J.Y., Chen, Q., and Peng, L.M. (2005) Fabrication and electrical and mechanical properties of carbon nanotube interconnections. *Advanced Functional Materials*, **15** (11), 1825–1831.

[118] San Paulo, A., Bokor, J., Howe, R.T. *et al*. (2005) Mechanical elasticity of single and double clamped silicon nanobeams fabricated by the vapor–liquid-solid method. *Applied Physics Letters*, **87** (5), 053111.

[119] Jin, Q., Wang, Y.L., Li, T. *et al*. (2008) A MEMS device for *in-situ* TEM test of SCS nanobeam. *Science in China Series E: Technological Sciences*, **51** (9), 1491–1496.

[120] Gravier, S., Coulombier, M., Safi, A. *et al.* (2009) New on-chip nanomechanical testing laboratory – applications to aluminum and polysilicon thin films. *Journal of Microelectromechanical Systems*, **18** (3), 555–569.

[121] Wang, M., Kaplan-Ashiri, I., Wei, X. *et al.* (2008) *In situ* TEM measurements of the mechanical properties and behavior of WS_2 nanotubes. *Nano Research*, **1** (1), 22–31.

[122] Agraït, N., Yeyati, A.L., and van Ruitenbeek, J.M. (2003) Quantum properties of atomic-sized conductors. *Physics Reports*, **377** (2–3), 81–279.

[123] Beche, A., Rouviere, J.L., Clement, L., and Hartmann, J.M. (2009) Improved precision in strain measurement using nanobeam electron diffraction. *Applied Physics Letters*, **95** (12), 123114.

[124] Liu, F., Collazo, R., Mita, S. *et al.* (2008) Direct observation of inversion domain boundaries of GaN on c-sapphire at sub-ångstrom resolution. *Advanced Materials*, **20** (11), 2162–2165.

[125] Vaughn, J.M. and Kordesch. M.E. (2009) *In situ* electron diffraction characterization of the deformation of nanobelts: gallium oxide. *Journal of Vacuum Science & Technology A: Vacuum, Surfaces, and Films*, **27** (4), 1058–1061.

[126] Qin, L.-C. (2006) Electron diffraction from carbon nanotubes. *Reports on Progress in Physics*, **69** (10), 2761.

[127] Richter, G., Hillerich, K., Gianola, D.S. *et al.* (2009) Ultrahigh strength single crystalline nanowhiskers grown by physical vapor deposition. *Nano Letters*, **9** (8), 3048–3052.

[128] Kis, A., Csanyi, G., Salvetat, J.P. *et al.* (2004) Reinforcement of single-walled carbon nanotube bundles by intertube bridging. *Nature Materials*, **3** (3), 153–157.

[129] Krasheninnikov, A.V. and Banhart, F. (2007) Engineering of nanostructured carbon materials with electron or ion beams. *Nature Materials*, **6** (10), 723–733.

[130] Nanofactory Instruments AB. (2011) http://www.nanofactory.com/Page.asp?nav=Nanomaterials&id=1 (accessed July 2012).

9

Engineering Nano-Probes for Live-Cell Imaging of Gene Expression

Gang Bao[1], Brian Wile[1], and Andrew Tsourkas[2]

[1]*Georgia Institute of Technology and Emory University, USA*
[2]*University of Pennsylvania, USA*

9.1 Introduction

The ability to image specific ribonucleic acid (RNA) in living cells in real time can provide essential information on RNA synthesis, processing, transport, and localization. Visualizing the dynamics of RNA expression and localization in response to external stimuli will offer unprecedented opportunities for advancement in molecular biology, disease pathophysiology, drug discovery, and medical diagnostics. There is increasing evidence to suggest that RNA molecules have a wide range of functions in living cells, including physically conveying and interpreting genetic information, catalyzing essential reactions, providing structural support for molecular machines, and silencing genes. These functions are realized through control of the expression level and stability, both temporally and spatially, of specific RNAs in a cell. Therefore, determining the dynamics and localization of RNA molecules in living cells will significantly impact molecular biology and medicine.

Of particular interest is the fluorescent imaging of specific messenger RNAs (mRNAs) in living cells. As shown schematically in Figure 9.1, for eukaryotic cells a pre-mRNA molecule is synthesized in the cell nucleus. After processing, such as splicing and polyadenylation, the mature mRNAs are transported from the cell nucleus to specific sites in the cytoplasm. The mRNAs are then translated by ribosomes before being degraded by RNases. The limited lifetime of mRNA enables a cell to alter protein synthesis rapidly in

Nano and Cell Mechanics: Fundamentals and Frontiers, First Edition. Edited by Horacio D. Espinosa and Gang Bao.
© 2013 John Wiley & Sons, Ltd. Published 2013 by John Wiley & Sons, Ltd.

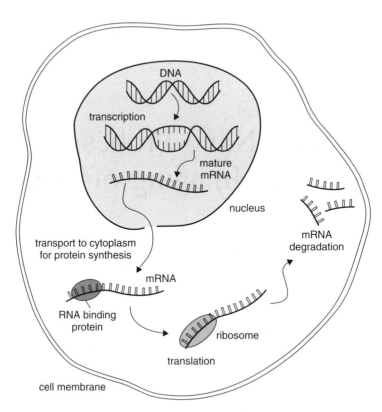

Figure 9.1   The mRNA life cycle. Messenger RNA (mRNA) that encodes the chemical "blueprint" for a protein is synthesized (transcribed) from a DNA template and the pre-mRNA is processed (spliced) to have mature mRNA, which is transported to specific locations in the cell cytoplasm. The coding information carried by mRNA is used by ribosomes to produce proteins (translation). After a certain amount of time, the message is degraded. mRNAs are almost always complexed with RNA-binding proteins to form ribonucleoprotein (RNP) molecules

response to its changing needs. mRNA is always complexed with RNA-binding proteins to form a ribonucleoprotein (RNP), which has significant implications to the live-cell imaging of mRNAs.

Many *in vitro* methods have been developed to provide a relative (mostly semi-quantitative) measure of gene expression level within a cell population using purified DNA or RNA obtained from cell lysate. These methods include *polymerase chain reaction* (PCR) [1], Northern hybridization (or Northern blotting) [2], expressed sequence tag (EST) [3], serial analysis of gene expression (SAGE) [4], differential display [5], and DNA microarrays [6]. These technologies, combined with the rapidly increasing availability of genomic data for numerous biological entities, present exciting possibilities for understanding human health and disease. For example, pathogenic and carcinogenic sequences are increasingly being used as clinical markers for diseased states. However, using *in vitro* methods to detect and identify foreign or mutated nucleic acids is often difficult in a clinical setting due to the low abundance of diseased cells in blood, sputum, and stool samples. Further, these methods cannot reveal the spatial and temporal variation of RNA within a single cell.

Labeled linear oligonucleotide (ODN) probes have been used to study intracellular mRNA via *in-situ* hybridization (ISH) [7], in which cells are fixed and permeabilized to increase the probe delivery efficiency. Unbound probes are removed by washing to reduce background and achieve specificity [8]. To enhance the signal level, multiple probes targeting the same mRNA can be used [7]. However, fixation agents and associated chemicals can have a considerable effect on signal level [9] and the integrity of certain organelles, such as mitochondria. Fixation of cells by either cross-linking or denaturing agents and the use of proteases in ISH assays may prevent the obtaining of an accurate description of intracellular mRNA localization. It is also difficult to obtain a dynamic picture of gene expression in cells using ISH methods.

In order to detect RNA molecules in living cells, the probes need to recognize RNA targets with high specificity, convert target recognition directly into a measurable signal, and differentiate between true and false-positive signals. This is especially important for low-abundance genes and clinical samples containing a small number of diseased cells. It is important for the probes to quantify low gene expression levels with high accuracy, and have fast kinetics in tracking alterations in gene expression in real time. For detecting genetic alterations such as mutations, insertions, and deletions, the ability to recognize single nucleotide polymorphisms (SNPs) is essential. To achieve this optimal performance, it is necessary to have a good understanding of the structure–function relationship of the probes, probe stability, and RNA target accessibility in living cells. It is also necessary to achieve efficient cellular delivery of probes.

In the remaining sections, we review commonly used fluorescent probes for RNA detection in living cells and discuss the critical biological and engineering issues, including probe design, target accessibility, cellular delivery of probes, and detection sensitivity, specificity, and signal-to-background ratio. Emphasis is placed on the design and application of molecular beacons, although some of the issues are common to other ODN probes. To help readers in engineering, key biological and technical terminologies used are summarized in Table 9.1.

9.2 Molecular Probes for RNA Detection

Several classes of molecular probes have been developed for RNA detection in living cells, including: (1) tagged linear ODN probes; (2) ODN hairpin probes; and (3) probes using fluorescent protein as a reporter. Although probes composed of full-length RNAs (mRNA or nuclear RNA) tagged with a fluorescent or radioactive reporter have been used to study the intracellular localization of RNA [10–12], these probes are not discussed here since they cannot be used to measure the expression level of specific RNAs in living cells.

9.2.1 Fluorescent Linear Probes

One of the simplest methods used to visualize endogenous RNA in single living cells involves introducing fluorescently labeled antisense linear ODN probes (Figure 9.2a) into cells [13–17]. In general, this approach requires the use of multiple probes targeting the same RNA transcript so that a high local concentration of hybridized probes can

Table 9.1   Descriptions of key biological and technical terminologies

Deoxyribonucleic acid (DNA)	A DNA molecule is a nucleic acid that contains the genetic instructions used in the development and functioning of all known living organisms and some viruses. It is a long polymer made from repeating nucleotides including adenine (A), quanine (G), cytosine (C), and thymine (T). Typically, G pairs with C and A pairs with T, according to Watson–Crick base pairing.
Ribonucleic acid (RNA)	RNA is a biologically important type of molecule that consists of a long chain of nucleotide units. Each nucleotide contains a ribose sugar, with carbons numbered 1′ through 5′. A base is attached to the 1′ position, generally adenine (A), cytosine (C), quanine (G), or uracil (U). Typically, G pairs with C and A pairs with U.
Oligonucleotide (ODN)	An ODN is a short nucleic acid polymer, typically with 20 or fewer bases. Because ODNs readily bind to their respective complementary nucleotide, they are often used as probes for detecting DNA or RNA.
Polymerase chain reaction (PCR)	PCR is a technique to amplify a single or few copies of a piece of DNA across several orders of magnitude, generating thousands to millions of copies of a particular DNA sequence. Almost all PCR applications employ a heat-stable DNA polymerase which enzymatically assembles a new DNA strand from DNA building blocks, the nucleotides, by using single-stranded DNA as a template and DNA ODNs (DNA primers) which are required for initiation of DNA synthesis.
Fluorescence *in-situ* hybridization (FISH)	FISH is a cytogenetic technique used to detect and localize the presence or absence of specific DNA sequences on chromosomes. FISH uses fluorescent probes that bind to only those parts of the DNA that show a high degree of sequence similarity. FISH is often used for finding specific features in DNA for use in genetic counseling, medicine, and species identification. FISH can also be used to detect and localize specific mRNAs within tissue samples.
Ribonucleoprotein (RNP)	RNP is a nucleoprotein that contains RNA and protein; that is, it is an association that combines ribonucleic acid and protein together.
Fluorescence resonance energy transfer (FRET)	FRET is the nonradiative transfer of energy from one molecule (called a donor) to another (called the acceptor) based on the proximity of the two molecules (typically <10 nm) and their spectral overlap. It is a widely applied tool for optical sensing and studies of biomolecular interaction.
Single nucleotide polymorphism (SNP)	A single-nucleotide polymorphism (SNP) is a DNA sequence variation occurring when a single nucleotide (A, T, C or G) in the genome (or other shared sequence) differs between members of a species (or between paired chromosomes in an individual).
Fluorophore	A fluorophore, in analogy to a chromophore, is a component of a molecule which causes a molecule to be fluorescent. It is typically a functional group in a molecule which will absorb energy of a specific wavelength and re-emit energy at a different (but equally specific) wavelength.

(continued overleaf)

Table 9.1   (*continued*)

Quencher	A quencher is a substance that absorbs excitation energy from a fluorophore; a dark quencher dissipates the energy as heat, thus eliminating a fluorophore's ability to fluoresce.
Codon	Codon is a sequence of three adjacent nucleotides constituting the genetic code that specifies the insertion of an amino acid in a specific structural position in a polypeptide chain during the synthesis of proteins.
Expressed sequence tag (EST)	A short portion of a gene that can often be used to identify whole genes.
Serial analysis of gene expression (SAGE)	A technique used to analyze the messenger RNA population in a sample of interest. The output typically provides quantitative information on the number of times each particular RNA is observed.
Representational difference analysis (RDA)	A technique used to find differences in RNA expression between two different cell samples.
Suppression subtractive hybridization (SSH)	A technique that is used to find differentially expressed genes in two different cell samples. Specifically, identifies unique genes with low expression, while highly expressed genes are suppressed.
Small nuclear RNA (snRNA)	A class of small RNA molecules that are found in the nucleus of eukaryotic cells and are involved in activities such as RNA splicing, regulation of transcription factors, and maintaining telomeres.
Green fluorescent protein (GFP)	A protein isolated from the jellyfish *Aequore victorea* that fluoresces green when exposed to blue light.
Streptolysin O (SLO)	Pore-forming endotoxin that binds to cell membranes.
Cell-penetrating peptide (CPP)	Short polycationic peptides that can facilitate the cellular uptake of attached cargo (e.g., ODNs, peptides, drugs, proteins, and nanoparticles).

be seen above the high background of unbound probes [16]. Alternatively, RNAs that exhibit a high local concentration within the nucleus can also be visualized, such as 28S ribosomal RNA in the nucleoli, poly(A) RNA in speckles, and U3 small nuclear RNA (snRNA) in Cajal bodies [16,17]. Linear ODN probes cannot be used to detect RNA in the cytoplasm, partly due to their rapid accumulation in the nucleus after delivery using microinjection. Further, linear probes may lack detection specificity, since a partial match between the probe and target sequences could induce probe hybridization to RNA molecules of multiple genes. Specificity can be improved by introducing backbone and/or base modifications that improve the ODN probe's affinity for mRNA and simultaneously shortening the probe to increase the energy penalty for a single base mismatch; however, this will result in reduced probe selectivity. In other words, the probability of having the probes hybridizing to nontarget RNAs in the cell with the same consensus sequence increases. These shortcomings have prevented the use of fluorescently labeled ODN probes in most live cell RNA imaging applications.

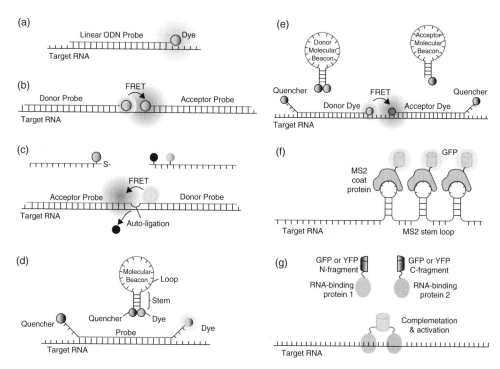

Figure 9.2 Illustrations of fluorescent probes for live-cell RNA detection. (a) Tagged linear oligonucleotide (ODN) probes. (b) Linear FRET probes in which two linear ODN probes have respectively a donor and an acceptor fluorophore that form a FRET pair. (c) Auto-ligation FRET probes. The fluorescence of the donor is initially quenched. Upon binding of the two probes to adjacent sites on the same RNA, the quencher is displaced and the ligation brings the donor and acceptor fluorophores together, resulting in a FRET signal. (d) Molecular beacons are dual-labeled stem–loop ODN hairpin probes with a reporter fluorophore at one end and a quencher molecule at the other end. (e) Dual-FRET donor and acceptor molecular beacons hybridize to adjacent regions on the same mRNA target, resulting in a FRET signal. (f) Probes using the coat protein of the bacterial phage MS2 fused with GFP (MS2–GFP). The MS2–GFP complexes bind to multiple hairpin sequences in the 3' UTR region of an mRNA, giving rise to a high signal compared with the background. (g) Probes based on fragment complementation of fluorescent protein. When two RNA-binding proteins, each carrying a fluorescent protein fragment, bind to adjacent sites on the same RNA, a fluorescence signal is generated due to fragment complementation of the fluorescent protein

9.2.2 Linear FRET Probes

As shown schematically in Figure 9.2b, linear fluorescence resonance energy transfer (FRET) probes utilize two linear ODNs that are fluorescently labeled at their 5' and 3' ends with donor and acceptor fluorophores, respectively, forming a FRET pair [18,19]. These probes are designed to hybridize to adjacent regions on a nucleic acid target such that the two fluorophores are brought into close proximity only when both probes are hybridized to the same RNA target. Excitation of the donor fluorophore leads to sensitized emission of acceptor fluorescence, producing a FRET signal indicative of target detection.

Since a FRET signal is only generated when both probes hybridize to adjacent sequences on target RNA, this method provides a novel way to differentiate between target recognition and background fluorescence from unbound probes. This wavelength shifting results in a significant improvement in signal-to-background ratio compared with fluorescently labeled linear probes [18]. The dual linear-probe approach may still exhibit a high background signal due to direct excitation of the acceptor and emission detection of the donor fluorescence; however, careful selection of donor and acceptor fluorophores can result in a signal-to-background of approximately 10:1. It has been reported that this can be increased to 20:1 by using two donor fluorophores in some instances [20].

An alternative approach that has been used to improve the sensitivity of RNA detection with linear FRET probes involved measuring the decay of acceptor fluorescence using a time-resolved method [19]. Specifically, when both donor and acceptor ODNs were bound to target RNA, the acceptor exhibited a significantly longer fluorescence lifetime than unhybridized probes did. The fluorescence decay of the acceptor was also much slower than autofluorescence, allowing it to be easily distinguished from background signal. It was found that time-resolved microscopy could be used to detect as few as ~900 probe–RNA hybrids in a single HeLa cell, compared with 10 000 hybrids when a conventional fluorescence microscope was used [18,19].

9.2.3 Quenched Auto-ligation Probes

Quenched auto-ligation FRET (QUAL-FRET) probes solve several problems inherent in linear or linear-FRET ODN probes by combining FRET principles with fluorescence quenching. In QUAL-FRET, one ODN is labeled with a FRET acceptor (e.g., Cy5) on the 5′ end and a nucleophile on the 3′ end. A second ODN is labeled with a FRET donor (e.g., FAM) on the 3′ end and an electrophilic dabsyl quencher on the 5′ end (Figure 9.2c) [21]. In the absence of target, the fluorescence of the donor fluorophore is quenched by the dabsyl quencher and the acceptor is not directly excited. Upon binding of the two probes to adjacent sites on the same RNA, the nucleophilic group displaces the dabsyl group via nucleophilic substitution reaction. This results in the formation of the ligated product with the FRET donor and acceptor held in close proximity. Excitation of the donor fluorophore therefore generates a detectable FRET signal. The irreversibility of the ligated product provides both favorable and unfavorable outcomes, depending on the application. Constant expression of an mRNA leads to the accumulation of ligated product and a corresponding increase in signal. This could allow for the detection of low copy number RNA that would not be detected otherwise. The disadvantage of using an irreversible probe is that it is not possible to monitor a down-regulation in gene expression.

Since ligation is required to generate the FRET signal, nonspecific signal is unlikely to occur even if both half probes were bound to a given nucleic acid binding protein. This is because the reactive ends are required to be held in very close proximity for a set period of time in order for the nucleophilic substitution reaction to occur. Unfortunately, it is expected that this time requirement also limits the detection of some RNAs. For example, if RNAs are rapidly degraded, there may not be sufficient time for auto-ligation to occur.

The selectivity and specificity of QUAL probes is expected to be similar to linear FRET probes; however, the signal-to-background is improved due to the quenching of donor fluorescence. In other words, fluorescence from unhybridized donor probes does not contribute to the background fluorescence in the acceptor channel even if there is a spectral overlap in donor and acceptor emission. Similar to the other two-probe methods, QUAL probes also require the identification of large stretches of RNA with little or no secondary structure.

9.2.4 Molecular Beacons

ODN hairpin probes have been the most widely adopted class of probes for live cell RNA imaging to date. As illustrated in Figure 9.2d, these probes are typically labeled at one end with a reporter fluorophore and at the other end with a quencher to form "molecular beacons." Molecular beacons are designed to form a stem–loop hairpin structure in the absence of a complementary target so that fluorescence of the fluorophore is quenched. Hybridization with the target nucleic acid opens the hairpin and physically separates the fluorophore from quencher, allowing a fluorescence signal to be emitted upon excitation (Figure 9.2d). This enables a molecular beacon to function as a sensitive probe with a high signal-to-background ratio. Under optimal conditions, the fluorescence intensity of molecular beacons can increase by >200-fold upon binding to their targets [22]. The ability to transduce target recognition *directly* into a fluorescence signal with high signal-to-background ratio has allowed molecular beacons to enjoy a wide range of biological and biomedical applications, including real-time monitoring of PCR, genotyping and mutation detection, multiple analyte detection, assaying for nucleic acid cleavage in real time, cancer cell detection, studying viral infection, and monitoring RNA expression and localization in living cells [13,16,23–35].

As illustrated in Figure 9.2d, a conventional molecular beacon has four essential components: a loop, stem, fluorophore, and quencher. The loop usually consists of 15–25 nucleotides and is selected to have a unique antisense targeting sequence. The stem, formed by two complementary short-arm sequences, is typically four to six bases long and chosen to be independent of the target sequence (Figure 9.2d).

Figure 9.3 compares the salient features of molecular beacons with fluorescence *in-situ* hybridization (FISH) probes. Molecular beacons are dual-labeled hairpin probes of 15–25 nucleotides, while FISH probes are dye-labeled linear ODNs of 40–50 nucleotides. The molecular-beacon-based approach has the advantage of detecting RNA in live cells without the need of cell fixation and washing. However, it requires cellular delivery of the probes and has low target accessibility (discussed in later sections). FISH assays enable facile probe design owing to better target accessibility in fixed cells. Much like a snapshot compared with a video, however, FISH assays provide an instantaneous image of RNA localization which may be altered from normal cell function due to the fixation and permeabilization process.

Figure 9.4 shows an example of molecular-beacon-based fluorescent images of the distribution of β-actin mRNA in live fibroblast cells visualized using molecular beacons with $2'$-O-methyl ribonucleotide backbones [33]. Each beacon was complexed with streptavidin to prevent nuclear localization of the beacon after microinjection. As shown in Figure 9.4a and b respectively, tetramethylrhodamine (TMR)-labeled β-actin mRNA-specific molecular beacons gave strong signal (TMR channel), while Texas-Red-labeled control molecular beacons show only weak background signal (Texas Red channel). Figure 9.4c shows the ratio image obtained by dividing the fluorescence intensity of TMR by that of Texas

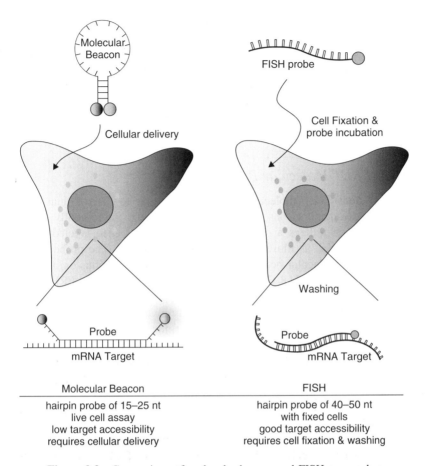

Figure 9.3 Comparison of molecular beacon and FISH approaches

Figure 9.4 Imaging the distribution of β-actin mRNA in living fibroblasts using molecular beacons. TMR-labeled β-actin mRNA-specific molecular beacons and Texas-Red-labeled control molecular beacons complexed with streptavidin were microinjected into cells. The fluorescence signals in the TMR (a) and Texas Red (b) channels were detected, and a ratio image (c) was obtained by dividing the fluorescence intensity of TMR by that of Texas Red at every pixel in the image, indicating the localization of β-actin mRNA. The color of each pixel in (c) reflects the value of the ratio, with warmer colors representing higher ratios and cooler colors representing lower ratios. Refer to Plate 6 for the colored version

Red at every pixel in the image, indicating clearly the localization of β-actin mRNA in fibroblast cells.

The use of a single type of reporter dye on each molecular beacon allows multiple, optically distinct, molecular beacons to be visualized simultaneously (i.e., multiplexing) [23,26]. This important attribute could potentially be taken advantage of to highlight the orchestration between various gene expression patterns in living cells. In fact, several groups have already demonstrated the feasibility of simultaneously imaging multiple genes in single living cells with molecular beacons [34,35].

9.2.5 Dual-FRET Molecular Beacons

The concept of dual-FRET molecular beacons is similar to that of dual-FRET linear probes, except molecular beacons are used in place of the fluorescently labeled linear ODNs. This approach utilizes a pair of molecular beacons each labeled with a donor and acceptor fluorophore, as illustrated in Figure 9.2e [30,31]. The probe sequences are chosen such that the molecular beacons hybridize to adjacent regions on a single nucleic acid target, similar to the dual-FRET linear probes. The resulting FRET signal (i.e., sensitized emission of the acceptor fluorophore) upon probe hybridization serves as a positive signal, which can be readily discerned from non-FRET false-positive signals due to probe degradation and nonspecific probe opening. Dual-FRET molecular beacons, therefore, combine the low background signal and high specificity of molecular beacons with the ability of two-probe FRET assays to differentiate between true target recognition and false-positive signals. Interestingly, it has been demonstrated that, upon hybridization to nucleic acid targets, dual-FRET molecular beacons can provide a better signal-to-background ratio than the single molecular beacon approach when appropriate FRET pairs are selected [30,31].

In the conventional molecular beacon design, the stem sequence is typically independent of the target sequence (Figure 9.2d); however, dual-FRET molecular beacons can be designed such that all the bases of one arm of the stem (to which a fluorophore is conjugated) are complementary to the target sequence, thus participating in both stem formation and target hybridization (shared-stem molecular beacons) [36] (Figure 9.2e). The advantage of this shared-stem design is to help fix the position of the fluorophore that is attached to the stem arm, limiting its degree-of-freedom of motion, and increasing the FRET in the dual-FRET molecular beacon design.

Dual-FRET molecular beacons have been used to detect K-ras, survivin, and Oct4 mRNAs in HDF, MIAPaCa-2, and H1 cells, respectively [31,37]. Interestingly, K-ras mRNAs seemed to localize with a filamentous pattern in HDF cell, whereas survivin mRNAs seemed to localize in a nonsymmetrical pattern within MIAPaCa-2 cells, often to one side of the nucleus of the cell, as shown in Figure 9.5. mRNA localization in living cells is believed to be closely related to post-transcriptional regulation of gene expression, but much remains to be seen if such localization indeed targets a protein to its site of function by producing the protein "right on the spot." Using a dual-beacon approach, the transport and localization of oskar mRNA in *Drosophila melanogaster* oocytes has also been visualized [29]. In this work, molecular beacons with $2'$-O-methyl backbone were delivered into cells using microinjection and the migration of oskar

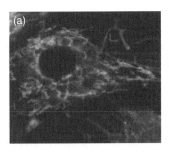
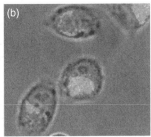

Figure 9.5 Imaging mRNA localization in HDF and MIAPaCa-2 cells using dual-FRET molecular beacons targeting respectively K-ras and survivin mRNA. (a) Fluorescence images of K-ras mRNA in stimulated HDF cells. Note the filamentous K-ras mRNA localization pattern. (b) A fluorescence image of survivin mRNA localization in MIAPaCa-2 cells. Note that survivin mRNAs often localize to one side of the nucleus of the MIAPaCa-2 cells. Refer to Plate 7 for the colored version

mRNAs was tracked in real time, from the nurse cells where they are produced to the posterior cortex of the oocyte where they are localized. Clearly, the direct visualization of specific mRNAs in living cells with molecular beacons will provide important insight into the intracellular trafficking and localization of RNA molecules.

As with other dual probes for live-cell RNA detection, although dual-FRET molecular beacons exhibit an improved signal-to-background and selectivity compared with single molecular beacons, this approach does require large stretches of RNA with little or no secondary structure that can accommodate the hybridization of two probes. This can be challenging, particularly when trying to identify single nucleotide polymorphisms, which significantly constrains the selection of target sites.

9.2.6 Fluorescent Protein-Based Probes

Fluorescent proteins have enjoyed widespread applications in biological studies since the introduction of a stabilized form of GFP in 1994. Inserting the DNA sequence encoding GFP into the 5′ end of the sequence for a protein of interest enables specific "tagging" of the protein. This strategy allows for a fluorescence signal to be emitted at the same location as the protein of interest while avoiding the often treacherous issue of delivering probes across the plasma membrane of a cell. However, this approach is not applicable to directly tagging a specific mRNA. An indirect method has been developed recently in which tandem repeats of a hairpin structure are inserted into the target RNA sequence in the 3′ untranslated region (3′ UTR), and GFP-tagged MS2 protein, a coat protein of the bacterial phage MS2, is co-expressed [38,39]. The GFP–MS2 binds specifically to the hairpin structure added to the mRNA of interest, thus generating a tag for it, as shown in Figure 9.2f [40]. In this approach, the binding of multiple GFP–MS2 fusion proteins to single RNA transcripts results in a high local concentration of GFP (i.e., bright fluorescent speckles) that can be seen above the high background of unbound GFP fusions – similar to the mechanism by which tagged linear ODN probes detect intracellular RNA. The GFP–MS2 approach has been used to track the localization and dynamics of RNA in living cells with single-molecule sensitivity and has, thus, provided a great deal of insight into RNA processing, localization, and transport [41,42].

Recently, a method to improve the signal-to-background of fluorescent-protein-based imaging of RNA has been introduced, based on the concept of protein fragment complementation (PFC) [43]. PFC refers to the rational dissection of a protein (or enzyme) into two inactive fragments that fold into the complete protein and regain functionality when held in close proximity. In the case of fluorescent proteins, the complementation of the two fluorescent protein fragments results in the regeneration of an optical signal. As illustrated in Figure 9.2g, for RNA imaging, two RNA-binding proteins are expressed in living cells, each containing a fragment of a split GFP (or YFP). Binding of the two RNA-binding proteins to adjacent sequences in an mRNA strand drives the association of the two GFP fragments, inducing GFP complementation and activation of a fluorescence signal. An optical signal was not observed unless the target mRNA was expressed and the GFP fragments were brought into close proximity upon binding the RNA binding site [43]. This approach clearly improves upon the signal-to-background signal compared with the GFP–MS2-based techniques that use full-length GFP, although the hybridization of the protein fragments requires more time and is not as bright as a single GFP molecule. Both approaches require extensive genetic modification and so are unsuitable for endogenous RNA detection as well as therapeutic and diagnostic applications.

Another PFC approach that enables the detection of endogenous RNA in living cells involves the co-expression of GFP fragments fused to genetically modified Pumilio homology domains (PUM-HDs) [44]. PUM-HD is composed of an array of eight peptide elements that recognize specific eight-base RNA sequences, with each element specifically recognizing a single base. The specificity of PUM-HD was tailored by changing the RNA-recognizing amino acid residues within the PUM-HD elements. As a result, two different PUM-HDs were created that recognize a 16-base sequence of mitochondrial RNA encoding NADH dehydrogenase subunit 6 (ND6). Upon targeting the split GFP–PUM-HD fusions into the mitochondrial matrix, it was possible to image the localization of ND6 mitochondrial RNA in real time.

The most significant advantage of a fluorescent-protein-based approach is that the probes do not have to be introduced into the cell, but can be expressed in the cytoplasm using various DNA-delivery vehicles. Further, the background signal is quite low, since there is very little fluorescent signal unless the RNA-binding proteins are bound to the target mRNA. The split-GFP method, however, may have difficulties in tracking the dynamics of RNA expression in real time, since the reconstitution of the fluorescent protein from the split fragments typically takes 0.5–4 h, during which the RNA expression level may change. Further, GFP, once it is reconstituted, does not readily dissociate, making it difficult to monitor the down-regulation of gene expression. GFP is also not particularly bright compared with most commercial fluorophores. Therefore, it can be difficult to detect faint signals above autofluorescence. Fluorescent-protein-based probes are also often criticized because the binding of large bulky protein to RNA could interfere with normal RNA transport. This is particularly true when multiple GFP molecules are bound to a single RNA transcript. Transfection efficiency could also be a major concern in the GFP-based approaches, in that usually only a few percent of the cells express the fluorescent proteins following transfection. This limits the application of the split-GFP methods in detecting diseased cells using mRNA as a biomarker for the disease.

9.3 Probe Design, Imaging, and Biological Issues

Molecular beacons provide an enticing opportunity to optically track some of the most elusive events in molecular biology. Despite the complexity of interpreting molecular beacon results, molecular beacons themselves are relatively simple molecules to synthesize. Designing molecular beacons involves understanding a series of trade-offs between the specificity, sensitivity, and signal-to-background for beacons. In the following sections we discuss how the major aspects of a molecular beacon (probe sequence, hairpin structure, and fluorophore/quencher selection) determine beacon performance.

9.3.1 Specificity of Molecular Beacons

The most important property of a molecular beacon is its specificity, which can be loosely quantified as the difference in melting temperature between perfectly complementary hybrids and hybrids with single base mismatches. Molecular beacons typically exhibit a higher specificity for perfectly complementary nucleic acid targets than linear ODN probes do. It has been shown that properly designed molecular beacons can readily discriminate between targets that differ by as little as a single nucleotide [45,46]. The reason is the energy penalty associated with unwinding the molecular beacon stem reduces binding to mismatched targets because it is less energetically favorable. In experiments where the detection of SNPs is required, the specificity of molecular beacons can be improved by increasing the stem length. A longer stem provides a wider set of conditions over which molecular beacons can discriminate between two targets. This can be attributed to the enhanced stability of the molecular beacon stem–loop structure and the resulting smaller free energy difference between closed (unbound) molecular beacons and molecular beacon–target duplexes, which generates a condition where a single-base mismatch reduces the energetic preference of probe–target binding. A longer stem also increases the signal-to-background ratio, since the more stable hairpin conformation reduces the probability of stem opening due to Brownian fluctuations and results in more efficient quenching of the fluorescent dye.

The type and number of nucleotides in a molecular beacon's loop must be carefully designed for the chosen RNA target and the set of experimental conditions [36,45]. For example, in live-cell studies it is desirable for the melting temperature of perfectly complementary hybrids to be above 37 °C and the melting temperature for hybrids with single base mismatches to be less than 37 °C (Figure 9.6). Both of these melting temperatures will increase with increasing loop length and decreasing stem length. If the melting temperatures are too high, it would not be possible to discriminate between perfectly complementary and mismatched targets under physiological conditions. On the other hand, if the melting temperatures are too low, a large number of molecular beacons may open under physiological conditions, leading to a high level of background signal. The melting temperature of a molecular beacon can be tailored by changing its stem–loop structure, as demonstrated in Figure 9.7a. Changing the probe length of a molecular beacon may also influence the rate of hybridization, as demonstrated by Figure 9.7b, but generally to a lesser extent than changing the stem length. While both the stability of the hairpin probe and its ability to discriminate targets over a wider range of temperatures increase with increasing stem length, this is accompanied by a decrease in hybridization on-rate

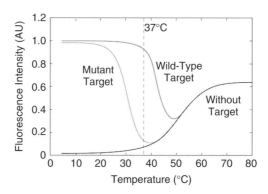

Figure 9.6   Typical molecular beacon thermal denaturation profiles. With wild-type (complementary) targets, molecular beacons emit a maximal signal at low temperatures, indicating that the molecular beacons are bound to target; as temperature increases, the molecular beacons melt away from the target. With mutant targets, the melting temperature of the molecular beacons is reduced. The difference between the "wild-type target" and "mutant target" curves over a range of temperatures represents the "window of discrimination" between wild-type and mutant targets. In live-cell studies it is desirable for the melting temperature of perfectly complementary hybrids to be above 37 °C and the melting temperature for hybrids with single base mismatches to be less than 37 °C

constant, as shown in Figure 9.7b. For example, molecular beacons with a four-base stem had an on-rate constant up to 100 times greater than molecular beacons with a six-base stem.

In addition to melting temperature considerations, the characteristics of the target sequence itself must be unique to guarantee specificity. The Basic Local Alignment Search Tool (BLAST) developed by the National Center for Biotechnology Information (NCBI) [47] or similar software can be used to select multiple target sequences of 15–25 bases that are unique for the target RNA. For any target sequence selected there might be multiple genes in the mammalian genome that have sequences differing by only a few bases. BLAST allows the selection of beacon target sequences that are not only unique for the specific RNA of interest, but also with minimal partial match between beacon sequence and the sequence of other RNAs. It is also necessary to select a target sequence with a balanced G–C content (the percentage of G–C pairs). A G–C mismatch carries a much larger energy penalty, enhancing specificity; however, the prevalence of C–G islands in the genome makes beacons with a high G–C content more likely to have off-target interactions. Therefore, it is important to select a G–C-balanced and unique target sequence to ensure the specificity of molecular beacons.

Several approaches can be taken to validate the signal specificity in live cells. The most common approach is to up- or down-regulate the expression level of a specific RNA and compare molecular-beacon-based imaging in live cells with RT-PCR data quantifying the RNA expression level. Complications may arise when the approach used to change the RNA expression level in living cells has an effect on multiple genes, leading to some ambiguity even when PCR and beacon results match. A common way to down-regulate the level of a single mRNA in live cells is to use small interfering RNA (siRNA) treatment, which typically leads to >80% reduction of the specific mRNA level when the siRNA protocol is properly optimized.

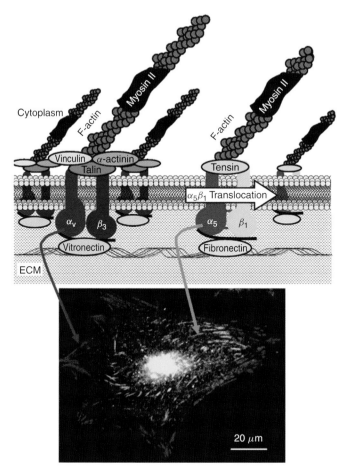

Plate 1 Various adhesions in two dimensions, with a graphical description of focal adhesions and fibrillar adhesions. Reprinted with permission from [45]

Nano and Cell Mechanics: Fundamentals and Frontiers, First Edition. Edited by Horacio D. Espinosa and Gang Bao.
© 2013 John Wiley & Sons, Ltd. Published 2013 by John Wiley & Sons, Ltd.

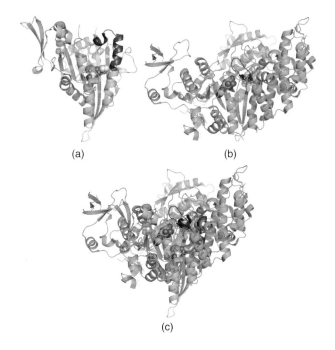

(a)   (b)

(c)

Plate 2   Structures of kinesin-5 and myosin-II bound to inhibitors

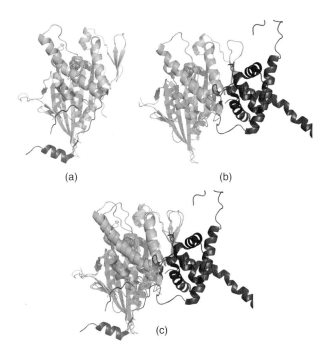

(a)   (b)

(c)

Plate 3   Regulation and calcium-dependent activation of KCBP

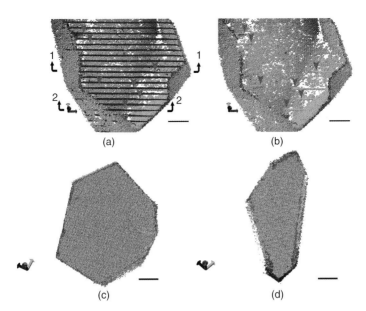

Plate 4 Dislocation nucleation in a representative grain in nanotwinned Cu with grain size of 20 nm and twin boundary spacing of 1.25 nm. Atomic views of dislocation structures (pointed by gray arrows). (a) Twin boundaries are painted in red. (b) Twin boundaries are invisible. (c, d) Top views of the cross-section marked by 1 and 2, respectively, in (a). The scale bars are 5 nm

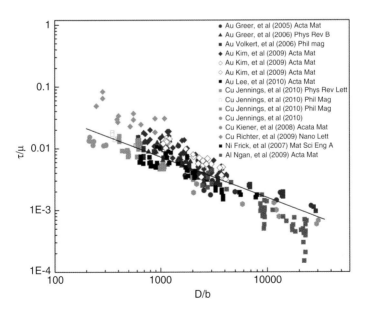

Plate 5 Shear strength normalized by shear modulus as a function of pillar diameter normalized by Burgers vector for a wide range of non-pristine fcc metals. Reprinted with permission from [1], © 2011 Elsevier

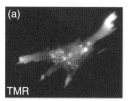 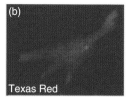 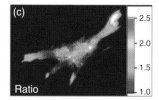

Plate 6 Imaging the distribution of β-actin mRNA in living fibroblasts using molecular beacons. TMR-labeled β-actin mRNA-specific molecular beacons and Texas-Red-labeled control molecular beacons complexed with streptavidin were microinjected into cells. The fluorescence signals in the TMR (a) and Texas Red (b) channels were detected, and a ratio image (c) was obtained by dividing the fluorescence intensity of TMR by that of Texas Red at every pixel in the image, indicating the localization of β-actin mRNA. The color of each pixel in (c) reflects the value of the ratio, with warmer colors representing higher ratios and cooler colors representing lower ratios

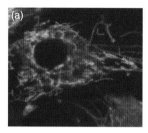 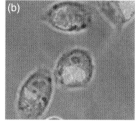

Plate 7 Imaging mRNA localization in HDF and MIAPaCa-2 cells using dual-FRET molecular beacons targeting respectively K-ras and survivin mRNA. (a) Fluorescence images of K-ras mRNA in stimulated HDF cells. Note the filamentous K-ras mRNA localization pattern. (b) A fluorescence image of survivin mRNA localization in MIAPaCa-2 cells. Note that survivin mRNAs often localize to one side of the nucleus of the MIAPaCa-2 cells

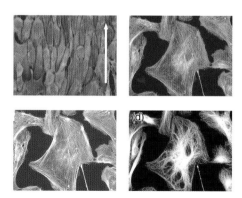

Plate 8 (a) Image of human umbilical vein endothelial cells (HUVECs) subjected to continuous shear stress, noting how the cells have aligned with the applied flow. Accordingly, filamentous actin fibers, stained green with phalloidin, also oriented to the direction of flow. The nuclei are stained blue with Hoechst. (b) In contrast, actin fibers, here stained green with phalloidin, in HUVECs under static conditions exhibit no such orientation. Filamentous actin is shown separately in (c). The microtubules are stained with anti-α-tubulin and appear red, shown separately in (d); while the nucleus has been stained blue with Hoechst. Photographs by Yumiko Sakurai

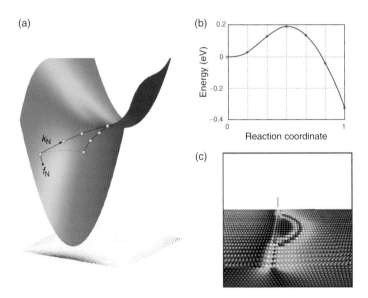

Plate 9 The free-end NEB method enables an efficient reaction pathway calculation of extended defects in the large system. (a) Illustration of the method showing one end of the elastic band is fixed and the other is freely moved along an energy contour. (b) An example of the converged MEP from the free-end NEB method calculation; the end node is pinned at 0.3 eV below the initial state. (c) The corresponding saddle-point structure of a dislocation loop bowing out from a mode II crack tip (the upper half crystal is removed for clarity)

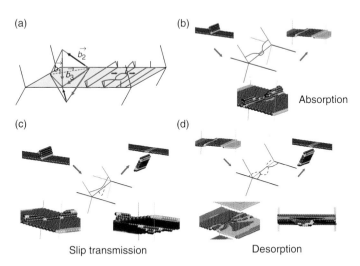

Plate 10 The free-end NEB modeling of twin-boundary-mediated slip transfer reactions [15]. (a) Schematics of dislocation–interface reactions based on double Thompson tetrahedra, showing different combinations of incoming and outgoing dislocations (the Burgers vectors with the same color) at a coherent TB. Atomic structures of the initial, saddle-point, and final states are shown for the slip transfer reactions, including (b) absorption, (c) direct transmission, and (d) desorption

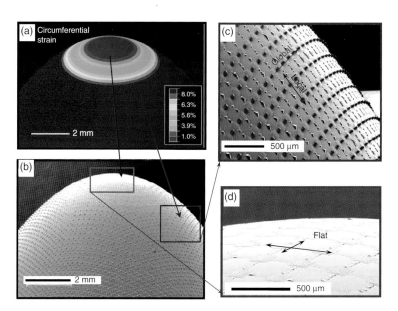

Plate 11 (a) Distribution of circumferential strain for a parabolic elastomeric transfer element given by the mechanics model. (b–d) Buckling patterns observed in experiments. Reprinted with permission from [51], © 2010 The Royal Society of Chemistry

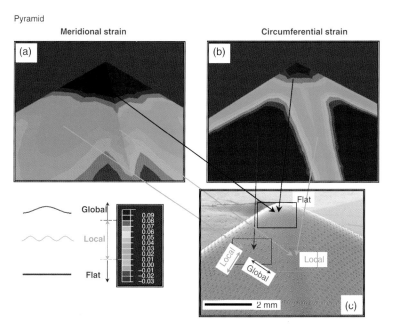

Plate 12 (a, b) Strain distributions in the meridional and circumferential directions around the pyramidal PDMS membrane. The corresponding buckling modes observed in experiments are shown in (c) for comparison. Reprinted with permission from [30], © 2009 John Wiley & Sons

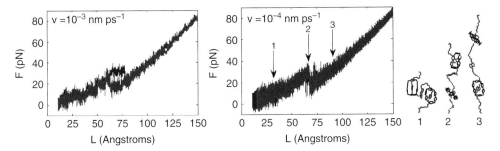

Plate 13 Simulated force–extension behavior of a [3] rotaxane immersed in a high dielectric medium at 300 K obtained using MD and the MM3 [66] force field. The plots show the force–extension curves during stretching (blue) and retraction (red) using pulling speeds v of 10^{-3} and 10^{-4} nm ps^{-1}. The cantilever employed has a soft spring constant of $k = 1.1$ pN Å^{-1}. Snapshots of structures (labels 1, 2 and 3) encountered during pulling are shown on the far right

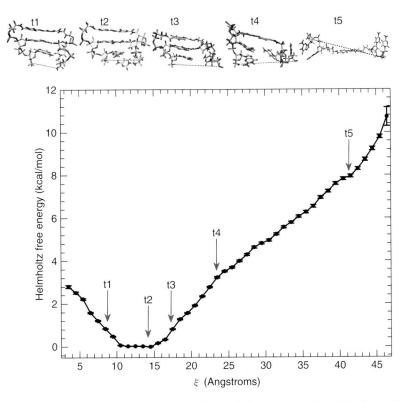

Plate 14 Potential of mean force along the end-to-end distance coordinate for the azobenzene-capped DNA hairpin. Snapshots of structures adopted during the unfolding are shown in the top panel. The error bars correspond to twice the standard deviation obtained from a bootstrapping analysis

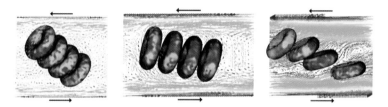

Plate 15 The shear of a four-RBC cluster at low (left), intermediate (middle), and high (right) shear rates S. The vectors represent the fluid velocity field

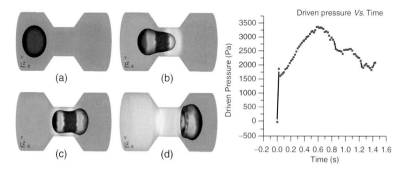

Plate 16 A 3D simulation of a single cell squeezing through a capillary vessel (left). The history of the driven pressure during the squeezing process (right). The cell diameter is 1.2 times the diameter of the constriction

Plate 17 Model of RBCs with a diameter of $7-8\,\mu$m (top, left), NPs of various shapes and sizes ranging between 10 and 1000 nm (top, middle), and a WBC with a diameter of $20\,\mu$m (top, right). Fluid domain ($40 \times 40 \times 200\,\mu$m^3) with RBCs (red), leukocytes (white), and NPs (green)

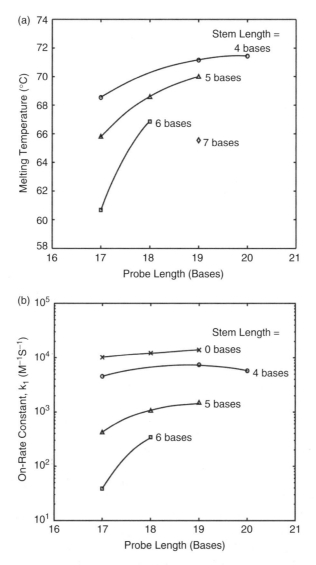

Figure 9.7 Structure–function relations of molecular beacons. (a) Melting temperatures for molecular beacons with different structures in the presence of target. (b) The rate constant of hybridization k_1 (on-rate constant) for molecular beacons with various probe and stem lengths hybridized to their complementary targets

9.3.2 Fluorophores, Quenchers, and Signal-to-Background

Selecting an appropriate dye–quencher pair yields important benefits for the signal-to-background ratio, multiplexing ability, and fluorescence quantification of molecular beacons. For example, a systematic study on a wide range of fluorophore–quencher combinations showed that the quenching efficiency (contact quenching) could vary between

57% and 98% [48]. Quenching efficiencies could potentially be further improved by either using inorganic quenchers, such as gold, or by incorporating multiple quenchers into a single molecular beacon [49,50]. Alternatively, the signal intensity of molecular beacons can be increased by using quantum dots or photoluminescent polymers [51–53]. Initial studies have suggested that quantum-dot-based molecular beacons only exhibit a signal-to-background ratio of 6 : 1 due to inefficient quenching. Inefficient quenching is unlikely to be an issue for photoluminescent polymers, which exhibit a superquenching effect. When the fluorescence of any single repeat unit is quenched, the entire polymer chain responds in the same fashion. Long polymer chains can, therefore, be used to provide an amplified fluorescent signal that can be modulated by a single quencher. However, nonspecific interaction of the photoluminescent polymers prevents them from widespread application in live-cell RNA detection assays [53].

A more conventional way to increase signal-to-background ratio is to use multiple beacons to target the same RNA molecule. As an example, molecular beacons were designed to target a sequence in the genome of bovine respiratory syncytial virus (bRSV) that has three exact repeats [54]. Figure 9.8 shows the molecular beacon signal indicating the spreading of viral infection at days 1, 3, 5, and 7 post-infection (PI), which demonstrates the ability of molecular beacons to monitor and quantify in real time the viral infection process. Molecular beacons were further used to image the viral genomic RNA (vRNA) of human RSV (hRSV) in live Vero cells, revealing the dynamics of filamentous virion egress, and providing insight into how viral filaments bud from the plasma membrane of the host cell [55].

In addition to creating brighter fluorescent labels or adding multiple labels to each mRNA, the excitation and emission peaks of the labels are also important. Red-shifted fluorophores can be used to improve signal-to-background in live cells by eliminating the interfering effects that result from autofluorescence. It is also possible to use lanthanide chelate as the donor in a dual-FRET probe assay and perform time-resolved measurements to dramatically increase the signal-to-background ratio [30].

As fluorescent imaging strategies have advanced, there has been a general trend towards more quantitative measurements of fluorescent signals, with the ultimate goal being absolute quantification. The absolute quantification of fluorescence could allow the exact number of fluorophores within a compartment/cell to be quantified and correspondingly allow the number of target genes, proteins, or enzymes to be quantified. However, factors such as nonspecific protein interactions and pH could have a dramatic effect on the fluorescence intensity of some fluorophores. Therefore, if accurate fluorescence measurements are desirable, it is necessary to select fluorescent labels that are insensitive to their environment. Recently, it has been shown that although the fluorescence intensity of a few fluorophores (e.g., fluorescein) was highly susceptible to the intracellular environment, other fluorophores (e.g., Dylight 649, Alexa647, and Alexa750) were insensitive to the intracellular environment [56].

9.3.3 Target Accessibility

A critical issue in molecular beacon design is target accessibility. It is well known that proteins are constantly bound to functional mRNA molecules in living cells, forming a ribonucleoprotein (RNP). Further, an mRNA molecule often has double-stranded portions

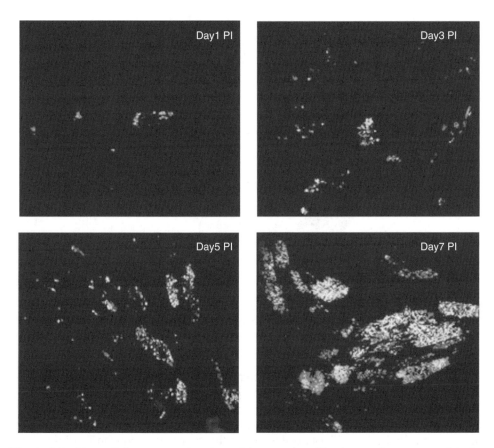

Figure 9.8   Live-cell fluorescence imaging of the genome of bovine respiratory syncytial virus (bRSV) using molecular beacons shows the spreading of infection in host cells at days 1, 3, 5, and 7 post-infection (PI). Primary bovine turbinate cells were infected by a clinical isolate of bRSV, CA-1, with a viral titer of $2 \times 10^{3.6}$ TCID$_{50}$/ml. Molecular beacons were designed to target several repeated sequences of the gene-end-intergenic-gene-start signal within the bRSV genome, with a signal-to-noise ratio of 50–200

and forms secondary (folded) structures (Figure 9.9). It is therefore necessary to avoid targeting mRNA sequences that are double stranded or occupied by RNA-binding proteins. These sites require the molecular beacon to compete off the RNA-binding protein or RNA strand in order to hybridize to the target. This competition is hypothesized to contribute to a lack of signal when certain molecular beacons designed for a specific mRNA are delivered to living cells. Although predictions of mRNA secondary structure can be made using software such as *Beacon Designer* (www.premierbiosoft.com) and *mfold* (http://www.bioinfo.rpi.edu/applications/mfold/old/dna/), they may be inaccurate due to limitations of the biophysical models used and a limited understanding of protein–RNA interaction. For each target mRNA, therefore, it may be necessary to synthesize multiple unique molecular beacon sequences along the target RNA and test them in living cells to select the best target sequence.

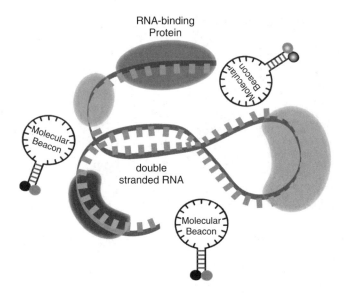

Figure 9.9 A schematic illustration of a segment of the target mRNA with a double-stranded portion and RNA-binding proteins. A molecular beacon has to compete off an mRNA strand or RNA-binding protein(s) in order to hybridize to the target

To uncover the possible molecular beacon design rules, the accessibility of BMP-4 mRNA was studied using different beacon targets for the same mRNA sequence [57]. Specifically, molecular beacons were designed to target the start codon, the termination codon, siRNA sites, anti-sense ODN probe sites, and random sites. All the target sequences were run through the BLAST database to ensure that they were unique to BMP-4 mRNA. Of the eight molecular beacons designed to target BMP-4 mRNA, it was found that only two beacons resulted in a strong signal inside cells. The positive beacons targeted the start codon region and the termination codon region. It was also found that shifting the target sequence of a molecular beacon by just a few bases towards the 3' or 5' ends would significantly reduce the fluorescence signal from beacons in a live-cell assay. This indicates that the target accessibility is quite sensitive to the location of the targeted sequence. These results, together with molecular beacons validated previously, suggest that the start codon and termination codon regions and the exon–exon junctions are often more accessible than other locations in an mRNA.

9.4 Delivery of Molecular Beacons

One of the critical steps in the accurate detection of RNA molecules in living cells is the efficient delivery of synthetic probes into the cytoplasm. ODN-based probes are generally prevented from gaining access to the cytoplasm due to the cell membrane [58]. Once the probes enter the cells successfully the fraction of probes that are free to hybridize intracellular RNA also becomes a concern. Numerous studies have shown that molecular beacons are rapidly sequestered into the nucleus once introduced into cells; however, there have been an equal number of studies that did not observe this pattern of intracellular distribution. It is not clear whether these differences are cell-line dependent, dependent on the method of delivering molecular beacons, or due to some other variable. Nonetheless, several methods have been introduced to prevent nuclear sequestration. Specifically, molecular beacons have been conjugated to large proteins (or nanoparticles) that prevent their passage through nuclear pores and they have been linked to tRNAs that drive nuclear export [33,59–61]. Common techniques that have been used to deliver ODN-based RNA imaging probes into live cells include microinjection, polycationic molecules (such as liposomes and dendrimers), electroporation, cell-penetrating peptides (CPPs) or steptolysin O (SLO).

9.4.1 Microinjection

Microinjection is, perhaps, one of the most direct methods for ensuring the delivery of ODN probes into the cytoplasm of live cells. This method is advantageous because it removes the question of whether beacons have crossed the cell membrane or not. Numerous studies have delivered molecular beacons to a small number of cells and used microscopy to determine the subcellular localization of the probes [62]. Microinjection is unsuitable for population-based confirmation of beacon results due to the relatively low number of cells that can be analyzed at any given time. Further, microinjection can often be damaging to the cell and may interfere with normal cell function. Because of these drawbacks, many investigators use alternative methods that result in higher delivery throughput and less physical damage than microinjection.

9.4.2 Cationic Transfection Agents

Cationic transfection agents can be used to deliver molecular beacons into living cells. These transfection agents generally form lipoplexes with molecular beacons, which can sometimes stabilize molecular beacons in their hairpin confirmation. While these agents have been found to be effective in some studies [34], others have found that many commercial transfection agents result in punctate fluorescent patterns that appear to be indicative of endosomal entrapment [32]. ODN probes that enter into the endosomal/lysosomal pathway are rapidly degraded by nucleases and the acidic environment [63]. Consequently, even when transfection methods allow for endosomal/lysosomal escape, only 0.01–10% of the probes retain their functionality [64]. Furthermore, probe delivery via the endocytic pathway typically takes 2–4 h. This long time period required increases the likelihood of molecular beacon degradation due to cellular nucleases.

9.4.3 Electroporation

To avoid the deleterious effects associated with endosomal entrapment, methods such as electroporation have been used to deliver ODNs directly into the cytoplasm of living cells. Although electroporation has traditionally been associated with low cell viability, recent advances in electroporation technology, such as the ability to perform electroporation in microliter-volume spaces (e.g., pipette tips), has led to a reduction in the many harmful events associated with this process, including heat generation, metal ion dissolution, pH variation, and oxide formation. This microliter-volume electroporation process is known as microporation. Recently, it was shown that microporation could lead to the uniform cytosolic distribution of ODN probes in live cells with a transfection efficiency of 93% and an average viability of 86% [60]. A unique advantage of microporation is that delivery of ODN probes takes seconds, so cells can be analyzed immediately for RNA content. Conversely, a potential disadvantage is that most electroporation techniques require cells to be detached from the culture surface. Therefore, it is several hours before the cells re-adhere to cell culture plate surfaces and RNA localization can be imaged.

9.4.4   Chemical Permeabilization

Another non-endocytic delivery method is toxin-based cell membrane permeabilization. One popular reagent is SLO, which is a pore-forming bacterial toxin that has been used as a simple and rapid means of introducing ODNs into eukaryotic cells [65–67]. SLO binds as a monomer to cholesterol and oligomerizes into a ring-shaped structure to form pores of approximately 25–30 nm in diameter, allowing the influx of both ions and macromolecules. An essential feature of this technique is that the toxin-based permeabilization is reversible [68]. This can be achieved by introducing ODNs with SLO under serum-free conditions and then removing the mixture and adding normal media with serum and calcium [66,68]. Since cholesterol composition varies with cell types, the permeabilization protocol needs to be optimized for each cell type by varying temperature, incubation time, cell number, and SLO concentration. Typically, RNA localization can be assessed 30 min to 2 h following the introduction of ODN probes into cells using SLO-based delivery.

9.4.5   Cell-Penetrating Peptide

CPPs have been used to introduce proteins, nucleic acids, and other biomolecules into living cells [69–71]. Among the family of peptides with membrane translocating activity are antennapedia, HSV-1 VP22, and the HIV-1 Tat peptide. To date, the most widely used peptides are the HIV-1 Tat peptide and its derivatives, owing to their small size and high delivery efficiency. The Tat peptide is rich in cationic amino acids, especially arginines, which are very common in many of the CPPs. However, the exact mechanism for CPP-induced membrane translocation remains elusive.

A wide variety of cargos have been delivered to living cells both in cell culture and in tissue using CPPs [72,73]. For example, Allinquant et al. [74] linked antennapedia peptide to the 5' end of DNA ODNs (with biotin on the 3' end) and incubated both peptide-linked ODNs and ODNs alone with cells. By detecting biotin using streptavidin–alkaline phosphatase amplification, it was found that the peptide-linked ODNs were internalized very efficiently into all cell compartments compared with control ODNs. No indication of endocytosis was found. Similar results were obtained by Troy et al. [75], with a 100-fold increase in antisense delivery efficiency when ODNs were linked to antennapedia peptides. Recently, Tat peptides were conjugated to molecular beacons using different linkages (Figure 9.10a and b); the resulting peptide-linked molecular beacons were delivered into living cells to target GAPDH and survivin mRNAs [32]. At relatively low concentrations,

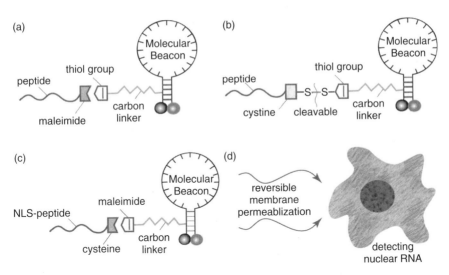

Figure 9.10   A schematic of peptide-linked molecular beacons. (a) A peptide-linked molecular beacon using the thiol-maleimide linkage in which the quencher-arm of the molecular beacon stem is modified by adding a thiol group which can react with a maleimide group placed to the C terminus of the peptide to form a direct, stable linkage. (b) A peptide-linked molecular beacon with a cleavable disulfide bridge in which the peptide is modified by adding a cysteine residue at the C terminus which forms a disulfide bridge with the thiol-modified molecular beacon. This disulfide bridge design allows the peptide to be cleaved from the molecular beacon by the reducing environment of the cytoplasm. (c) A schematic illustration of the design of a peptide-linked molecular beacon and its delivery into cell nucleus. The NLS peptide is covalently linked to the molecular beacon using a modified nucleotide in its quencher arm. The NLS linked molecular beacons are delivered into the cytoplasm first using streptolysin O (SLO), and the NLS peptide actively transports the probes into the nucleus of a living cell

peptide-linked molecular beacons were internalized into living cells within 30 min with nearly 100% efficiency. Peptide-based delivery did not interfere with molecular beacon binding to survivin or GAPDH, since similar levels of fluorescence and a similar pattern of localization was seen in cells with beacons delivered using alternative means. Peptide-linked molecular beacons show impressive potential as an all-in-one molecule capable of self-delivery, targeting, and reporting in live cells.

Peptide-linked molecular beacons can also be delivered using an SLO-based approach to target RNA molecules in the cell nucleus by attaching a nuclear localization signal (NLS) peptide to a molecular beacon. Molecular beacons designed to target snRNAs U1 and U2 were linked to NLS peptides and delivered to cells using the SLO-based reversible membrane permeabilization method (Figure 9.10c). The small nucleolar RNA U3 was delivered into the nuclei of live HeLa cells, and the localization and co-localization (U1 and U2) of these nuclear RNAs was imaged [76]. This delivery method can potentially be used to image transcriptional and post-transcriptional processing of RNAs in the nucleus of living cells.

9.5 Engineering Challenges and Future Directions

Nanostructured molecular probes such as molecular beacons have the potential to address a wide range of applications that require sensitive detection of genomic sequences. For example, molecular beacons are used as a tool for the detection of single-stranded nucleic acids in homogeneous *in vitro* assays [77,78]. Surface-immobilized molecular beacons used in microarray assays allow for the high-throughput parallel detection of nucleic acid targets while avoiding the difficulties associated with PCR-based labeling [77,79]. Another novel application of molecular beacons is the detection of double-stranded DNA targets using PNA "openers" that form triplexes with the DNA strands [80]. Further, proteins can be detected by synthesizing "aptamer molecular beacon," which, upon binding to a protein, undergoes a conformational change that results in the restoration of fluorescence [81,82].

The most exciting application of nanostructured ODN probes, however, is living-cell gene expression detection. Molecular beacons can detect endogenous mRNA in living cells with high specificity, sensitivity, and signal-to-background ratio. Thus, molecular beacons have the potential to provide a powerful tool for laboratory and clinical studies of gene expression *in vivo*. For example, molecular beacons can be used in high-throughput cell-based assays to quantify and monitor the dose-dependent changes of specific mRNA expression in response to different drug leads. The ability of molecular beacons to detect and quantify the expression of specific genes in living cells will also facilitate disease studies, such as viral infection detection and cancer diagnosis.

There are a number of challenges in detecting and quantifying RNA expression in living cells. In addition to issues of probe design and target accessibility, quantifying gene expression in living cells in terms of mRNA copy-number per cell poses a significant challenge. It is necessary to distinguish true signal from background signals, to determine the fraction of mRNA molecules hybridized with probes, and to quantify the possible self-quenching effect of the reporter, especially when mRNA is highly localized. Since the fluorescence intensity of the reporter may be altered by the intracellular environment, it is also necessary to create an internal control by, for example, injecting a known quantity of additional fluorescently labeled ODNs into the same cells and obtaining the corresponding fluorescence intensity. Further, unlike in RT-PCR studies, where the mRNA expression is averaged over a large number of cells (usually over 1 million), in optical imaging of mRNA expression in living cells only a relatively small number of cells (typically less than 1000) are observed. Therefore, the average copy number per cell may change with the total number of cells observed due to the (often large) cell-to-cell variation of mRNA expression.

Another issue in living-cell gene detection using hairpin ODN probes is the possible effect of probes on normal cell function, such as protein expression. As revealed in antisense therapy research, complementary pairing of a short segment of an exogenous ODN to mRNA can have a profound impact on protein expression levels and even cell fate. For example, tight binding of the probe to the translation start site may block mRNA translation. Binding of a DNA probe to mRNA can also trigger RNase H-mediated mRNA degradation. However, the probability of eliciting antisense effects with hairpin probes may be very low, because low concentrations of probes (<200 nM) are used for mRNA detection in contrast to the high concentrations (typically 20 μM; [67]) employed in antisense experiments. Furthermore, antisense effects are generally unable to be observed for at least 4 h, whereas visualization of mRNA with hairpin probes requires less than 2 h after delivery. Well-designed experiments should carry out a systematic study of possible antisense effects, especially for molecular beacons with $2'$-O-methyl backbone, which may bind to mRNA for longer periods of time.

As a new approach for *in vivo* gene detection, the nanostructured probes can be further developed to have enhanced sensitivity and a wider range of applications. Hairpin ODN probes with near-infrared (NIR) dye as the reporter combined with peptide-based delivery have the potential to detect specific RNAs in tissue samples, animals, or even humans. It is also possible to use lanthanide chelate as the donor in a dual-FRET probe assay and perform time-resolved measurements to dramatically increase the signal-to-noise ratio, thus achieving high sensitivity in detecting low-abundance genes. Although very challenging, the development of these and other nanostructured ODN probes will significantly enhance our ability to image, track, and quantify gene expression *in vivo*, and provide a powerful tool for basic and clinical studies of human health and disease.

There are many possibilities for nanostructured ODN probes to become clinical tools for disease detection and diagnosis. For example, molecular beacons could be used to perform cell-based early cancer detection using clinical samples, including blood, saliva, and other bodily fluid. In this case, cells in the clinical sample are separated and molecular beacons designed to target specific cancer genes are delivered to the cytoplasm for detecting mRNAs of the cancer biomarker genes. Cancer cells having a high level of the target mRNAs (such as survivin) or mRNAs with specific mutations that cause cancer (such as K-ras codon 12 mutations) would show a high level of fluorescence signal, while normal cells would show just low background signal. In this approach, the target mRNAs would not be diluted compared with the approaches using cell lysate. Thus, a molecular-beacon-based assay has the potential to positively identify cancer cells in a clinical sample with high specificity and sensitivity. It may also be possible to detect cancer cells *in vivo* by using NIR-dye-labeled molecular beacons in combination with endoscopy. Although there remain significant challenges, imaging methods using nanostructured probes have a great potential in becoming a powerful clinical tool for disease detection and diagnosis.

Acknowledgments

This work was supported by the National Heart Lung and Blood Institute of the NIH as a Program of Excellence in Nanotechnology award (HHSN268201000043C to G.B.) and an NCI grant (CA116102 to A.T.). This work was also supported by the NSF (BES-0616031 to A.T.) and the American Cancer Society (RSG-07-005-01 to A.T.).

References

[1] Saiki, R.K., Scharf, S., Faloona, F. *et al.* (1985) Enzymatic amplification of beta-globin genomic sequences and restriction site analysis for diagnosis of sickle cell anemia. *Science*, **230**, 1350–1354.

[2] Alwine, J.C., Kemp, D.J., Parker, B.A. *et al.* (1979) Detection of specific RNAs or specific fragments of DNA by fractionation in gels and transfer to diazobenzyloxymethyl paper. *Methods in Enzymology*, **68**, 220–242.

[3] Adams, M.D., Dubnick, M., Kerlavage, A.R. *et al.* (1992) Sequence identification of 2,375 human brain genes. *Nature*, **355**, 632–634.

[4] Velculescu, V.E., Zhang, L., Vogelstein, B., and Kinzler, K.W. (1995) Serial analysis of gene expression. *Science*, **270**, 484–487.

[5] Liang, P. and Pardee, A.B. (1992) Differential display of eukaryotic messenger RNA by means of the polymerase chain reaction. *Science*, **257**, 967–971.

[6] Schena, M., Shalon, D., Davis, R.W., and Brown, P.O. (1995) Quantitative monitoring of gene expression patterns with a complementary DNA microarray. *Science*, **270**, 467–470.

[7] Bassell, G.J., Powers, C.M., Taneja, K.L., and Singer, R.H. (1994) Single mRNAs visualized by ultrastructural *in situ* hybridization are principally localized at actin filament intersections in fibroblasts. *Journal of Cell Biology*, **126**, 863–876.

[8] Buongiorno-Nardelli, M. and Amaldi, F. (1970) Autoradiographic detection of molecular hybrids between RNA and DNA in tissue sections. *Nature*, **225**, 946–948.

[9] Behrens, S., Fuchs, B.M., Mueller, F., and Amann, R. (2003) Is the *in situ* accessibility of the 16S rRNA of *Escherichia coli* for Cy3-labeled oligonucleotide probes predicted by a three-dimensional structure model of the 30S ribosomal subunit? *Applied and Environmental Microbiology*, **69**, 4935–4941.

[10] Huang, Q. and Pederson, T. (1999) A human U2 RNA mutant stalled in 3′ end processing is impaired in nuclear import. *Nucleic Acids Research*, **27**, 1025–1031.

[11] Glotzer, J.B., Saffrich, R., Glotzer, M., and Ephrussi, A. (1997) Cytoplasmic flows localize injected oskar RNA in *Drosophila* oocytes. *Current Biology*, **7**, 326–337.

[12] Jacobson, M.R. and Pederson, T. (1998) Localization of signal recognition particle RNA in the nucleolus of mammalian cells. *Proceedings of the National Academy of Sciences of the United States of America*, **95**, 7981–7986.

[13] Dirks, R.W., Molenaar, C., and Tanke, H.J. (2001) Methods for visualizing RNA processing and transport pathways in living cells. *Histochemistry and Cell Biology*, **115**, 3–11.

[14] Dirks, R.W. and Tanke, H.J. (2006) Advances in fluorescent tracking of nucleic acids in living cells. *Biotechniques*, **40**, 489–496.

[15] Molenaar, C., Abdulle, A., Gena, A. *et al*. (2004) Poly(A)⁺ RNAs roam the cell nucleus and pass through speckle domains in transcriptionally active and inactive cells. *Journal of Cell Biology*, **165**, 191–202.

[16] Molenaar, C., Marras, S.A., Slats, J.C. *et al*. (2001) Linear 2′ O-methyl RNA probes for the visualization of RNA in living cells. *Nucleic Acids Research*, **29**, E89.

[17] Paillasson, S., Van De Corput, M., Dirks, R.W. *et al*. (1997) *In situ* hybridization in living cells: detection of RNA molecules. *Experimental Cell Research*, **231**, 226–233.

[18] Tsuji, A., Koshimoto, H., Sato, Y. *et al*. (2000) Direct observation of specific messenger RNA in a single living cell under a fluorescence microscope. *Biophysical Journal*, **78**, 3260–3274.

[19] Tsuji, A., Sato, Y., Hirano, M. *et al*. (2001) Development of a time-resolved fluorometric method for observing hybridization in living cells using fluorescence resonance energy transfer. *Biophysical Journal*, **81**, 501–515.

[20] Okamura, Y., Kondo, S., Sase, I. *et al*. (2000) Double-labeled donor probe can enhance the signal of fluorescence resonance energy transfer (FRET) in detection of nucleic acid hybridization. *Nucleic Acids Research*, **28**, E107.

[21] Silverman, A.P. and Kool, E.T. (2005) Quenched autoligation probes allow discrimination of live bacterial species by single nucleotide differences in rRNA. *Nucleic Acids Research*, **33**, 4978–4986.

[22] Tyagi, S. and Kramer, F.R. (1996) Molecular beacons: probes that fluoresce upon hybridization. *Nature Biotechnology*, **14**, 303–308.

[23] Tyagi, S., Bratu, D.P., and Kramer, F.R. (1998) Multicolor molecular beacons for allele discrimination. *Nature Biotechnology*, **16**, 49–53.

[24] Li, J.J., Geyer, R., and Tan, W. (2000) Using molecular beacons as a sensitive fluorescence assay for enzymatic cleavage of single-stranded DNA. *Nucleic Acids Research*, **28**, E52.

[25] Sokol, D.L., Zhang, X., Lu, P., and Gewirtz, A.M. (1998) Real time detection of DNA.RNA hybridization in living cells. *Proceedings of the National Academy of Sciences of the United States of America*, **95**, 11538–11543.

[26] Vet, J.A., Majithia, A.R., Marras, S.A. *et al*. (1999) Multiplex detection of four pathogenic retroviruses using molecular beacons. *Proceedings of the National Academy of Sciences of the United States of America*, **96**, 6394–6399.

[27] Kostrikis, L.G., Tyagi, S., Mhlanga, M.M. *et al*. (1998) Spectral genotyping of human alleles. *Science*, **279**, 1228–1229.

[28] Piatek, A.S., Tyagi, S., Pol, A.C. *et al.* (1998) Molecular beacon sequence analysis for detecting drug resistance in Mycobacterium tuberculosis. *Nature Biotechnology*, **16**, 359–363.

[29] Bratu, D.P., Cha, B.J., Mhlanga, M.M. *et al.* (2003) Visualizing the distribution and transport of mRNAs in living cells. *Proceedings of the National Academy of Sciences of the United States of America*, **100**, 13308–13313.

[30] Tsourkas, A., Behlke, M.A., Xu, Y., and Bao, G. (2003) Spectroscopic features of dual fluorescence/luminescence resonance energy-transfer molecular beacons. *Analytical Chemistry*, **75**, 3697–3703.

[31] Santangelo, P.J., Nix, B., Tsourkas, A., and Bao, G. (2004) Dual FRET molecular beacons for mRNA detection in living cells. *Nucleic Acids Research*, **32**, e57.

[32] Nitin, N., Santangelo, P.J., Kim, G. *et al.* (2004) Peptide-linked molecular beacons for efficient delivery and rapid mRNA detection in living cells. *Nucleic Acids Research*, **32**, e58.

[33] Tyagi, S. and Alsmadi, O. (2004) Imaging native beta-actin mRNA in motile fibroblasts. *Biophysical Journal*, **87**, 4153–4162.

[34] Peng, X.H., Cao, Z.H., Xia, J.T. *et al.* (2005) Real-time detection of gene expression in cancer cells using molecular beacon imaging: new strategies for cancer research. *Cancer Research*, **65**, 1909–1917.

[35] Medley, C.D., Drake, T.J., Tomasini, J.M. *et al.* (2005) Simultaneous monitoring of the expression of multiple genes inside of single breast carcinoma cells. *Analytical Chemistry*, **77**, 4713–4718.

[36] Tsourkas, A., Behlke, M.A., and Bao, G. (2002) Structure–function relationships of shared-stem and conventional molecular beacons. *Nucleic Acids Research*, **30**, 4208–4215.

[37] King, F.W., Liszewski, W., Ritner, C., and Bernstein, H.S. (2011) High-throughput tracking of pluripotent human embryonic stem cells with dual fluorescence resonance energy transfer molecular beacons. *Stem Cells and Development*, **20**, 475–484, doi: 10.1089/scd.2010.0219.

[38] Brodsky, A.S. and Silver, P.A. (2002) Identifying proteins that affect mRNA localization in living cells. *Methods*, **26**, 151–155.

[39] Fusco, D., Accornero, N., Lavoie, B. *et al.* (2003) Single mRNA molecules demonstrate probabilistic movement in living mammalian cells. *Current Biology*, **13**, 161–167.

[40] Bertrand, E., Chartrand, P., Schaefer, M. *et al.* (1998) Localization of ASH1 mRNA particles in living yeast. *Molecular Cell*, **2**, 437–445.

[41] Shav-Tal, Y., Darzacq, X., Shenoy, S.M. *et al.* (2004) Dynamics of single mRNPs in nuclei of living cells. *Science*, **304**, 1797–1800.

[42] Haim, L., Zipor, G., Aronov, S., and Gerst, J.E. (2007) A genomic integration method to visualize localization of endogenous mRNAs in living yeast. *Nature Methods*, **4**, 409–412.

[43] Valencia-Burton, M., McCullough, R.M., Cantor, C.R., and Broude, N.E. (2007) RNA visualization in live bacterial cells using fluorescent protein complementation. *Nature Methods*, **4**, 421–427.

[44] Ozawa, T., Natori, Y., Sato, M., and Umezawa, Y. (2007) Imaging dynamics of endogenous mitochondrial RNA in single living cells. *Nature Methods*, **4**, 413–419.

[45] Tsourkas, A., Behlke, M.A., Rose, S.D., and Bao, G. (2003) Hybridization kinetics and thermodynamics of molecular beacons. *Nucleic Acids Research*, **31**, 1319–1330.

[46] Bonnet, G., Tyagi, S., Libchaber, A., and Kramer, F.R. (1999) Thermodynamic basis of the enhanced specificity of structured DNA probes. *Proceedings of the National Academy of Sciences of the United States of America*, **96**, 6171–6176.

[47] States, D.J., Gish, W., and Altschul, S.F. (1991) Improved sensitivity of nucleic acid database searches using application-specific scoring matrices. *Methods*, **3**, 66–70.

[48] Marras, S.A., Kramer, F.R., and Tyagi, S. (2002) Efficiencies of fluorescence resonance energy transfer and contact-mediated quenching in oligonucleotide probes. *Nucleic Acids Research*, **30**, e122.

[49] Dubertret, B., Calame, M., and Libchaber, A.J. (2001) Single-mismatch detection using gold-quenched fluorescent oligonucleotides. *Nature Biotechnology*, **19**, 365–370.

[50] Yang, C.J., Lin, H., and Tan, W. (2005) Molecular assembly of superquenchers in signaling molecular interactions. *Journal of the American Chemical Society*, **127**, 12772–12773.

[51] Kim, J.H., Morikis, D., and Ozkan, M. (2004) Adaptation of inorganic quantum dots for stable molecular beacons. *Sensors and Actuators B: Chemical*, **102**, 315–319.

[52] Kim, Y., Sohn, D., and Tan, W. (2008) Molecular beacons in biomedical detection and clinical diagnosis. *International Journal of Clinical and Experimental Pathology*, **1**, 105–116.

[53] Kushon, S.A., Ley, K.D., Bradford, K. *et al.* (2002) Detection of DNA hybridization via fluorescent polymer superquenching. *Langmuir*, **18**, 7245–7249.

[54] Santangelo, P., Nitin, N., LaConte, L. *et al.* (2006) Live-cell characterization and analysis of a clinical isolate of bovine respiratory syncytial virus, using molecular beacons. *Journal of Virology*, **80**, 682–688.

[55] Santangelo, P.J. and Bao, G. (2007) Dynamics of filamentous viral RNPs prior to egress. *Nucleic Acids Research*, **35**, 3602–3611.

[56] Chen, A.K., Cheng, Z., Behlke, M.A., and Tsourkas, A. (2008) Assessing the sensitivity of commercially available fluorophores to the intracellular environment. *Analytical Chemistry*, **80**, 7437–7444.

[57] Rhee, W.J., Santangelo, P.J., Jo, H., and Bao, G. (2007) Target accessibility and signal specificity in live-cell detection of BMP-4 mRNA using molecular beacons. *Nucleic Acids Research*, **36**, e30.

[58] Chen, A.K., Rhee, W.J., Bao, G., and Tsourkas, A. (2011) Delivery of molecular beacons for live-cell imaging and analysis of RNA. *Methods in Molecular Biology*, **714**, 159–174.

[59] Chen, A.K., Behlke, M.A., and Tsourkas, A. (2007) Avoiding false-positive signals with nuclease-vulnerable molecular beacons in single living cells. *Nucleic Acids Research*, **35**, e105.

[60] Chen, A.K., Behlke, M.A., and Tsourkas, A. (2008) Efficient cytosolic delivery of molecular beacon conjugates and flow cytometric analysis of target RNA. *Nucleic Acids Research*, **36**, e69.

[61] Mhlanga, M.M., Vargas, D.Y., Fung, C.W. *et al.* (2005) tRNA-linked molecular beacons for imaging mRNAs in the cytoplasm of living cells. *Nucleic Acids Research*, **33**, 1902–1912.

[62] Chen, A.K., Behlke, M.A., and Tsourkas, A. (2009) Sub-cellular trafficking and functionality of 2′-*O*-methyl and 2′-*O*-methyl-phosphorothioate molecular beacons. *Nucleic Acids Research*, **37**, e149.

[63] Price, N.C. and Stevens, L. (1999) *Fundamentals of Enzymology: The Cell and Molecular Biology of Catalytic Proteins*, 3rd edn, Oxford University Press.

[64] Dokka, S. and Rojanasakul, Y. (2000) Novel non-endocytic delivery of antisense oligonucleotides. *Advanced Drug Delivery Reviews*, **44**, 35–49.

[65] Giles, R.V., Ruddell, C.J., Spiller, D.G. *et al.* (1995) Single base discrimination for ribonuclease H-dependent antisense effects within intact human leukaemia cells. *Nucleic Acids Research*, **23**, 954–961.

[66] Barry, M.A. and Eastman, A. (1993) Identification of deoxyribonuclease II as an endonuclease involved in apoptosis. *Archives of Biochemistry and Biophysics*, **300**, 440–450.

[67] Giles, R.V., Spiller, D.G., Grzybowski, J. *et al.* (1998) Selecting optimal oligonucleotide composition for maximal antisense effect following streptolysin O-mediated delivery into human leukaemia cells. *Nucleic Acids Research*, **26**, 1567–1575.

[68] Walev, I., Bhakdi, S.C., Hofmann, F. *et al.* (2001) Delivery of proteins into living cells by reversible membrane permeabilization with streptolysin-O. *Proceedings of the National Academy of Sciences of the United States of America*, **98**, 3185–3190.

[69] Snyder, E.L. and Dowdy, S.F. (2001) Protein/peptide transduction domains: potential to deliver large DNA molecules into cells. *Current Opinion in Molecular Therapeutics*, **3**, 147–152.

[70] Wadia, J.S. and Dowdy, S.F. (2002) Protein transduction technology. *Current Opinion in Biotechnology*, **13**, 52–56.

[71] Becker-Hapak, M., McAllister, S.S., and Dowdy, S.F. (2001) TAT-mediated protein transduction into mammalian cells. *Methods*, **24**, 247–256.

[72] Wadia, J.S. and Dowdy, S.F. (2005) Transmembrane delivery of protein and peptide drugs by TAT-mediated transduction in the treatment of cancer. *Advanced Drug Delivery Reviews*, **57**, 579–596.

[73] Brooks, H., Lebleu, B., and Vives, E. (2005) Tat peptide-mediated cellular delivery: back to basics. *Advanced Drug Delivery Reviews*, **57**, 559–577.

[74] Allinquant, B., Hantraye, P., Mailleux, P. *et al.* (1995) Downregulation of amyloid precursor protein inhibits neurite outgrowth *in vitro*. *Journal of Cell Biology*, **128**, 919–927.

[75] Troy, C.M., Derossi, D., Prochiantz, A. *et al.* (1996) Downregulation of Cu/Zn superoxide dismutase leads to cell death via the nitric oxide-peroxynitrite pathway. *Journal of Neuroscience*, **16**, 253–261.

[76] Nitin, N. and Bao, G. (2008) NLS peptide conjugated molecular beacons for visualizing nuclear RNA in living cells. *Bioconjugate Chemistry*, **19**, 2205–2211.

[77] Liu, X. and Tan, W. (1999) A fiber-optic evanescent wave DNA biosensor based on novel molecular beacons. *Analytical Chemistry*, **71**, 5054–5059.

[78] Kambhampati, D., Nielsen, P.E., and Knoll, W. (2001) Investigating the kinetics of DNA–DNA and PNA–DNA interactions using surface plasmon resonance-enhanced fluorescence spectroscopy. *Biosensors and Bioelectronics*, **16**, 1109–1118.

[79] Steemers, F.J., Ferguson, J.A., and Walt, D.R. (2000) Screening unlabeled DNA targets with randomly ordered fiber-optic gene arrays. *Nature Biotechnology*, **18**, 91–94.

[80] Kuhn, H., Demidov, V.V., Coull, J.M. *et al*. (2002) Hybridization of DNA and PNA molecular beacons to single-stranded and double-stranded DNA targets. *Journal of the American Chemical Society*, **124**, 1097–1103.

[81] Hamaguchi, N., Ellington, A., and Stanton, M. (2001) Aptamer beacons for the direct detection of proteins. *Analytical Biochemistry*, **294**, 126–131.

[82] Yamamoto, R., Baba, T., and Kumar, P.K. (2000) Molecular beacon aptamer fluoresces in the presence of Tat protein of HIV-1. *Genes to Cells*, **5**, 389–396.

10

Towards High-Throughput Cell Mechanics Assays for Research and Clinical Applications

David R. Myers,[1,2] Daniel A. Fletcher,[3] and Wilbur A. Lam[1,2]
[1]Aflac Cancer Center and Blood Disorders Service of Children's Healthcare of Atlanta and Emory University School of Medicine, USA
[2]Georgia Institute of Technology and Emory University, USA
[3]University of California at Berkeley, USA

10.1 Cell Mechanics Overview

A number of excellent review papers have been published examining cell mechanics [1]; mechanical models of cells and molecules [2], the role of cell mechanics in disease [3,4]; tools used to study cell mechanics [5,6]; as well as tools used to apply forces to cells [7]. This chapter, which is aimed at those new to cell mechanics, seeks to complement the existing body of work by examining the clinical applicability of current tools used to study cell mechanics. In general, these tools measure quantitative mechanical properties of cells, including modulus of elasticity, viscoelastic properties, adhesion, and forces created by cells. Here, we consider whether current methods can be adapted to high-throughput measurements in clinics for use in screening for or diagnosis of diseases known to be associated with changes in mechanical properties at the single-cell level.

Decades ago, studies found that the mechanical properties of cells played a central role in cardiovascular and hematologic diseases [8]; however, increasing numbers of studies show that cell mechanics is intimately linked to other disease [3,4], such as cancer and malaria. The key barrier to further study has been the ability to successfully screen large numbers of cells for altered mechanical properties. Importantly, pathologic cells responsible for causing diseases often begin as small subpopulations amid larger cell populations. As such, this chapter is written to provide a review of existing cell mechanical techniques

Nano and Cell Mechanics: Fundamentals and Frontiers, First Edition. Edited by Horacio D. Espinosa and Gang Bao.
© 2013 John Wiley & Sons, Ltd. Published 2013 by John Wiley & Sons, Ltd.

and to specifically discuss these techniques in the context of use for large numbers of precise mechanical measurements. The ideal clinical assay would be able to mechanically interrogate and measure the response of single cells at high throughput to enable isolation and statistical comparisons of cellular subpopulations of different mechanical properties. Furthermore, the platform developed should be translatable to clinical settings, easy to use, and rapid, providing physicians with new quantitative data that can be used to help guide treatment of diseases and pathologic conditions.

We begin the chapter with a brief overview of cell mechanics and how forces both applied by and applied to cells affect their behavior. The next portion of the chapter focuses on bulk and single-cell mechanical assays and highlights the key needs for high-throughput cell mechanics. Bulk mechanical assays measure the average cell responses of large populations and are unable to achieve single-cell resolution. Single-cell mechanical assays, on the other hand, provide detailed information on individual cells, but typically cannot measure large numbers of cells in a reasonable time frame.

We then examine several recently developed devices that are capable of large-scale, individual-cell mechanical measurements. Some of these devices have been developed to study a specific cell mechanical phenomenon and would need further refinement before being used for high-throughput cell mechanical measurements in a clinical setting. Other devices have achieved high-throughput mechanical analysis of single cells and have been successfully used to examine the role of cell mechanics in certain disease states. However, there remains a need for new assays to identify and quantify clinically relevant mechanical properties of single cells. We close with a review of diseases which are known to be affected by cell mechanics and a summary of key needs for future high-throughput cell mechanical-assays.

10.1.1 Cell Cytoskeleton and Cell-Sensing Overview

Cells are inherently mechanical, physically interacting with their local microenvironment in a number of different manners. Cells may exert stresses on their local environment but also respond to forces in the local microenvironment. An early example of this was observed with endothelial cells which changed shape and orientation in the direction of a constant fluidic shear stress [9], as shown in the comparison of Figure 10.1a and b. Subsequent studies determined that shear stress altered genetic expression in endothelial cells when compared with static conditions [10]. Cells are also capable of undergoing large deformations. For example, erythrocytes (red blood cells) have diameters of $7-8\,\mu m$, and regularly squeeze through capillaries with diameters of $3\,\mu m$. This correlates to approximately 100% elastic deformation [1].

When examining the mechanics of cells, the key factor affecting the mechanical properties of the cells are the concentration and architecture of different components present in the cell cytoskeleton. The cell cytoskeleton is a viscoelastic material and is responsible for the cell shape, holding cell organelles, assisting in mitosis, creating mechanical force, and transducing extracellular mechanical and biochemical stimuli to the cell cytoplasm. The main structural components of the cytoskeleton are the actin filaments, intermediate filaments, and microtubules, as shown in Figure 10.2. The cell cytoskeleton is a dynamically changing structure that is affected by different internal biochemical cues, changes in genetic expression, and by a variety of external factors, such as signals from

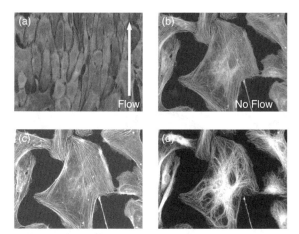

Figure 10.1   (a) Image of human umbilical vein endothelial cells (HUVECs) subjected to continuous shear stress, noting how the cells have aligned with the applied flow. Accordingly, filamentous actin fibers, stained green with phalloidin, also oriented to the direction of flow. The nuclei are stained blue with Hoechst. (b) In contrast, actin fibers, here stained green with phalloidin, in HUVECs under static conditions exhibit no such orientation. Filamentous actin is shown separately in (c). The microtubules are stained with anti-α-tubulin and appear red, shown separately in (d); while the nucleus has been stained blue with Hoechst. Photographs by Yumiko Sakurai. Refer to Plate 8 for the colored version

the extracellular matrix [3]. When the cell moves or responds to external stimuli, the cytoskeleton is actively changing and remodeling.

Actin filaments are twisted double strands composed of actin subunits, and are approximately 7 nm in diameter and can form organized structures several centimeters in length in muscle cells, as shown in Figure 10.1c. Actin filaments are involved with cell and muscle contraction, with altering the cell shape, and with movement of the cell [11]. Muscle contraction is produced from the sliding of actin filaments in response to forces generated by the protein myosin. Actin filaments and the networks formed from them are important not only in muscle contraction, but also in cell motility, cell division, cytokinesis, cell signaling, vesicle and organelle movement, cell shape, and other cellular functions.

Intermediate filaments are a family of polymers that form rope-like strands, composed of various proteins depending on the tissue type, and are shown in Figure 10.1d. They are approximately 8–12 nm in diameter and range from 10 to 100 μm in length. They are primarily responsible for the cell shape [11] and provide tensile strength for the cell. Intermediate filaments cross-link with microtubules, the nuclear membrane, and actin to improve the stability of the cell. Intermediate filaments also connect adjacent cells through desmosomes, which are specialized structures which facilitate cell-to-cell adhesion. Intermediate filaments may be stretched to several times (250% tensile strain) their original length [12]. This unique stretching property is possibly due to the hierarchical structure of the proteins [13], which adds unique mechanical properties to the cell.

Microtubules are highly dynamic polymers formed from α- and β-tubulin dimers. The tubulin dimers polymerize end to end in protofilaments that are assembled into hollow cylindrical filaments. Microtubules are involved in maintaining cell structure, providing

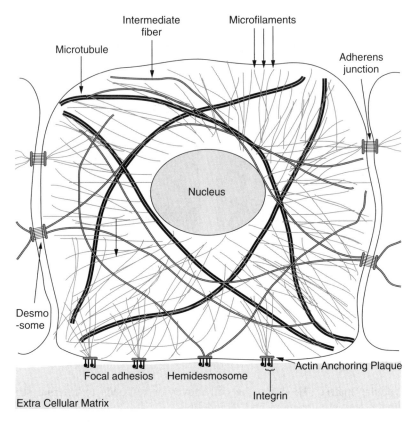

Figure 10.2 The cell cytoskeleton is primarily responsible for the mechanical properties of the cell. The cytoskeleton is a very dynamic structure which is constantly changing with externally applied mechanical and chemical signals. The cell primarily receives mechanical and chemical information as well as sends information through integrins

platforms for intracellular transport and mitosis, as well as other cellular processes. During these processes, motor proteins, such as kinesin and dynein, bind to microtubule polymers.

Integrins are cell receptors which are involved in cell attachment and cell signaling. They are capable of communication in two directions, both sensing attachment for the cell and presenting status information to other cells. Integrins are formed by two different macromolecules and contain an α and β subunit. Mammalian cells have 18 α and 8 β subunits. Each combination of subunits is associated with different adhesions and cell signaling conditions.

Focal adhesions are dynamic structures which attach the cell to substrates and serve as anchoring locations for actin filaments. They are composed of over 90 components [14], and include the following notable constituents: integrins, actin, vinculin, talin, tensin, and paxilin. Focal adhesions along with other adhesion sites such as focal complexes are thought to be the main surface-sensing organelles. Focal adhesions are important in a number of cell mechanical events, including regulating actin assembly and cytoskeleton signaling [15]. They are also affected by applied mechanical forces [16], and applied force appears to be necessary for early-stage development of focal adhesions [15,17].

Adherens junctions are protein complexes that appear in endothelial cells, which are important in cell–cell adhesion. Other cells have a similar structure termed the fascia adherens. Adherens junctions are composed of cadherins as well as α, β, and δ (p120) catenins. Cadherins are named for their adhesive and calcium-dependent properties, "*ca*lcium-dependent *adhe*sion." They are transmembrane proteins which are essential to the adhesive properties of adherens junctions.

The response of cells to mechanical stimuli is termed mechanotransduction. This area has been the subject of several excellent reviews [18–20], but a brief overview is presented here. Cell mechanotransduction is broadly classified as either passive or active. In the passive case, cells biological respond to forces such as shear, extension, and compression via protein activation and gene expression. In the active case, the cell is capable of probing the local microenvironment, determining the stiffness, topography, and ligand density. Several well-known mechanosensing pathways, the Rho/ROCK signaling pathway, stretch-activated channels, and force-induced protein unfolding are covered in detail by Holle and Engler [18].

Finally, many models have been developed to explain the behavior of cells under applied mechanical loads. One is the tensegrity model [21–23], in which cells stabilize their structure through "continuous tension" as opposed to "continuous compression, as used in a stone arch" [24]. Specifically, the actin filaments and intermediate filaments support tension, while the microtubules and extracellular matrix adhesion support compressive loads. Other models view the cell as a solid continuum or as a liquid-filled membrane [6]. Other authors have reported on scaling laws which apply to a large number of cells and indicate cell mechanical behaviors similar to soft glassy materials [25–27] and have also identified a fluidization response seen when stretching cells [28]. These models provide a framework for describing the cell as a dynamically changing mechanical structure capable of applying forces, moving, and physically remodeling the cytoskeleton in response to both mechanical and chemical stimuli.

10.1.2   Forces Applied by Cells

Cells are inherently mechanical and may apply mechanical forces to the local extracellular microenvironment and other cells. For decades, researchers have observed that cells appear to be applying forces to the local microenvironment. However, until recently it has been difficult to measure the forces of individual cells. Some of the earliest work involved the use of 1 μm thickness silicone-rubber sheets [29], in which cells placed on the sheets caused local wrinkles, enabling estimates of individual cell forces.

Cell movement and cell contraction, such as that seen by muscle cells, fibroblasts, or platelets, is due to acto-myosin-based contraction. In this arrangement, myosin, a type of molecular motor, converts chemical energy into mechanical movement and by sliding over actin filaments decreases the overall length of the structure and exerts contractile forces. Single-cell studies of platelets [30] and myoblasts [31] have shown that the force produced by cells is dependent on the local stiffness of the environment. In these experiments, cells or platelets were connected to cantilever structures of varying mechanical stiffness and the subsequent force applied by the cell was measured. Increasing the local stiffness of the microenvironment caused cells to exert larger contraction forces. Noting that a single filament of actin in conjunction with eight heads of myosin contracts similarly with respect to stiffness changes [32], this offers evidence that the actin–myosin unit is indeed sensing and adapting to the local microenvironment. Recently, follow-up work has suggested that stiffness and not force is responsible for the cell response [33].

The dynamics of cell motility, or the ability of cells to move, is a major focus of cell mechanics. Changes in how cells move have large effects in embryogenesis, inflammation, wound healing, and tissue engineering. In addition, alterations in cell motility have been observed to be associated with the metastatic potential of cancer cells [34]. Cell movement begins with extensions of the membrane due to polymerization of actin filaments. The membrane subsequently adheres to the local substrate, and acto-myosin contraction in the rear of the cell moves the cell forward. Recently, a mechanical model linking the motility of the cell with the actin network and membrane tension has been proposed and shown to have good agreement with experimental observations [35].

Recent experiments have been able to identify how cells apply forces during collective cell migration. Contrary to previous theories, it was discovered the cells move in a push–pull mechanism during collective cell migration, experiencing both tensile and contractile forces during migration [36]. Cell movement was in the direction of maximum local stress, where traction forces were highly anisotropic, termed "plithotaxis." This research also showed that intracellular mechanical communication was dominated by direct cell–cell adhesion contacts [37].

The forces applied by cells are complex in nature, and several models and theories have been proposed. For example, with the collective cell migration measurements, some theories proposed that cells were simply pulled or pushed by groups of cells. However, one of the key insights to collective cell motion is that cells must create internal contractile forces to remain adherent to the substrate. The adhesion molecules only become active and sticky when stress is applied [37]. As such, forces applied by cells need to meet two constraints: first, the forces must maintain cell adhesion and, second, maintain cell motion.

10.1.3 Cell Responses to Force and Environment

Mechanical forces have been shown to be important in organizing tissues [38], and by extension individual cells. Forces applied to cells may have a range of effects in health and disease. For example, continuously stretching axons [39,40] promotes growth, enabling lengths of up to 10 cm. However, when rates of stretch become too high, the axons disconnect. Similarly, central nervous system injury damage is hypothesized to be caused by an initial mechanical event followed by an extended neurochemical or cellular cascade [41]. Furthermore, mechanical cues have been shown to affect signal pathways [21], and may play a role in gene regulation [42].

In general, the stiffness of the cell is much lower than typically encountered in other materials. Stiffness is typically reported by the modulus of elasticity or Young's modulus. The modulus of elasticity describes how much a material will deform for a given applied stress and has units of pascals (Pa, or N m^{-2}). Larger numbers indicate that the material will move very little with an applied stress. Steel and aluminum have moduli of roughly 200×10^9 Pa and 70×10^9 Pa respectively. Materials such as rubber typically fall within the range of $(10–100) \times 10^6$ Pa. However, numbers for cells are typically 1000 times smaller than rubbers. For comparison, shaving foams have an elastic modulus of 10–1000 Pa [43], which is very close to that of measured cells.

Early research found that the viability of cells could be controlled with geometric constraints [44,45]. By patterning islands of fibronectin on a substrate, it was found that cells would undergo cell death, or apoptosis, when the fibronectin area was small. Furthermore, there was a reduction in DNA synthesis on small islands. To further evaluate the mechanism for the cell death, dots of fibronectin were patterned on the surface with varying spacing and total area [44]. This enabled researchers to decouple the total area of fibronectin from final size of the cell once attached to the surface. In doing so, it was discovered that apoptosis was dependent on the cell area and not the total amount of fibronectin on the surface.

In the case of stem cells, differentiation may be altered by changing the local substrate stiffness. Body tissues have a range of stiffnesses; for example, brain tissues tend to be softer than bone tissues. When mesenchymal stem cells were cultured on collagen-coated gels mimicking brain, muscle, and bone, the stem cells committed to neurogenic, myogenic, and osteogenic lineages [46]. Since the only difference in the experiments was the stiffness of the collagen gels, it appears that the cells are capable of sensing the local microenvironment stiffness, and they must be actively pulling on the substrate and responding to the different levels of force needed to move the substrate.

Applying a pulse of strain to cells will cause the cells to promptly soften, termed fluidization. In experiments, typical strains applied were up to 10%, and after hundreds of seconds the cells would return to the original stiffness [28]. This unique response was observed in a variety of cell types and even when treated with pharmacological agents to affect the cell response. These experiments add to existing theories that indicate cytoskeleton dynamics are governed by mechanical strain stiffening [47,48] and active reinforcement [49,50]. Based on experimental findings, the authors propose a power law to explain the temporal response of cells to a strain pulse [28].

10.1.4   General Principles of Combined Mechanical and Biological Measurements

Many traditional mechanical measurements are made via physical contact. One of the most common measurements is of the modulus of elasticity of a structure. In the simplest form, the measurement is accomplished by applying a known force to a structure and measuring the resulting deformation. Softer structures will deform more for a given force. Typically, data are reported in the more universal numbers of stress versus strain to make comparisons simpler. Stress is a measure of the force applied over a known area, and provides an easier number to compare with other experiments. Strain is a measurement of the amount of deformation relative to the original size of the structure.

While this basic technique of applying a force and measuring a deflection seems trivial, it is used in a number of different situations. In essence, all the methods discussed later in this chapter are variations on the same theme. Each method applies the force and measures the deflection differently. The challenge is finding a high-throughput method of applying forces and measuring the deflections on individual cells.

In order to truly quantify aspects of cell mechanics, large numbers of measurements are needed to compensate for the noisy nature of biological systems. Individual cells show a great deal of variation when exposed to the same forces. The measurement system must also reproduce the sensitivity and resolution of the existing single-cell techniques which have enabled this research area. This combination of sensitivity and high throughput has been a difficult challenge to meet, but recent advances in micro- and nano-fabrication techniques enable researchers to build highly specialized tools capable of interacting with cells one by one.

Ideally, mechanical measurements should integrate with existing biological measurements, coupling with them to provide rich and more diverse data to quantitatively determine how cell mechanics and cells interact. In general, biological measurements of cells are based on optical and/or biochemical approaches. Optically, cells are observed using microscopy, through a number of different techniques, such as phase contrast, fluorescence, confocal, two-photon, and differential interference contrast.

In designing high-throughput devices, however, the key need is for the device itself to be transparent within wavelengths used by the microscopy method, typically wavelengths in the visual, infrared, and ultraviolet (UV) range. It is also important to consider the focal depth of the microscope objective used in each of the techniques. Certain microscope techniques, such as confocal, place larger constraints on how far away the cell may be from the lens objective, constraining the mechanical device design.

10.2 Bulk Assays

This section will briefly provide an overview of the existing bulk cell measurement techniques that are used to study cell mechanics. This section will inform the reader of the ways in which cells may be mechanically interrogated and will provide the key advantages and limitations of the existing techniques. Ideally, knowledge of these assays will act as a starting point for future high-throughput single-cell mechanical assays. Bulk assays provide information about a large number of cells, but do not track or monitor individual cells.

10.2.1 Microfiltration

Microfiltration is used to sterilize fluids by removal of microorganisms or to concentrate cell populations by capturing cells from fluids using a filter. The filter has pore sizes much smaller than the cells of interest. However, this technique was adapted to measure the deformability of cells by choosing larger pore sizes which allowed cells to squeeze through the pores. For example, with red blood cells, the typical cell diameter is 6–8 μm, and a good starting pore size for deformation studies is 5 μm [51]. Microfilters provide an experimental platform which can simulate the geometric size scale of capillaries and were used in hematologic studies. Furthermore, they are inexpensive, widely available, and could be easily integrated with existing equipment. Subsequent cell mechanic assays use the same fundamental principle of pushing a cell through a pore, originally used in microfiltration techniques.

Depending on the experimental setup, it is possible to either measure the number of cells trapped in the filter or the time cells take to pass through the filters. In both cases, the cell filtration is dictated by pore geometry, flow rates, cell adhesion, and cell mechanical properties. Figure 10.3 shows the experimental setup and equipment used for both measurement strategies.

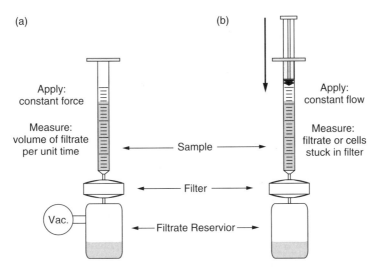

Figure 10.3   Microfiltration techniques are based on readily available, off-the-shelf equipment. (a) The volume of material passing through the filter is measured; often useful for whole blood studies. (b) The number of cells captured in the filter is measured by counting cells both before and after passing through the filter at a constant flow rate

To determine how many cells are stuck in a filter, one simply measures how many cells are in solution both before and after passing through the microfilter. To increase consistency, the cells are flown through the filter at a predetermined set rate. This technique was used to measure the filtration properties of stimulated neutrophils through microfilters. The microfiltration experiments found that up to 80% of neutrophils become stuck in filters [52]. In combination with cell stiffness experiments, this experiment provided evidence that increased cell stiffness could directly lead to neutrophil accumulation in capillaries.

Alternatively, the time in which cells take to pass through the filter may be measured. In this type of experiment, a constant force is applied to cells, such as one created by a pressure differential as described by Reid *et al*. [51]. In this method, either the flow time or the volume of filtrate passed through the filter in a preset amount of time could be used to compare the deformability of cells. This method is often useful in measuring blood samples, and it was linked to a reduction in cell deformability in patients with peripheral vascular disease who experienced rest pain as compared with patients with intermittent pain [8]. Other clinical studies have shown that filterability of red blood cells and white blood cells was reduced in patients with acute myocardial infarction [53,54]. Other studies used microfiltration to measure blood cells from a variety of species [55].

One challenge involves the use of multiple cell types with the same filter, such as filtering whole blood; white blood cells could cause measurement artifacts [56]. White blood cells are much less deformable and larger than red blood cells and often clog the pores transiently [57]. Of the white blood cells, the least deformable to the most deformable cells were monocytes, granulocytes, and lymphocytes, respectively [58]. For examples of different models and preparation methods used for microfiltration techniques involving multiple cells, the interested reader is referred to Refs [57,59].

There are several issues with microfiltration techniques which rely on pre-manufactured filters with a distribution of pore sizes and directions. Commercial filers may contain multi-branching channels, which create conditions in which the filter pore size is much larger than the cell size. In this circumstance, cells pack into the channel and create a central column, allowing cells in the middle of the pore to slip past the other cells at the edge of the pore, affecting measurement results [60]. Another study has shown that the shape of the cell or bacteria highly affects the filterability and passage through pores [61]. Also, cell stiffness and cell adhesion cannot be uncoupled in this assay, leading to more causes of potential experimental artifact. Furthermore, only bulk measurements can be made, and average cell properties for a given sample can be measured. Despite these limitations, researchers have leveraged microfiltration to learn much about cell biomechanics over the decades.

Microfiltration is difficult to adapt for use in diagnostics in the clinical arena. As mentioned in the studies above, there are a number of different pathologic phenomena which can affect the deformability of cells. When diagnosing a patient, microfiltration will only inform the physician that something has changed in the patient's blood deformability, but gives little insight into the origin of this change and, thus, what disease process is involved.

10.2.2 Rheometry

Rheometry takes advantage of tools originally developed to measure liquid properties and uses them to apply a shear stress to a group of cells. The design takes advantage of Couette flow, which occurs in liquids bounded by a moving and nonmoving wall. The velocity profile of a fluid moving between two infinite flat plates, in which one plate is moving relative to the other, is given by

$$u(y) = \frac{u_0}{h} y \tag{10.1}$$

where u_0 (m s^{-1}) is the velocity of the top moving plate, h (m) is the height between the plates, and y (m) is the distance up from the bottom, stationary plate. The velocity will linearly increase from zero at the stationary plate to the u_0, the velocity of the moving plate. This velocity profile is shown in Figure 10.4a for the cylindrical rheometer. For the cone-and-plate rheometer shown in Figure 10.4b, both the fluid velocity and height change linearly such that a homogeneous shear stress is still created. Detailed equations for the cone-and-plate rheometer are given by Bussolari et al. [62]. For the flat-plate condition, the corresponding shear stress is given by

$$\tau(y) = \mu \frac{\partial u}{\partial y} = \mu \frac{u_0}{h} \tag{10.2}$$

where μ (Pa s) is the dynamic viscosity of the fluid. As such, in the cylindrical rheometer, the shear stress is constant at all locations in the fluid. Corresponding equations are given for the cone and plate by Bussolari et al. [62]. There are a number of different operational regimes; for example, counter-rotating plates may be used [63]. Both rheometers offer an interesting platform to study cells, since they provide very controlled shear stresses. Many studies use rheometers to apply forces to cells, and measure the biochemical changes to

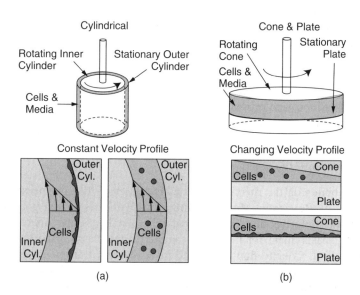

Figure 10.4   Two separate types of rheometers adapted for applying constant shear stress to both adherent and nonadherent cells: (a) cylindrical rheometer; (b) cone-and-plate rheometer

cells which experience constant shear stress. Early work showed that, under constant shear stress, endothelial cells remodeled their cytoskeleton to align with the predominant flow direction [9]. Furthermore, these devices are capable of working with both adherent and nonadherent cells.

One of the early studies examined the rheological properties of blood and blood cells. Using the rheometer in conjunction with an optical setup, it was possible to view an elongation in red blood cells while shear stress was applied [63]. For ease, a cell elongation parameter E was created. Note: this should not be confused with Young's modulus, also denoted E. Circular disks correspond to the value of $E = 0$, and may typically go as high as 0.7 before shear rupture for red blood cells. The elongation parameter is dependent on the cell length L and width W, and is given as

$$E = \frac{L - W}{L + W} \tag{10.3}$$

By combining the visual measurement of cells with well-controlled, repeatable applications of shear stresses, later experiments were able to measure the stiffness of red blood cells when affected by the malaria parasite *Plasmodium falciparum*, finding that the deformability of the cells is dependent on the length of time in which the parasite has infected the cell [64].

Rheometry does have several drawbacks. First, the microscope objective is fixed, so it is difficult to track individual nonadherent cells. In visualizing the cells, as is the case with all light microscopes, the best linear or two-point resolution is around 0.3 μm [65], although the ability to detect displacements or track edges does offer improved resolution. Other authors have argued that this technique requires a fair amount of operator skill [65], making it difficult to adapt to clinical use. As a clinical tool, rheometry would best be suited to blood measurements, since the setup time for measurements would be fairly minimal. However, since the only metric measured by this technique is the blood cell elongation, it is difficult for physicians to diagnose the source of displayed symptoms. Overall, rheometry provided a leap forward in research, since it applies forces in a very controlled fashion, and provides more quantitative data about cell deformation than microfiltration does.

10.2.3 Ektacytometry

Ektacytometry builds on the rheometer and adds a laser diffraction measurement, originally developed by Bessis and Mohandus [66]. This work has similar goals to rheology, and measures the deformability of cells, but relies on lasers and optical scattering to create a quick measurement for the entire sample. In the classic experiments by Bessis and Mohandus [66], Couette flow is created in concentric, transparent cylinders. Laser light passed through the cylinders is scattered into a "diffraction" pattern, as shown in Figure 10.5. The intensity of scattering is dependent on the number of scattered particles, and the shape of the diffraction pattern is dependent on the shapes of the scattering particles. Circular diffraction patterns indicate that the cells are circular, whereas elliptical diffraction patterns indicate that the cells are elliptical, as shown in Figure 10.5.

To measure the diffraction pattern, photodiodes are mounted perpendicular to one another along a fixed radius from a center point, as shown in Figure 10.5. The diffraction

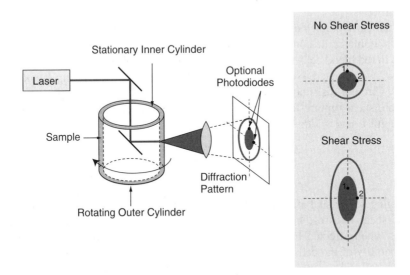

Figure 10.5 Ektacytometry begins with a cylindrical rheometer but uses a novel optical measurement setup to help automate measurements. As cells deform under shear stress, the diffraction pattern changes from a circular to elliptical shape. The eccentricity of the diffraction pattern directly correlates to the eccentricity of the cells. By mounting photodiodes at positions (1) and (2) it is possible to get a single number for the eccentricity of the cells

pattern is projected onto the photodiodes, and centered along the same center point as the radius. By measuring the voltages of each photodiode, V_1 and V_2, a similar metric to the one used in rheology is found, called the deformability index (DI). This was shown to have good correlation, giving DI numbers which match E within 5% [65]:

$$DI = \frac{V_1 - V_2}{V_1 + V_2} \tag{10.4}$$

While this technique greatly streamlined the measurement of the deformability of large groups of cells, one of the primary challenges is determining what is causing the change in deformability, as it is with the rheometer. For example, reduced deformability of red blood cells can be caused by reduced surface-to-volume ratios, increased viscosity, and increased membrane stiffnesses [67]. Also, this tool has limited use in studying adherent cells, which would require releasing cells from the surface of interest. Unfortunately, this potentially changes the mechanical properties of the cell.

As such, this tool is well suited to laboratory settings, in which the cells of interest are carefully controlled and changed, and is still used to this extent [68]. Unlike other rheometry-based cell mechanical assays, this assay has been applied clinically to measure and diagnose a hematologic disease called hereditary spherocytosis, a disorder of the red cell membrane that leads to decreased deformability under different osmotic conditions [69,70]. Overall, ektacytometry provided a clever adaption to rheometry, simplifying the use and calculation of cell deformability.

10.2.4 Parallel-Plate Flow Chambers

Similar to rheometers, parallel-plate flow chambers are designed to apply shear stresses to cells. The typical setup uses a gasket placed on a glass slide, and is vacuum held to the slide with a top inlet and outlet port. Fluid may be passed into the ports using a number of techniques, including pressure, syringe pumps, or a peristaltic pump [71]. This technique has several advantages, including simpler equipment, ease of access to the cultured cells and medium, and ease of viewing with microscopy. Furthermore, since it incorporates fluidic ports, media may be sampled and exchanged quite easily.

One area of interest for flow chambers, as well as more traditional rheometers, is the pulsatile and reverse flow, specifically in endothelial cells. In general, sustained high shear stress from laminar flow upregulates genes which protect against atherosclerosis, whereas low shear stress upregulates genes which promote inflammation and plaque build-up [71]. However, these cells exhibit different responses to no flow, constant flow, and changing flow conditions. Different endothelial cell types have also been shown to react differently to similar conditions [72].

Parallel-plate flow chambers are similar to rheology, in that they are primarily used in conjunction with optical microscopy. As with rheology, measurements of cell deformation, mechanical properties, or adhesion are done optically. However, since the equipment is easy to disassemble and use, experiments examining the interaction between multiple cell types are easier to perform. For example, endothelial cells may be grown on a glass slide and then assembled into the parallel-plate flow chamber, and studies examining the interaction between these cells and blood cells may be performed.

10.3 Single-Cell Techniques

Single-cell techniques represent powerful methods for measuring mechanical properties of cells. However, all of the techniques require a fair amount of operator skill and may only measure a single cell at a time. These techniques represent the standard of comparison for novel high-throughput mechanical assays. Ideally, the same underlying physics used in the devices discussed in this section could be scaled to measure many cells at a time.

10.3.1 Micropipette Aspiration

In this technique, cells are deformed into a micropipette, originally reported in 1954 by Mitchison and Swann [73]. Glass pipettes with diameters smaller than the cell of interest are brought in contact with the target cell. By applying controlled pressures and by measuring the amount of deformation into the pipette, it is possible to quantitatively measure the material properties of cells, such as the modulus of elasticity. By measuring the rate of deformation, it is possible to quantify additional mechanical phenomena such as creep [60]. Pressures are typically controlled with exceptional precision by changing the height of a fluid reservoir with a micrometer, as shown in Figure 10.6a.

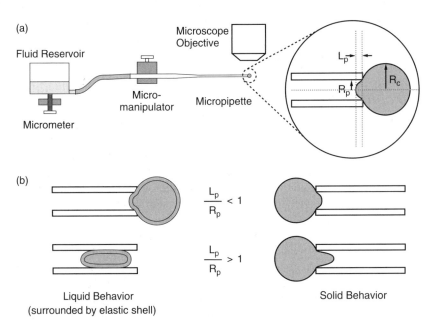

Figure 10.6 (a) A typical micropipette aspiration setup in which the pressure in the micropipette is controlled by changing the height of a fluid reservoir, giving exceptional control of the pressure. A micromanipulator is used to position the micropipette for capturing and visualizing the cell underneath the microscope objective. (b) Two common models for cell behavior are the liquid behavior, in which a cell is surrounded by an elastic shell, and the solid behavior, in which the cell is modeled as a homogeneous elastic solid. In the case of the liquid behavior, the cell will experience an unstable behavior when the aspirated portion of the cell is longer than the pipette radius ($L_p/R_p > 1$), and will be sucked into the pipette

Mechanical parameters such as viscosity and modulus of elasticity are determined by assuming that the cell fits a continuum model. Two popular models are to assume that the cell is a homogeneous elastic solid or assume that the cell is a liquid surrounded by an elastic cortical shell [6]. For example, chondrocytes have been shown to behave using the elastic solid model [74], whereas granulocytes have been shown to behave using the elastic cortical shell model [75,76]. It should be noted that, within these models, the red blood cell has an ability to hold a unique shape which is associated with a shear rigidity, and a separate model is used [60,77]. Both liquid-type and solid-type behaving cells act similarly when the projection into the micropipette is less than or equal to the pipette radius [6]. However, liquid cells will be sucked into the pipette with increasing pressure past this point, whereas solid cells will simply continue to deform further, as shown in Figure 10.6b.

The equations for the liquid and solid cells are given below. Note how different mechanical terms are appropriate for each behavior type. For example, in the liquid case a cortical tension is calculated, whereas in the solid case a modulus of elasticity is calculated. In both cases, ΔP is the change in pressure:

$$\Delta P = 2T_c \left(\frac{1}{R_p} - \frac{1}{R_c} \right) \tag{10.5}$$

where T_c is the cortical tension, R_c is the radius of the cell outside the pipette, and ΔP is the suction pressure. The equation may be used to calculate the critical pressure ΔP_c, which occurs when $L_p / R_p = 1$ [6]:

$$\Delta P = \frac{2\pi}{3} E \frac{L_p}{R_p} \varphi \tag{10.6}$$

where E is the modulus of elasticity of the solid and φ is a term which depends weakly on the ratio of the thickness of the pipette wall to the radius of the pipette. Note: the modulus of elasticity E should not be confused with the elongation parameter used in rheology. A typical value for φ is given as 2.1 [78]. It is possible to equate the cortical tension to the modulus of elasticity, as well as to find terms for the viscosity, but this is beyond the scope of this chapter and the interested reader is directed to Hochmuth [6].

The resolution of the micropipette is governed by the ability to track edges of cells and by the minimum force that may be applied to deform the cell. Currently, the edges of cells may be tracked within micropipettes with an accuracy of 50 nm [79]. The minimum suction pressure is practically given as $0.1 - 0.2 \, pN \, \mu m^{-2}$, due to drift limitations [6]. The maximum pressure is on the order of $96 \, nN \, \mu m^{-2}$ and is limited by the vapor pressure of water at room temperature [6].

One unique aspect of micropipette aspiration is that it typically deforms cells in an opposite direction to other techniques. The cell surface is extended into a pipette rather than depressed inwards by some other means. As such, this technique has found widespread use in mechanical studies of cells, since performing a measurement will simultaneously immobilize the cell. One interesting application of micropipettes is to megakaryocytes, which create platelets through extensions called proplatelets. When subjected to pressure, the megakaryocyte membrane flows into the micropipette, and the resulting structure in the pipette is similar to proplatelets. Under continued suction, platelet-like fragments break off the membrane extensions. By using this technique, it was possible to study proplatelets under varying conditions [80].

An extension of the micropipette aspiration technique is the biomembrane force probe [81]. Here, a cell or cell-size membrane capsule or a red blood cell is held in place with a micropipette. In this case the cell or membrane capsule acts as a mechanical spring which can be tuned over a large range. By applying larger suction pressures on the micropipette, the cell membrane becomes more stretched, increasing the spring constant. As such, the forces applied by the probe may be as small as tens of femtonewtons (fN) to 1 nN (nanonewtons) [81]. Using this probing technique it was possible to measure a variety of molecular bond forces, showing the effect of different loading rates on bond lifetimes and strength [82].

Building upon this work, techniques have been developed which enable the measurements of receptor–ligand binding kinetics [83]. One implementation, called the adhesion frequency assay, involves taking two cells and contacting them together and then pulling them apart. After the contact event, it is possible to see the cells deform due to receptor–ligand binding. This binding will not occur every time and is treated as a statistical event. The experimenter simply counts the number of adhesions per total cell contacts and can change the contact area and duration of the adhesion event. These data are then correlated with a model of the receptor–ligand interaction to determine kinetic rate constants [84]. A separate experiment uses thermal fluctuations in conjunction with a biomembrane force probe to measure kinetics [85]. Using both the cell adhesion frequency assay and the thermal fluctuation assay, it was possible to measure the kinetics of t-cell receptors and peptide-major histocompatibility complexes, which are responsible for identifying pathogens and self-antigens in the immune system [86].

10.3.2 Atomic Force Microscopy

Atomic force microscopy is a powerful technique which uses specialized piezoelectric elements to move a scanning cantilever with nanometer precision (Figure 10.7). In essence,

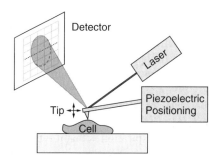

Figure 10.7 Atomic force microscopy is capable of high force resolution and "feels" the cell or substrate of interest, either making single measurements or by tapping across a surface and taking repeated measurements. The high precision is achieved through a combination of piezoelectric control of the cantilever, a high precision micro-fabricated cantilever, and the ability to measure small deflections using a laser system. The laser diffracts off the cantilever, and the deflection is determined by measuring the intensity of light above and below a set plane

an atomic force microscope (AFM) "feels" the surface below the cantilever tip, and it is capable of reporting back information related to the mechanical properties and surface topography. To make force measurements, the system relies on a cantilever with a known spring constant. The cantilever will deflect linearly with applied force over small amplitudes. By measuring the deflection, it is possible to determine the exact force applied. Measurements on the order of piconewtons (pN) are possible. This high precision has enabled measurements of many different biological structures, including microfilaments [12]. As with micropipette aspiration, it is necessary to assume a mechanical model of the cell in order to extract parameters such as the modulus of elasticity and viscoelastic properties from an indentation measurement. The Hertzian mechanics equation has been applied to both adherent and nonadherent cells [87,88].

When choosing a cantilever, the spring constant and tip radius dictate the resolution of the AFM. It is important to specify a spring constant such that there will be a measurable deflection for the measured force of interest. The spring constant of a cantilever is a function of the material modulus of elasticity and the geometric dimensions. For example, the spring constant for a rectangular cantilever is

$$k = \frac{F}{\delta} = \frac{Etw^3}{4L^3} \tag{10.7}$$

where F is the applied force, δ is the deflection, E is the modulus of elasticity, t is the cantilever thickness, w is the width, and L is the total length of the cantilever. In general, very small and thin cantilevers are needed to measure the forces of interest, such as those associated with cells or studying material surfaces. These cantilevers are typically created using microfabrication processes, since these enable a high uniformity and geometric precision.

Microfabrication also enables the creation of very small cantilever tips, on the order of nanometers, enabling high spatial resolution. There will still be some variation between each cantilever, so each new cantilever must be calibrated prior to experimentation, although several methods exist to perform this calibration [89]. Finally, it is important to keep in mind that the AFM cantilever will still be very stiff in comparison with the cell and, as such, very small forces exerted by the cell may not be measured [6]. Other probes, such as the bio-force membrane probe or laser tweezing, may be more appropriate.

The AFM is capable of measuring both adherent [87] and nonadherent cells. It is possible to measure nonadherent cells [90] directly, although this is challenging and often results in low-throughput conditions. In creating an experimental procedure or fixture, the key challenge is creating something which will not change the cell before the measurement is made. For example, one may biochemically attach a cell to a surface, but the mechanical properties will undoubtedly change during that process, introducing membrane tension and deformation. One recent adaptation involves the use of microwells to hold nonadherent cells in place [88,91]. This technique was later used to show that increased leukemia cell stiffness was associated with leucostasis, in which leukemia cells cause vital organ damage and microvasculature damage through cell accumulation [92].

It is also possible to have an AFM measure the contractile force of a cell, rather than simply push on a cell surface [30]. Recent experiments on platelets used fibrinogen-coated AFM cantilevers to cause platelet activation on contact [30]. Platelets attached between the cantilever and glass substrate began contracting, pulling the cantilever towards the underlying substrate. One key advantage to using the AFM for this experiment was the high level of control over the position of the cantilever. In particular, it was possible to write control algorithms which enabled the cantilever to maintain a constant distance away from the substrate, while still delivering force information. This enabled the researchers to determine that platelets generate high forces despite their small volume and show that the contractile force is dependent on the local stiffness of the microenvironment.

Modifications to the AFM can be made to aid in visualizing the experimental samples, such as the recently developed side-view AFM [93]. This was particularly helpful in the aforementioned experiment, as it enabled the researchers to better position the platelets between an AFM tip and substrate. Most AFM setups view both the AFM cantilever and measurement substrate from directly above or below. This can make visualizing small objects, such as platelets, attached between the cantilever and substrate difficult.

The combination of cost, slow data acquisition, and operator skill has prevented the AFM from becoming a standard clinical diagnostic tool. Although there have been some interesting studies of cancer in a clinical setting [94,95], sample sizes are small. Each sample requires a fair amount of time for the measurement, and measurements must be made serially, limiting the total number of samples which may be measured in a given time frame. The measurement time associated with each sample depends on how many spatial measurements are needed for each cell. Taking a single measurement speeds up the process, but does not utilize the exceptional capabilities of the AFM. With regard to the tip itself, unwanted materials may stick to the tip, complicating measurements by requiring a mid-experiment tip change. In particular, it is helpful to have a skilled operator who is aware of potential measurement errors and nuances associated with the use of the AFM.

10.3.3 Microplate Stretcher

A similar device to the AFM has been developed termed the microplate stretcher. In this technique, rectangular bars of glass are pulled to different cantilever dimensions, suitable for cell manipulation [96]. By using glass, this material set works well with optical microscopy and biochemical techniques. The microplates may be used for compressing, pulling, and rupturing cells, as well as serving as a base for aspiration methods. In measuring forces, the system works similar to an AFM by using a cantilever with a known spring constant and measures the resulting displacement, once force is applied. Systems are capable of measuring forces between 1 and 1000 nN [96], although force calibrations must be done manually using an existing object of known spring constant. Later studies expanded the use of microplates to rheological measurements of cells [97].

Microplates offer an interesting alternative to AFM measurements. The cantilevers are fairly inexpensive and simply fabricated using a common micro-pipetter glass puller. While the span of forces is not as large as AFM, measurements of whole cell contraction, adhesion, and traction experiments may be performed. For example, microplates were used to measure the contractile response of myoblasts in different mechanical microenvironments [31] and to confirm that the stiffness, and not the force, altered the cells' contractile properties [33].

10.3.4   Optical Tweezers

Optical trapping was originally reported by Ashkin and coworkers [98,99], who also pioneered the use of the technique with living cells [100]. The technique uses radiation pressure to trap and move dielectric particles. Typically, a laser is used in conjunction with a microscope objective, which focuses the laser onto the sample plane. The key to successful optical trapping lies in the use of a beam which has an intensity gradient. As light is passed through the particle, it is refracted and changes direction, imparting some momentum to the particle. The gradient intensity ensures that the particle will "feel" different forces as it traverses through the beam, with a tendency to return the particle to the center of the beam, as shown in Figure 10.8.

Since a force will be applied when the particle is displaced from the center of the beam, the trap itself may be modeled as a spring. Hence, by measuring the displacement of a particle from the center of the beam, it is possible to calculate the applied force, just as in an AFM. Similarly, the precision of the trap depends on the associated spring constant and the resolution at which displacements may be measured. As the forces applied by light are quite small, the spring constant of an optical trap is much lower than that of an AFM. Combined with the ability to make sub-nanometer displacement measurements, optical traps are frequently used to measure the properties of individual biological molecules [101–104].

To measure the mechanical properties of both cells and single molecules, dielectric spheres are typically attached to the object of interest, and manipulated using laser tweezers [105,106]. In some instances it is preferable to hold the cell with a micropipette and use a single optical trap to manipulate the dielectric sphere. The interested reader is directed to Moffitt *et al*. [107], where additional detailed information on the use of optical tweezers and traps is included. Using these techniques, a number of mechanical properties have been measured, including the shear modulus [108], nonlinear elastic properties [105], and viscoelastic properties [105]. To measure the viscoelastic and nonlinear properties, large forces of up to 400 pN have been created to stretch cells [109]. In all cases, to determine these properties, an *a priori* mechanical model of the deformation is needed.

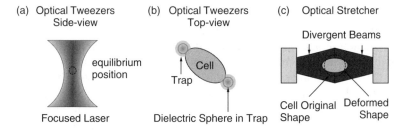

Figure 10.8   (a) Shows the side view of a focused laser optical trap and the equilibrium position of a particle. Since a restoring force will be applied to the particle when it is displaced from the equilibrium position, the optical trap acts as a spring. By knowing the spring constant of the trap, small displacements of the particle may be measured and the applied force may be deduced. (b) How optical tweezers may be used to apply known forces to cells using attached dielectric spheres. (c) An optical stretcher which purposefully defocuses the beam to apply a force to the cell membrane, causing subsequent stretching

Optical traps have recently been used to probe the cell mechanics of microorganisms and blood cells in the study of disease [110–114]. More recently, researchers have modified this platform to enable higher throughput, as discussed below.

10.4 Existing High-Throughput Cell Mechanical-Based Assays

Several assays capable of measuring mechanical properties of multiple cells, while still maintaining single-cell resolution, have been developed. The following reviews these high-throughput cell mechanical assays.

10.4.1 Optical Stretchers

One particular implementation of optical trapping utilizes a combination of optical fibers and microfluidics to create a high-throughput optical trapping cell mechanical system, termed the optical stretcher [115,116]. In this technique, rather than a focused laser beam, two opposed and slightly divergent beams are used to trap cells (Figure 10.8c). The opposing beams keep the cell in a constant spatial position. However, most of the momentum transfer occurs at the surface of the cell [117], causing deformations of the cell membrane. By measuring the deformations optically, the modulus of elasticity or viscoelastic properties of the cell may be determined. Also, the force is applied with no mechanical contact, avoiding potential adhesion issues.

The divergent beams have the added benefit of enabling higher laser powers. One of the key concerns in using optical trapping is damage to cells and biological molecules through high-intensity radiation damage and generation of free radicals. Since the beams are not converging on the sample, the laser intensities experienced by the cells are much lower, approximately 400 times lower [117]. As such, much higher forces may be generated without damaging the cell, compared with optical tweezers, and the force ranges are between those of optical tweezers and AFM measurements [115].

The opposing beams also enable high-throughput studies and a high level of automation. By integrating a fluidic delivery system around the optical fibers used for the laser, a technique for delivering large numbers of cells was created. Furthermore, the cells did not need to be perfectly aligned in the channels prior to optical deformations, since they would be aligned by the measurement process itself. This system was used to study the progression of mouse and human cell lines from normal to cancerous based on changes in deformability. Furthermore, it was possible to determine when some cells became metastatic [118]. Later studies showed that cancer cells could stretch up to five times more than normal cells [119] and that, among cancer cells, metastatic cancer cells were distinguishable from nonmetastatic cancer cells. Recently, microfluidic channels have been integrated with the system [120,121] and used to study migration of cells [122]. Optical stretchers have also been used to measure the viscoelastic properties of cells [123,124]. Other work has focused on simplifying the system to a single beam [125].

There are several limitations when using the optical stretcher. The first is that the cell deformation is dependent on both the optical and mechanical properties of the cell. While there is some variation in the cell optical properties in previous reports [118], these are not contributing to a significant error in the measurement. Second, in order to use this system, adherent cells must first be detached from their substrate. Typical methods of detachment involve digesting proteins expressed on the surface of the cell and could have an effect on the cell mechanics. Finally, although the system is a high-throughput system, there are some time constraints. It is estimated that the system is capable of measuring one cell per second [115], but current best demonstrations are at 1–1.7 cells per minute [120].

10.4.2 Traction Force Microscopy via Bead-Embedded Gels

Cells exert traction forces on the extracellular matrix (ECM) to which they are adhered through integrin-mediated adhesions. Regulation of these forces is crucial in various processes, including cell motility and remodeling of the surrounding tissue and ECM. Several methods have been developed for quantitative measurements of these cellular traction forces. These methods have largely involved developing calibrated, compliant substrates to which adherent cells exert surface stresses. By imaging the deformations of these substrates due to cell traction and taking account of the mechanical properties of each substrate, the cellular traction forces can be calculated.

One method uses continuous gels (e.g., polyacrylamide or polydimethylsiloxane (PDMS)) embedded with fluorescent beads as fiduciary markers to visualize the gel displacement field induced by cellular traction in which the bead displacement is proportional to the force applied [126,127]. Combined with other microscopy methods, this technique was recently used to measure the migration of sheets of cells [36]. One of the key advances in this technique was the use of a second bead layer which does not move, creating a fiduciary layer for subsequent imaging. Each subsequent image may be aligned to this nonmoving layer, reducing errors in bead displacement associated with movement of the microscope stage or drift. This technique has recently been scaled to three dimensions with the cells themselves embedded in the gel [128,129].

10.4.3 Traction Force Microscopy via Micropost Arrays

Another method, pioneered in the Chen group [130], uses the mechanical properties of large numbers of repeating posts, which act in a cantilever-like fashion to measure cell traction forces. The posts are oriented vertically and cells are placed on top of the posts, as shown in Figure 10.9a. When a cell attached to the end of the post applies a traction force, the post will deflect. Since the cell needs to be attached to the post, this technique is best suited to adherent cells. The amount of deflection is directly related to the spring constant of the post, which can be tuned by changing the geometrical and material properties. The spring constant of a post with a force applied at the end of the post is given by [130]

$$k_{post} = \frac{F}{\delta} = \frac{3EI}{L^3} \tag{10.8}$$

where E is the modulus of elasticity, I is the moment of inertia of the post, and L is the length of the post. The microposts are created using microfabrication techniques and create highly uniform and repeatable structures with tunable geometry. As such, it is possible to create a number of different shapes, even creating gradients of shapes, as shown in Figure 10.9b. The elliptical shapes create anisotropic stiffness, in which the posts have a lower spring constant in one direction versus another. This was used to determine

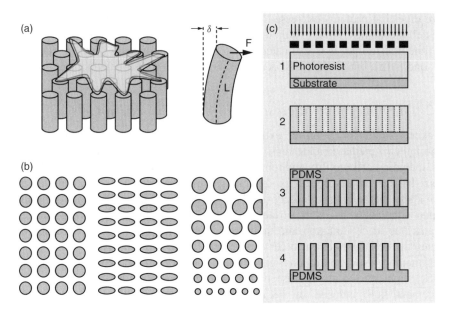

Figure 10.9 (a) A cell on top of a micropost array will deflect the posts, indicating the traction force exerted by the cell. The cells are typically transparent, enabling optical measurements of the deflection. (b) A few of the micropost patterns which can be created to tailor the stiffness experienced by the cell, including an anisotropic pattern (ellipse) and gradient stiffnesses (changing size). (c) The fabrication process for micropost arrays is straightforward. Holes are patterned in photoresist using standard photolithography. The photoresist serves as a template for the micropost arrays, which are typically created from PDMS

that endothelial cells grow and migrate along the direction of greatest stiffness [131], termed "durotaxis." Similarly, changing the radius of the microposts can create a stiffness gradient and was used to further investigate cell migration towards stiffer areas [132].

While there are different techniques and methods to creating microposts [133], the most basic approach is shown in Figure 10.9c. Here, photoresist is spun onto a substrate, typically on a silicon or glass wafer. The thickness of the photoresist is controlled by the spin speed and determines the final height of the posts. Next, a mask selectively covers the photoresist, allowing UV light onto the photoresist. Exposure to UV light chemically changes the photoresist compared with the unexposed photoresist, enabling selective removal of this photoresist using solvents. The remaining photoresist is then baked to harden the structure. Finally, PDMS is poured onto the resulting structure and placed under a vacuum to remove any air bubbles. The entire assembly is cured in an oven; the PDMS is then peeled off the structure, creating the micropost array. This PDMS pouring, curing, and removal may be repeated to created many structures from a single mold, and is termed soft lithography [134,135].

Some recent work has incorporated cobalt nanowires into the microposts, enabling deflection of the posts in magnetic fields. Magnetic nanowires were mixed into PDMS, poured into the mold, and cured. As such, the distribution of nanowires was random and only some of the posts contained nanowires. Nevertheless, the ability to apply forces to the micropost array did reveal that applying a step force led to increased focal adhesion size at the site of application. Furthermore, when forces were applied by the magnetic posts, cells would experience a loss in contractility, either suddenly or gradually over several minutes [136]. This unique response highlights the need to be able to measure individual cells. More information on the unique response of individual cells and subpopulations is needed to better understand the link between cell mechanics and disease.

Micropost traction force microscopy is not without limitations. Below $1\,\mu m$, fabrication and imaging of the arrays becomes difficult. As such, cells must be large enough to interact with at least several posts for meaningful data, rendering small cells ($<5\,\mu m$) difficult to analyze. In addition, although the spatial resolution of the force measurement is well defined by the geometry of the sensor array and enables the force calculation to be straightforward, this geometry also puts constraints on the cellular adhesions and introduces topographical cues that may alter normal cellular physiology and confound interpretation of data. Bead-embedded gels with continuous contact area better approximate physiological conditions. However, calculating the traction force field of the cell from bead displacements can be computationally rigorous and often involves complex computational models [126,137–139]. Overall, both traction force microscopy methods are unique, in that they passively measure the mechanical forces applied by cells.

10.4.4 Substrate Stretching Assays

Substrate stretching devices are best suited for applying forces to adherent cells and examining the biochemical and biophysical response. Cells are cultured on a flexible substrate in which axial strain is applied [140,141]. Alternatively, it is possible to use pressures to deform the substrate [142]. One interesting area of study has been the effect of applying strain in cycles [143] and watching how cells remodel in the presence of chemical agents known to affect their cytoskeleton. While it is possible to measure the forces applied by cells using substrate stretching assays, there are much better techniques that provide additional spatial information, such as the aforementioned micropost array.

10.4.5 Magnetic Twisting Cytometry

Magnetic twisting cytometry presents a unique opportunity to apply forces to cells in a different manner from other techniques. As such, it has provided unique insight into the cell and cell mechanics, since forces may be applied to local areas of the cell without causing large changes in the cell shape. To perform magnetic twisting cytometry, magnetic beads are chemically modified, mixed with cells, and become attached to the cell surface. Once attached, a magnetic field is applied to the beads, causing a twisting motion. This enables the application of highly localized torques on the cell cytoskeleton, which is difficult to achieve with other techniques. Furthermore, specific binding sites on the cell surface may be targeted through the selection of the chemical linkers used on the magnetic bead.

This technique was developed by Wang *et al*. to determine that transmembrane ECM receptors provide a molecular path for mechanical signal transfer [21]. Later uses applied time-varying magnetic fields to measure the cell rheological properties as a function of the frequency dependence of the applied force [25]. By imaging the beads individually and applying oscillatory forces, the local elastic modulus and loss modulus associated with the bead could be determined. These modulus relationships all followed a weak power-law dependence, even after treatment with biological antagonists, suggesting an underlying mechanism. Later studies showed that the bead coating and type also had an effect on the measured modulus, but again the cells followed a unique power law [144,145]. Other work has found changes in the gene expression with applied twisting forces [146].

While the forces applied to the cells using this method are unique in their torsional manner and, furthermore, occur at specific binding sites, there is limited spatial control over which binding sites are used. Similarly, there is no guarantee that each cell will possess the same number of magnetic beads, making it difficult to determine if biological changes in the cells are linked to the number of beads attached to each cell. Also, this technique is practically limited to adherent cells, since applying a magnetic field to nonadherent cells will also cause the cell itself to move.

10.4.6 Microfluidic Pore and Deformation Assays

Microfluidic pore and deformation assays use similar principles of operation to microfiltration methods discussed earlier, but they solve many of the issues surrounding microfiltration and are able to report individual cell data. Even in the simplest case of simply filtering cells, microfluidics have been able to show improved capture of rare cells for later analysis, such as circulating tumor cells [147] or hematopoietic stem cells [148]. The basic unit of a microfluidic system is a channel and is often created with soft lithography. Soft lithography techniques are similar to those used for micropost creation, and are briefly described here. Photoresist is spun onto a wafer and patterned using a mask and UV light in the shape of a channel. PDMS is then poured over the patterned photoresist, cured, and removed. However, unlike the microposts, the PDMS mold needs to be subsequently bonded to a substrate such as glass or more PDMS to complete the channel. This is done by treating both surfaces that are to be bonded with an oxygen plasma and subsequently placing the surfaces in contact with one another. Once completed, fluidic connections are rather straightforward. By simply piercing holes in the cured PDMS and inserting tubing of a slightly larger size, it is possible to take advantage of an interference fit. Note that the holes for the fluidic connections should be created before bonding for best results. PDMS is also permeable to gases, enabling exchange of carbon dioxide and oxygen with the surrounding environment. By using microfabrication, it is possible to scale the system rapidly to measure large numbers of cells simultaneously.

Microfluidic cell deformability measurements are typically performed in two ways. One method involves using interference or contact-based measurement, in which cells interact with a nonmoving surface, shown in Figure 10.10. Using microfabrication techniques, a microfluidic channel or structure that is slightly smaller than the cell of interest is first created. The cell will physically touch the wall of the channel and be moved through the channel with applied pressure. By using transparent materials to make the device, such as PDMS, it is possible to view the cell passing through the microchannel directly. A number of different parameters may be measured for the cell in the channel, including entry time [149], transit velocity [149], and total transit time [150].

Many techniques are available to measure the transit time, including optoelectronic [151,152], resistive pulse detection [153–155], and optical microscopy [150]. One advantage to the optical system is the simplicity of the device. Microscopy in

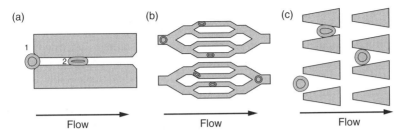

Figure 10.10 (a) The basic unit for microfluidic cell deformation is a channel slightly smaller than the cell of interest. Multiple measurements may be made, including entry time, transit velocity, and total transit time. (b) Shows a high-throughput cell deformation device which repeats the basic cell deformation channel and adds interconnects. In this device, many cells may be measured simultaneously, enabling high-throughput measurements. (c) Alternative geometries which may be used to deform the cells [169]

conjunction with automated image analysis algorithms and software (e.g., MATLAB) creates a high-throughput system capable of measuring cells with minimal operator supervision. Such a system was recently shown to measure between 50 and 100 cells per minute [150]. More importantly, the system was able to measure populations of cell transit times, showing unique populations of leukemia cells which were affected by different drug treatments [150].

The other method involves using the varying physical response of different cells under fluidic flows. Since microfabrication techniques offer precise control over the fluidic channel geometry, unique fluidic properties and forces may be used to sort and test cells. For example, curved channels may be designed to create inertial forces which sort and order particles and cells within microchannels [156]. Later work focused on using inertial forces to separate cells by size and deformability [157], or scaling the system to remove bacteria from diluted blood [158]. Other systems have induced pressure gradients through unique channel geometries to separate particles based on size [159], which in principle could be applied to examine cells. When compared with interference-based systems, these devices are less prone to clogging, which is advantageous when performing continuous cell separation.

One of the advantages of microfluidic pore and deformation assays is that they can be used to create a first-order approximation of the microvasculature [160,161]. As such, cell behavior occurring in the device has the possibility of occurring in physiological conditions as well. For example, in examining the effect of chemotherapy on leukemia cell stiffness, it was possible to use a microfluidic deformation-based device to visualize dead leukemia cells becoming stuck seven times more often than live cells [162], offering insight into the effects of chemotherapy. Another interesting example found that unexpected events may happen with red blood cells infected with malaria. In one instance, a red blood cell burst after passing through a channel. The same study also showed that a healthy red blood cell was able to weave through a blockage caused by infected red blood cells [163]. Although more *in vivo* work would be needed to confirm such findings, it does offer new insight into possible mechanisms for the spread of the malaria parasite, as one rarely envisions cells bursting after passing through a channel.

Microfluidic deformability assays as high-throughput cell mechanical devices are well suited for clinical applications. They are capable of measuring large numbers of individual cells and provide population statistics. These population statistics provide clinicians with more information, enabling detection of mechanical cellular subpopulations, which may be associated with disease. Furthermore, devices are readily transferable to the clinical setting, requiring minimal operator skill. Currently, microfluidic platforms function as research-enabling tools to study cell mechanics of disease. As research progresses and sufficient data prove the clinical utility of these devices, the role of microfluidics will likely transition from research method to diagnostic assay.

10.5   Cell Mechanical Properties and Diseases

The sensitivity of single-cell measurements has proven to be extremely clinically useful, as pathologic processes originate at the single-cell level [164,165]. To date, the only clinically approved high-throughput assay with single-cell resolution is flow cytometry, which quantifies cell surface protein expression and signaling activity and has widespread use in clinical hematology and immunology [166]. However, flow cytometry cannot measure cellular mechanical properties, as the variance in cellular mechanical properties within a given population is vast and single mechanically altered cells are theoretically sufficient to induce pathology [167,168]. For example, metastatic cancer cells, those which begin the spread of cancer across the body, are shown to be 70% softer and have a very narrow standard deviation compared with healthy body tissue and that initially, the pathologic subpopulation is rare and grows over time [94]. Thus, a clinical need exists for improved single-cell mechanical analysis.

Future research into cells and cell mechanics needs to expand the number of cells which can be measured for an individual experiment. For convenience, the techniques presented in this chapter are summarized in Table 10.1. Ideally, a review of these methods will inspire the reader to create new high-throughput mechanical devices. Given an overview of both the history and mechanisms used to measure the mechanical properties of cells, it is possible to build on the existing techniques to create new devices. The current high-throughput cell deformability devices are in essence an extension of the microfiltration assays, but vastly improve on the original and provide new capabilities. In a similar manner, it should be possible to build on the other bulk and single-cell assays to create high-throughput devices. For example, building on the AFM, one could envision creating an array of simple micro-cantilevers capable of measuring the contraction of certain cells. Or, in considering the magnetic bead assays, many additional functionalities could be created with better control of magnetic beads, especially if the beads could be controlled individually. Finally, one could envision a micropost array in which each post could be separately actuated to probe the cell.

One future challenge lies in understanding the underlying mechanisms of how cell mechanics is involved in the pathophysiology of disease. Only after that is achieved can cell mechanics-based platforms be applied in the clinical setting. Several excellent review papers have examined the role of cell mechanics in disease and vascular injury [1,3,4]. Malaria, cancer, and atherosclerosis have all been previously demonstrated to have a definite link to cell mechanics. Some of the recent findings will be summarized below along with several future areas for inquiry.

Table 10.1 Summary of techniques used for the mechanical analysis of cells

Technique	Cell types	Type of information	Advantages	Limitations	Resolution/force limits
Microfiltration	Suspended cells	Number of cells stuck in filter, transit times of cells	Inexpensive, easily integrated with other equipment, little specialized training needed	Bulk assay, pore uniformity, cell shape changes results, hard to interpret results for multicell systems	Pore size, shape, and branching dictated by manufacturing tolerances
Rheometry	Adherent cells, suspended cells	Deformation of cells under shear (optical measurement)	Controlled application of shear force, possible to get individual cell information	Bulk assay, difficult to use in clinical setting, cell deformability metric may be influenced by many factors	Light microscope resolution limits, two-point resolution limit is 0.3 μm; see micropipette for more info
Ektacytometry	Suspended cells	Diffraction pattern based on deformation of cells under shear	Controlled application of shear force, simplified readout	Bulk assay, cell deformability metric may be influenced by many factors	Diffraction limits for groups of cells in suspension
Parallel plate flow chambers	Adherent cells		Controlled application of shear force	Bulk assay, does not necessarily provide mechanical data	This technique is best suited to applying forces to cells
Micropipette aspiration	Adherent cells, suspended cells	Deformation of cell into micropipette (correlates to membrane or cell stiffness)	Individual cell information, unique information about cell membrane, cell immobilized during measurement (pulls on cell)	Low throughput, difficult to use in clinical setting	Resolution governed by light microscope limits (best reported is 0.025 μm) and limits on smallest suction pressure $(\text{pN } \mu\text{m}^{-2})$

Table 10.1 (*continued*)

Technique	Cell types	Type of information	Advantages	Limitations	Resolution/ force limits
Atomic force microscopy	Adherent cells, suspended cells	Surface topography of cell, deformation of cell at different locations, deformation of entire cell, contraction, adhesion rupture	Individual cell information, high spatial resolution, surface topography measurements possible	Low throughput, difficult to use in clinical setting	Piconewton measurements possible, has large dynamic force range
Microplate stretcher	Adherent cells	Contraction, deformation of entire cell, adhesion rupture	Individual cell information, alternative to AFM, fairly inexpensive	Low throughput, need force calibrations before measurements, difficult to use in clinical setting	Force range 1–1000 nN
Optical tweezers	Adherent cells, suspended cells	Deformation of cell when pulled by point forces (beads attached to cell)	Individual cell information	Low throughput, potential radiation damage, generation of free radicals, difficult to use in clinical setting	Sub piconewton measurements possible, spring constant $\sim$pN nm^{-1}, maximum applied force limited (400 pN best reported)
Optical stretcher	Suspended cells	Deformation of cell when deformed by divergent laser beam	Individual cell information, improved throughput (1 cell/min), reduced radiation concerns	Cell deformation dependent on optical and mechanical properties of cells, difficult to use in clinical setting	Force range between optical tweezers and AFM

Assay	Cell type	Force application	Advantages	Disadvantages	Notes
Traction force microscopy (bead-embedded gels)	Adherent cells	Forces applied by cells when attached to surface	Individual cell information, better physiological approximation than microposts	Low throughput, models used to calculate applied forces can be computationally intensive, traction forces not linked to clinical diagnostics	Force limit is function of gel stiffness as well as microscope resolution
Traction force microscopy (micropost arrays)	Adherent cells	Forces applied by cells when attached to surface	Individual cell information, simplified calculation of applied forces	Low throughput, cell traction forces not linked to clinical diagnostic	Force limit is function of micropost stiffness and geometry, as well as microscope resolution, typically nN resolution
Substrate stretching assays	Adherent cells	Applies force to cells attached to flexible substrate	Well suited to cyclical studies of repeated application and removal of force	Measurement of forces applied by cells is limited, measurement of cell properties limited, difficult to use in clinical settings	Large range of forces may be applied, enough to rupture cell
Magnetic twisting cytometry	Adherent cells	Forces applied by magnetic beads which may be chemically attached to cells	Individual cell information, application of force to chemically targeted cell areas	Difficult to use in clinical settings, not linked to clinical needs	Locally applies forces to specific locations
Microfluidic pore and deformation assay	Suspended cells	Number of cells stuck in microfluidic channels, transit times of cells, sorting based on deformability of cells	High throughput, individual cell information, inexpensive, improved pore size and shape compared with microfiltration, good tolerances enable unique fluid physics	More research needed to prepare for clinical use, whole unfiltered blood is rarely used, often need sample prep	Pore size, shape dictated by lithographic processes

Cancer research is a large area of research, especially given the large number and types of cancer that exist and the complex nature of the disease itself. In the past decade, cell mechanics researchers have found that cancer cells are not as stiff as typical healthy cells [94,118,149]. The current challenge lies in finding a high-throughput cell mechanics device that is capable of measuring large numbers of cells and is capable of identifying cancer cells, which are often rare, with high specificity and selectivity. Alternatively, another area of research involves matching cancer symptoms, such as leukostasis, with biophysical phenomenon, such as stiffening cells [92]. In either case, the cells which are of interest are a small subset of a normal healthy population of cells.

Malaria is similar to cancer, in that the cells of interest are a small subset of a normal healthy population of cells. When infected by *P. falciparum*, red blood cells experience an increase in stiffness dependent on the stage of the infection [64]. This has been measured in several different ways, including an optical stretcher [170] and deformability-based microfluidics [169]. Furthermore, experiments have been able to inform biophysical modeling of infected cells [171]. One of the consequences of malaria is an increased adhesiveness of red blood cells which may sequester in the microvasculature of major organs. These studies form the biophysical basis of malaria pathophysiology, and further work will determine whether cell mechanics has a role in discovering new therapeutic targets for this dreaded infectious disease.

A large body of research on endothelial cells [71] has found that endothelial cells are highly affected by the shear stress conditions surrounding them. These studies were among the first to prove that the physical environment can directly alter the biological function of cells and, as such, pioneered the field of cell mechanics. More recent studies have subsequently shown that different flow patterns, including pulsatile and reversing flow, also alter endothelial cell physiology. Specifically, high shear stress created by sustained laminar flow on endothelial cells causes upregulation of genes and proteins that protect against atherosclerosis. However, few studies have focused on large-scale studies of individual endothelial cells to determine the population variation, examining the potential role of unique subpopulations.

Although the majority of studies investigating cell mechanics and disease involve endothelial dysfunction in cardiovascular disease, malaria, and cancer, cell mechanics is potentially implicated in numerous other biological and pathologic processes. Recent work has shown that cell mechanics likely plays significant roles in inflammation [172,173], embryogenesis [174], nerve injury and regrowth [175], sickle cell disease [176], and stem cell differentiation [46,177,178]. Novel and innovative high-throughput techniques to investigate cell mechanics in the context of these biological and pathologic processes are needed to advance our basic understanding of human physiology and develop new diagnostic assays and therapies for disease.

References

[1] Bao, G. and Suresh, S. (2003) Cell and molecular mechanics of biological materials. *Nature Materials*, **2**, 715–725.
[2] Zhu, C., Bao, G., and Wang, N. (2000) Cell mechanics: mechanical response, cell adhesion, and molecular deformation. *Annual Review of Biomedical Engineering*, **2**, 189–226, doi: 10.1146/annurev.bioeng.2.1.189.

[3] Suresh, S. (2007) Biomechanics and biophysics of cancer cells. *Acta Materialia*, **55**, 3989–4014, doi:10.1016/j.actamat.2007.04.022.

[4] Suresh, S., Spatz, J., Mills, J.P. *et al.* (2005) Connections between single-cell biomechanics and human disease states: gastrointestinal cancer and malaria. *Acta Biomaterialia*, **1**, 15–30, doi: 10.1016/j.actbio.2004.09.001.

[5] Kim, D.-H., Wong, P.K., Park, J. *et al.* (2009) Microengineered platforms for cell mechanobiology. *Annual Review of Biomedical Engineering*, **11**, 203–233, doi: 10.1146/annurev-bioeng-061008-124915.

[6] Hochmuth, R.M. (2000) Micropipette aspiration of living cells. *Journal of Biomechanics*, **33**, 15–22.

[7] Brown, T.D. (2000) Techniques for mechanical stimulation of cells *in vitro*: a review. *Journal of Biomechanics*, **33**, 3–14, doi: 10.1016/s0021-9290(99)00177-3.

[8] Reid, H.L., Dormandy, J.A., Barnes, A.J. *et al.* (1976) Impaired red cell deformability in peripheral vascular disease. *The Lancet*, **307**, 666–668, doi: 10.1016/s0140-6736(76)92778-1.

[9] Dewey, J.C.F., Bussolari, S.R., Gimbrone, J.M.A., and Davies, P.F. (1981) The dynamic response of vascular endothelial cells to fluid shear stress. *Journal of Biomechanical Engineering*, **103**, 177–185, doi: 10.1115/1.3138276.

[10] McCormick, S.M., Eskin, S.G., McIntire, L.V. *et al.* (2001) DNA microarray reveals changes in gene expression of shear stressed human umbilical vein endothelial cells. *Proceedings of the National Academy of Sciences of the United States of America*, **98**, 8955–8960, doi: 10.1073/pnas.171259298.

[11] Audesirk, T., Audesirk, G., and Byers, B.E. (2000) *Life on Earth*, 2nd edn., Prentice-Hall.

[12] Herrmann, H., Bar, H., Kreplak, L. *et al.* (2007) Intermediate filaments: from cell architecture to nanomechanics. *Nature Reviews. Molecular Cell Biology*, **8**, 562–573.

[13] Qin, Z., Kreplak, L., and Buehler, M.J. (2009) Hierarchical structure controls nanomechanical properties of vimentin intermediate filaments. *PLoS ONE*, **4**, e7294.

[14] Zaidel-Bar, R., Itzkovitz, S., Ma'ayan, A. *et al.* (2007) Functional atlas of the integrin adhesome. *Nature Cell Biology*, **9**, 858–867.

[15] Geiger, B., Spatz, J.P., and Bershadsky, A.D. (2009) Environmental sensing through focal adhesions. *Nature Reviews. Molecular Cell Biology* **10**, 21–33.

[16] Riveline, D., Zamir, E., Balaban, N.Q. *et al.* (2001) Focal contacts as mechanosensors. *The Journal of Cell Biology*, **153**, 1175–1186, doi: 10.1083/jcb.153.6.1175.

[17] Galbraith, C.G., Yamada, K.M., and Sheetz, M.P. (2002) The relationship between force and focal complex development. *The Journal of Cell Biology*, **159**, 695–705, doi: 10.1083/jcb.200204153.

[18] Holle, A.W. and Engler, A.J. (2011) More than a feeling: discovering, understanding, and influencing mechanosensing pathways. *Current Opinion in Biotechnology*, **22**, 648–654, doi: 10.1016/j.copbio.2011.04.007.

[19] Discher, D.E., Janmey, P., and Wang, Y.-L. (2005) Tissue cells feel and respond to the stiffness of their substrate. *Science*, **310**, 1139–1143, doi: 10.1126/science.1116995.

[20] Chen, C.S. (2008) Mechanotransduction – a field pulling together? *Journal of Cell Science*, **121**, 3285–3292, doi: 10.1242/jcs.023507.

[21] Wang, N., Butler, J., and Ingber, D. (1993) Mechanotransduction across the cell surface and through the cytoskeleton. *Science*, **260**, 1124–1127, doi: 10.1126/science.7684161.

[22] Wang, N., Naruse, K, Stamenović, D. *et al.* (2001) Mechanical behavior in living cells consistent with the tensegrity model. *Proceedings of the National Academy of Sciences of the United States of America*, **98**, 7765–7770, doi: 10.1073/pnas.141199598.

[23] Ingber, D.E. (2003) Tensegrity II. *How structural networks influence cellular information processing networks. Journal of Cell Science*, **116**, 1397–1408, doi: 10.1242/jcs.00360.

[24] Ingber, D.E. (2003) Tensegrity I. *Cell structure and hierarchical systems biology. Journal of Cell Science*, **116**, 1157–1173, doi: 10.1242/jcs.00359.

[25] Fabry, B., Maksym, G.N., Butler, J.P. *et al.* (2001) Scaling the microrheology of living cells. *Physical Review Letters*, **87**, 148102.

[26] Gunst, S.J. and Fredberg, J.J. (2003) The first three minutes: smooth muscle contraction, cytoskeletal events, and soft glasses. *Journal of Applied Physiology*, **95**, 413–425, doi: 10.1152/japplphysiol.00277.2003.

[27] Deng, L., Trepat, X., Butler, J.P. *et al.* (2006) Fast and slow dynamics of the cytoskeleton. *Nature Materials*, **5**, 636–640.

[28] Trepat, X., Deng, L., An, S.S. *et al.* (2007) Universal physical responses to stretch in the living cell. *Nature*, **447**, 592–595.

[29] Harris, A., Wild, P., and Stopak, D. (1980) Silicone rubber substrata: a new wrinkle in the study of cell locomotion. *Science*, **208**, 177–179, doi: 10.1126/science.6987736.

[30] Lam, W.A., Chaudhuri, O., Crow, A. *et al*. (2011) Mechanics and contraction dynamics of single platelets and implications for clot stiffening. *Nature Materials*, **10**, 61–66.

[31] Mitrossilis, D., Fouchard, J., Guiroy, A. *et al*. (2009) Single-cell response to stiffness exhibits muscle-like behavior. *Proceedings of the National Academy of Sciences of the United States of America*, **106**, 18243–18248, doi: 10.1073/pnas.0903994106.

[32] Debold, E.P., Patlak, J.B., and Warshaw, D.M. (2005) Slip sliding away: load-dependence of velocity generated by skeletal muscle myosin molecules in the laser trap. *Biophysical Journal*, **89**, L34–L36, doi: 10.1529/biophysj.105.072967.

[33] Mitrossilis, D., Fouchard, J., Pereira, D. *et al*. (2010) Real-time single-cell response to stiffness. *Proceedings of the National Academy of Sciences of the United States of America*, **107**, 16518–16523, doi: 10.1073/pnas.1007940107.

[34] Lauffenburger, D.A. and Horwitz, A.F. (1996) Cell migration: a physically integrated molecular process. *Cell*, **84**, 359–369, doi: 10.1016/s0092-8674(00)81280-5.

[35] Keren, K., Pincus, Z., Allen, G.M. *et al*. (2008) Mechanism of shape determination in motile cells. *Nature*, **453**, 475–480.

[36] Trepat, X., Wasserman, M.R., Angelini, T.E. *et al*. (2009) Physical forces during collective cell migration. *Nature Physics*, **5**, 426–430.

[37] Gov, N. (2011) Cell mechanics: moving under peer pressure. *Nature Materials*, **10**, 412–414.

[38] Fung, Y.C. (1988) *Biomechanics: Mechanical Properties of Living Tissues*, Springer-Verlag.

[39] Smith, D.H., Wolf, J.A., and Meaney, D.F. (2001) A new strategy to produce sustained growth of central nervous system axons: continuous mechanical tension. *Tissue Engineering*, **7**, 131–139, doi: 10.1089/107632701300062714.

[40] Pfister, B.J., Iwata, A., Meaney, D.F., and Smith, D.H. (2004) Extreme stretch growth of integrated axons. *The Journal of Neuroscience*, **24**, 7978–7983, doi: 10.1523/jneurosci.1974-04.2004.

[41] Morrison, B., Saatman, K.E., Meaney, D.F., and McIntosh, T.K. (1998) *In vitro* central nervous system models of mechanically induced trauma: a review. *Journal of Neurotrauma*, **15**, 911–928, doi: 10.1089/neu.1998.15.911.

[42] Chicurel, M.E., Singer, R.H., Meyer, C.J., and Ingber, D.E. (1998) Integrin binding and mechanical tension induce movement of mRNA and ribosomes to focal adhesions. *Nature*, **392**, 730–733.

[43] Kealy, T., Abram, A., Hunt, B., and Buchta, R. (2008) The rheological properties of pharmaceutical foam: implications for use. *International Journal of Pharmaceutics*, **355**, 67–80, doi: 10.1016/j.ijpharm.2007.11.057.

[44] Chen, C.S., Mrksich, M., Huang, S. *et al*. (1997) Geometric control of cell life and death. *Science*, **276**, 1425–1428, doi: 10.1126/science.276.5317.1425.

[45] Huang, S. and Ingber, D.E. (1999) The structural and mechanical complexity of cell-growth control. *Nature Cell Biology*, **1**, E131–E138.

[46] Engler, A.J., Sen, S., Sweeney, H.L., and Discher, D.E. (2006) Matrix elasticity directs stem cell lineage specification. *Cell*, **126**, 677–689, doi: 10.1016/j.cell.2006.06.044.

[47] Gardel, M.L., Shin, J.H., MacKintosh, F.C. *et al*. (2004) Elastic behavior of cross-linked and bundled actin networks. *Science*, **304**, 1301–1305, doi: 10.1126/science.1095087.

[48] Storm, C., Pastore, J.J., MacKintosh, F.C. *et al*. (2005) Nonlinear elasticity in biological gels. *Nature*, **435**, 191–194.

[49] Matthews, B.D., Overby, D.R., Mannix, R., and Ingber, D.E. (2006) Cellular adaptation to mechanical stress: role of integrins, Rho, cytoskeletal tension and mechanosensitive ion channels. *Journal of Cell Science*, **119**, 508–518, doi: 10.1242/jcs.02760.

[50] Vogel, V. and Sheetz, M. (2006) Local force and geometry sensing regulate cell functions. *Nature Reviews. Molecular Cell Biology*, **7**, 265–275.

[51] Reid, H.L., Barnes, A.J., Lock, P.J. *et al*. (1976) A simple method for measuring erythrocyte deformability. *Journal of Clinical Pathology*, **29**, 855–858, doi: 10.1136/jcp.29.9.855.

[52] Worthen, G.S., Schwab, B., Elson, E.L., and Downey, G.P. (1989) Mechanics of stimulated neutrophils: cell stiffening induces retention in capillaries. *Science*, **245**, 183–186.

[53] Waldenlind, L., Edlund, B.L., Hulting, J. *et al*. (1988) Decreased red cell filterability in patients with acute myocardial infarction. *Acta Medica Scandinavica*, **224**, 225–229, doi: 10.1111/j.0954-6820.1988.tb19365.x.

[54] Nash, G.B., Christopher, B., Morris, A.J., and Dormandy, J.A. (1989) Changes in the flow properties of white blood cells after acute myocardial infarction. *British Heart Journal*, **62**, 329–334, doi: 10.1136/hrt.62.5.329.

[55] Kikuchi, Y., Sato, K., Ohki, H., and Kaneko, T. (1992) Optically accessible microchannels formed in a single-crystal silicon substrate for studies of blood rheology. *Microvascular Research*, **44**, 226–240, doi:10.1016/0026-2862(92)90082-z.

[56] Rampling, M. (1985) Haemorheology and blood flow in diabetes and limb ischaemia. *Journal of the Royal Society of Medicine*, **78**, 596–598.

[57] Chien, S., Schmalzer, E.A., Lee, M.M. *et al.* (1983) Role of white blood cells in filtration of blood cell suspensions. *Biorheology*, **20**, 11–27.

[58] Nash, G.B., Jones, J.G., Mikita, J., and Dormandy, J.A. (1988) Methods and theory for analysis of flow of white cell subpopulations through micropore filters. *British Journal of Haematology*, **70**, 165–170, doi: 10.1111/j.1365-2141.1988.tb02458.x.

[59] Nash, G.B., Jones, J.G., Mikita, J. *et al.* (1988) Effects of preparative procedures and of cell activation on flow of white cells through micropore filters. *British Journal of Haematology*, **70**, 171–176, doi: 10.1111/j.1365-2141.1988.tb02459.x.

[60] Evans, E. and La Celle, P. (1975) Intrinsic material properties of the erythrocyte membrane indicated by mechanical analysis of deformation. *Blood*, **45**, 29–43.

[61] Wang, Y., Hammes, F., Düggelin, M., and Egli, T. (2008) Influence of size, shape, and flexibility on bacterial passage through micropore membrane filters. *Environmental Science and Technology*, **42**, 6749–6754, doi: 10.1021/es800720n.

[62] Bussolari, S.R., Dewey, C.F., and Gimbrone, M.A. (1982) Apparatus for subjecting living cells to fluid shear stress. *Review of Scientific Instruments*, **53**, 1851–1854.

[63] Fischer, T., Stohr-Lissen, M., and Schmid-Schonbein, H. (1978) The red cell as a fluid droplet: tank tread-like motion of the human erythrocyte membrane in shear flow. *Science*, **202**, 894–896, doi: 10.1126/science.715448.

[64] Cranston, H., Boylan, C.W., Carroll, G.L. *et al.* (1984) *Plasmodium falciparum* maturation abolishes physiologic red cell deformability. *Science*, **223**, 400–403, doi: 10.1126/science.6362007.

[65] Groner, W., Mohandas, N., and Bessis, M. (1980) New optical technique for measuring erythrocyte deformability with the ektacytometer. *Clinical Chemistry*, **26**, 1435–1442.

[66] Bessis, M. and Mohandas, N. (1975) Diffractometric method for mearurement of cellular deformability. *Blood Cells*, **1**, 307–313.

[67] Mohandas, N., Clark, M.R., Jacobs, M.S., and Shohet, S.B. (1980) Analysis of factors regulating erythrocyte deformability. *The Journal of Clinical Investigation*, **66**, 563–573.

[68] Streekstra, G.J., Dobbe, J.G.G., and Hoekstra, A.G. (2010) Quantification of the fraction poorly deformable red blood cells using ektacytometry. *Optics Express*, **18**, 14173–14182.

[69] Chasis, J.A. and Mohandas, N. (1986) Erythrocyte membrane deformability and stability: two distinct membrane properties that are independently regulated by skeletal protein associations. *Journal of Cell Biology*, **103**, 343–350.

[70] Pautard, B., Feo, C., Dhermy, D. *et al.* (1988) Occurrence of hereditary spherocytosis and beta thalassaemia in the same family: globin chain synthesis and visco diffractometric studies. *British Journal of Haematology*, **70**, 239–245.

[71] Chiu, J.-J. and Chien, S. (2011) Effects of disturbed flow on vascular endothelium: pathophysiological basis and clinical perspectives. *Physiological Reviews*, **91**, 327–387, doi: 10.1152/physrev.00047.2009.

[72] Sung, H.-J., Yee, A., Eskin, S.G., and McIntire, L.V. (2007) Cyclic strain and motion control produce opposite oxidative responses in two human endothelial cell types. *American Journal of Physiology: Cell Physiology* **293**, C87–C94, doi: 10.1152/ajpcell.00585.2006.

[73] Mitchison, J.M. and Swann, M.M. (1954) The mechanical properties of the cell surface. *Journal of Experimental Biology*, **31**, 443–460.

[74] Jones, W., Ting-Beall, H.P., Lee, G.M. *et al.* (1999) Alterations in the Young's modulus and volumetric properties of chondrocytes isolated from normal and osteoarthritic human cartilage. *Journal of Biomechanics*, **32**, 119–146.

[75] Evans, E. and Yeung, A. (1989) Apparent viscosity and cortical tension of blood granulocytes determined by micropipet aspiration. *Biophysical Journal*, **56**, 151–211, doi: 10.1016/s0006-3495(89)82660-8.

[76] Yeung, A. and Evans, E. (1989) Cortical shell-liquid core model for passive flow of liquid-like spherical cells into micropipets. *Biophysical Journal*, **56**, 139–188, doi: 10.1016/s0006-3495(89)82659-1.

[77] Evans, E.A. (1973) New membrane concept applied to the analysis of fluid shear- and micropipette-deformed red blood cells. *Biophysical Journal*, **13**, 941–954, doi: 10.1016/s0006-3495(73)86036-9.

[78] Theret, D.P., Levesque, M.J., Sato, M. *et al.* (1988) The application of a homogeneous half-space model in the analysis of endothelial cell micropipette measurements. *Journal of Biomechanical Engineering*, **110**, 190–199, doi: 10.1115/1.3108430.

[79] Shao, J.-Y., Ting-Beall, H.P., and Hochmuth, R.M. (1998) Static and dynamic lengths of neutrophil microvilli. *Proceedings of the National Academy of Sciences of the United States of America*, **95**, 6797–6802.

[80] Shin, J.-W., Swift, J., Spinler, K.R., and Discher, D.E. (2011) Myosin-II inhibition and soft 2D matrix maximize multinucleation and cellular projections typical of platelet-producing megakaryocytes. *Proceedings of the National Academy of Sciences*, **108**, 11458–11463, doi: 10.1073/pnas.1017474108.

[81] Evans, E., Ritchie, K., and Merkel, R. (1995) Sensitive force technique to probe molecular adhesion and structural linkages at biological interfaces. *Biophysical Journal*, **68**, 2580–2587.

[82] Merkel, R., Nassoy, P., Leung, A. *et al.* (1999) Energy landscapes of receptor-ligand bonds explored with dynamic force spectroscopy. *Nature*, **397**, 50–53.

[83] Chen, W., Zarnitsyna, V., Sarangapani, K. *et al.* (2008) Measuring receptor–ligand binding kinetics on cell surfaces: from adhesion frequency to thermal fluctuation methods. *Cellular and Molecular Bioengineering*, **1**, 276–288, doi: 10.1007/s12195-008-0024-8.

[84] Chesla, S.E., Selvaraj, P., and Zhu, C. (1998) Measuring two-dimensional receptor–ligand binding kinetics by micropipette. *Biophysical Journal*, **75**, 1553–1572.

[85] Chen, W., Evans, E.A., McEver, R.P., and Zhu, C. (2008) Monitoring receptor–ligand interactions between surfaces by thermal fluctuations. *Biophysical Journal*, **94**, 694–701.

[86] Huang, J., Zarnitsyna, V.I., Liu, B. *et al.* (2010) The kinetics of two-dimensional TCR and pMHC interactions determine T-cell responsiveness. *Nature*, **464**, 932–936.

[87] Radmacher, M., Fritz, M., Kacher, C.M. *et al.* (1996) Measuring the viscoelastic properties of human platelets with the atomic force microscope. *Biophysical Journal*, **70**, 556–567.

[88] Rosenbluth, M.J., Lam, W.A., and Fletcher, D.A. (2006) Force microscopy of nonadherent cells: a comparison of leukemia cell deformability. *Biophysical Journal*, **90**, 2994–3003.

[89] Lévy, R. and Maaloum, M. (2002) Measuring the spring constant of atomic force microscope cantilevers: thermal fluctuations and other methods. *Nanotechnology*, **13**, 33.

[90] Radmacher, M. (2002) Measuring the elastic properties of living cells by AFM, in *Atomic Force Microscopy in Cell Biology* (eds. B.P. Jena and H.J.K. Hörber), *Methods in Cell Biology*, vol. **68**, Academic Press, pp. 67–90.

[91] Ng, L., Hung H.,H., Sprunt A. *et al.* (2007) Nanomechanical properties of individual chondrocytes and their developing growth factor-stimulated pericellular matrix. *British Journal of Haematology*, **40**, 1011–1023.

[92] Lam, W.A., Rosenbluth, M.J., and Fletcher, D.A. (2008) Increased leukaemia cell stiffness is associated with symptoms of leucostasis in paediatric acute lymphoblastic leukaemia. *British Journal of Haematology*, **142**, 497–501, doi: 10.1111/j.1365-2141.2008.07219.x.

[93] Chaudhuri, O., Parekh, S.H., Lam, W.A., and Fletcher, D.A. (2009) Combined atomic force microscopy and side-view optical imaging for mechanical studies of cells. *Nature Methods*, **6**, 383–387.

[94] Cross, S.E., Jin, Y.-S., Rao, J., and Gimzewski, J.K. (2007) Nanomechanical analysis of cells from cancer patients. *Nature Nanotechnology*, **2**, 780–783.

[95] Cross, S.E., Jin, Y.-S., Tondre, J. *et al.* (2008) AFM-based analysis of human metastatic cancer cells. *Nanotechnology*, **19**, 384003.

[96] Thoumine, O., Ott, A., Cardoso, O., and Meister, J.-J. (1999) Microplates: a new tool for manipulation and mechanical perturbation of individual cells. *Journal of Biochemical and Biophysical Methods*, **39**, 47–62, doi: 10.1016/s0165-022x(98)00052-9.

[97] Desprat, N., Guiroy, A., and Asnacios, A. (2006) Microplates-based rheometer for a single living cell. *Review of Scientific Instruments*, **77**, 055111.

[98] Ashkin, A. (1970) Acceleration and trapping of particles by radiation pressure. *Physical Review Letters*, **24**, 156.

[99] Ashkin, A., Dziedzic, J.M., Bjorkholm, J.E., and Chu, S. (1986) Observation of a single-beam gradient force optical trap for dielectric particles. *Optics Letters*, **11**, 288–290.

[100] Ashkin, A. and Dziedzic, J. (1987) Optical trapping and manipulation of viruses and bacteria. *Science*, **235**, 1517–1520, doi: 10.1126/science.3547653.

[101] Bustamante, C., Bryant, Z., and Smith, S.B. (2003) Ten years of tension: single-molecule DNA mechanics. *Nature*, **421**, 423–427.

[102] Mehta, A.D., Rief, M., Spudich, J.A. *et al.* (1999) Single-molecule biomechanics with optical methods. *Science*, **283**, 1689–1695, doi: 10.1126/science.283.5408.1689.

[103] Kim, J., Zhang, C.-Z., Zhang, X., and Springer, T.A. (2010) A mechanically stabilized receptor–ligand flex-bond important in the vasculature. *Nature*, **466**, 992–995.

[104] Jakobi, A.J., Mashaghi, A., Tans, S.J., and Huizinga, E.G. (2011) Calcium modulates force sensing by the von Willebrand factor A2 domain. *Nature Communications*, **2**, 385.

[105] Mills, J.P., Qie, L., Dao, M. *et al.* (2004) Nonlinear elastic and viscoelastic deformation of the human red blood cell with optical tweezers. *Mechanics and Chemistry of Biosystems*, **1**, 169–180.

[106] Dao, M., Lim, C.T., and Suresh, S. (2003) Mechanics of the human red blood cell deformed by optical tweezers. *Journal of the Mechanics and Physics of Solids*, **51**, 2259–2280, doi: 10.1016/j.jmps.2003.09.019.

[107] Moffitt, J.R., Chemla, Y.R., Smith, S.B., and Bustamante, C. (2008) Recent advances in optical tweezers. *Annual Review of Biochemistry*, **77**, 205–228, doi: 10.1146/annurev.biochem.77.043007.090225.

[108] Henon, S., Lenormand, G., Richert, A., and Gallet, F. (1999) A new determination of the shear modulus of the human erythrocyte membrane using optical tweezers. *Biophysical Journal*, **76**, 1145–1151, doi: 10.1016/S0006-3495(99)77279-6.

[109] Lim, C.T., Dao, M., Suresh, S. *et al.* (2004) Large deformation of living cells using laser traps. *Acta Materialia*, **52**, 1837–1845, doi: 10.1016/j.actamat.2003.12.028.

[110] Yang, B.W., Mu, Y.H., Huang, K.T. *et al.* (2010) The evaluation of interaction between red blood cells in blood coagulation by optical tweezers. *Blood Coagulation and Fibrinolysis*, **21**, 505–510, doi: 10.1097/MBC.0b013e328339cc5d.

[111] Tam, J.M., Castro, C.E., Heath, R.J.W. *et al.* (2010) Control and manipulation of pathogens with an optical trap for live cell imaging of intercellular interactions. *PLoS ONE*, **5**, e15215, doi: 10.1371/journal.pone.0015215.

[112] Dharmadhikari, J., Roy, S., Dharmadhikari, A. *et al.* (2004) Torque-generating malaria-infected red blood cells in an optical trap. *Optics Express*, **12**, 1179–1184.

[113] Brandão, M.M., Saad, S.T., Cezar, C.L. *et al.* (2003) Elastic properties of stored red blood cells from sickle trait donor units. *Vox Sanguinis*, **85**, 213–215.

[114] Brandão, M.M., Fontes, A., Barjas-Castro, M.L. *et al.* (2003) Optical tweezers for measuring red blood cell elasticity: application to the study of drug response in sickle cell disease. *European Journal of Haematology*, **70**, 207–211.

[115] Guck, J., Ananthakrishnan, R., Mahmood, H. *et al.* (2001) The optical stretcher: a novel laser tool to micromanipulate cells. *Biophysical Journal*, **81**, 767–784.

[116] Lincoln, B., Wottawah, F., Schinkinger, S. *et al.* (2007) High-throughput rheological measurements with an optical stretcher, in *Cell Mechanics* (eds. Y.-L. Wang and D.E. Discher), *Methods in Cell Biology*, vol. **83**, Academic Press, pp. 397–423.

[117] Guck, J., Ananthakrishnan, R., Moon, T.J. *et al.* (2000) Optical deformability of soft biological dielectrics. *Physical Review Letters*, **84**, 5451.

[118] Guck, J., Schinkinger, S., Lincoln, B. *et al.* (2005) Optical deformability as an inherent cell marker for testing malignant transformation and metastatic competence. *Biophysical Journal*, **88**, 3689–3698.

[119] Lincoln, B., Erickson, H.M., Schinkinger, S. *et al.* (2004) Deformability-based flow cytometry. *Cytometry Part A*, **59**, 203–209.

[120] Lincoln, B., Schinkinger, S., Travis, K. *et al.* (2007) Reconfigurable microfluidic integration of a dual-beam laser trap with biomedical applications. *Biomedical Microdevices*, **9**, 703–710, doi: 10.1007/s10544-007-9079-x.

[121] Bellini, N., Vishnubhatla, K.C., Bragheri, F *et al.* (2010) Femtosecond laser fabricated monolithic chip for optical trapping and stretching of single cells. *Optics Express*, **18**, 4679–4688.

[122] Lautenschläger, F., Paschke, S., Schinkinger, S. *et al.* (2009) The regulatory role of cell mechanics for migration of differentiating myeloid cells. *Proceedings of the National Academy of Sciences of the United States of America*, **106**, 15696–15701, doi: 10.1073/pnas.0811261106.

[123] Wottawah, F., Schinkinger, S., Lincoln, B. *et al.* (2005) Optical rheology of biological cells. *Physical Review Letters*, **94**, 098103.

[124] Wottawah, F., Schinkinger, S., Lincoln, B. *et al.* (2005) Characterizing single suspended cells by optorheology. *Acta Biomaterialia*, **1**, 263–271, doi: 10.1016/j.actbio.2005.02.010.

[125] Sraj, I. , Eggleton, C.D., Jimenez, R. *et al*. (2010) Cell deformation cytometry using diode-bar optical stretchers. *Journal of Biomedical Optics*, **15**, 047010.

[126] Gardel, M.L., Sabass, B., Ji, L. *et al*. (2008) Traction stress in focal adhesions correlates biphasically with actin retrograde flow speed. *Journal of Cell Biology*, **183**, 999–1005, doi: 10.1083/jcb.200810060.

[127] Sabass, B., Gardel, M.L., Waterman, C.M., and Schwarz, U.S. (2008) High resolution traction force microscopy based on experimental and computational advances. *Biophysical Journal*, **94**, 207–220, doi: 10.1529/biophysj.107.113670.

[128] Legant, W.R., Miller, J.S., Blakely, B.L. *et al*. (2010) Measurement of mechanical tractions exerted by cells in three-dimensional matrices. *Nature Methods*, **7**, 969–971, doi: 10.1038/nmeth.1531.

[129] Franck, C., Maskarinec, S.A., Tirrell, D.A., and Ravichandran, G. (2011) Three-dimensional traction force microscopy: a new tool for quantifying cell-matrix interactions. *PLoS ONE*, **6**, e17833, doi: 10.1371/journal.pone.0017833.

[130] Tan, J.L., Tien, J., Pirone, D.M. *et al*. (2003) Cells lying on a bed of microneedles: an approach to isolate mechanical force. *Proceedings of the National Academy of Sciences of the United States of America*, **100**, 1484–1489, doi: 10.1073/pnas.0235407100.

[131] Saez, A., Ghibaudo, M., Buguin, A. *et al*. (2007) Rigidity-driven growth and migration of epithelial cells on microstructured anisotropic substrates. *Proceedings of the National Academy of Sciences of the United States of America*, **104**, 8281–8286, doi: 10.1073/pnas.0702259104.

[132] Sochol, R.D., Higa, A.T., Janairo, R.R.R. *et al*. (2011) Unidirectional mechanical cellular stimuli via micropost array gradients. *Soft Matter*, **7**, 4606–4609.

[133] Sniadecki, N.J. and Chen, C.S. (2007) Microfabricated silicone elastomeric post arrays for measuring traction forces of adherent cells, in *Cell Mechanics* (eds. Y.-L. Wang and D.E. Discher), *Methods in Cell Biology*, vol. 83, Academic Press, pp. 313–328.

[134] Rogers, J.A. and Nuzzo, R.G. (2005) Recent progress in soft lithography. *Materials Today*, **8**, 50–56, doi: 10.1016/s1369-7021(05)00702-9.

[135] Xia, Y. and Whitesides, G.M. (1998) Soft lithography. *Annual Review of Materials Science*, **28**, 153–184, doi: 10.1146/annurev.matsci.28.1.153.

[136] Sniadecki, N.J., Anguelouch, A., Yang, M.T. *et al*. (2007) Magnetic microposts as an approach to apply forces to living cells. *Proceedings of the National Academy of Sciences of the United States of America*, **104**, 14553–14558, doi: 10.1073/pnas.0611613104.

[137] Yang, Z., Lin, J.S., Chen, J., and Wang, J.H. (2006) Determining substrate displacement and cell traction fields--a new approach. *Journal of Theoretical Biology*, **242**, 607–616, doi: 10.1016/j.jtbi.2006.05.005.

[138] Stricker, J., Sabass, B., Schwarz, U.S., and Gardel, M.L. (2010) Optimization of traction force microscopy for micron-sized focal adhesions. *Journal of Physics: Condensed Matter*, **22**, 194104, doi: 10.1088/0953-8984/22/19/194104.

[139] Huang, J., Peng, X., Qin, L. *et al*. (2009) Determination of cellular tractions on elastic substrate based on an integral Boussinesq solution. *Journal of Biomechanical Engineering*, **131**, 061009, doi: 10.1115/1.3118767.

[140] Kurpinski, K., Park, J., Thakar, R.G., and Li, S. (2006) Regulation of vascular smooth muscle cells and mesenchymal stem cells by mechanical strain. *Molecular and Cellular Biomechanics*, **3**, 21–34.

[141] Kurpinski, K. and Li, S. (2007) Mechanical stimulation of stem cells using cyclic uniaxial strain. *Journal of Visualized Experiments*, **2007** (6), e242.

[142] Ellis, E.F., McKinney, J.S., Willoughby, K.A. *et al*. (1995) A new model for rapid stretch-induced injury of cells in culture: characterization of the model using astrocytes. *Journal of Neurotrauma*, **12**, 325–339.

[143] Wang, J.H., Goldschmidt-Clermont, P., and Yin, F.C. (2000) Contractility affects stress fiber remodeling and reorientation of endothelial cells subjected to cyclic mechanical stretching. *Annals of Biomedical Engineering*, **28**, 1165–1171.

[144] Puig-de-Morales, M., Millet, E., Fabry, B. *et al*. (2004) Cytoskeletal mechanics in adherent human airway smooth muscle cells: probe specificity and scaling of protein–protein dynamics. *American Journal of Physiology: Cell Physiology*, **287**, C643–C654, doi: 10.1152/ajpcell.00070.2004.

[145] Bursac, P., Lenormand, G., Fabry, B. *et al*. (2005) Cytoskeletal remodelling and slow dynamics in the living cell. *Nature Materials*, **4**, 557–561.

[146] Chen, J., Fabry, B., Schiffrin, E.L., and Wang, N. (2001) Twisting integrin receptors increases endothelin-1 gene expression in endothelial cells. *American Journal of Physiology: Cell Physiology*, **280**, C1475–C1484.

[147] Nagrath, S., Sequist, L.V., Maheswaran, S. *et al.* (2007) Isolation of rare circulating tumour cells in cancer patients by microchip technology. *Nature*, **450**, 1235–1239.

[148] Schirhagl, R., Fuereder, I., Hall, E.W. *et al.* (2011) Microfluidic purification and analysis of hematopoietic stem cells from bone marrow. *Lab on a Chip*, **11**, 3130–3135.

[149] Hou, H., Li, Q.S., Lee, G.Y. *et al.* (2009) Deformability study of breast cancer cells using microfluidics. *Biomedical Microdevices*, **11**, 557–564, doi: 10.1007/s10544-008-9262-8.

[150] Rosenbluth, M.J., Lam, W.A., and Fletcher, D.A. (2008) Analyzing cell mechanics in hematologic diseases with microfluidic biophysical flow cytometry. *Lab on a Chip*, **8**, 1062–1070.

[151] Kiesewetter, H., Dauer, U., Teitel, P., Schmid-Schonbein, H., and Trapp, R. (1982) The single erythrocyte rigidometer (SER) as a reference for RBC deformability. *Biorheology*, **19**, 737–753.

[152] Seiffge, D. (1984) Dependency of red blood cell passage time on pore geometry in the single-pore erythrocyte rigidometer (SER). *Biorheology. Supplement: The Official Journal of the International Society of Biorheology*, **1**, 245–247.

[153] Frank, R.S. and Hochmuth, R.M. (1987) An investigation of particle flow through capillary models with the resistive pulse technique. *Journal of Biomechanical Engineering*, **109**, 103–109, doi: 10.1115/1.3138650.

[154] Frank, R.S. and Tsai, M.A. (1990) The behavior of human neutrophils during flow through capillary pores. *Journal of Biomechanical Engineering*, **112**, 277–282, doi: 10.1115/1.2891185.

[155] Frank, R. (1990) Time-dependent alterations in the deformability of human neutrophils in response to chemotactic activation. *Blood*, **76**, 2606–2612.

[156] Di Carlo, D., Irimia, D., Tompkins, R.G., and Toner, M. (2007) Continuous inertial focusing, ordering, and separation of particles in microchannels. *Proceedings of the National Academy of Sciences of the United States of America*, **104**, 18892–18897, doi: 10.1073/pnas.0704958104.

[157] Hur, S., Henderson-MacLennan, N., McCabe, E., and Di Carlo, D. (2011) Deformability-based cell classification and enrichment using inertial microfluidics. *Lab on a Chip*, **11**, 912–932, doi:10.1039/c0lc00595a.

[158] Mach, A. and Di Carlo, D. (2010) Continuous scalable blood filtration device using inertial microfluidics. *Biotechnology and Bioengineering*, **107**, 302–313, doi: 10.1002/bit.22833.

[159] Choi, S. and Park, J.-K. (2007) Continuous hydrophoretic separation and sizing of microparticles using slanted obstacles in a microchannel. *Lab on a Chip*, **7**, 890–897, doi: 10.1039/b701227f.

[160] Shevkoplyas, S.S., Gifford, S.C., Yoshida, T., and Bitensky, M.W. (2003) Prototype of an *in vitro* model of the microcirculation. *Microvascular Research*, **65**, 132–136, doi: 10.1016/s0026-2862(02)00034-1.

[161] Shevkoplyas, S.S., Yoshida, T., Gifford, S.C., and Bitensky, M.W. (2006) Direct measurement of the impact of impaired erythrocyte deformability on microvascular network perfusion in a microfluidic device. *Lab on a Chip*, **6**, 914–920.

[162] Lam, W.A., Rosenbluth, M.J., and Fletcher, D.A. (2007) Chemotherapy exposure increases leukemia cell stiffness. *Blood*, **109**, 3505–3508, doi: 10.1182/blood-2006-08-043570.

[163] Shelby, J.P., White, J., Ganesan, K. *et al.* (2003) A microfluidic model for single-cell capillary obstruction by *Plasmodium falciparum*-infected erythrocytes. *Proceedings of the National Academy of Sciences of the United States of America*, **100**, 14618–14622, doi: 10.1073/pnas.2433968100.

[164] Nolan, G.P. (2006) Deeper insights into hematological oncology disorders via single-cell phospho-signaling analysis. *Hematology: The American Society of Hematology Education Program*, 123–127, 509.

[165] Davey, H.M. and Kell, D.B. (1996) Flow cytometry and cell sorting of heterogeneous microbial populations: the importance of single-cell analyses. *Microbiological Reviews*, **60**, 641–696.

[166] Irish, J.M., Kotecha, N., and Nolan, G.P. (2006) Mapping normal and cancer cell signalling networks: towards single-cell proteomics. *Nature Reviews. Cancer*, **6**, 146–155.

[167] Aprelev, A., Weng, W., Zakharov, M. *et al.* (2007) Metastable polymerization of sickle hemoglobin in droplets. *Journal of Molecular Biology*, **369**, 1170–1174.

[168] Bagge, U., Amundson, B., and Lauritzen, C. (1980) White blood cell deformability and plugging of skeletal muscle capillaries in hemorrhagic shock. *Acta Physiologica Scandinavica*, **108**, 159–163.

[169] Bow, H., Pivkin, I.V., Diez-Silva, M. *et al.* (2011) A microfabricated deformability-based flow cytometer with application to malaria. *Lab on a Chip*, **11**, 1065–1073.

[170] Mauritz, J.M.A., Tiffert, T., Seear, R. *et al.* (2010) Detection of *Plasmodium falciparum*-infected red blood cells by optical stretching. *Journal of Biomedical Optics*, **15**, 030517.

[171] Fedosov, D.A., Caswell, B., Suresh, S., and Karniadakis, G.E. (2011) Quantifying the biophysical characteristics of *Plasmodium-falciparum*-parasitized red blood cells in microcirculation. *Proceedings of the National Academy of Sciences of the United States of America*, **108**, 35–39, doi: 10.1073/pnas.1009492108.

[172] Wayman, A.M., Chen, W., McEver, R.P., and Zhu, C. (2010) Triphasic force dependence of E-selectin/ligand dissociation governs cell rolling under flow. *Biophysical Journal*, **99**, 1166–1174, doi: 10.1016/j.bpj.2010.05.040.

[173] Gupta, V.K., Sraj, I.A., Konstantopoulos, K., and Eggleton, C.D. (2010) Multi-scale simulation of L-selectin-PSGL-1-dependent homotypic leukocyte binding and rupture. *Biomechanics and Modeling in Mechanobiology*, **9**, 613–627, doi: 10.1007/s10237-010-0201-2.

[174] Guevorkian, K., Gonzalez-Rodriguez, D., Carlier, C. *et al.* (2011) Mechanosensitive shivering of model tissues under controlled aspiration. *Proceedings of the National Academy of Sciences of the United States of America*, **108**, 13387–13392, doi: 10.1073/pnas.1105741108.

[175] Laplaca, M.C. and Prado, G.R. (2010) Neural mechanobiology and neuronal vulnerability to traumatic loading. *Journal of Biomechanics*, **43**, 71–78, doi: 10.1016/j.jbiomech.2009.09.011.

[176] Barabino, G.A., Platt, M.O., and Kaul, D.K. (2010) Sickle cell biomechanics. *Annual Review of Biomedical Engineering*, **12**, 345–367, doi: 10.1146/annurev-bioeng-070909-105339.

[177] Li, D., Zhou, J., Chowdhury, F. *et al.* (2011) Role of mechanical factors in fate decisions of stem cells. *Regenerative Medicine*, **6**, 229–240, doi: 10.2217/rme.11.2.

[178] Tenney, R.M. and Discher, D.E. (2009) Stem cells, microenvironment mechanics, and growth factor activation. *Current Opinion in Cell Biology*, **21**, 630–635, doi: 10.1016/j.ceb.2009.06.003.

11

Microfabricated Technologies for Cell Mechanics Studies

Sri Ram K. Vedula[1], Man C. Leong[2], and Chwee T. Lim[1,3]
[1]*National University of Singapore, Singapore*
[2]*NUS Graduate School for Integrative Sciences, Singapore and*
[3]*Department of Bioengineering, Singapore*

11.1 Introduction

The area of cell mechanics refers to how the mechanical responses and properties of cells and the microenvironment within which they exist (e.g., extracellular matrix (ECM)) are altered and/or regulated under various physiological and pathological conditions. Back in the nineteenth century, Julius Wolff described the active remodeling of bone in response to alteration in mechanical loading, suggesting an intricate relationship between mechanical forces and tissue function. At the cellular level, altered mechanical properties are manifested in cells during disease progression. In diseases such as malaria and cancer, cell size, shape, and deformability change due to changes in both composition and organization of the subcellular components.

Different micromechanical and physical cues in the microenvironment (e.g., geometrical patterns, topography, and substrate stiffness) can directly influence and regulate various biological processes. For example, the cells which are spatially confined by their physical environment and are rounded have a tendency to undergo cell death, whereas cell growth is associated with well-spread cells [1]. Spatial confinements imposed by the microenvironment have also been reported to affect the shape [2], proliferation [3], and differentiation [4] of mesenchymal stem cells (MSCs); adipogenesis is associated with round and less-spread stem cells, whereas stiff extracellular environment promotes cell spreading and osteogenesis [5].

Cells are in constant interaction with their mechanical environment. They have the ability to sense, react, and generate forces in their respective tissue context. For example, under high physiological shear stress as seen in laminar flows, endothelial cells exhibit

Nano and Cell Mechanics: Fundamentals and Frontiers, First Edition. Edited by Horacio D. Espinosa and Gang Bao.
© 2013 John Wiley & Sons, Ltd. Published 2013 by John Wiley & Sons, Ltd.

preferred orientation toward the direction of fluid flow, have low proliferation, and are less susceptible to cell death [6]. In contrast, endothelial cells under nonlaminar flow become randomly aligned when exposed to shear stress of $1.5\,\mathrm{dyn\,cm^{-2}}$ ($1\,\mathrm{dyn} = 10\,\mu\mathrm{N}$) for 16 h, with a considerable increase in DNA synthesis [7]. Understanding how these extracellular mechanical or physical cues regulate cellular functions requires techniques which allow us to manipulate and characterize these micromechanical factors *in vitro*. For example, the use of various microfabrication tools has substantially helped us to characterize in great detail the role of cell mechanics in regulating a number of important biological processes, such as cell adhesion, migration, spreading, division, and differentiation to name a few. On the other hand, the use of microfluidic devices allows us to mimic the application of shear stresses to cells in their native environment. In this chapter, we review three important applications of microfabricated technologies to understand cell mechanics: micropatterned substrates, micropillars, and microfluidic devices.

11.2 Microfabrication Techniques

Microfabrication technology originates from the semiconductor industry, and was traditionally used to generate miniature structures on different substrates, such as the integrated chips in various electronics. In more recent years, this technology gained popularity in biology as there has been a need to engineer substrates with precise geometry or topography. Many studies now focus on the interaction of cells with one another or their extracellular micro-environment, and they require well-controlled environments to characterize and stimulate cells. In this section, we will highlight commonly used microfabrication techniques which allow facile and repeatable design of substrate for the study of cell mechanics. Choice of the technique depends on the type of pattern, size of pattern, and availability of equipment, among others. For example, soft lithography offers a low-cost solution in producing patterns with up to 100 nm resolution [8] but is still less preferred for producing very closely spaced features.

11.2.1 Photolithography and Soft Lithography

Photolithography is a process in which patterns on a photomask are transferred to a substrate coated with a thin layer of photoresist utilizing light (typically ultraviolet (UV) light) [9]. Photomasks are usually made of fused silica or quartz material and the pattern itself is obtained by depositing chromium metal on the photomask. Photoresists are light-sensitive compounds that undergo changes in their molecular structures when exposed to light of specific wavelengths. They are spin coated on substrates (e.g., SiO_2 films on silicon wafers) to form layers that can vary from tens of micrometers to several millimeters in thickness. There are two different types of photoresists: positive and negative. Positive photoresists become soluble on exposure to light and can be washed off with developer solution in a subsequent step, while negative photoresists work in the opposite way. UV light exposure causes polymerization of the negative photoresist, after which it becomes insoluble. One of the most important steps in photolithography is the alignment of the photomask. The mask needs to be aligned to and in physical contact with the wafer before the UV exposure. The patterns created from the insoluble photoresist can be used as such or further etched using various etching agents (Figure 11.1).

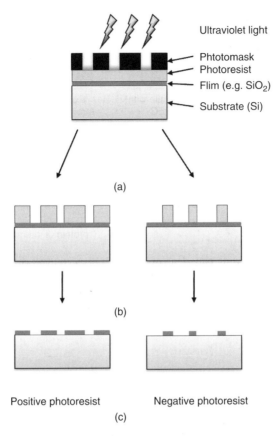

(a)

(b)

Positive photoresist   Negative photoresist

(c)

Figure 11.1 A schematic showing various steps in photolithography process: (a) exposure to UV light renders the exposed photoresist soluble or insoluble depending on whether it is a positive or negative photoresist; (b) removal of soluble photoresist using a developer solution; (c) etching and stripping of photoresist to finally expose the film

Alternatively, the same technique can be applied directly to generate patterns on substrates. Glass substrates coated with poly-L-lysine–polyethylene glycol (PLL–PEG) can be directly patterned by exposing them to UV light through a mask. UV exposure causes cleavage of the PEG chains of the coating. As a result, the regions exposed to UV lose the PLL–PEG coating and are able to bind proteins (e.g., fibronectin; Figure 11.2) [10].

Soft lithography is an inexpensive and effective method for generating patterns using elastomeric stamps (prepared from polydimethylsiloxane, PDMS). PDMS is a type of silicon polymer, widely used in biomedical applications and soft lithography. It is viscous when uncured but is suitable for replicating features up to hundreds of nanometers. PDMS stamps are usually molded from a "master" template containing the desired patterns (Figure 11.3). The masters can be prepared using photolithography, electron-beam etching, or micromachining. Alternatively, to prevent breakage or wearing out of the master silicon mold due to repeated usage, some prefer to work with PDMS replicas of

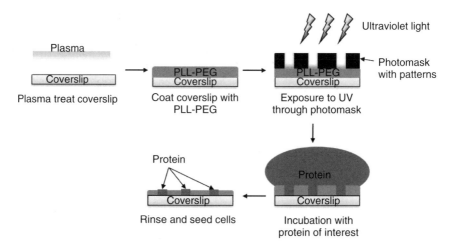

Figure 11.2   Schematic showing patterning of PLL–PEG-coated glass coverslips using UV treatment. The surface of the glass coverslip is activated by plasma treatment, followed by incubation with PLL–PEG. The photomask is brought into physical contact with the coverslip, after which the coverslip is irradiated through the photomask. Protein of interest is added to the surface of the coverslip and the region exposed to the UV light would be coated with the protein

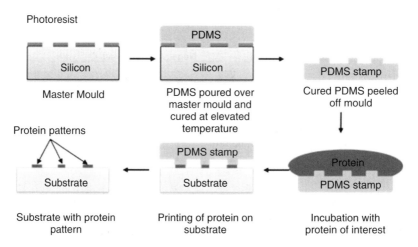

Figure 11.3   Schematic of microcontact printing process. PDMS stamps are casted from the silicon masters containing the desired patterns. Fully cured stamps are then inked with the protein of interest and are brought into contact with the substrate of interest (e.g., petri dish or glass). Unstamped regions are backfilled with Pluronic or PLL–PEG to render them cell repulsive

the master mold. The PDMS mold can be created from the PDMS stamps by reverse molding.

The patterns existing on the stamp can then be transferred to the desired substrate (plastic or glass) using different methods like microcontact printing (μCP) or microtransfer molding (μTM), etc. [11]. In the case of microcontact printing, the surface of elastomeric stamps containing the features is coated with an ink of interest and then dried appropriately before the stamp is brought into contact with the substrate. The stamps are subjected to mechanical stresses such as thermal expansion/contraction of the stamp (during curing and cooling of the elastomer), capillary forces (during the drying of the ink), and stress during the imprinting, and this could limit the size or shape of the features to be imprinted. For example, submicrometer or high aspect-ratio features on the stamp will easily collapse or buckle during the drying of the ink or during imprinting, resulting in poor transfer of the desired patterns [12,13].

More recently, by combining with nanolithography, it has been applied to generate arrays of discrete protein-binding regions. The precise control of spacing and density of the arrays can provide significant insights into how spatial distribution of ligands can affect the behavior of cells [14].

11.2.2   Microphotopatterning (μPP)

Microphotopatterning is a fairly recent technique developed to generate different patterns on glass substrates. This method utilizes local photo ablation of poly(vinyl) alcohol (PVA) thin films deposited on glass using a two-photon excitation guided through a point-scanning confocal microscope to generate the desired patterns [15]. The PVA film prevents protein adsorption, while the photo-ablated region pattern is protein adhesive. There are two main advantages of this method: (a) it does not require physical stamps for pattern generation and (b) it allows the patterning of different proteins of submicrometer resolution very close to one another (Figure 11.4).

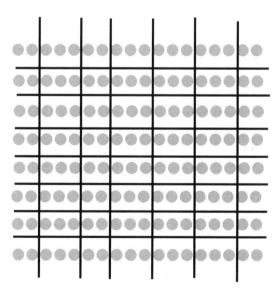

Figure 11.4   Two different proteins (fibrinogen (lines) and vitronectin (dots)) are patterned using μPP [15]

11.3 Applications to Cell Mechanics

A wide range of cellular functions and processes are influenced by mechanical cues in the immediate environment of the cells. These include rigidity of the ECM, fluid shear flow, and patterns and gradients of proteins in the ECM (haptotaxis), among others. In this section we describe how microfabricated technologies have been applied to study various cellular processes.

11.3.1 Micropatterned Substrates

Micropatterned substrates provide a very simple way of altering and regulating cell shape, spreading and migration in a controlled and systematic manner. These substrates can be obtained using either microcontact printing or photo patterning techniques. Micropatterned substrates have been used to investigate the role of cell mechanics in regulating several cellular processes, as described below.

Cell Differentiation

The ability to control and understand differentiation of stem cells into cells of different lineages is currently one of the most important areas of research because of its obvious impact on regenerative medicine. While several studies have established the roles of various chemicals and growth factors as means to induce differentiation of stem cells into specific cell types, recent studies suggest that altering the biomechanical cues (e.g., matrix elasticity, seeding density) can also be used to control differentiation [16]. Since any significant change in the density of the cells also changes the shape and spreading area of the cells as well as intercellular contact (which can alter downstream signaling pathways), it is difficult to separate the individual contribution of each of these cues using traditional experiments. However, using microcontact printing, it is possible to restrict the cells to a predetermined shape and size by varying the patterns on the stamps. Furthermore, such a system is high throughput, since it allows thousands of single cells to be analyzed for differentiation. It has been observed that the cell shape and spreading area are indeed independent biomechanical cues that regulate the differentiation of human mesenchymal stem cells (hMSCs). hMSCs seeded on small islands of fibronectin (area $\sim 1024\,\mu m^2$) surrounded by a nonadhesive area spread very little and attained a round shape. A large percentage of these cells stained positive for lipid stain but almost no cells stained positive for alkaline phosphatase (marker for bone differentiation), suggesting that they had differentiated specifically into adipocytes. On the other hand, hMSCs that were seeded on large fibronectin islands (area $\sim 10\,000\,\mu m^2$) spread out well to completely occupy the whole fibronectin region. A large number of these well-spread cells stained positive for alkaline phosphatase but not lipid, showing specific differentiation of hMSCs into osteoblasts [17]. Cells seeded on fibronectin islands of intermediate area showed a mixed population of adipocytes and osteoblasts. Furthermore, it was also shown that hMSCs on large fibronectin islands differentiated into adipocytes in the presence of ROCK inhibitor Y-27632, suggesting that RhoA-mediated increase in intracellular acto-myosin tension was involved in regulating differentiation of hMSCs in response to altered spreading.

Recent studies using microcontact printing of various geometrical patterns (e.g., rectangle with different aspect ratios or pentagons with varying curvature of the edges) have shown that patterns that provide local shape cues (e.g., sharp curvatures or pointed features) increase intracellular acto-myosin tension and regulate the differentiation behavior of MSCs [18]. Accordingly, cells grown on rectangular patterns with high aspect ratio (4 : 1) showed a higher percentage of bone cells, while those grown on a square showed almost equal number of bone and fat cells even though the actual area for both the patterns was the same. Similarly, patterns of pentagons also altered the percentage of adipocytes and osteoblasts depending on the presence of sharp local features. Hence, pentagon patterns showing gentle curvature ("flower" pattern) showed a higher number of adipocytes, while those showing sharp vertices ("star" pattern) showed a higher number of osteoblasts even though both patterns had the same total area. Furthermore, it was found that the difference in the differentiation of the stem cells on such patterns was mediated by altered intracellular tension mediated by acto-myosin contraction. Cells treated with cytochalasin D or blebbistatin (both of which inhibit acto-myosin contractility) showed a tendency to differentiate into fat cells even on "star" patterns, while those treated with nocodazole (inhibitor of microtubules that also causes increased intracellular tension) showed a tendency to form osteoblasts on "flower" patterns.

Cell Division and Polarization

The axis along which a cell divides determines the position and fate of the daughter cells; hence, it is important to understand how biomechanical cues imposed by ECM proteins regulate it. Microcontact printed patterns of fibronectin provide a simple way to regulate cell shape and observe how this influences the cell division axis (Figure 11.5). It was observed that, on high aspect-ratio rectangles, cells had a tendency to divide perpendicular to the long axis of the rectangle, while on circular patterns there was no preference to divide along any particular axis. Interestingly, on triangular patterns and "L"-shaped patterns, while the cell shapes were similar, the axis of division was much more constrained and sharply distributed parallel to the hypotenuse. This suggests that while cell shape played an important role in determining the division axis, it is not the only parameter regulating it. Further analysis using nocodazole wash-out experiments on these patterns strongly suggests that the spatial distribution of ECM determines cortical actin

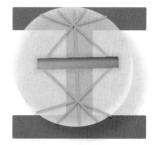

Figure 11.5 Cartoon showing the use of microcontact-printed ECM protein patterns (red) to study their influence on cell division axis [19]. Nucleus and spindle fibers are depicted in blue and green, respectively

dynamics and heterogeneity, which in turn regulates the axis of cell division [19,20]. Disruption of the cortical heterogeneity using tyrosine kinase inhibitors led to spindle mis-positioning.

Polarization of cell-motility machinery determines the direction of cell migration and, hence, dictates various processes such as wound healing and morphogenesis. Polarization is associated with characteristic redistribution of intracellular organelles. For example, the Golgi apparatus and centrosomes relocate to the front end, while the nucleus and stress fibers are present at the rear end. The formation of new cell–ECM adhesions in the protruding lamellipodia of cells at the edge of a wound has been proposed to be one of the major driving forces in establishing polarization. Conventional scratch wound assays, however, do not provide a suitable experimental model to delineate the role of cell–cell adhesion in regulating the migratory polarization of cells. Since microcontact printed patterns provide an experimental setup that allows us to systematically regulate the extent of intercellular adhesion formation, prevent cells from forming new cell–ECM adhesions and inhibit cell migration, it has been possible to investigate the exclusive role of intercellular adhesion in inducing cell polarization. Using these patterns, it has been observed that initiation of intercellular adhesion mediated by E-cadherins can induce cell polarization [21]. On such patterns, the nucleus was located near the cell–cell contact, while the centrosome was located away from the adhesion. Significantly higher ruffling activity was also observed away from the intercellular contact. When the constraints imposed by the patterns were removed by electrodesorption, polarized cells migrated away from each other, suggesting that polarization can in turn regulate cell migration [22]. Furthermore, E-cadherin-adhesion-mediated polarization required intact actin cytoskeleton as well as Cdc42 activity.

Cell Migration

Cell migration is influenced not only by chemical signals, but also by a variety of biomechanical cues in their microenvironment (e.g., topography, geometrical constraints, ECM protein distribution patterns). Microfabrication techniques have provided us with an opportunity to understand how these biomechanical cues regulate cell migration. For example, it has been observed that the Golgi apparatus is positioned behind the nucleus in cells migrating on microcontact-printed narrow fibronectin patterns [23]. Similarly, it has been shown using microphotopatterned substrates that migration of cells on very narrow patterns (one-dimensional migration) is faster than that on two-dimensional surfaces. Furthermore, the migration on such narrow channels has been observed to be much more dependent on myosin II than migration in two dimensions is [24]. Another interesting application of microcontact printing to study cell migration has been to use functionalized surfaces comprising alternating strips of collagen IV and adjustable concentrations of E-cadherin molecules [25]. It was observed that cells migrated on collagen IV but not on surfaces functionalized with only E-cadherin. Furthermore, the lamellipodial activity of the cells was also significantly altered on such patterned substrates, demonstrating a crosstalk between E-cadherin- and integrin-mediated adhesions.

Collective cell migration underlies several important biological processes, such as wound healing, gastrulation, and cancer metastasis. Scratch wound assay remains the most important *in vitro* assay to study collective cell migration. However, this method has several disadvantages, like disruption of the ECM coating on which the cells are migrating, difficulty in uncoupling the influence of biomechanical cues (e.g., providing a free surface) from other possible biochemical cues (release of chemicals from damaged and/or permeabilized cells), significant sample-to-sample variation depending on cell densities, difficulty in controlling the wound size, etc. Recently, a microfabrication technique has been applied to study wound healing under much more controlled conditions. In this method, a microfabricated soft elastic "microstencil" is first placed on a cell culture surface [26]. Once cells grow to confluence within the stencil, the stencil is lifted off, resulting in multiple model wounds for observation.

11.3.2 Micropillared Substrates

These are typically cast out of PDMS and consist of arrays of pillars whose stiffness can be controlled by altering their diameter, height, and curing conditions. The tips of these micropillars can be functionalized by stamping ECM proteins (most often fibronectin) on the top to allow the attachment and migration of cells [27].

Characterizing Cell Substrate Traction Forces

Micropillar substrates provide a very convenient way to characterize the traction forces exerted by cells on the substrate by monitoring the extent to which they deflect (Figure 11.6) [27,28]. They have been used extensively to study traction forces exerted

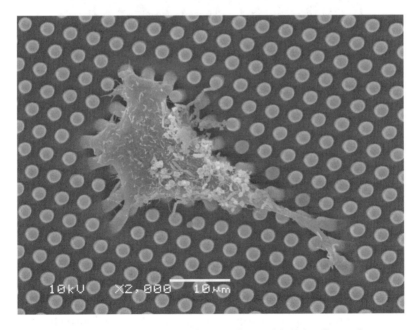

Figure 11.6 Scanning electron microscopy image of an epithelial cell growing on an array of micropillars and deflecting them

by epithelial cells sheets [29,30] as well as single migrating cells. In contrast to the polyacrylamide gels with embedded fluorescent beads or thin silicone rubber sheets traditionally used to study traction forces exerted by cells, micropillared substrates provide several distinct advantages. First, computation of the forces from the deflected pillars is relatively straightforward. Second, the deflection of a pillar is localized and not transmitted across to other pillars. Third, the deflection of the pillars is almost instantaneous, giving a much better temporal resolution. However, one of the biggest disadvantages of these micropillared substrates is that they direct and dictate the formation of the focal adhesions on top of the pillars, which could independently affect cell spreading and migration. Furthermore, it is necessary to make sure that the fibronectin is localized to the top of the pillar, else the cells can dip in between the pillars, leading to errors in estimation of the forces.

Intercellular Adhesion Forces

There are several methods to quantify and characterize the intercellular adhesion forces between cells in a suspended state (e.g., dual micropipette assay, atomic force microscopy) [31,32]. However, it is difficult to quantify the adhesion forces between two cells that remain anchored to the substrate. Recently, micropillared substrates have been successfully applied to address this question. Here, soft microfabricated pillars are first microcontact printed with "bowtie" patterns of fibronectin (Figure 11.7) [33]. Cells are subsequently

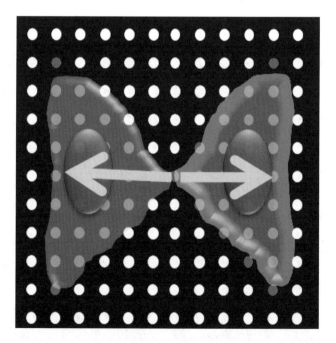

Figure 11.7   Two cells forming an intercellular adhesion on a bowtie pattern (red) on a micropillar array. The force of intercellular adhesion in this case is equal and opposite to the sum of the forces exerted by each single cell on the micropillars [33]

allowed to adhere and spread on these patterns and the cell substrate traction forces are determined from the deflections of the micropillars. Since cells exist in a state of quasi-static equilibrium (i.e., the sum of the traction forces exerted by a single cell is zero), the intercellular adhesion force between two cells is equal and opposite to the sum of the traction forces exerted by a single cell. Using this technique, it was shown that mechanical tugging force regulated the size of intercellular adhesion.

In an alternative application of micropillared substrates to study intercellular adhesion, the tops of micropillars were functionalized with purified recombinant N-cadherin molecules. Cells attached to these substrates and exerted significant traction forces [34]. Further experiments conducted using micropillars of varying stiffness functionalized with N-cadherins showed that cells spread better on stiffer pillars than on softer ones. Formation and maturation of the N-cadherin contacts was also better on stiffer substrates. Traction forces exerted by the cells on the micropillars through the N-cadherin contacts also increased with increase in the stiffness of the micropillars. Together, these results suggested that N-cadherin-mediated adhesions exhibit force-dependent strengthening that is characteristic of several other mechanosensor modules in the cell.

Substrates with Anisotropic Rigidity

As referred to in the previous sections, substrate rigidity has been shown to play a dominant role in regulating cytoskeleton reorganization and migration (durotaxis). Most of these studies, though, have been performed at macroscopic scales using polyacrylamide gels [35]. However, micropillars can be used to generate substrates with microscopic stiffness gradients. The underlying principle is to fabricate micropillars that are "oval" in cross-section. Such pillars show higher bending stiffness along the long axis than along the short axis, leading to anisotropic stiffness of the substrate at the microscopic scale. Islands of epithelial cells grown on these substrates showed a preferential overall orientation along the long axis of the pillars (i.e., along the stiffest direction). The actin cytoskeleton and focal adhesion also aligned along the long axis of the pillars [36]. Further experiments suggested that their orientation was not due to any contact guidance cues provided by the pillars, since cells did not show any preferred orientation on glass substrates stamped with the same patterns. Following scattering of the cells induced by hepatocyte growth factor, single cells migrating on the oval pillars showed a more directed migration (along the stiffest direction), while cells on pillars with circular cross-section (without any stiffness gradient) exhibited a random-walk migration. This clearly indicated that the cells were able to sense the microscopic anisotropic gradient of stiffness of the substrate.

Cell Migration in Three-Dimensional Micropatterns

It is well recognized that cell migration in three dimensions displays different characteristics than those observed in two dimensions. Micropillar substrates can also be used as an alternative system to study cell migration in a more three-dimensional environment. In this case, both the top and the whole length of the micropillars is covered by fibronectin. When micropillars are sparsely spaced (spacing varying from 4 to 12 μm), cells first adhere to the tops but progressively get impaled by the micropillars [37]. In contrast to stress fibers normally observed in cells grown on a two-dimensional substrate, cells adhering in between the micropillars formed thick actin bundles in the regions where they surrounded the micropillars. Furthermore, forces exerted by the cells could be measured by measuring the deflection of the micropillars.

Magnetic Micropillars for Local Force Application on Cells

The general understanding of the effect of mechanical forces on cells has traditionally been obtained by subjecting cells to macroscopic stretching. Methods to apply localized forces to cells at the microscopic scale include optical traps, magnetic microbeads, and micropipettes. Recently, micropillars have also been modified to apply localized forces to cells. The underlying principle here is to incorporate magnetic nanowires or magnetic aggregates inside PDMS or acrylamide micropillars [38,39]. Such micropillars can be deflected by the application of a strong external homogeneous or gradient magnetic field. Unlike other systems of localized force application, magnetic micropillars provide a platform for simultaneous force application and measurement.

Cell Differentiation

Apart from allowing us to quantify traction forces exerted by cells on the substrate, micropillared substrates also allow us to tune the apparent stiffness of the substrate sensed by the cells. The bending stiffness of the micropillars varies directly as the fourth power of its radius and inversely with the cube of its height, allowing us to explore a wide range of stiffness values. Furthermore, unlike the use of hydrogels and polyacrylamide gels, micropillared substrates can decouple the effect of stiffness from adhesive and surface properties of the substrate [40]. hMSCs grown on micropillared substrates of varying rigidities did not show any differentiation. However, after the addition of a bipotential medium (stimulating both adipogenesis and osteogenesis), hMSCs on the stiffer micropillars had a tendency to differentiate into osteocytes while those on soft micropillars predominantly differentiated into adipocytes. This suggested that, while the rigidity of the substrate by itself did regulate differentiation behavior of the stem cells, the balance of hMSC fates was shifted in response to stimulating medium.

11.3.3 Microfluidic Devices

Several cell types, in their physiological environment, are subjected to shear stress acting on them. For example, endothelial cells lining the inner wall of the blood vessels are constantly subjected to shear stresses exerted by the flowing blood. Other cell types (e.g., red and white blood cells) are subjected to high deformation forces as they migrate through the narrow blood vessels and capillaries. Understanding the response of cells to such "shear" and "squeezing" forces is essential to understand the pathogenesis of several diseases. Microfluidic devices provide a simple and elegant *in vitro* substitute for studying the effect of shear stress and deformation forces exerted on cells as they flow through narrow channels.

Probing the Elastic Properties of Diseased Cells

Microfabricated channels of various aspect ratios provide a useful way to study the elasticity and deformability of cells. Understanding the deformability of cells is of great interest in several diseased states (e.g., malaria and cancer). In malaria, red blood cells infected by the malarial parasite show increased stiffness as well as adhesiveness to endothelial cells lining the blood vessels. This results from the expression of proteins synthesized by the parasite that are exported to the red-cell membrane. These proteins interact with various molecules expressed on endothelial cells, leading to increased adhesiveness and sequestration of the infected red blood cells in peripheral blood vessels. Some of the synthesized proteins also cross-link the spectrin network, leading to stiffening of the red-cell membrane. Such stiffening of the cell membrane and the presence of the nucleated parasite prevents the red blood cells from flowing through narrow capillaries.

Microfabricated channels can be used to understand how the ability of red blood cells to deform and squeeze through narrow capillaries is affected at different stages of infection by the malarial parasite. Using such microchannels, it has been shown that later stages of infection by the parasite decrease the ability of red blood cells to flow through narrow capillaries compared with healthy red blood cells [41]. Furthermore, endothelial cells can be grown inside the microfluidic channels to study the effect of increased adhesiveness to endothelial cells. Such microfluidic channels can be used as a platform for high-throughput testing of drugs that can interfere with the ability of the parasite to induce cell stiffness and stickiness.

Another interesting application of microfabricated channels is to study the deformability of cancer cells. Metastasis of cancer cells from the primary tumor site requires them to enter local blood vessels and flow in the bloodstream to distant organs where they have to exit the blood vessels to form secondary tumors. Micropipette aspiration and atomic force microscopy indentation experiments suggest that metastatic breast cancer cells are more deformable than benign breast cancer cells. Microfluidic channels can be used as direct evidence to test whether such differences in deformability can indeed translate into better ability of cancer cells to flow through narrow blood vessels. Indeed, such experiments do show that the time taken for squeezing into the microchannels and the speed of the flow through the microchannels for benign breast cancer cells (MCF-10A) was significantly different from that of maglinant breast cancer cells (MCF-7), suggesting that the latter were more deformable (Figure 11.8) [42].

Deformability and Size-Based Approach to Cell Sorting

Microfabricated channels can be efficiently applied as a means to detect, enrich, and sort cells based on their size and deformability. For example, malaria-infected red blood cells tend to marginate to the walls of the microfluidic channels due to their stiffness and can subsequently be separated using different outlets (Figure 11.9a) [43]. Similarly, circulating tumor cells (especially of epithelial origin, which tend to be stiffer than red and white blood cells) in the blood can be separated from other blood cells in microdevices that combine micropillars, magnetic micropatterned substrates, and shear-modulated channels

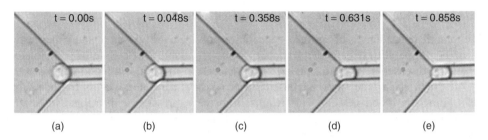

(a)   (b)   (c)   (d)   (e)

Figure 11.8 A series of images showing the entry of a breast cancer cell into a microfluidic channel. The cell underwent significant deformation as it entered the channel [42]

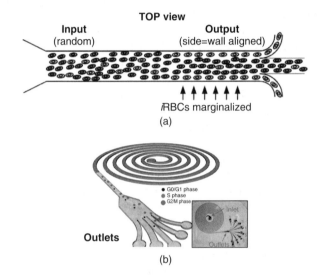

Figure 11.9 (a) Schematic of the microfluidic device to sort infected red blood cells. The stiffer infected cells tend to marginalize towards the walls of the channel, allowing them to be sorted at the outlet [43]; (b) microfluidic device with spiral channels to sort stem cells based on their size [44]. Reproduced by permission of The Royal Society of Chemistry

with microfluidic flow [45–47]. Recently, spiral channels have been used to sort cells into different stages of cell cycle based on their size (Figure 11.9b) [44].

Probing Cell Adhesion

Microfluidic devices can also be used to understand and characterize cell adhesion to other cells and ligands. For example, malaria-infected red blood cells express certain proteins on their surface that make them sticky and highly adherent to endothelial cells lining the blood vessels. Similarly, white blood cells tend to adhere strongly to endothelial cells during inflammation. Microfluidic devices can be used to visualize these adhesion processes in real time and in great detail [48]. High-speed imaging of these processes also allows the biophysical characterization of these various interactions. Furthermore, these devices can also be used for high-throughput screening of various drugs that interfere with the adhesion process.

11.4 Conclusions

The use of microfabrication technologies described above has made it possible to successfully address various questions related to cell mechanics, regenerative medicine, diagnostic devices, molecular immunodiagnosis, and microbiology [49,50]. Apart from their ability to address various biological questions that were previously difficult to address and not amenable to experimental verification, microfabricated technologies provide two distinct advantages to traditional approaches: miniaturization of devices and high-throughput analysis. Miniaturization can revolutionize translational medicine by providing point-of-care devices for rapid diagnosis in regions that may not have sophisticated facilities. In addition, it can offer cost-effective solutions by reducing the amount of chemicals or reagents, the time required for each testing and the easily controllable microenvironment. High-throughput analysis using such devices would be a boon to the pharmaceutical industry, in particular, allowing large-scale screening of various drugs under reproducible and well-controlled conditions. In conclusion, microfabricated technologies that exploit various aspects of cell mechanics are destined to play a much larger role in both basic and translational research.

References

[1] Chen, C.S., Mrksich, M., Huang, S. *et al.* (1997) Geometric control of cell life and death. *Science*, **276** (5317), 1425–1428.

[2] McBride, S.H., Falls, T., and Knothe Tate, M.L. (2008) Modulation of stem cell shape and fate B: mechanical modulation of cell shape and gene expression. *Tissue Engineering Part A*, **14** (9), 1573–1580.

[3] Riddle, R.C., Taylor, A.F., Genetos, D.C., and Donahue, H.J. (2006) MAP kinase and calcium signaling mediate fluid flow-induced human mesenchymal stem cell proliferation. *American Journal of Physiology: Cell Physiology*, **290** (3), C776–C784.

[4] Arnsdorf, E.J., Tummala, P., Kwon, R.Y., and Jacobs, C.R. (2009) Mechanically induced osteogenic differentiation – the role of RhoA, ROCKII and cytoskeletal dynamics. *Journal of Cell Science*, **122** (4), 546–553.

[5] McBeath, R., Pirone, D.M., Nelson, C.M. *et al.* (2004) Cell shape, cytoskeletal tension, and RhoA regulate stem cell lineage commitment. *Developmental Cell*, **6** (4), 483–495.

[6] García-Cardeña, G., Comander, J., Anderson, K.R. *et al.* (2001) Biomechanical activation of vascular endothelium as a determinant of its functional phenotype. *Proceedings of the National Academy of Sciences of the United States of America*, **98** (8), 4478–4485.

[7] Davies, P.F., Remuzzi, A., Gordon, E.J. *et al.* (1986) Turbulent fluid shear stress induces vascular endothelial cell turnover *in vitro*. *Proceedings of the National Academy of Sciences of the United States of America*, **83** (7), 2114–2117.

[8] Renault, J.P., Bernard, A., Bietsch, A. *et al.* (2002) Fabricating arrays of single protein molecules on glass using microcontact printing. *The Journal of Physical Chemistry B*, **107** (3), 703–711.

[9] Khan Malek, C. (1991) A review of microfabrication technologies: application to X-ray optics. *Journal of X-Ray Science and Technology*, **3** (1), 45–67.

[10] Reymann, A.C., Martiel, J.L., Cambier, T. *et al.* (2010) Nucleation geometry governs ordered actin networks structures. *Nature Materials*, **9** (10), 827–832.

[11] Xia, Y. and Whitesides, G.M. (1998) Soft lithography. *Annual Review of Materials Science*, **28** (1), 153–184.

[12] Chandra, D. and Yang, S. (2009) Capillary-force-induced clustering of micropillar arrays: is it caused by isolated capillary bridges or by the lateral capillary meniscus interaction force? *Langmuir*, **25** (18), 10430–10434.

[13] Sharp, K.G., Blackman, G.S., Glassmaker, N.J. *et al*. (2004) Effect of stamp deformation on the quality of microcontact printing: theory and experiment. *Langmuir*, **20** (15), 6430–6438.

[14] Schvartzman, M., Palme, M., Sable, J. *et al*. (2011) Nanolithographic control of the spatial organization of cellular adhesion receptors at the single-molecule level. *Nano Letters*, **11** (3), 1306–1312.

[15] Doyle, A.D. (2009) Generation of micropatterned substrates using micro photopatterning. *Current Protocols in Cell Biology*, **45**, 10.15.1–10.15.35.

[16] Engler, A.J., Sen, S., Sweeney, H.L., and Discher, D.E. (2006) Matrix elasticity directs stem cell lineage specification. *Cell*, **126** (4), 677–689.

[17] McBeath, R., Pirone, D.M., Nelson, C.M. *et al*. (2004) Cell shape, cytoskeletal tension, and RhoA regulate stem cell lineage commitment. *Developmental Cell*, **6** (4), 483–495.

[18] Kilian, K.A., Bugarija, B., Lahn, B.T., and Mrksich, M. (2010) Geometric cues for directing the differentiation of mesenchymal stem cells. *Proceedings of the National Academy of Sciences of the United States of America*, **107** (11), 4872–4877.

[19] Thery, M., Racine, V., Pépin, A. *et al*. (2005) The extracellular matrix guides the orientation of the cell division axis. *Nature Cell Biology*, **7** (10), 947–953.

[20] Thery, M., Jiménez-Dalmaroni, A., Racine, V. *et al*. (2007) Experimental and theoretical study of mitotic spindle orientation. *Nature*, **447** (7143), 493–496.

[21] Desai, R.A., Gao, L., Raghavan, S. *et al*. (2009) Cell polarity triggered by cell–cell adhesion via E-cadherin. *Journal of Cell Science*, **122** (Pt 7), 905–911.

[22] Jiang, X., Bruzewicz, D.A., Wong, A.P. *et al*. (2005) Directing cell migration with asymmetric micropatterns. *Proceedings of the National Academy of Sciences of the United States of America*, **102** (4), 975–978.

[23] Pouthas, F., Girard, P., Lecaudey, V. *et al*. (2008) In migrating cells, the Golgi complex and the position of the centrosome depend on geometrical constraints of the substratum. *Journal of Cell Science*, **121** (Pt 14), 2406–2414.

[24] Doyle, A.D., Wang, F.W., Matsumoto, K. *et al*. (2009) One-dimensional topography underlies three-dimensional fibrillar cell migration. *Journal of Cell Biology*, **184** (4), 481–490.

[25] Borghi, N., Lowndes, M., Maruthamuthu, V. *et al*. (2010) Regulation of cell motile behavior by crosstalk between cadherin- and integrin-mediated adhesions. *Proceedings of the National Academy of Sciences of the United States of America*, **107** (30), 13324–13329.

[26] Poujade, M., Grasland-Mongrain, E., Hertzog, A. *et al*. (2007) Collective migration of an epithelial monolayer in response to a model wound. *Proceedings of the National Academy of Sciences of the United States of America*, **104** (41), 15988–15993.

[27] Tan, J.L., Tien, J., Pirone, D.M. *et al*. (2003) Cells lying on a bed of microneedles: an approach to isolate mechanical force. *Proceedings of the National Academy of Sciences of the United States of America*, **100** (4), 1484–1489.

[28] Sniadecki, N.J. and Chen, C.S. (2007) Microfabricated silicone elastomeric post arrays for measuring traction forces of adherent cells. *Methods in Cell Biology*, **83**, 313–328.

[29] Saez, A., Anon, E., Ghibaudo, M. *et al*. (2010) Traction forces exerted by epithelial cell sheets. *Journal of Physics: Condensed Matter*, **22** (19), 194119.

[30] Du Roure, O., Saez, A., Buguin, A. *et al*. (2005) Force mapping in epithelial cell migration. *Proceedings of the National Academy of Sciences of the United States of America*, **102** (7), 2390–2395.

[31] Benoit, M., Gabriel, D., Gerisch, G. *et al*. (2000) Discrete interactions in cell adhesion measured by single-molecule force spectroscopy. *Nature Cell Biology*, **2** (6), 313–317.

[32] Lim, C.T., Zhou, E.H., Li, A. *et al*. (2006) Experimental techniques for single cell and single molecule biomechanics. *Materials Science and Engineering C*, **26** (8), 1278–1288.

[33] Liu, Z., Tan, J.L., Cohen, D.M. *et al*. (2010) Mechanical tugging force regulates the size of cell–cell junctions. *Proceedings of the National Academy of Sciences of the United States of America*, **107** (22), 9944–9949.

[34] Ganz, A., Lambert, M., Saez, A. *et al*. (2006) Traction forces exerted through N-cadherin contacts. *Biology of the Cell*, **98** (12), 721–730.

[35] Lo, C.M., Wang, H.B., Dembo, M. *et al*. (2000) Cell movement is guided by the rigidity of the substrate. *Biophysical Journal*, **79** (1), 144–152.

[36] Saez, A., Ghibaudo, M., Buguin, A. *et al*. (2007) Rigidity-driven growth and migration of epithelial cells on microstructured anisotropic substrates. *Proceedings of the National Academy of Sciences of the United States of America*, **104** (20), 8281–8286.

[37] Ghibaudo, M., Di Meglio, J.-M., Hersen, P., and Ladoux, B. (2011) Mechanics of cell spreading within 3D-micropatterned environments. *Lab on a Chip*, **11** (5), 805–812.

[38] Sniadecki, N.J., Lamb, C.M., Liu, Y. *et al*. (2008) Magnetic microposts for mechanical stimulation of biological cells: fabrication, characterization, and analysis. *Review of Scientific Instruments*, **79** (4), 044302.

[39] Le Digabel, J., Biais, N., Fresnais, J. *et al*. (2011) Magnetic micropillars as a tool to govern substrate deformations. *Lab on a Chip*, **11** (15), 2630–26306.

[40] Fu, J., Wang, Y.K., Yang, M.T. *et al*. (2010) Mechanical regulation of cell function with geometrically modulated elastomeric substrates. *Nature Methods*, **7** (9), 733–736.

[41] Shelby, J.P., White, J., Ganesan, K. *et al*. (2003) A microfluidic model for single-cell capillary obstruction by *Plasmodium falciparum*-infected erythrocytes. *Proceedings of the National Academy of Sciences of the United States of America*, **100** (25), 14618–14622.

[42] Hou, H.W., Li, Q.S., Lee, G.Y. *et al*. (2009) Deformability study of breast cancer cells using microfluidics. *Biomedical Microdevices*, **11** (3), 557–564.

[43] Hou, H.W., Bhagat, A.A., Chong, A. *et al*. (2010) Deformability based cell margination – a simple microfluidic design for malaria-infected erythrocyte separation. *Lab on a Chip*, **10** (19), 2605–2613.

[44] Lee, W.C., Bhagat, A.A.S., Huang, S. *et al*. (2011) High-throughput cell cycle synchronization using inertial forces in spiral microchannels. *Lab on a Chip*, **11** (7), 1359–1367.

[45] Tan, S.J., Yobas, L., Lee, G.Y. *et al*. (2009) Microdevice for the isolation and enumeration of cancer cells from blood. *Biomedical Microdevices*, **11** (4), 883–892.

[46] Saliba, A.E., Saias, L., Psychari E. *et al*. (2010) Microfluidic sorting and multimodal typing of cancer cells in self-assembled magnetic arrays. *Proceedings of the National Academy of Sciences of the United States of America*, **107** (33), 14524–14529.

[47] Bhagat, A.A.S., Hou, H.W., Li, L. *et al*. (2011) Pinched flow coupled shear-modulated inertial microfluidics for high-throughput rare blood cell separation. *Lab on a Chip*, **11** (11), 1870–1878.

[48] Antia, M., Herricks, T., and Rathod, P.K. (2008) Microfluidic approaches to malaria pathogenesis. *Cellular Microbiology*, **10** (10), 1968–1974.

[49] Weibel, D.B., Diluzio, W.R., and Whitesides, G.M. (2007) Microfabrication meets microbiology. *Nature Reviews. Microbiology*, **5** (3), 209–218.

[50] Li, H.Y., Dauriac, V., Thibert, V. *et al*. (2010) Micropillar array chips toward new immunodiagnosis. *Lab on a Chip*, **10** (19), 2597–2604.

Part Four

Modeling

12

Atomistic Reaction Pathway Sampling: The Nudged Elastic Band Method and Nanomechanics Applications

Ting Zhu[1], Ju Li[2], and Sidney Yip[2]

[1]*Georgia Institute of Technology, USA*
[2]*Massachusetts Institute of Technology, USA*

12.1 Introduction

Two of the central recurring themes in nanomechanics are strength and plasticity [1–3]. They are naturally coupled because plastic deformation is a major strength-determining mechanism and understanding the resistance to deformation (strength) is a principal motivation for studying plasticity. Many phenomena of interest in mechanics can be discussed in the framework of microstructural evolution of a system where defects like cracks and dislocations are formed and evolve interactively. Microstructure evolution at the nanoscale is particularly relevant from the standpoint of probing unit processes of deformation, such as advancement of a crack front by a lattice unit or propagation of a dislocation core by a lattice vector. These atomic-level details can reveal the mechanisms of deformation, which are the essential inputs to describing microstructure evolution at the mesoscale – the next length and time scales. This hierarchical relation is the essence of multiscale modeling and simulation paradigm [4–6].

The purpose of this chapter is to discuss the atomistic approach to describe the evolution of crystalline defects [1–3,6]. We focus on a method, which is becoming widely used, that allows one to track the microstructure evolution through the sampling of a

Nano and Cell Mechanics: Fundamentals and Frontiers, First Edition. Edited by Horacio D. Espinosa and Gang Bao.
© 2013 John Wiley & Sons, Ltd. Published 2013 by John Wiley & Sons, Ltd.

minimum-energy path (MEP) [7–9]. This path describes the variation of the energy of the system as it moves through a unit process reaction from an initial to a final state. The sampling also produces a reaction coordinate which is a collective coordinate that can be used to label system atomic configurations along the reaction pathway. The determination of the MEP, therefore, provides two pieces of valuable information: the saddle point energy of the reaction, which is the activation energy required for the reaction to occur, and the atomic configurations at the saddle point, which can reveal the molecular mechanism associated with the unit process. Several case studies are then discussed to illustrate the applicability of the sampling method, known as the nudged elastic band (NEB) [7–9], and the MEP that it can produce for each unit process. For microstructure evolution over extended time intervals the system still can be characterized by reaction pathways, although the concept of a unit process is no longer appropriate. For these problems, other atomistic sampling methods can be used to determine the appropriate transition-state pathway (TSP) trajectories which are effectively an ordered sequence of MEPs. Such studies are only emerging. While they are beyond the scope of the present discussion, we will nevertheless provide a brief outlook on some recent developments [10].

The chapter is laid out as follows. Motivation and brief introductions to reaction pathway sampling, the NEB method of determining the MEP, and transition-state theory of activated processes are covered in Section 12.1. Section 12.2 deals with the NEB method as applied to stress-driven unit processes. Section 12.3 describes the atomistic results for several scenarios: dislocation emission at a crack tip, interaction between a silica nanorod and a water molecule, twinning effects at the nanoscale, temperature and strain-rate sensitivity, and size effects. Section 12.4 is a concluding outlook on microstructure evolution at longer times from the standpoint of atomistic sampling methods and challenging problems in the area of materials ageing.

12.1.1 Reaction Pathway Sampling in Nanomechanics

We are primarily concerned here with the modeling and simulation of unit processes in nanoscale deformation. The nucleation and evolution of defects will be described in terms of reaction pathways with specified initial and final states. We will focus on a particular atomistic method of sampling (determined by simulation) the defect reaction pathway, a method that is known as the NEB [7–9]. With this method one can obtain the MEP, which describes the variation in system energy along a reaction coordinate. The determination of the MEP allows one to find the energy barrier for the reaction. Since a one-to-one relation exists between the reaction coordinate and the atomic configurations of the system, the MEP also allows us to study the atomic configurations at the saddle point. Such details are needed to establish the unit mechanism of deformation.

12.1.2 Extending the Time Scale in Atomistic Simulation

Atomistic simulations, of which molecular dynamics (MD) is the primary method, are known for their time-scale limitations. The limitations usually appear in one of two ways. The explicit limitation is the time interval that a simulation can cover, while a more implicit limitation is the rate of change that one can impose externally on the system.

MD studies, as direct simulation of Newtonian dynamics, cannot extend over times longer than nanoseconds without using acceleration techniques [11] in conjunction with transition-state theory (see Section 12.1.3). MD gives directly the time responses of the system to external perturbations (imposed in discrete steps); the effective rates of perturbation associated with the simulation are invariably orders of magnitude higher than what can be practically imposed in experiments or in nature. The reason is simply because the basic time step in the integration of Newton's equations of motion is restricted to femtoseconds. The extreme rate of perturbation of all MD simulations, therefore, calls into question the physical meaning of the mechanisms revealed by simulation in comparison with rate-dependent responses of systems studied experimentally.

12.1.3 Transition-State Theory

Many of the inelastic deformations in solids, including dislocation slip, twinning, and phase transformation, occur by the thermally activated processes of atomic rearrangement. According to transition-state theory [12,13], the rate of a thermally activated process can be estimated according to

$$v = v_0 \exp\left(-\frac{Q(\sigma, T)}{k_B T}\right) \tag{12.1}$$

where v_0 is the trial frequency, k_B is the Boltzmann constant, and Q is the activation free energy whose magnitude is controlled by the local stress σ and temperature T.

To develop a quantitative sense of Equation (12.1), it is instructive to consider some numbers. The physical trial frequency v_0 is typically on the order of 10^{11} s^{-1}, as dictated by the atomic vibration. In order for a thermally activated process observable in a typical laboratory experiment, the rate v should be of the order of say 10^{-2} s^{-1}, so that the activation energy needs to be around $30k_B T$. As such, a thermally activated process with an energy barrier of ~ 0.7 eV would be relevant to the laboratory experiment at room temperature (the corresponding thermal energy $k_B T \approx 1/40$ eV).

Clearly, the activation free energy Q in Equation (12.1) is a quantity of central importance for determining the kinetic rate within transition-state theory. Its value depends on the specific activation processes of atomic rearrangement, and is a function of stress, temperature, and system size, and so on. To a first approximation, Q can be estimated by the 0 K energy barrier, which can be effectively evaluated by using the NEB method.

12.2 The NEB Method for Stress-Driven Problems

12.2.1 The NEB method

The NEB method is a chain-of-states approach of finding the MEP on the potential energy surface (PES). In configuration space, each point represents one configuration of atoms in the system. This configuration has a 0 K potential energy, which can be evaluated by using the empirical interatomic potential or first-principles method [5]. The PES is the surface of the potential energy of each point in configuration space. In general, there are basins, ridges, local minima and saddle points on the PES. The MEP is the lowest energy path for a rearrangement of a group of atoms from one stable configuration to another; that is, from one local energy minimum to another [7]. The potential energy maximum along the MEP is the saddle-point energy which gives the activation energy barrier; that is, Q in Equation (12.1).

In an NEB calculation, the initial and final configurations should be first determined by using energy minimization. Then, a discrete band, consisting of a finite number of replicas of the system, is constructed. These replicas can be generated by linear interpolation between the two end states. Every two adjacent replicas are connected by a spring, mimicking an elastic band made up of beads and springs. Beads in the band can be equally spaced in a relaxation process due to the spring forces. To solve the problems of corner cutting and sliding down that often arise with the plain elastic band method, a force projection is needed; this is what is referred to as the "nudging" operation [7]. With appropriate relaxation, the band converges to the MEP. Figure 12.1 illustrates the NEB method by showing an elastic band before and after relaxation on the energy contour of a model system with two degrees of freedom.

The algorithm of the NEB method involves the following basic steps. According to Henkelman *et al*. [9], let us denote $\mathbf{R}_i$ as the atomic coordinates in the system at replica i. Given an estimate of the unit tangent to the path at each replica $\mathbf{t}_i$, the force on each replica contains the parallel component of the spring force and perpendicular component of the potential force:

$$\mathbf{F}_i = -\nabla E(\mathbf{R}_i)|_\perp + (\mathbf{F}_i^s \cdot \mathbf{t}_i)\mathbf{t}_i \tag{12.2}$$

where $\nabla E(\mathbf{R}_i)$ is the gradient of the energy with respect to the atomic coordinates in the system at replica i, $\nabla E|_\perp$ is the component of the gradient perpendicular to $\mathbf{t}_i$ and it can

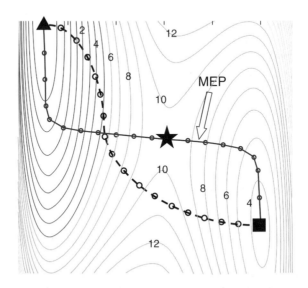

Figure 12.1 An illustration of the NEB method by using the energy contour of a model system with two degrees of freedom. The dashed line represents an initially guessed MEP that links the initial (triangle) and final (square) states of local energy minima; the solid curve is the converged MEP passing through the saddle point (star)

be obtained by subtracting out the parallel component:

$$\nabla E(\mathbf{R}_i)|_\perp = \nabla E(\mathbf{R}_i) - [\nabla E(\mathbf{R}_i) \cdot \mathbf{t}_i]\mathbf{t}_i \qquad (12.3)$$

In Equation (12.2), $\mathbf{F}_i^s$ is the spring force acting on replica i and it can be evaluated according to

$$\mathbf{F}_i^s = k(|\mathbf{R}_{i+1} - \mathbf{R}_i| - |\mathbf{R}_i - \mathbf{R}_{i-1}|)\mathbf{t}_i \qquad (12.4)$$

where k is the spring constant. Henkelman *et al.* [9] also discussed in detail how to estimate the tangent $\mathbf{t}_i$, minimize the force $\mathbf{F}_i$, and find the saddle point by the climbing image method, and so on.

The converged MEP is usually plotted as the energy, relative to the initial state, versus reaction coordinate. The latter can be defined in the following sense. Each replica on the MEP is a specific configuration in a $3N$ configurational hyperspace, where N is the number of movable atoms in the simulation. For each replica one calculates the hyperspace arc length

$$l \equiv \int_{\mathbf{R}_0^{3N}}^{\mathbf{R}_i^{3N}} \sqrt{d\mathbf{R}^{3N} \cdot d\mathbf{R}^{3N}} \qquad (12.5)$$

between the initial state $\mathbf{R}_0^{3N}$ and the state of the replica $\mathbf{R}_i^{3N}$. The normalized reaction coordinate s can be calculated according to $s = l/l_0$, where l_0 denotes the hyperspace arc length between the initial and final states.

Example: Vacancy Migration in Cu

Let us consider the application of the NEB method to a simple problem of migration of a single vacancy in an otherwise perfect crystal of Cu. The system is initially a cube of face-centered-cubic (FCC) lattice with 500 atoms. The side length of the cube is $5a_0$, where a_0 is the equilibrium lattice constant, 3.615 Å. A vacancy is generated by simply removing one atom from the perfect lattice. The vacant site in the initial (i) and final (f) configurations differs by a displacement of $a_0/\sqrt{2}$ in the $\langle 110 \rangle$ direction. Prior to the NEB calculation, both the initial and final configurations are fully relaxed to the zero stress under periodic boundary conditions, such that they correspond to two local energy minima on the zero-stress PES. The embedded atom potential (EAM) of Cu [14] is used in calculations.

Figure 12.2 shows the converged MEP from an NEB calculation, giving a vacancy migration barrier of 0.67 eV, consistent with experimental measurement of 0.71 eV [14]. The atomic configurations of the initial, saddle point, and final states are shown in Figure 12.2. Incidentally, the reaction coordinate in this case can be physically considered as the displacement of the vacancy (normalized by $a_0/\sqrt{2}$) in the $\langle 110 \rangle$ direction from the initial to final state, while it has been mathematically calculated according to the hyperspace arc length along the MEP (Equation (12.5)).

12.2.2 The Free-End NEB Method

The NEB method is effective in finding the MEP of a highly localized activation process that involves the rearrangement of a small number of atoms, such as lattice and surface

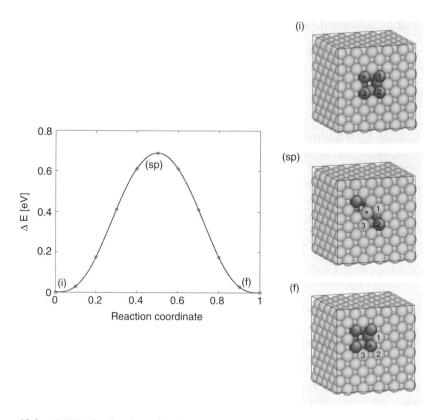

Figure 12.2   MEP of migration of a single vacancy in an fcc Cu lattice. Also shown are the atomic configurations of the initial (i), saddle point (sp), and final (f) states. The migration of such a vacancy in the ⟨110⟩ direction can be better visualized in terms of displacing an atom (adjacent to the vacant site) in the opposite direction; that is, the starred atom moving in between (i) and (f)

diffusion. However, it is inefficient to probe the activation process of extended three-dimensional (3D) defects such as dislocations and crack fronts, typically involving the collective motion of a large group of atoms and thus requiring a large model system with a long reaction path. This issue becomes more significant when the energy landscape is highly tilted by the applied load, such that the saddle point becomes closer to the initial state. Under such a condition, the plain NEB method becomes highly inefficient. That is, although only a small portion of the path close to the initial state is actually needed for finding the saddle point, a large number of replicas, and thus computations, have to be used to ensure the sufficient nodal densities for mapping out the long path between the saddle point and final state. To improve the computational efficiency, a free-end NEB method [15] has been developed.

The idea of the free-end NEB method is to reduce the modeled path length. This is realized by cutting short the elastic band and meanwhile allowing the end of the shortened band to move freely on an energy iso-surface close to the beginning of the band. The free-end NEB method contrasts with the plain one that requires the fixed final state at an energy minimum far from the beginning. To appreciate the need for allowing the final state to move freely on an energy iso-surface, we note that if the initial state (node 0) is a local minimum and the finial state (node n) is not, but fixed during relaxation, then the NEB algorithm can behave badly. That is, node $n - 1$ moves along its potential gradient except the component parallel to the path direction. If the fixed node n is not chosen to be right on the MEP, node $n - 1$ will droop down and end up having much lower energy than node n. In the relaxation process it will drag all the path nodes along due to the spring force, which makes the quality of the NEB mesh degrades with time.

Consider, for example, a process of dislocation nucleation with the saddle-point energy of around 0.2 eV (taking the energy of node 0 zero). Usually, it would be sufficient if final node's energy is -0.3 eV, since this means the final node is already in another attraction basin on the PES. Instead of seeking energy minimization, the free-end NEB algorithm requires that the final node's energy stays constant on the energy iso-surface of -0.3 eV. As shown schematically in Figure 12.3a, the band swings to improve nodal density around the saddle point. Figure 12.3b shows a converged MEP for a model problem of a dislocation loop bowing out from a mode II crack tip, with the saddle-point configuration shown in Figure 12.3c. Here, the final state is fixed at 0.3 eV below the initial state. The free-end NEB method captures the saddle point with only seven replicas along the band, thus significantly improving the computational efficiency.

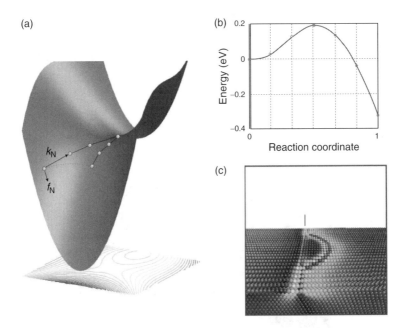

Figure 12.3 The free-end NEB method enables an efficient reaction pathway calculation of extended defects in the large system. (a) Illustration of the method showing one end of the elastic band is fixed and the other is freely moved along an energy contour. (b) An example of the converged MEP from the free-end NEB method calculation; the end node is pinned at 0.3 eV below the initial state. (c) The corresponding saddle-point structure of a dislocation loop bowing out from a mode II crack tip (the upper half crystal is removed for clarity). Refer to Plate 9 for the colored version

12.2.3   Stress-Dependent Activation Energy and Activation Volume

Increasing the applied stress will generally increase the thermodynamic driving force, reduce the energy barrier, and, accordingly, increase the rate of a thermally activated process [16]. Consider, for example, the nucleation of a dislocation loop from the surface of a Cu nanowire, Figure 12.4a. Suppose we begin to apply an axial stress σ in incremental steps. Initially, a dislocation would not form spontaneously because the driving force is not sufficient to overcome the nucleation resistance. What does this mean? Consider an initial configuration of a perfect wire and a final configuration with a fully nucleated partial dislocation; that is, states "i" and "f" in Figure 12.4b, respectively. At low loads (e.g., $\sigma_1 < \sigma_{cr}$ in Figure 12.4c), the initial configuration (white circle) has a lower energy than the final configuration (black circle). They are separated by an energy barrier (gray circle) with the saddle-point (sp) configuration shown in Figure 12.4b. At this load level, the nucleation is thermodynamically unfavorable because the energy of the final state is higher than the initial one. As the load increases, the system will be driven toward the final state, such that the nucleation becomes favorable thermodynamically when $\sigma_2 > \sigma_{cr}$. One can regard the overall energy landscape as being tilted toward the final state with a corresponding reduction in the energy barrier – compare the saddle-point states (gray circles) in Figure 12.4c. As the load increases further, the biasing becomes stronger. So long as the barrier remains finite, the state of a prefect wire will not move out of its initial basin without additional activation, such as from thermal fluctuations. When the load reaches the point where the nucleation barrier disappears altogether, the wire is then unstable at the initial configuration. It follows that a dislocation will nucleate instantaneously without any thermal activation. This is the athermal load threshold, denoted by σ_{ath} in Figure 12.4c.

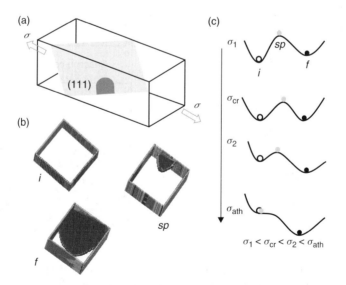

Figure 12.4   Effects of the stress-dependent activation energy. (a) Schematic of surface dislocation nucleation in a nanowire at a given applied load. (b) Atomic structure of initial (i), saddle-point (sp) and final (f) states of surface nucleation. (c) Energy landscape at different applied loads; white circle represents the initial state (i), black circle is the final state (f), and gray circle corresponds to the saddle-point (sp) state in between

A concept related to the stress dependence of activation energy is activation volume, defined as

$$\Omega = -\frac{\partial Q(\sigma, T)}{\partial \sigma} \tag{12.6}$$

Physically, the activation volume corresponds to the amount of material (i.e., volume of atoms) in the high-energy state in a thermally activated process. As such, it measures the individualistic and collective nature of transition. During thermal activation, the stress does work on the activation volume to assist the transition by reducing the effective energy barrier. To reflect this stress-work effect, the rate formula of Equation (12.1) is commonly rewritten as

$$v = v_0 \exp\left(-\frac{Q_0 - \sigma\Omega}{k_B T}\right) \tag{12.7}$$

where Q_0 is the activation energy at zero stress.

Different rate processes can have drastically different characteristic activation volumes; for example, $\Omega \approx 0.1b^3$ for lattice diffusion versus $\Omega \approx 1000b^3$ for the Orowan looping of a dislocation line across the pinning points in coarse-grained metals, where b is the Burgers vector length. As a result, the activation volume can serve as an effective kinetic signature of deformation mechanism. This is illustrated by the schematic in Figure 12.5, where the activation volume corresponds to the slope of an activation energy curve plotted as a function of stress. Notice that while the activation volume generally varies with stress, it is often treated as a constant when the rate or stress change is not large. Suppose the two competing processes have the same activation energy (indicated by the short-dash line in Figure 12.5) giving the same rate of transition, one can use the activation volume to identify the operative one in an experiment or a simulation. Furthermore, Figure 12.5 indicates that for the two processes with different activation volumes, the process with a higher energy barrier at low stresses may change to have a lower barrier at high stresses.

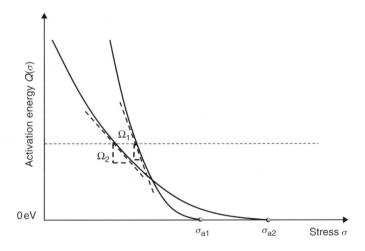

Figure 12.5 Schematic of the stress-dependent activation energy for two competing thermally activated processes. They have different activation volumes (Ω_1 versus Ω_2) and athermal threshold stresses (σ_{a1} versus σ_{a2})

This cross-over of energy barriers often underlies the switching of the rate-controlling mechanism in experiments.

In experiments, the activation volume can be determined by measuring the strain-rate sensitivity. Consider, as an example, uniaxial tension of a polycrystalline specimen. The empirical power-law relation is often used to represent the measured stress σ versus strain rate $\dot{\varepsilon}$ response:

$$\frac{\sigma}{\sigma_0} = \left(\frac{\dot{\varepsilon}}{\dot{\varepsilon}_0}\right)^m \tag{12.8}$$

where σ_0 is the reference stress, $\dot{\varepsilon}_0$ is the reference strain rate, and m is the nondimensional rate-sensitivity index, which generally lies in between 0 and 1 ($m = 0$ gives the rate-independent limit and $m = 1$ corresponds to the linear Newtonian flow). The apparent activation volume Ω^* is conventionally defined as

$$\Omega^* \equiv \sqrt{3}k_{\mathrm{B}}T\frac{\partial \ln\dot{\varepsilon}}{\partial \sigma} \tag{12.9}$$

Combining Equations (12.8) and (12.9), one can readily show that m is related to Ω^* by [17]

$$m = \sqrt{3}\frac{k_{\mathrm{B}}T}{\sigma\Omega^*} \tag{12.10}$$

In Equations (12.9) and (12.10), the factor of $\sqrt{3}$ arises because, similar to the von Mises yield criterion, the normal stress σ is converted to the effective shear stress τ^* according to $\tau^* = \sigma/\sqrt{3}$. Since τ^* is related to the resolved shear stress on a single slip plane τ by $\tau^* = M/\sqrt{3}\tau$, where $M = 3.1$ is the Taylor factor, it follows that the true activation volume Ω associated with a unit process and the apparent activation volume Ω^* measured from a polycrystalline sample are related by

$$\Omega^* = \frac{\sqrt{3}}{M}\Omega \tag{12.11}$$

The activation volume and rate sensitivity provide a direct link between experimentally measurable plastic flow characteristics and underlying deformation mechanisms. However, this link can be complicated by such important factors as mobile dislocation density and strain hardening [18].

Finally, we note that the scalar activation volume in Equations (12.6) and (12.7) can be generalized to a definition of the activation volume tensor [19], when all the stress components are considered. Albeit broad implications of the tensorial activation volume, in this chapter we focus on the simple scalar activation volume for highlighting its physical characteristics and usefulness. Importantly, the activation volume can be determined by both experiment and atomistic modeling, thus providing a unique link in coupling the two approaches for revealing the rate-controlling deformation mechanisms [15,17].

12.2.4 Activation Entropy and Meyer–Neldel Compensation Rule

In order to provide a reasonable estimate of the absolute magnitude of the rate of a thermally activated process, one should also pay special attention to the effect of activation entropy $S(\sigma)$ [20–24]. The activation free energy $Q(\sigma, T)$ in Equation (12.1) can be decomposed into

$$Q(\sigma, T) = E(\sigma) - T S(\sigma) \tag{12.12}$$

where $E(\sigma)$ is the activation enthalpy that corresponds to the energy difference between the saddle point and initial equilibrium state on the $0\,\text{K}$ PES. Substitution of Equation (12.12) into Equation (12.1) leads to

$$v = \tilde{v}_0 \exp\left(-\frac{E(\sigma)}{k_{\text{B}} T}\right) \tag{12.13}$$

where

$$\tilde{v}_0 = v_0 \exp\left(\frac{S(\sigma)}{k_{\text{B}}}\right) \tag{12.14}$$

It should be emphasized that the pre-exponential factor $\tilde{v}_0$ in Equation (12.13) generally varies for different rate processes. More importantly, $\tilde{v}_0$ changes for the same kind of processes under different applied stresses as well. As seen from Equation (12.14), such a change arises due to the change of activation entropy $S(\sigma)$. There is a well-known empirical Meyer–Neldel compensation law or iso-kinetic rule [25], which suggests that $S(\sigma)$ is likely to correlate the activation energy by

$$S(\sigma) = \frac{E(\sigma)}{T_{\text{MN}}} \tag{12.15}$$

where T_{MN} denotes the Meyer–Neldel temperature constant.

The so-called "compensation" rule can be understood as follows: when the applied stress decreases at a constant temperature, the activation energy $E(\sigma)$ typically increases (see Figure 12.5), causing a decrease of the value of $\exp(-E(\sigma)/k_{\text{B}}T)$. However, according to Equation (12.15), the activation entropy $S(\sigma)$, and accordingly $\tilde{v}_0$, will increase with $E(\sigma)$. As a result, the rate of activation v in Equation (12.13) does not decrease as rapidly as one would expect from only considering $\exp(-E(\sigma)/k_{\text{B}}T)$. According to Yelon et al. [26], the Meyer–Neldel compensation law is obeyed in a wide range of kinetic processes, including annealing phenomena, electronic processes in amorphous semiconductors, trapping in crystalline semiconductors, conductivity in ionic conductors, ageing of insulating polymers, biological death rates, and chemical reactions.

The empirical linear relation between the activation energy and activation entropy in Equation (12.15) is surprisingly simple and effective. Such a remarkable connection has been explained earlier in the context of solid-state diffusion, and more generally through the role of multi-excitation entropy [27,28]. Intuitively, the activation process with large activation energy involves the collective motion of a group of atoms. This gives rise to a large number of ways in which the activation can be done through the multi-phonon processes; namely, a large entropy change between the saddle point and initial equilibrium state. On the other hand, the increasing number of activated atoms is also manifested through the increase of activation volume Ω defined by Equations (12.6) and (12.7).

Substitution of Equation (12.15) into Equation (12.12) gives

$$Q(\sigma, T) = E(\sigma) \left(1 - \frac{T}{T_{\text{MN}}}\right) \tag{12.16}$$

This relation furnishes a simple estimate, and interpretation, of the Meyer–Neldel temperature T_{MN}. Mott assumed T_{MN} to be the melting temperature of the crystal by noting that the activation free energy $Q(\sigma, T)$ should approach zero at the melting temperature [20]. This provides an explanation of the large pre-factor of the temperature exponential in the measured rate of grain-boundary slip in pure polycrystalline aluminum. Generally, T_{MN} can be treated as the local melting or disordering temperature [21]. More extensive discussions on the activation entropy and Meyer–Neldel compensation rule can be found in a recent atomistic study of surface dislocation nucleation by Hara and Li [23].

12.3 Nanomechanics Case Studies

The NEB method has been applied successfully to a wide range of nanomechanics problems. In this section we present the case studies to clarify the concepts and demonstrate the applications. These cases are taken mostly from our own work. In Section 12.3.1, the crack-tip dislocation emission is discussed to demonstrate the necessity of 3D simulation of energy barriers as opposed to two-dimensional (2D) simulation. In Section 12.3.2, the stress-mediated chemical reaction is presented to illustrate the competing reaction pathways, mediated by stress. In Section 12.3.3, the dislocation and interface interaction is studied in nanotwinned Cu. This subsection highlights how to bridge the laboratory experiments with atomistic simulation of the rate-controlling mechanisms in terms of rate sensitivity and activation volume. In Section 12.3.4, surface dislocation nucleation in a nanowire is studied to predict the temperature and strain-rate dependence of strength limit and yield stress spanning a wide range of loading conditions. Finally, in Section 12.3.5 we present the size effects on fracture, with an emphasis on the difference of energy barriers under stress- versus strain-controlled loadings.

12.3.1 Crack Tip Dislocation Emission

An important problem in the study of the mechanical behavior of materials is to understand the ductile–brittle transition of fracture. This has been studied in terms of competition between dislocation emission and cleavage bond breaking at an atomically sharp crack tip [29]. To evaluate the rate of dislocation emission from a crack tip, one needs to determine the saddle-point configuration and energy barrier of a dislocation loop nucleating from a crack tip. This has been studied by using various approaches, including the dislocation line model [29], the semi-analytic model based on the Peierls concept (i.e., periodic relation between shear stress and atomic shear displacement) [30], and the numerical approach of the boundary integral [31]. Application of the NEB method to this problem makes possible a direct atomistic determination of the saddle-point configuration and energy barrier of dislocation nucleation [32].

Consider, as an example, a crack in an FCC single crystal of Cu. The simulation cell consists of a cracked cylinder cut from the crack tip, with radius $R = 80\,\text{Å}$. The straight crack front, lying on a (111) plane, runs along the [110] direction. The cracked system is subjected to a mode I load of the stress intensity factor $K_I = 0.44\,\text{MPa m}^{1/2}$. Atoms within $5\,\text{Å}$ of the outer surface are fixed according to a prescribed atomic displacement, and all the other atoms are fully relaxed. The embedded atom (EAM) potential of Cu is

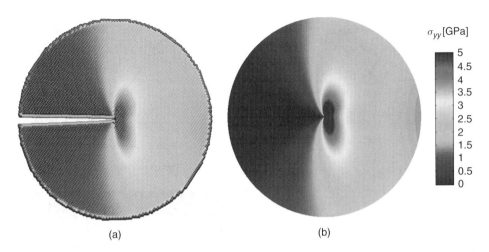

σ_{yy}[GPa]

Figure 12.6   Distribution of stress σ_{yy} near a crack tip in Cu, subjected to an applied stress intensity factor $K_I = 0.44$ MPa m$^{1/2}$ [32]. (a) Atomistic result from energy minimization of a cracked lattice and the atomic stress is calculated by the virial formula. (b) Numerical result based on the analytic solution of the crack-tip stress field from continuum fracture mechanics

used in calculations. To validate the atomistic study, we obtained the consistent atomic calculation (Figure 12.6a) and analytic solution (Figure 12.6b) of the stress distribution near the crack tip.

It should be emphasized that the thermally activated dislocation emission is intrinsically a 3D process involving the growth of a small dislocation loop from the crack front. In other words, one cannot simply use a quasi-2D NEB calculation to determine the energy barrier of emission of a straight line parallel to the crack front, which will increase unphysically with thickness of the simulation cell. Figure 12.7a shows the schematics of emission of a 3D dislocation loop on a {111} slip plane inclined to the crack tip. In the NEB calculation, the initial state is the loaded crack system without dislocation, and the final state consists of a fully nucleated dislocation. Note that the atomic structure of the initial state corresponds to that in Figure 12.6, but replicated by 24 unit cells with a total length of 61 Å in the out-of-plane direction, along which a periodic boundary condition is imposed. Such a large thickness minimizes the interaction of a dislocation loop between neighboring supercells in the out-of-plane direction. The final state can be generated by a two-step operation: first, impose a high load to nucleate a dislocation instantaneously; second, unload the system by prescribing the same atomic positions of boundary atoms as the initial state and then relax the system.

The converged MEP from the NEB calculation is shown in Figure 12.7b, giving an energy barrier of 1.1 eV. Figure 12.7c shows the saddle-point atomic configuration of an emanating dislocation loop. From this structure, the shear displacement distribution on the slip plane is quantitatively extracted as shown in Figure 12.7d, which clearly demonstrates that the dislocation line is the boundary between slipped (red) and unslipped (blue) regions. The implications of those results on homogeneous versus heterogeneous dislocation nucleation at the crack tip are discussed by Zhu *et al.* [32].

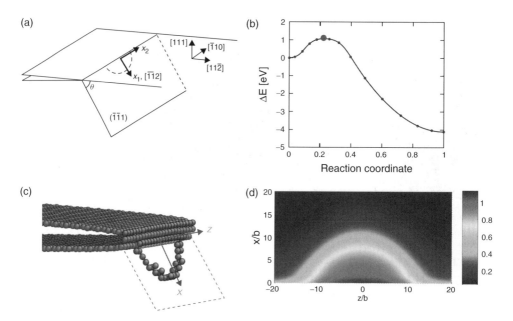

Figure 12.7 The NEB calculation of dislocation emission from a crack tip in single-crystal Cu [32]. (a) Orientations of the crack and the inclined {111} slip plane across which a dislocation loop nucleates. (b) MEP of emission of a 3D dislocation loop. (c) Saddle-point atomic structure. Atom color indicates coordination number, and atoms with perfect coordination ($N = 12$) are made invisible. (d) Contour plot of shear displacement distribution, normalized by the Burgers vector length $b = 1.476\,\text{Å}$, across the slip plane; this plot is extracted from (c)

12.3.2   Stress-Mediated Chemical Reactions

Chemical reaction rates in solids are known to depend on mechanical stress levels. This effect can be generally described in terms of a change of activation energy barrier in the presence of stress. A typical example is stress corrosion of silica (SiO_2) glass by water; the strength of the glass decreases with time when subjected to a static load in an aqueous environment [33]. The phenomenon, also known as delayed failure or static fatigue, essentially refers to the slow growth of pre-existing surface flaws as a result of corrosion by water in the environment. From a microscopic viewpoint, it is believed that the intrusive water molecules chemically attack the strained siloxane ($Si-O-Si$) bonds at the crack tip, causing bond rupture and formation of terminal silanol ($Si-OH$) groups which repel each other at the conclusion of the process [34]. This molecular-level mechanism, intrinsically, governs the macroscopic kinetics of quasi-static crack motion.

Stress corrosion of silica by water is studied by exploring the stress-dependent PES computed at the level of molecular orbital theory [35]. Figure 12.8a shows that an ordered silica nanorod with clearly defined nominal tensile stress is constructed to model a structural unit of the stressed crack tip. Using the NEB method, we are able to explicitly map out families of reaction pathways, parameterized by the continuous nominal stress. Figure 12.8b shows that three competing hydrolysis reaction pathways are determined, each involving

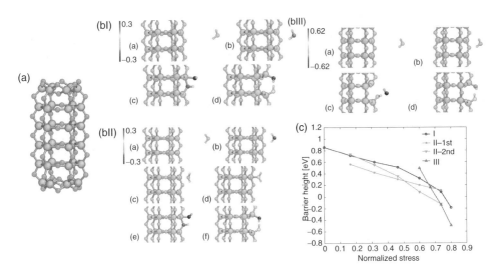

Figure 12.8 Stress-dependent molecular reaction pathways between an SiO_2 nanorod and a single water molecule [35]. (a) Structure of a fully relaxed SiO_2 nanorod. (b) MEPs of reactions, involving the characteristic initiation step: (I) water dissociation, (II) molecular chemisorption, and (III) direct siloxane bond rupture. Atoms are colored by charge variation relative to the initial configuration. (c) Comparison of the stress-dependent activation barriers for the three molecular mechanisms of hydrolysis

a distinct initiation step: (I) water dissociation, (II) molecular chemisorption, and (III) direct siloxane bond rupture.

Figure 12.8c compares the energy barriers as a function of stress for the three different mechanisms of hydrolysis reaction. Evidently, the tensile stress will reduce the activation energy barrier for any specific reaction mechanism. More importantly, as the relative barrier height of different mechanisms changes with an increase of stress, the switching of rate-limiting steps will occur either within one type of reaction pathway (e.g., the second reaction mechanism) or among different reaction mechanisms. The three reactions dominate at low, intermediate, and high stress levels, respectively.

12.3.3 Bridging Modeling with Experiment

As an important type of nanostructured metal, ultrafine crystalline Cu with nanoscale growth twins has attracted considerable attention in recent years. Experiments by Lu *et al*. [36] showed that nano-twinned Cu exhibited an unusual combination of ultrahigh strength and reasonably good tensile ductility, while most nanostructured metals have high strength and low ductility. As an essential step toward understanding the mechanistic origins of such extraordinary mechanical properties, the experiment showed that the nano-twinned Cu increases the rate sensitivity ($m \approx 0.02$) by up to an order of magnitude relative to microcrystalline metals with grain size in the micrometer range, and a concomitant decrease in the activation volume by two orders of magnitude (e.g., down to $\Omega \approx 20b^3$ when twin lamellae are approximately 20 nm thick) [37].

We have studied the mechanistic origin of decreased activation volume and increased rate sensitivity in the nano-twinned system [15]. Slip transfer reactions are simulated between a lattice dislocation and a coherent twin boundary (TB), involving dislocation absorption and desorption into the TB and slip transmission across the TB. The dislocation–interface reactions have previously been studied by MD simulations [38]. To overcome the well-known time-scale limitation of molecular dynamics, the free-end NEB method is used to determine the MEPs of the above-mentioned slip transfer reactions (Figure 12.9), such that atomistic predictions of yield stress, activation volume, and rate sensitivity can be directly compared with measurements from laboratory experiments on long time scales (seconds to minutes); see Table 12.1. The modeling predictions are consistent with experimental measurements, thereby showing that the slip transfer reactions are the rate-controlling mechanisms in nano-twinned Cu.

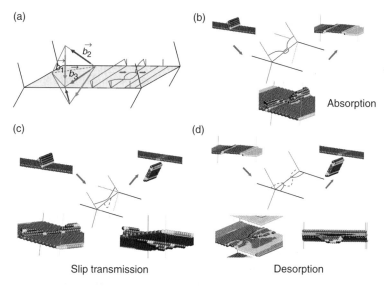

Figure 12.9 The free-end NEB modeling of twin-boundary-mediated slip transfer reactions [15]. (a) Schematics of dislocation–interface reactions based on double Thompson tetrahedra, showing different combinations of incoming and outgoing dislocations (the Burgers vectors with the same color) at a coherent TB. Atomic structures of the initial, saddle-point, and final states are shown for the slip transfer reactions, including (b) absorption, (c) direct transmission, and (d) desorption. Refer to Plate 10 for the colored version

Table 12.1 The free-end NEB calculations predict activation energy, activation volume, and yield stress for nano-twinned Cu, consistent with experimental measurements [15]

	Athermal stress threshold σ_{ath}	Activation volume Ω^*	Strain-rate sensitivity m
Uniaxial tension experiment	~1 GPa		
Nanoindentation experiment	>700 MPa	$12–22b^3$	0.025–0.036
Atomistic calculation	780 MPa	$24–44b^3$	0.013–0.023

12.3.4 Temperature and Strain-Rate Dependence of Dislocation Nucleation

Direct MD simulations have been widely used to explore the temperature and strain-rate dependence of defect nucleation. However, MD is limited to exceedingly high stresses and strain rates. To overcome this limitation, the statistical models have been developed that integrate transition-state theoretical analysis and reaction pathway modeling [21,24]. Such models require the atomistic input of the stress-dependent energy barriers of defect nucleation, which can be calculated by using the NEB method.

Consider, as an example, surface nucleation in a Cu nanopillar (Figure 12.10a) under a constant applied strain rate. Because of the probabilistic nature of the thermally activated nucleation processes, the nucleation stress has a distribution even if identical samples are used. The most probable nucleation stress is defined by the peak of the frequency distribution of nucleation events. Specifically, the statistical distribution of nucleation events is the product of a nucleation rate that increases in time and a likelihood of pillar

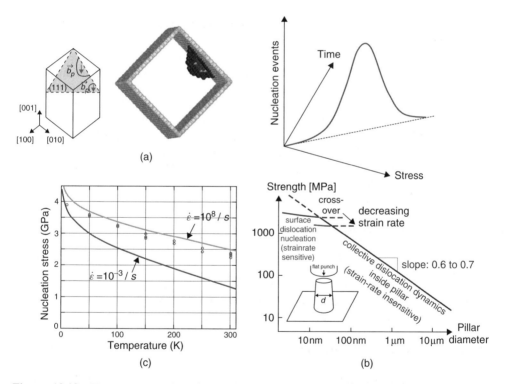

(a)

(c)

(b)

Figure 12.10 Temperature and strain-rate dependence of surface dislocation nucleation [21]. (a) Nucleation of a partial dislocation from the side surface of a single-crystal Cu nanowire under uniaxial compression. (b) Under a constant strain rate, a peak of nucleation events arises because of the two competing effects: the increasing nucleation rate with time (stress) and decreasing survival probability without nucleation. (c) Nucleation stress as a function of temperature and strain rate from predictions (solid lines) and direct MD simulations (circles). (d) Illustration of the surface effect on the rate-controlling process and the size dependence of yield strength in micro- and nano-pillars of diameter d under compression

survival without nucleation that decreases with time. These two competing effects lead to a maximum at a specific time (stress), as illustrated in Figure 12.10b. The nucleation stress, therefore, represents the most likely moment of nucleation under a particular loading rate. It is not a constant.

Based on the above nucleation statistics-based definition, we have developed a nonlinear theory of the most probable nucleation stress as a function of temperature and strain rate. Here, the nonlinearity arises primarily because of the nonlinear stress dependence of activation energy, which has been numerically determined using the NEB calculations [21]. A key result from these calculations is that the activation volumes associated with a surface dislocation source are in a characteristic range of $1-10b^3$, which is much lower than that of the bulk dislocation processes ($100-1000b^3$). The physical effect of such small activation volumes can be clearly seen from a simplified linear version of the theory, giving an analytic formula of the nucleation stress:

$$\sigma = \sigma_a - \frac{k_B T}{\Omega} \ln \frac{k_B T N \nu_0}{E \dot{\varepsilon} \Omega} \tag{12.17}$$

where σ_a is the athermal stress of instantaneous nucleation, E is Young's modulus, and N is the number of equivalent nucleation sites on the surface. Notice that the nucleation stress σ has a temperature scaling of $T \ln T$, and the activation volume Ω appears outside the logarithm, such that a small Ω associated with a surface source should lead to sensitive temperature and strain-rate dependence of nucleation stress, as quantitatively shown in atomistic simulations; see Figure 12.10c. In nano-sized volumes, surface dislocation nucleation is expected to dominate, as supported by recent experiment. As shown schematically in Figure 12.10d, the strength mediated by surface nucleation should provide an upper bound to the size–strength relation in nanopillar compression experiments. This upper bound is strain-rate sensitive because of the small activation volume of surface nucleation at ultra-high stresses.

12.3.5 Size and Loading Effects on Fracture

In the study of brittle fracture at the nanometer scale, questions often arise:

1. Is the classical theory of Griffith's fracture still applicable?
2. What is the influence of the discreteness of the atomic lattice?
3. Do the stress- and strain-controlled loadings make a difference?

These questions can be directly addressed by the NEB calculations of energetics of nano-sized cracks [39].

Recall that, in the Griffith theory of fracture [40], one considers a large body with a central crack of length $2a$; see Figure 12.11a, for example. Suppose the system is subject to an average tensile stress σ. This load can be imposed by either a fixed displacement or a constant force at the far field. Relative to the uncracked body under the same load (e.g., fixed displacement), the elastic energy decrease due to the formation of a crack of length $2a$ is $\pi \sigma^2 a^2 / E$ per unit thickness (where E is Young's modulus) and the corresponding increase of surface energy is $4\gamma a$, where γ is the surface energy density. As a result, the total energy change is $U(a) = 4\gamma a - p\sigma^2 a^2 / E$. The critical crack length of Griffith

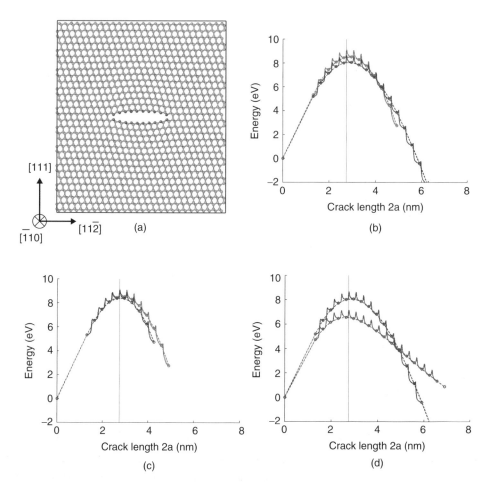

Figure 12.11 Size and loading effects on nanoscale fracture in single-crystal Si [39]. (a) Relaxed atomic structure with a central nanocrack at its critical length of Griffith fracture, $2a_{cr} = 3$ nm; the size of simulation cell is 9.1×10.1 nm^2. (b) The system's energy as a function of crack length under stress-controlled (red) and strain-controlled (blue) loading conditions; the size of simulation cell is 18.3×20.1 nm^2, doubling the width and height of the cell shown (a). (c) Sample-size effect on system's energy for stress-controlled fracture, showing the energy of the crack system in the cells of 18.3×20.1 nm^2 (red) and 9.1×10.1 nm^2 (brown). (d) Sample-size effect on system's energy for strain-controlled fracture, showing the energy of the crack system in the cells of 18.3×20.1 nm^2 (blue) and 9.1×10.1 nm^2 (green). Reprinted with permission from [39]. © 2009 Elsevier

fracture $2a_{cr}$ is defined in terms of the condition when $U(a)$ reaches the maximum, giving $a_{cr} = 2\gamma E/(\pi\sigma^2)$. Note that the displacement/strain-controlled and force/stress-controlled loadings give the same formula of $U(a)$ and a_{cr} in the classical fracture theory.

Figure 12.11b shows the atomistic calculations of $U(a)$ for single-crystal Si based on the Stillinger–Weber (SW) interatomic potential [41]. In the figure, a circle represents the energy of a metastable state with a nominal crack length $2a$ given by the number of broken bonds times the lattice spacing. The envelope curves connecting circles give $U(a)$ under stress-controlled (red) and strain-controlled (blue) loadings. The Griffith crack length can be determined by the maximum of $U(a)$, giving $2a_{cr} = 2.8$ nm. Alternatively, one can evaluate the Griffith crack length using the material constants of E and γ calculated from the SW potential, giving $2a_{cr} = 2.74$ nm, as indicated by vertical lines in Figure 12.11b–d. The agreement between the two methods for predicting the critical crack length (with a difference of less than one atomic spacing of 0.33 nm) suggests that the Griffith formula is applicable to the nanoscale fracture.

Energy barriers of crack extension arise because of the lattice discreteness, leading to the so-called lattice trapping effect [42]. In Figure 12.11b, each spike-like curve linking adjacent circles gives the MEP of breaking a single bond at the crack tip; that is, unit crack extension by one lattice spacing. The maximum of each MEP gives the energy barrier of bond breaking. Such MEPs manifest the corrugation of the atomic-scale energy landscape of the system due to the lattice discreteness. As a result, a crack can be locally "trapped" in a series of metastable states with different crack lengths and crack-tip atomic structures. The time-dependent kinetic crack extension then corresponds to the transition of the system from one state to another via thermal activation.

Figure 12.11 also demonstrates the effects of system size and loading method on the nanoscale brittle fractures. It is seen from Figure 12.11b that, when the system size is about 10 times larger than the crack size, the curves of $U(a)$ are close for stress-controlled (red) and strain-controlled (blue) fracture. Comparison of Figure 12.11c and d indicates that $U(a)$ for strain-controlled fracture is much more sensitive to the reduction of system size than the stress-controlled fracture. This is because the average stress in a sample under strain control can significantly change with crack length in small systems (i.e., the sample size is less than 10 times the crack length).

12.4 A Perspective on Microstructure Evolution at Long Times

A longstanding and still largely unresolved question in multiscale materials modeling and simulation is the connection between atomic-level deformation processes at the nanoscale and the overall system behavior seen on macroscopic time scales. Two special cases have been examined recently that may serve to elaborate on this issue [10]. One is the viscous relaxation in glassy liquids, where the shear viscosity increases sharply with supercooling. This is a phenomenon of temperature-dependent stress fluctuations, which is becoming amenable to atomistic simulations due to the development of a method for sampling microstructure evolutions at long times [43–45]. The other case is the slow structural deformation in solids, the phenomenon of anelastic stress relaxation and defect dynamics [46]. Although physically quite different, both problems involve microstructure developments that span many unit processes. The corresponding trajectories in these problems may be called TSP trajectories [43], to distinguish them from the MEPs discussed in Sections 12.2 and 12.3. In this chapter we have seen several atomistic studies where NEB is applicable, in which case one can evolve the system using transition-state theory and a knowledge of the activation energy of a particular transition. Since the traditional NEB method requires knowing the initial and final states of a reaction, tracking the system evolution over an extended time interval will require extension of the NEB methodology.

The search for an atomistic method capable of reaching the macro time scale has motivated the development of a number of simulation techniques, among them being hyperdynamics [47] and metadynamics [48,49], which are methods designed to accelerate the sampling of rare events. By the use of history-dependent bias potentials, metadynamics can enable the efficient sampling of the potential-energy landscapes, leading to the determination of the activation barriers and the associated rate constants through the transition-state theory. Here, we will briefly mention a metadynamics-based method which seems to be promising in dealing with deformation and reaction problems in nanomechanics.

12.4.1 Sampling TSP Trajectories

A procedure has recently been implemented that was designed to simulate situations where the system is trapped in a deep local minimum and, therefore, requires long times to get out of the minimum and continue exploration of the phase space [43]. The idea is to lift a system of particles out of an arbitrary potential well by a series of activation–relaxation steps. The algorithm is an adaptation of the metadynamics method originally devised to drive a system from its free-energy minima [48,49]. The particular procedure of activation–relaxation will be described here only schematically; for details, the reader should consult the original work [43].

Suppose our system starts out at a local energy minimum E_1^m, as shown in Figure 12.12a. To push the system out of its initial position a penalty function is imposed and the system is allowed to relax and settle into a new configuration (Figure 12.12b). The two-step process of activation and relaxation is repeated until the system moves to an adjacent local minimum, indicated in Figure 12.12c. As the process continues, the previous local minima are not visited because the penalty functions providing the previous activations are not removed; the system is, therefore, encouraged always to

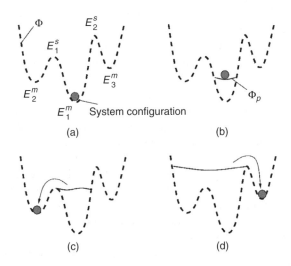

Figure 12.12 Schematic illustration explaining the autonomous basin climbing (ABC) method [43]. Dashed and solid lines indicate original PES and penalty potential, respectively. Penalty functions push the system out of a local minimum to a neighboring minimum by crossing the lowest saddle barrier. Reprinted with permission from [43]. © 2009 American Institute of Physics

sample new local minima, illustrated in Figure 12.12d. The sequence of starting from an initial local minimum E_1^m to cross a saddle point E_1^s to reach a nearby local minimum E_2^m, and so on, thus generates a TSP trajectory; an example of three local minima and two saddle points is depicted in Figure 12.12. We will refer to this algorithm as the autonomous barrier climbing (ABC) method. The most distinguishing feature of metadynamics is the way in which the bias potential is formulated. The bias potential is history dependent, in that it depends on the previous part of the trajectory that has been sampled [48,49].

12.4.2 Nanomechanics in Problems of Materials Ageing

We have recently started to explore how ABC could be used to provide an atomic-level explanation of mechanical behavior of materials governed by very slow microstructure evolution, involving time scales well beyond the reach of traditional atomistic methods [10]. Two problems were studied; one has to do with the temperature variation of the shear viscosity of glassy liquids [43–45] and the other concerns the relaxation behavior in solids undergoing creep deformation [46]. We believe there are other complex materials behavior problems where the understanding of long-time systems behavior may benefit from sampling of microscale processes of molecular rearrangements and collective inter-actions. As examples of a class of phenomena that could be broadly classified as materials ageing, we briefly introduce here an analogy between viscous flow and creep on the one hand and stress-corrosion cracking (SCC) and cement setting on the other hand [10]. All are challenges for atomistic modeling and simulation studies.

Figure 12.13 shows four selected functional behaviors of materials: temperature vari-ation of viscosity of supercooled liquids, stress variation of creep strain rate in steel, stress variation of crack speed in a glass, and time variation of the shear modulus of a cement paste. Each is a technologically important characteristic of materials noted for slow microstructure evolution. It is admittedly uncommon to bring them together to suggest a common issue in sampling slow dynamics. Qualitatively speaking, these prob-lems illustrate the possible extensions of the case studies that have been discussed in Section 12.3.

To bring out the commonality among the apparently different physical behaviors in Figure 12.13, we first note the similar appearance between Figure 12.13a and b, where the data are plotted against inverse temperature and stress, respectively. The presence of two distinct stages of variation may be taken to indicate two competing atomic-level responses. In the viscosity case we know the high-T regime of small η and the low-T stage of large η are governed by continuous collision dynamics and barrier hopping, respectively. One may expect analogous interplay between dynamical processes occurring at different spatial and temporal scales in the case of stress-driven creep. The variation of strain rate with stress and temperature, seen in Figure 12.13b, is a conventional way to characterize structural deformation.

The two stages are a low-stress strain rate, usually analyzed as a power-law exponent, and a high-stress high strain-rate regime. Attempts to explain creep on the basis of an atomic-level mechanism of dislocation climb are in its infancy [53]; the usefulness of TSP methods such as ABC is suggested by recent studies of self-interstitial [54] and vacancy clusters [55].

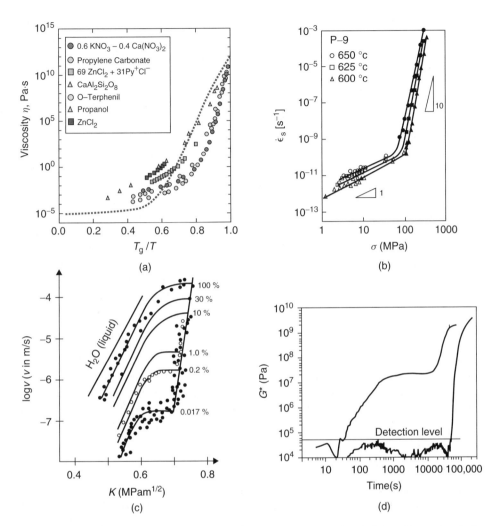

Figure 12.13 A collection of materials' behaviors illustrating the defining character of slow dynamics. (a) Temperature variation of shear viscosity of supercooled liquids [50]. Reprinted with permission from [56]. © 1988 Elsevier; (b) stress variation of strain rate in P-91 steel [51]. Reprinted with permission from [57]. © 2005 Maney Publishing; (c) stress loading variation of crack speed in soda-lime (NaOH) glass at various humidities [33]. Reprinted with permission from [33]. © 1967 Wiley; (d) time variation of shear modulus in hardening of Portland cement paste (water/cement ratio of 0.8) measured by ultrasound propagation [52]. Reprinted with permission from [58]. © 2004 EDP Sciences

Viscosity and creep are phenomena where the system microstructure evolves through cooperative rearrangements or lattice defect interactions without chemical (compositional or stoichiometric) changes. In contrast, the stress variation of the crack velocity in Figure 12.13c shows the classic three-stage behavior of SCC [56]. One can distinguish in the data a corrosion-dominated regime at low stress (stage I), followed by a plateau (stage II) where the crack maintains its velocity with increasing stress, and the onset of a rapid rise at high stress (stage III) where stress effects now dominate. Relative to Figure 12.13a and b, the implication here is that a complete understanding of SCC requires an approach where chemistry (corrosion) and stress effects are treated on an equal footing. This is an extension of the case study discussed in Section 12.3.2. The extension of ABC to reactions involving bond breaking and formation, which have been previously studied by MD simulation, is an area for further work.

Another reason we have for showing the different behaviors together in Figure 12.13 is to draw attention to a progression of microstructure evolution complexity, from viscosity to creep, to SCC, and to cement setting. In Figure 12.13d one sees another classic three-stage behavior in the hardening of cement paste [52]. The hydration or setting curve is known to everyone in the cement science community; on the other hand, an explanation in terms of molecular mechanisms remains elusive. A microstructure model of calcium silicate hydrate, the binder phase of cement, has recently been established [57] which could serve as starting point for the study of viscosity, creep deformation, and even the setting characteristics of cement. If viscosity or glassy dynamics is the beginning of this progression where the challenge lies in the area of statistical mechanics, followed by creep in the area of solid mechanics (mechanics of materials), then SCC and cement setting may be regarded as future challenges in the emerging area of chemo-mechanics.

Our brief outlook on viscous relaxation and creep deformation points to a direction for extending reaction pathway sampling to flow and deformation of matter in chemical and biomolecule applications. It seems to us that the issues of materials ageing and degradation in extreme environments should have fundamental commonality across a spectrum of physical and biological systems. Thus, one can expect that simulation-based concepts and algorithms allowing one to understand temporal evolution at the systems level in terms of molecular interactions and cooperativity will have enduring interest.

References

[1] Li, J. (2007) The mechanics and physics of defect nucleation. *MRS Bulletin*, **32**, 151–159.

[2] Zhu, T., Li, J., Ogata, S., and Yip, S. (2009) Mechanics of ultra-strength materials. *MRS Bulletin*, **34**, 167–172.

[3] Zhu, T. and Li, J. (2010) Ultra-strength materials. *Progress in Materials Science*, **55**, 710–757.

[4] Yip, S. (2003) Synergistic science. *Nature Materials*, **2**, 3–5.

[5] Yip, S. (2005) *Handbook of Materials Modeling*, Springer, Berlin.

[6] Yip, S. (2010) Multiscale materials, in *Multiscale Methods* (ed. J. Fish), Oxford University Press, New York, pp. 481–511.

[7] Jonsson, H., Mills, G., and Jacobsen, K.W. (1998) Nudged elastic band method for finding minimum energy paths of transitions, in *Classical and Quantum Dynamics in Condensed Phase Simulations* (eds. B.J. Berne, G. Ciccotti, and D.F. Coker), World Scientific, pp. 385–404.

[8] Henkelman, G., Uberuaga, B.P., and Jonsson, H. (2000) A climbing image nudged elastic band method for finding saddle points and minimum energy paths. *Journal of Chemical Physics*, **113**, 9901–9904.

[9] Henkelman, G., Johannesson, G., and Jonsson, H. (2000) Methods for finding saddle points and minimum energy paths, in *Progress on Theoretical Chemistry and Physics* (ed. S.D. Schwartz), Kluwer Academic, pp. 269–300.

[10] Kushima, A., Eapen, J., Li, J. *et al.* (2011) Time scale bridging in atomistic simulation of slow dynamics: viscous relaxation and defect mobility. *European Physical Journal B*, **82**, 271–293.

[11] Voter, A.F., Montalenti, F., and Germann, T.C. (2002) Extending the time scale in atomistic simulation of materials. *Annual Review of Materials Research*, **32**, 321–346.

[12] Vineyard, G.H. (1957) Frequency factors and isotope effects in solid state rate processes. *Journal of Physics and Chemistry of Solids*, **3**, 121–127.

[13] Weiner, J.H. (2002) *Statistical Mechanics of Elasticity*, Dover, New York.

[14] Mishin, Y., Farkas, D., Mehl, M.J., and Papaconstantopoulos, D.A. (1999) Interatomic potentials for monoatomic metals from experimental data and *ab initio* calculations. *Physical Review B*, **59**, 3393–3407.

[15] Zhu, T., Li, J., Samanta, A. *et al.* (2007) Interfacial plasticity governs strain rate sensitivity and ductility in nanostructured metals. *Proceedings of the National Academy of Sciences of the United States of America*, **104**, 3031–3036.

[16] Zhu, T., Li, J., and Yip, S. (2005) Nanomechanics of crack front mobility. *Journal of Applied Mechanics, Transactions ASME*, **72**, 932–935.

[17] Asaro, R.J. and Suresh, S. (2005) Mechanistic models for the activation volume and rate sensitivity in metals with nanocrystalline grains and nano-scale twins. *Acta Materialia*, **53**, 3369–3382.

[18] Lu, L., Zhu, T., Shen, Y.F. *et al.* (2009) Stress relaxation and the structure size-dependence of plastic deformation in nanotwinned copper. *Acta Materialia*, **57**, 5165–5173.

[19] Cahn, J.W. and Nabarro, F.R.N. (2001) Thermal activation under shear. *Philosophical Magazine A*, **81**, 1409–1426.

[20] Mott, N.F. (1948) Slip at grain boundaries and grain growth in metals. *Proceedings of the Physical Society of London*, **60**, 391–394.

[21] Zhu, T., Li, J., Samanta, A. *et al.* (2008) Temperature and strain-rate dependence of surface dislocation nucleation. *Physical Review Letters*, **100**, 025502.

[22] Warner, D.H. and Curtin, W.A. (2009) Origins and implications of temperature-dependent activation energy barriers for dislocation nucleation in face-centered cubic metals. *Acta Materialia*, **57**, 4267–4277.

[23] Hara, S. and Li, J. (2010) Adaptive strain-boost hyperdynamics simulations of stress-driven atomic processes. *Physical Review B*, **82**, 184114.

[24] Ryu, S., Kang, K., and Cai, W. (2011) Entropic effect on the rate of dislocation nucleation. *Proceedings of the National Academy of Sciences of the United States of America*, **108**, 5174–5178.

[25] Meyer, W. and Neldel, H. (1937) Über die Beziehungen zwischen der Energiekonstanten ε und der Mengenkonstanten a der Leitwerts-Temperaturformel bei oxydischen Halbleitern. *Zeitschrift für Technische Physik*, **12**, 588–593.

[26] Yelon, A., Movaghar, B., and Branz, H.M. (1992) Origin and consequences of the compensation (Meyer–Neldel) law. *Physical Review B*, **46**, 12244–12250.

[27] Yelon, A. and Movaghar, B. (1990) Microscopic explanation of the compensation (Meyer–Neldel) rule. *Physical Review Letters*, **65**, 618–620.

[28] Yelon, A., Movaghar, B., and Crandall, R.S. (2006) Multi-excitation entropy: its role in thermodynamics and kinetics. *Reports on Progress in Physics*, **69**, 1145–1194.

[29] Rice, J.R. and Thomson, R. (1974) Ductile versus brittle behavior of crystals. *Philosophical Magazine*, **29**, 73–97.

[30] Rice, J.R. and Beltz, G.E. (1994) The activation-energy for dislocation nucleation at a crack. *Journal of the Mechanics and Physics of Solids*, **42**, 333–360.

[31] Xu, G., Argon, A.S., and Ortiz, M. (1995) Nucleation of dislocations from crack tips under mixed-modes of loading – implications for brittle against ductile behavior of crystals. *Philosophical Magazine A*, **72**, 415–451.

[32] Zhu, T., Li, J., and Yip, S. (2004) Atomistic study of dislocation loop emission from a crack tip. *Physical Review Letters*, **93**, 025503.

[33] Wiederhorn, S.M. (1967) Influence of water vapor on crack propagation in soda-lime glass. *Journal of the American Ceramic Society*, **50**, 407–414.

[34] Michalske, T.A. and Freiman, S.W. (1982) A molecular interpretation of stress-corrosion in silica. *Nature*, **295**, 511–512.

[35] Zhu, T., Li, J., Lin, X., and Yip, S. (2005) Stress-dependent molecular pathways of silica–water reaction. *Journal of the Mechanics and Physics of Solids*, **53**, 1597–1623.

[36] Lu, L., Shen, Y.F., Chen, X.H. *et al.* (2004) Ultrahigh strength and high electrical conductivity in copper. *Science*, **304**, 422–426.

[37] Lu, L., Schwaiger, R., Shan, Z.W. *et al.* (2005) Nano-sized twins induce high rate sensitivity of flow stress in pure copper. *Acta Materialia*, **53**, 2169–2179.

[38] Jin, Z.H., Gumbsch, P., Ma, E. *et al.* (2006) The interaction mechanism of screw dislocations with coherent twin boundaries in different face-centred cubic metals. *Scripta Materialia*, **54**, 1163–1168.

[39] Huang, S., Zhang, S.L., Belytschko, T. *et al*. (2009) Mechanics of nanocrack: fracture, dislocation emission, and amorphization. *Journal of the Mechanics and Physics of Solids*, **57**, 840–850.

[40] Lawn, B. (1993) *Fracture of Brittle Solids*, Cambridge University Press, Cambridge.

[41] Stillinger, F.H. and Weber, T.A. (1985) Computer-simulation of local order in condensed phases of silicon. *Physical Review B*, **31**, 5262–5271.

[42] Thomson, R., Hsieh, C., and Rana, V. (1971) Lattice trapping of fracture cracks. *Journal of Applied Physics*, **42**, 3154–3160.

[43] Kushima, A., Lin, X., Li, J. *et al*. (2009) Computing the viscosity of supercooled liquids. *Journal of Chemical Physics*, **130**, 224504.

[44] Kushima, A., Lin, X., Li, J. *et al*. (2009) Computing the viscosity of supercooled liquids. II. Silica and strong–fragile crossover behavior. *Journal of Chemical Physics*, **131**, 164505.

[45] Li, J., Kushima, A., Eapen, J. *et al*. (2011) Computing the viscosity of supercooled liquids: Markov network model. *PLoS ONE*, **6**, e17909.

[46] Lau, T.T., Kushima, A., and Yip, S. (2010) Atomistic simulation of creep in a nanocrystal. *Physical Review Letters*, **104**, 175501.

[47] Voter, A.F. (1997) Hyperdynamics: accelerated molecular dynamics of infrequent events. *Physical Review Letters*, **78**, 3908–3911.

[48] Laio, A. and Parrinello, M. (2002) Escaping free-energy minima. *Proceedings of the National Academy of Sciences of the United States of America*, **99**, 12562–12566.

[49] Laio, A. and Gervasio, F.L. (2008) Metadynamics: a method to simulate rare events and reconstruct the free energy in biophysics, chemistry and material science. *Reports on Progress in Physics*, **71**, 126601.

[50] Angell, C.A. (1988) Perspective on the glass-transition. *Journal of Physics and Chemistry of Solids*, **49**, 863–871.

[51] Klueh, R.L. (2005) Elevated temperature ferritic and martensitic steels and their application to future nuclear reactors. *International Materials Reviews*, **50**, 287–310.

[52] Lootens, D., Hebraud, P., Lecolier, E., and Van Damme, H. (2004) Gelation, shear-thinning and shear-thickening in cement slurries. *Oil & Gas Science and Technology-Revue De L'Institut Francais Du Petrole*, **59**, 31–40.

[53] Kabir, M., Lau, T.T., Rodney, D. *et al*. (2010) Predicting dislocation climb and creep from explicit atomistic details. *Physical Review Letters*, **105**, 095501.

[54] Fan, Y., Kushima, A., and Yildiz, B. (2010) Unfaulting mechanism of trapped self-interstitial atom clusters in bcc Fe: a kinetic study based on the potential energy landscape. *Physical Review B*, **81**, 104102.

[55] Fan, Y., Kushima, A., Yip, S., and Yildiz, B. (2011) Mechanism of void nucleation and growth in bcc Fe: atomistic simulations at experimental time scales. *Physical Review Letters*, **106**, 125501.

[56] Ciccotti, M. (2009) Stress-corrosion mechanisms in silicate glasses. *Journal of Physics D: Applied Physics*, **42**, 214006.

[57] Pellenq, R.J.M., Kushima, A., Shahsavari, R. *et al*. (2009) A realistic molecular model of cement hydrates. *Proceedings of the National Academy of Sciences of the United States of America*, **106**, 16102–16107.

[58] Lootens, D., Hebraud, P., Lecolier, E., and Van Damme, H. (2004) Gelation, shear-thinning and shear-thickening in cement slurries. *Oil & Gas Science and Technology-Revue De L'Institut Francais Du Petrole*, 59:31.

13

Mechanics of Curvilinear Electronics

Shuodao Wang[1], Jianliang Xiao[2], Jizhou Song[3], Yonggang Huang[1], and John A. Rogers[4]

[1]*Northwestern University, USA*
[2]*University of Colorado, USA*
[3]*University of Miami, USA*
[4]*University of Illinois, USA*

13.1 Introduction

Biology is soft, elastic, and curved; electronics (using inorganic semiconductors) are not. Dominant forms of electronics are currently constrained to semiconductor technologies where *rigid*, *brittle* silicon-based materials are configured into *planar* layouts on wafers or plates of glass. A rapidly growing range of applications demands electronic systems that cannot be formed with hard, planar circuits that exist today. Research on mechanically unconventional forms of electronics began around 15 years ago through the development of polymer/organic semiconductors [1–8], mainly targeting applications such as paper-like displays [9,10]. The compliant organic semiconductors of those systems were key to their mechanical flexibility. However, owing to the high natural abundance of inorganic materials (especially silicon), together with their excellent reliability and high efficiency, many opportunities still exist for research into exploiting inorganics in unconventional geometry forms (ultrathin inorganics, nanotubes, and nanowires [11–13]), for electronic applications [14]. This research now extends to technically challenging opportunities in curved, bio-inspired designs, where the electronics is subject to large strains associated with stretching, compressing, twisting, and deforming into three-dimensional (3D) shapes [15–23], while maintaining high levels of electronic performance and reliability equal to established technology. Realization of curvilinear electronics will enable many innovative

Nano and Cell Mechanics: Fundamentals and Frontiers, First Edition. Edited by Horacio D. Espinosa and Gang Bao.
© 2013 John Wiley & Sons, Ltd. Published 2013 by John Wiley & Sons, Ltd.

application possibilities, such as curvilinear displays [24], electronic eye camera [25–27], smart surgical gloves [28], and structural health-monitoring devices [29].

Advanced strategies and procedures were recently introduced to conformally wrap two-dimensional planar electronic circuits onto arbitrarily curvilinear surfaces [30], enabling integration of conventional silicon-based electronics into the previously mentioned innovative systems of the future. The processes uses compressible circuit mesh structures (consisting of arrays of islands interconnected by narrow strips of conductive polyimide) and elastomeric transfer elements to wrap conformally curvilinear objects with complex shapes, such as the representative case of a conical surface illustrated in Figure 13.1. The steps include:

1. *Step A*. A thin, structured transfer element ($\sim$300 μm thick) is fabricated with an elastomer such as poly(dimethylsiloxane) (PDMS) by double casting and thermal curing against a target substrate, to replicate its surface geometry.
2. *Step B*. The resulting transfer element is radially stretched at the rim to form a drumhead membrane (top right frame).
3. *Step C*. The flat membrane contacts a prefabricated circuit with ultrathin, planar mesh geometry on silicon wafer, and then is removed to lift the circuit onto the membrane.
4. *Step D*. The radial tension is relaxed, thereby geometrically transforming the transfer element into the shape of the target object. During this process, the circuit on the

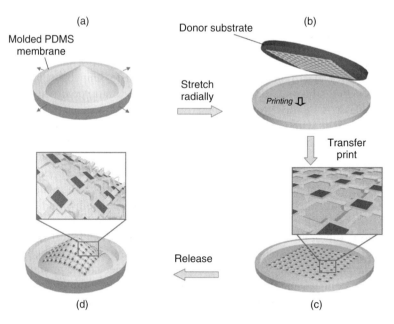

Figure 13.1 Schematic illustration for the use of compressible circuit mesh structures and elastomeric transfer element to wrap conformally curvilinear object. Reprinted with permission from [51], © 2010 The Royal Society of Chemistry

surface of the membrane follows the deformation, and the interconnect bridges buckle to adopt non-coplanar arch shapes in a way that accommodates the compressive forces and avoids significant strains in the circuit.

5. Finally, the target object is coated with a thin layer of adhesive and then the transfer element (with the circuit mesh on surface) delivers the device onto its surface.

Ko and coworkers [26,27,31] studied the buckling of a similar type of circuit mesh and showed that interconnect bridges always yield arch shapes when totally delaminated from

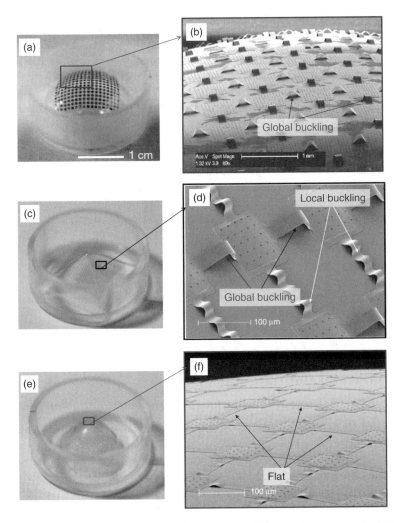

Figure 13.2 Experimentally observed buckling patterns of interconnect bridges in different areas of various shapes of curvilinear surfaces. Reprinted with permission from [26] © 2008 Nature Publishing Group, and [51] © 2010 The Royal Society of Chemistry

the transfer element (Figure 13.2a and b), which we refer to as *global buckling* in the present study. However, buckling of interconnect bridges on more complex shapes, such as a pyramid (Figure 13.2c), results in several different patterns: interconnect bridges along the circumferential direction buckle to a single arch (i.e., global buckling), but interconnect bridges along the meridional direction buckle to multiple, small arches and are partially delaminated from the transfer element, which we refer to as *local buckling* (Figure 13.2d). The interconnect bridges around the top of a parabola (Figure 13.2e–f) do not buckle at all; that is, *no buckling*. These different buckling modes, and the amount of compression induced during the wrapping processes described in Figure 13.1, result in different strains in the circuit mesh structures. Unlike the approaches of Vella *et al.* [32] and others [33–35] which investigate similar phenomena from a fracture mechanics point of view, the present study focuses on the buckling-driven delamination [36–40] observed in Figures 13.1 and 13.2. Here, the delamination is driven by the buckling of the stiff interconnect bridges, so the critical buckling condition must be satisfied prior to any out-of-plane deflection or interfacial delamination [40].

Mechanics models are developed to study the wrapping process, the buckling physics, and to estimate the strain in the brittle circuits. Section 13.2 gives analytically the strain distribution in both circumferential and meridional directions, for an arbitrary (axisymmetric) elastomeric transfer element stretched to flat (step B in Figure 13.1); the approaches in Section 13.2 are also used to track the positions of device islands of the circuits during the wrapping processes, demonstrated with its application to the fabrication of electronic eye cameras; the mechanisms for the different buckling behaviors (*global buckling*, *local buckling*, and *no buckling*) and a simple, analytical criterion separating different buckling modes are established in Section 13.3, which shows that the buckling patterns upon relaxation of tension observed in experiments (step D above) are determined by the strain distribution in the transfer element obtained in Section 13.2, and the work of adhesion between interconnect bridges and the transfer element; finally, based on the strain distribution from Section 13.2 and buckling patterns from Section 13.3, the strain in the stiff, brittle circuit mesh structure is determined analytically in Section 13.4.

13.2 Deformation of Elastomeric Transfer Elements during Wrapping Processes

13.2.1 Strain Distribution in Stretched Elastomeric Transfer Elements

For analysis of large deformations in curvilinear membranes (shells) [41–43], the mean and Gaussian curvatures [44,45] are widely used in the published literature, but such an approach usually cannot lead to simple, analytical solutions. However, for the special case of an axisymmetric transfer element, this section gives a simple, analytical model to obtain the strain in the stretched transfer element. Figure 13.3 illustrates this mechanics model, which corresponds to the same processes in Figure 13.1. This general strategy extends the model for a hemispherical target object (validated by the finite-element method) [26,46] to arbitrarily axisymmetric objects. The processes start with a thin elastomeric transfer element characterized by $R = R(z)$ in cylindrical coordinates (R, z), where $z \in (0, z_{max})$ (step A). This transfer element is radially stretched to a (nearly) flat state (step $\overline{AB}$), and

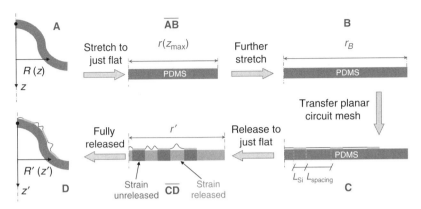

Figure 13.3 Schematic diagrams of the mechanics model for transferring compressible circuit mesh structures from planar arrays onto an arbitrarily axisymmetric curvilinear elastomeric transfer element. Reprinted with permission from [51], © 2010 The Royal Society of Chemistry

then further stretched to a flat, circular plate of radius r_B (step B). The prefabricated coplanar circuit mesh is then transfer printed [47–49] onto the transfer element (step C). Releasing of the radial tension relaxes the transfer element back to a new shape $R'(z')$ with height z'_{max} to be determined (step D), where (R', z') corresponds to the deformed position of original point (R, z), through an intermediate step $\overline{CD}$ that gives a (nearly) flat shape (similar to step $\overline{AB}$).

Finite-element analysis [26,46] has shown that, from step A to step $\overline{AB}$, the strain in the meridional direction is negligibly small, $\varepsilon_{meridional} \approx 0$, while the strain in the circumferential direction is much larger. In other words, during this process, the transfer element is mainly stretched in the circumferential direction. Therefore, a ring at the original height z $(0 \leq z \leq z_{max})$ in step A deforms (in step $\overline{AB}$) to a ring of the radius

$$r(z) = \int_0^z \sqrt{1 + \left[\frac{dR(y)}{dy}\right]^2}\, dy \tag{13.1}$$

which equals the arc length from 0 to z in step A. The circular plate of radius $r(z_{max})$ is then further stretched to radius of r_B in step B. This extra stretch imposes uniform strains in both meridional and circumferential directions, and eliminates compressive strain on the top surface of the membrane. The strains on the surface of elastomeric transfer element are given by [50]

$$\varepsilon_{circumferential} \approx \frac{r_B}{r(z_{max})}\frac{r(z)}{R(z)} - 1 \tag{13.2}$$

$$\varepsilon_{meridional} \approx \frac{r_B}{r(z_{max})} - 1 \tag{13.3}$$

Finite-element analysis is used to validate the above equations. For the parabolic elastomeric transfer element in Figure 13.2e, which has shape $R(z) = (7.58z)^{1/2}$ for $0 \leq z \leq 6.9$

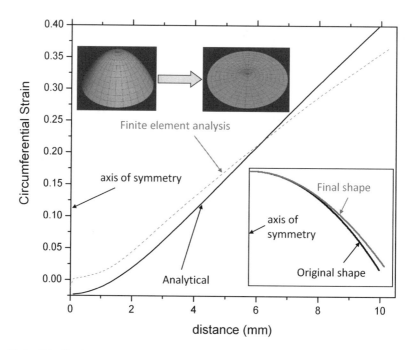

Figure 13.4 Distribution of circumferential strain for a parabolic elastomeric transfer element stretched to a flat state. The insets show the finite-element meshes for the initial and stretched configurations, as well as the initial and final shapes of the transfer element after the circuits are printed. Reprinted with permission from [51], © 2010 The Royal Society of Chemistry

(unit: mm) and thickness of 0.3 mm, Figure 13.4 shows that the distribution of circumferential strain given by Equation (13.2) agrees well with the finite-element analysis. The inset in Figure 13.4 shows the initial and the deformed meshes of the parabola transfer element when it is stretched to a flat state at $r_B = 10.3$ mm as in experiments. The meridional strain given by Equation (13.3) is very small, which is consistent with the small meridional strain ($\sim 2\%$) given by the finite-element method. Therefore, the simple, analytical expressions in Equations (13.2) and (13.3) provide good estimates of strains in the process of stretching arbitrarily axisymmetric elastomeric transfer elements to flat states (from steps A to B).

13.2.2 Deformed Shape of Elastomeric Transfer Elements

The circuit mesh is transferred to the flat elastomeric transfer element in step C. From steps C to $\overline{CD}$, the stretch in the transfer element is released except for that beneath the device islands, since the device islands (Young's modulus ~ 130 GPa for silicon [51]) are much stiffer than the elastomer (Young's modulus ~ 2 MPa for PDMS [52]). Therefore, the size of device islands L_{Si} remains unchanged, which gives their area fraction $f = NL_{Si}^2/\pi r_B^2$ at the flat state in step B, N being the number of islands in the circuit mesh. The average fractions of device islands and spacings along any direction are f and $1-f$, respectively. The radius of the circular plate in the intermediate step $\overline{CD}$ can be obtained by $r' = fr_B + (1-f)r(z_{max})$. Releasing the tension from steps $\overline{CD}$ to D is, similar to steps A to $\overline{AB}$, mainly in the circumferential direction. Applying similar analysis in both the circumferential and meridional directions results in [50]

$$R'(z') = fr_B \frac{r(z)}{r(z_{max})} + (1-f)R(z) \tag{13.4}$$

and

$$z' = \int_0^z \sqrt{(1-f)^2 + 2\frac{(1-f)fr_B}{r(z_{max})}\sqrt{1+\left[\frac{dR(y)}{dy}\right]^2}\left\{\sqrt{1+\left[\frac{dR(y)}{dy}\right]^2} - \frac{dR(y)}{dy}\right\}} \, dy \tag{13.5}$$

Equations (13.4) and (13.5) give the function R' for any given position (R, z) in the original shape. For $N = 3025$ (55×55) square device islands of $L_{Si} = 0.1$ mm on the parabolic transfer element stretched to $r_B = 10.3$ mm, the area fraction is $f = 0.0916$. The inset in Figure 13.4 shows that the initial shape $R(z)$ does not deform much because the area fraction f is small. The maximum height is reduced, also by a very small amount, from $z_{max} = 6.90$ mm to $z'_{max} = 6.64$ mm. The small change in shape helps to ensure conformal wrapping of the deformed transfer element onto the target object. As the area fraction (i.e., fill factor) f increases (e.g., $f > 50\%$), the final shape will be very different from the initial one, and could lead to extra deformations in the brittle circuits.

For the specific application in the fabrication of electronic eye cameras (with hemispherical shapes), the positions of the device islands (in this case, photodetectors) after the wrapping processes are important to the image quality and electronic performance of the camera. Wang and coworkers [26,46] used similar techniques to track the position of coplanar photodetector mesh (Figure 13.5a) during the wrapping processes described in Figure 13.1, for a hemispherical surface. Let R, φ_{max} (Figure 13.5b left frame, corresponds to step A in Figure 13.1) denote the radius and the spherical angle of the original shape of a hemispherical transfer element, respectively, and R' and φ'_{max} for the deformed shape (Figure 13.5b, right frame, corresponds to step D in Figure 13.1). The relations between the deformed and original shapes are given by [46]

$$R' = R(1-f)\left(1 + \frac{f}{1-f}\frac{r_B}{R\varphi_{max}}\right)^{3/2}, \qquad \varphi'_{max} = \frac{\varphi_{max}}{\sqrt{1 + \frac{f}{1-f}\frac{r_B}{R\varphi_{max}}}} \tag{13.6}$$

For the hemispherical shape shown in Figure 13.2a in experiments [46], the deformed shape ($R' = 13.9$ mm, $\varphi'_{max} = 58.1°$) obtained from Equation (13.6) agrees well with the finite-element analysis, which gives $R' = 13.8$ mm, $\varphi'_{max} = 55.5°$. The model is also used to track the positions of the device islands on the hemispherical surface. Let r_{pixel} denote the distance between the center of an island and the axis of symmetry on the circular plate in step C. The island moves radially towards the axis of symmetry from steps C to D; therefore, the spherical angle is given by

$$\varphi' = \varphi'_{max} \frac{r_{pixel}}{r_B} \tag{13.7}$$

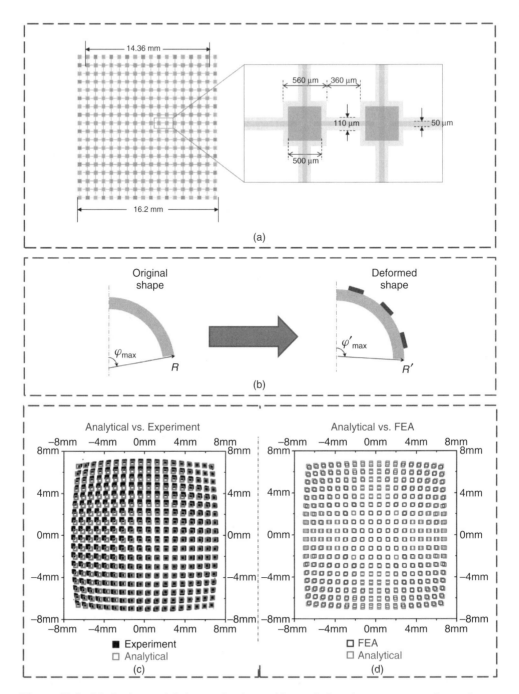

Figure 13.5 Mechanics model that tracks the positions of photodetectors on an electronic eye camera during the wrapping process: (a) the coplanar circuit geometry; (b) original and deformed shape of the hemispherical transfer element; (c, d) (top view) positions of photodetectors obtained by experiments, analytical model and finite-element analysis. Reprinted with permission from [26], © 2008 Nature Publishing Group, and [47], © 2009 American Institute of Physics

In the Cartesian coordinates, a photodetector with its center at $(x, y, 0)$ on the circular plate in step C moves onto to $[R', \varphi'_{max}(x^2 + y^2)^{1/2}/r, \tan^{-1}(y/x)]$ in step D. In experiments, the positions of the photodetectors are obtained by analyzing a top-view optical image, where each detector is noted as a black pad, with its center noted as a white dot in Figure 13.5c. Figure 13.5c also gives the top view of island positions (red square) generated by Equations (13.6) and (13.7), which agrees very well (<2% error) with experiments. The analytical results also agree well with the finite-element analysis using 3D continuum shell elements for the PDMS and conventional shell elements for the device islands, as shown in Figure 13.5d.

Equation (13.7) provides a simple way to design nonuniform pattern of circuits with coplanar geometry (step C), to achieve any desired pattern (e.g., uniform) on the deformed hemisphere (step D). The position at spherical coordinates (R', φ', θ') in step D has Cartesian coordinates $(r \cos\theta' \varphi'/\varphi'_{max}, r \sin\theta' \varphi'/\varphi'_{max}, 0)$ on the circular plate (i.e. the original coplanar circuit layouts) in step C, according to Equations (13.6) and (13.7). Furthermore, the original shape of the transfer element (step A) can be obtained analytically as $R = R'[1 - fr(R'\varphi'_{max})]^{3/2}/(1 - f)$ and $\varphi_{max} = \varphi'_{max}[1 - fr/(R'\varphi'_{max})]^{-1/2}$ from Equation (13.7) such that the final shape at step D has the desired radius R' and spherical angle φ'_{max}. The analysis provides a simple, analytical method to track the positions of the device islands on the surface of the transfer element, during the wrapping processes. This is important for applications where the positions of active device islands are critical for appropriate functioning of the curvilinear electronic system.

13.3 Buckling of Interconnect Bridges

Relaxation of the radially stretched elastomeric transfer element (with circuit mesh on its surface) leads to buckling of interconnect bridges. For each interconnect bridge, the tensile strain in step B (given analytically in Section 13.2) becomes a compressive strain upon relaxation of the stretch (from steps C to D). This compressive strain induces buckling of the interconnect bridges. Ko *et al.* [30] analyzed different buckling modes of interconnect bridges, which are summarized in the following. Their analysis on the buckling physics is combined with the strain distribution in Section 13.2 to determine the buckling patterns of interconnect bridges on complex curvilinear surfaces.

Let L and h denote the length and thickness of interconnect bridges, E the Young's modulus of the bridge, and γ the work of adhesion between the bridge and the transfer element. The interconnect bridge is subjected to the axial compressive strain ε (<0). Prior to buckling (*no buckling* mode) the bridge remains flat. The total potential energy U_{flat} is $U_{flat} = \frac{1}{2}EhL|\varepsilon|^2 - \gamma L$ [50].

For *global buckling*, the interconnect bridge buckles to an arch shape of $w = (A/2)[1 + \cos(2\pi x/L)]$ (the buckle amplitude A is to be determined). The out-of-plane displacement $w(x)$ satisfies the condition of vanishing displacement and slope at the two ends ($x = \pm L/2$). Energy minimization gives the buckle amplitude $A = (2L/\pi)(|\varepsilon| - \varepsilon_c)^{1/2}$, where $\varepsilon_c = \pi^2 h^2/(3L^2)$ is the critical Euler buckling strain. The total energy for global buckling is $U_{global} = EhL\varepsilon_c[|\varepsilon| - (\varepsilon_c/2)]$, which must be less than U_{flat} for global buckling to occur. This gives the critical strain for transition from flat to global buckling as [50]

$$\left(\frac{|\varepsilon|}{\varepsilon_c}\right)_{\text{flat}-\text{global}} = 1 + \sqrt{\frac{2\gamma}{Eh\varepsilon_c^2}} \tag{13.8}$$

For *local buckling*, the interconnect bridge buckles to multiple small sinusoidal arches, with amplitude a and wavelength l to be determined. The total potential energy is obtained as

$$U_{\text{local}} = \frac{\pi^2 Eh^3 L}{3l^2}\left(|\varepsilon| - \frac{\pi^2 h^2}{6l^2}\right) - \gamma L\left(1 - \frac{l}{L}\right)$$

where l could be solved from the equation

$$\frac{\pi^2 h^3}{3l^3}\left(\frac{\pi^2 h^2}{3l^2} - |\varepsilon|\right) + \frac{\gamma}{2EL} = 0$$

U_{local} must be less than U_{flat} for local buckling to occur, which gives the critical strain for transition from flat to local buckling [50]:

$$\left(\frac{|\varepsilon|}{\varepsilon_c}\right)_{\text{flat}-\text{local}} = 5\left(\frac{\gamma}{8Eh\varepsilon_c^2}\right)^{2/5} \tag{13.9}$$

The interfacial adhesion also plays a key role in determining the buckling patterns. For weak adhesion $\gamma/(8Eh\varepsilon_c^2) \leq 1$, the critical strain in Equation (13.9) (for flat–local transition) is larger than that in Equation (13.8) (for flat–global transition) such that global buckling occurs, and it never transitions to local buckling. For relatively strong adhesion $\gamma/(8Eh\varepsilon_c^2) > 1$, the critical strain in Equation (13.9) is smaller than that in Equation (13.8) such that local buckling occurs first. As the compressive strain $|\varepsilon|$ increases, global buckling occurs when $U_{\text{global}} < U_{\text{local}}$. This gives the critical strain for the transition from local to global buckling as

$$\left(\frac{|\varepsilon|}{\varepsilon_c}\right)_{\text{local}-\text{global}} = 1 + \frac{\gamma}{2Eh\varepsilon_c^2} \tag{13.10}$$

For the experimental case, the work of adhesion $\gamma = 0.16\,\text{J·m}^{-2}$ between PDMS [49,53,54] and the polyimide interconnect bridges ($E = 2.5$ GPa, $h = 1.4\,\mu\text{m}$, $L = 150\,\mu\text{m}$) corresponds to strong adhesion because $\gamma/(8Eh\varepsilon_c^2) = 70 \gg 1$. The normalized total potential energies are plotted in Figure 13.6, for no buckling, local buckling, and global buckling versus the normalized compressive strain $|\varepsilon|/\varepsilon_c$. Below the critical strain for local buckling 0.78% (Equation (13.9)), the total potential energy for no buckling is the lowest and the bridges remain flat. The critical strain 0.78% is much (27 times) larger than the Euler buckling strain $\varepsilon_c = 0.029\%$, due to the strong interfacial adhesion. Local buckling becomes the prevailing mode (with lowest energy) until the compressive strain reaches 8.0% (Equation (13.10)). For strain larger than 8.0%, global buckling then gives the lowest energy. The critical strains, 0.78% and 8.0% for local and global buckling, respectively, have been confirmed by experiments of one-dimensional arrays of islands and interconnect bridges [30] on a flat transfer element.

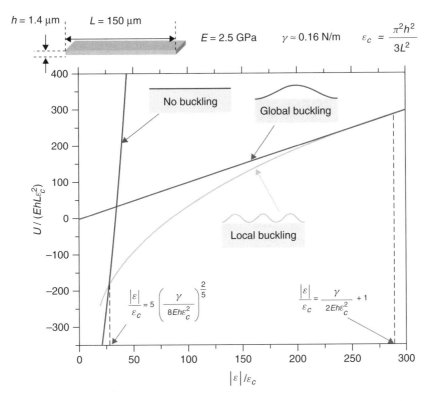

Figure 13.6 Comparison of the energy curves for global, local and no buckling modes. Reprinted with permission from [51], © 2010 The Royal Society of Chemistry

Figure 13.7 shows the deformation map for the different buckling modes (e.g., global, local, and no buckling), determined by range of compressive strain and work of adhesion. For weak adhesion $\gamma/(8Eh\varepsilon_c^2) \leq 1$, local buckling never occurs, and global buckling occurs once the normalized compressive strain reaches $1 + [2\gamma/(Eh\varepsilon_c^2)]^{1/2}$. For relatively strong adhesion $\gamma/(8Eh\varepsilon_c^2) > 1$, local buckling occurs when the compressive strain reaches $5[\gamma/(8Eh\varepsilon_c^2)]^{2/5}$, which is followed by global buckling at larger compressive strain of $1 + [\gamma/(2Eh\varepsilon_c^2)]$.

The results in Figure 13.7, with the strain distribution ε given in Section 13.2, provide a simple, robust way to determine the buckling patterns of interconnect bridges over different parts of curvilinear substrates (of arbitrary shapes), according to the following:

1. Use Equations (13.2) and (13.3) (or use the finite-element method for non-axisymmetric shapes) to determine the strain distribution ε along the meridional and circumferential directions on the curvilinear substrates.
2. Predict the buckling pattern (no buckling, local and global buckling) based on the deformation map Figure 13.7 and ε obtained above.
3. Use Equations (13.4) and (13.5) (or the finite-element method) to track the locations of interconnect bridges on the deformed shape of elastomeric transfer element.

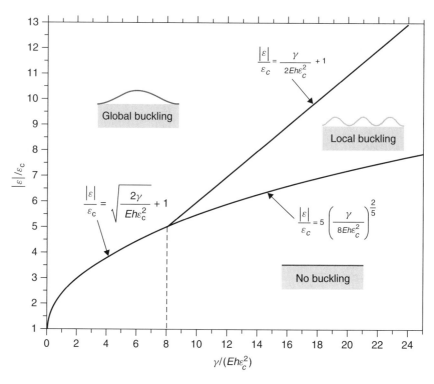

Figure 13.7   Deformation map for different buckling modes. Reprinted with permission from [30], © 2009 John Wiley and Sons

For example, Figure 13.8a shows analytically the circumferential strain distribution (from Section 13.2) for the parabolic elastomeric transfer element in Figure 13.2e (also Figure 13.4) stretched to $r_\mathrm{B} = 10.3$ mm. The contour values are set such that:

- red color is for the compressive strain larger than 8%, which predicts global buckling;
- blue color is for the compressive strain smaller than 1%, which predicts no buckling (flat interconnects); and
- all other colors are for compressive strain between 1% and 8%, which predicts local buckling.

The predicted buckling patterns for the parabolic transfer element, based on the above analysis, agree well with the experiment in Figure 13.8b. For example, Figure 13.8c clearly shows global buckling along the circumferential direction away from the peak, which is consistent with the red color in Figure 13.8a at the corresponding location. Figure 13.8d indicates no buckling in any direction around the peak, which corresponds to the blue color in Figure 13.8a. The strain in the meridional direction (not shown in Figure 13.8) is always small, which is consistent with local buckling (Figure 13.8c) away from the peak. The above approach also agrees well with experiments for many other complex shapes, such as pyramid, cone, or even golf ball shapes that have lots of dimples and edges [30], which are shown in Figures 13.9 13.10, and 13.11 [30], respectively.

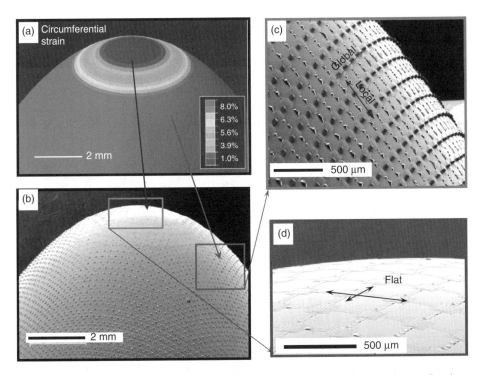

Figure 13.8 (a) Distribution of circumferential strain for a parabolic elastomeric transfer element given by the mechanics model. (b–d) Buckling patterns observed in experiments. Reprinted with permission from [51], © 2010 The Royal Society of Chemistry. Refer to Plate 11 for the colored version

13.4 Maximum Strain in the Circuit Mesh

For designs of curvilinear electronics, keeping the strain in the brittle circuits at a low level is one of the key issues for proper electronic functioning and reliability. Even for a large compressive strain ε on the curvilinear substrate, the strain in the circuit mesh is still very small because the interconnect bridges buckle to accommodate the compression. Based on the strain distribution from Section 13.2 and buckling patterns from Section 13.3, both analytical and finite-element models are developed in this section to determine the maximum strain in the island-interconnect structure of circuit mesh wrapped onto the deformed curvilinear surface (step D in Figure 13.1). The device islands consist of two layers of different materials: silicon (thickness h_{Si} and plane-strain modulus $\bar{E}_{Si}$) and polyimide (thickness h_{PI} and plane-strain modulus $\bar{E}_{PI}$). For an interconnect bridge subject to a large compressive strain ε (<0), global buckling will occur (see the analysis in Section 13.3). The maximum strain in interconnect bridges is given approximately by [31]

$$\varepsilon_{interconnect}^{max} \approx \frac{2\pi h}{L} \sqrt{|\varepsilon| - \varepsilon_c} \tag{13.11}$$

The device island is modeled as a composite plate of silicon and polyimide. The island has an equivalent bending stiffness $\overline{EI}_{island}$. The device island, sitting on

Pyramid

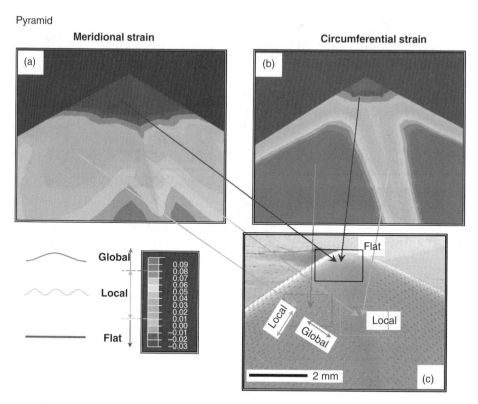

Figure 13.9 (a, b) Strain distributions in the meridional and circumferential directions around the pyramidal PDMS membrane. The corresponding buckling modes observed in experiments are shown in (c) for comparison. Reprinted with permission from [30], © 2009 John Wiley & Sons. Refer to Plate 12 for the colored version

the PDMS substrate, is subject to compressive force of $Eh\varepsilon_c$ and bending moment $M \approx \pi E h^3 \sqrt{|\varepsilon|/(1 + |\varepsilon|)}/3L$ resulting from buckled interconnect bridges. The strain due to the compressive force is negligible compared with that due to the bending. The maximum strain in the device islands is given by [31]

$$\varepsilon_{\text{island}}^{\max} = \frac{16\pi E h^3 h_{\text{Si}}}{(13\bar{E}_{\text{PI}} + \bar{E}_{\text{Si}})(h_{\text{Si}} + h_{\text{PI}})^3 L} \sqrt{\frac{|\varepsilon|}{1 + |\varepsilon|}} \qquad (13.12)$$

The finite-element method is used to validate the above approaches. Owing to symmetry, only one half of the device island and one half of the interconnect bridge are considered in the computational model. The polyimide interconnect bridge has Young's modulus $E = 2.5$ GPa, thickness $h = 1.4\,\mu\text{m}$, and half length $L/2 = 75\,\mu\text{m}$; the island is made of a layer of silicon (plane-strain modulus $\bar{E}_{\text{Si}} = 140$ GPa and thickness $h_{\text{Si}} = 700$ nm) and a layer of polyimide (plane-strain modulus $\bar{E}_{\text{PI}} = 2.83$ GPa and thickness $h_{\text{PI}} = 900$ nm),

Cone shape

Meridional strain   **Circumferential strain**

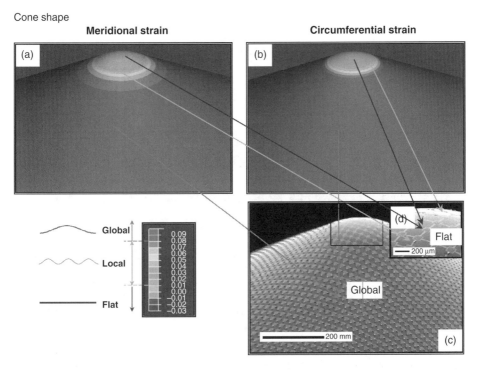

Figure 13.10 (a, b) Strain distributions in the meridional and circumferential directions for the conical-shape PDMS membrane. The corresponding buckling modes observed in experiments are shown in (c, d) for comparison. Reprinted with permission from [30], © 2009 John Wiley & Sons

and has a half length $50\,\mu\mathrm{m}$. The transfer element under the circuits (not shown in Figure 13.12b–d) has a thickness of $300\,\mu\mathrm{m}$ and plane-strain modulus of 2.7 MPa. For the stretch of elastomeric transfer element to $r_{\mathrm{B}} = 10.3$ mm as in experiments, the maximum strain $|\varepsilon|$ reaches 34.0% (given in Section 13.2) at the outermost point of the circuit mesh (9.62 mm from the center). Figure 13.12b–d shows the deformation and strain distribution of the circuit mesh, while Figure 13.12a gives the experimental image of the buckled shape. The finite-element analysis and the analytical expressions agree well: the maximum strain in the interconnect bridge is 3.2% (Figure 13.12d), which is very close to 3.4% given by Equation (13.11); the maximum strain in silicon is 0.16% (Figure 13.11c), which agrees with the 0.13% given by Equation (13.12). Therefore, Equations (13.11) and (13.12) provide simple and robust ways to estimate the strains in the circuits on curvilinear surfaces. These strains are much smaller than the frature strain of the corresponding materials ($\sim$1% for silicon and $\sim$7% for PI). No cracking or any other related mechanical failures is observed in the experiments (for even the most complicated shape of golf ball), which demonstrates the robustness of the techniques and promises its applications on complex biological systems in the future.

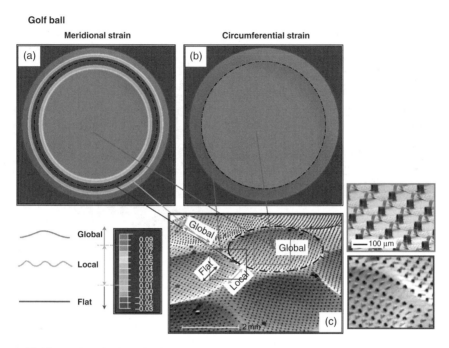

Figure 13.11 (a, b) Distributions in the meridional and circumferential directions around a golf ball dimple. The corresponding buckling modes observed in experiments are shown in (c). Reprinted with permission from [30]. © 2009 John Wiley & Sons

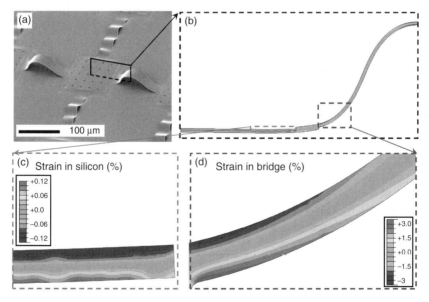

Figure 13.12 Buckled interconnect bridges and deformed device islands observed in experiments (a) and finite-element analysis (b). The strain distribution in the silicon part of the island and in the interconnect bridge are shown in (c) and (d). Reprinted with permission from [51], © 2010 The Royal Society of Chemistry

13.5 Concluding Remarks

An analytical model, validated by both finite-element analysis and experiments, is established for transfer of planar, silicon-based electronics circuits onto complex, curvilinear surfaces. The compressible circuits (consisting of silicon islands interconnected by narrow polyimide strips) have different buckling patterns, along different directions and over different parts of the curvilinear surface. It is shown that the adhesion between interconnect bridges and transfer elements, together with the compressive strain induced by wrapping, control the buckling pattern. The buckling patterns predicted by simple, analytical approaches agree well with the experiments. Maximum strains in the interconnect bridges and device islands are obtain analytically, and they agree well with the finite-element analysis. It is shown that even under large compressive strain, the strains in the brittle circuits remain low such that they do not exceed the fracture strains of corresponding materials. These outcomes provide a simple, robust method for the design and optimization for wrapping coplanar circuit mesh structure onto any curvilinear surface.

Acknowledgments

We acknowledge support from NSF Grant Nos. DMI-0328162, ECCS-0824129, OISE-1043143, and DOE, Division of Materials Sciences Grant No. DE-FG02-07ER46453. S. Wang gratefully acknowledges support from the Ryan Fellowship and the Northwestern University International Institute for Nanotechnology. J. Song acknowledges the supports from the Provost Award (University of Miami), the Ralph E. Powe Junior Faculty Enhancement Award (ORAU) and NSF Grant No. OISE1043161.

References

[1] Garnier, F., Hajlaoui, R., Yassar, A., and Srivastava, P. (1994) All-polymer field-effect transistor realized by printing techniques. *Science*, **265** (5179), 1684–1686.

[2] Bao, Z., Feng, Y., Dodabalapur, A. *et al.* (1997) High-performance plastic transistors fabricated by printing techniques. *Chemistry of Materials*, **9** (6), 1299–1301.

[3] Baldo, M.A., Thompson, M.E., and Forrest, S. R. (2000) High-efficiency fluorescent organic light-emitting devices using a phosphorescent sensitizer. *Nature*, **403** (6771), 750–753.

[4] Crone, B., Dodabalapur, A., Lin, Y.Y. *et al.* (2000) Large-scale complementary integrated circuits based on organic transistors. *Nature*, **403** (6769), 521–523.

[5] Loo, Y.-L., Someya, T., Baldwin, K.W. *et al.* (2002) Soft, conformable electrical contacts for organic semiconductors: high-resolution plastic circuits by lamination. *Proceedings of the National Academy of Sciences of the United States of America*, **99** (16), 10252–10256.

[6] Facchetti, A., Yoon, M.H., and Marks, T.J. (2005) Gate dielectrics for organic field-effect transistors: new opportunities for organic electronics. *Advanced Materials*, **17** (14), 1705–1725.

[7] Forrest, S.R., and Thompson, M.E. (2007) Introduction: organic electronics and optoelectronics. *Chemical Reviews*, **107** (4), 923–925.

[8] Sekitani, T., Noguchi, Y., Hata, K. *et al.* (2008) A rubberlike stretchable active matrix using elastic conductors. *Science*, **321** (5895), 1468–1472.

[9] Rogers, J.A., Bao, Z., Baldwin, K. *et al.* (2001) Paper-like electronic displays: large-area rubber-stamped plastic sheets of electronics and microencapsulated electrophoretic inks. *Proceedings of the National Academy of Sciences of the United States of America*, **98** (9), 4835–4840.

[10] Gelinck, G.H., Huitema, H.E.A., van Veenendaal, E. *et al.* (2004) Flexible active-matrix displays and shift registers based on solution-processed organic transistors. *Nature Materials*, **3** (2), 106–110.

[11] Sun, Y., and Rogers, J.A. (2007) Inorganic semiconductors for flexible electronics. *Advanced Materials*, **19** (15), 1897–1916.

[12] Cao, Q., and Rogers, J.A. (2009) Ultrathin films of single-walled carbon nanotubes for electronics and sensors: a review of fundamental and applied aspects. *Advanced Materials*, **21** (1), 29–53.

[13] Fan, Z., Ho, J.C., Takahashi, T. *et al.* (2009) Toward the development of printable nanowire electronics and sensors. *Advanced Materials*, **21** (37), 3730–3743.

[14] Rogers, J.A., Someya, T., and Huang, Y. (2010) Materials and mechanics for stretchable electronics. *Science*, **327** (5973), 1603–1607.

[15] Gray, D.S., Tien, J., and Chen, C.S. (2004) High-conductivity elastomeric electronics. *Advanced Materials*, **16** (5), 393–397.

[16] Hung, P., Jeong, K., Liu, G., and Lee, L. (2004) Microfabricated suspensions for electrical connections on the tunable elastomer membrane. *Applied Physics Letters*, **85** (24), 6051–6053.

[17] Lacour, S.P., Jones, J., Wagner, S. *et al.* (2005) Stretchable interconnects for elastic electronic surfaces. *Proceedings of the IEEE*, **93** (8), 1459–1467.

[18] Khang, D.-Y., Jiang, H., Huang, Y., and Rogers, J.A. (2006) A stretchable form of single-crystal silicon for high-performance electronics on rubber substrates. *Science*, **311** (5758), 208–212.

[19] Kim, D., Choi, W.M., Ahn, J. *et al.* (2008) Complementary metal oxide silicon integrated circuits incorporating monolithically integrated stretchable wavy interconnects. *Applied Physics Letters*, **93** (4), 044102.

[20] Kim, D.-H., Ahn, J.-H., Choi, W.M. *et al.* (2008) Stretchable and foldable silicon integrated circuits. *Science*, **320** (5875), 507–511.

[21] Kim, D.-H., Song, J., Choi, W.M. *et al.* (2008) Materials and noncoplanar mesh designs for integrated circuits with linear elastic responses to extreme mechanical deformations. *Proceedings of the National Academy of Sciences of the United States of America*, **105** (48), 18675–18680.

[22] Hsu, Y.-Y., Gonzalez, M., Bossuyt, F. *et al.* (2011) The effects of encapsulation on deformation behavior and failure mechanisms of stretchable interconnects. *Thin Solid Films*, **519** (7), 2225–2234.

[23] Kim, D.-H., Liu, Z., Kim, Y.-S. *et al.* (2009) Optimized structural designs for stretchable silicon integrated circuits. *Small*, **5** (24), 2841–2847.

[24] Crawford, G.P. (2005) *Flexible Flat Panel Displays*, John Wiley & Sons.

[25] Jin, H., Abelson, J., Erhardt, M., and Nuzzo, R. (2004) Soft lithographic fabrication of an image sensor array on a curved substrate. *Applied Physics Letters*, **22** (5), 2548–2551.

[26] Ko, H.C., Stoykovich, M.P., Song, J. *et al.* (2008) A hemispherical electronic eye camera based on compressible silicon optoelectronics. *Nature*, **454** (7205), 748–753.

[27] Shin, G., Jung, I., Malyarchuk, V. *et al.* (2010) Micromechanics and advanced designs for curved photodetector arrays in hemispherical electronic-eye cameras. *Small*, **6** (7), 851–856.

[28] Someya, T., Sekitani, T., Iba, S. *et al.* (2004) A large-area, flexible pressure sensor matrix with organic field-effect transistors for artificial skin applications. *Proceedings of the National Academy of Sciences of the United States of America*, **101** (27), 9966–9970.

[29] Nathan, A., Park, B., Sazonov, A. *et al.* (2000) Amorphous silicon detector and thin film transistor technology for large-area imaging of X-rays. *Microelectronics Journal*, **31** (11–12), 883–891.

[30] Ko, H.C., Shin, G., Wang, S. *et al.* (2009) Curvilinear electronics formed using silicon membrane circuits and elastomeric transfer elements. *Small*, **5** (23), 2703–2709.

[31] Song, J., Huang, Y., Xiao, J. *et al.* (2009) Mechanics of noncoplanar mesh design for stretchable electronic circuits. *Journal of Applied Physics*, **105** (12), 123516.

[32] Vella, D., Bico, J., Boudaoud, A. *et al.* (2009) The macroscopic delamination of thin films from elastic substrates. *Proceedings of the National Academy of Sciences of the United States of America*, **106** (27), 10901–10906.

[33] Cotterell, B. and Chen, Z. (2000) Buckling and cracking of thin films on compliant substrates under compression. *International Journal of Fracture*, **104** (2), 169–179.

[34] Gioia, G. and Ortiz, M. (1997) Delamination of compressed thin films, in *Advances in Applied Mechanics*, vol. 33 (eds. J.W. Hutchinson, and T.Y. Wu), Elsevier, pp. 119–192.

[35] Kendall, K. (1976) Preparation and properties of rubber dislocations. *Nature*, **261** (5555), 35–36.

[36] Moody, N.R., Reedy, E.D., Corona, E. *et al.* (2010) Buckle driven delamination in thin hard film compliant substrate systems. *EPJ Web of Conferences*, **6**, 40006.

[37] Hutchinson, J.W. (2001) Delamination of compressed films on curved substrates. *Journal of the Mechanics and Physics of Solids*, **49** (9), 1847–1864.

[38] Thouless, M.D., Hutchinson, J.W., and Liniger, E.G. (1992) Plane-strain, buckling-driven delamination of thin-films – model experiments and mode-II fracture. *Acta Metallurgica et Materialia*, **40** (10), 2639–2649.

[39] Hutchinson, J.W., Thouless, M.D., and Liniger, E.G. (1992) Growth and configurational stability of circular, buckling-driven film delaminations. *Acta Metallurgica et Materialia*, **40** (2), 295–308.

[40] Evans, A.G. and Hutchinson, J.W. (1984) On the mechanics of delamination and spalling in compressed films. *International Journal of Solids and Structures*, **20** (5), 455–466.

[41] Marder, M., Deegan, R.D., and Sharon, E. (2007) Crumpling, buckling, and crackling: elasticity of thin sheets. *Physics Today*, **60** (2), 33–38.

[42] Stumpf, H. and Makowski, J. (1987) On large strain deformations of shells. *Acta Mechanica*, **65** (1–4), 153–168.

[43] Taber, L.A. (1985) On approximate large strain relations for a shell of revolution. *International Journal of Non-Linear Mechanics*, **20** (1), 27–39.

[44] Yamauchi, H., Gumhold, S., Zayer, R., and Seidel, H.-P. (2005) Mesh segmentation driven by Gaussian curvature. *The Visual Computer*, **21** (8), 659–668.

[45] Jung, H.T., Lee, S.Y., Kaler, E.W. *et al.* (2002) Gaussian curvature and the equilibrium among bilayer cylinders, spheres, and discs. *Proceedings of the National Academy of Sciences of the United States of America*, **99** (24), 15318–15322.

[46] Wang, S., Xiao, J., Jung, I. *et al.* (2009) Mechanics of hemispherical electronics. *Applied Physics Letters*, **95** (18), 181912.

[47] Feng, X., Meitl, M.A., Bowen, A.M. *et al.* (2007) Competing fracture in kinetically controlled transfer printing. *Langmuir*, **23** (25), 12555–12560.

[48] Meitl, M.A., Zhu, Z.T., Kumar, V. *et al.* (2006) Transfer printing by kinetic control of adhesion to an elastomeric stamp. *Nature Materials*, **5** (1), 33–38.

[49] Huang, Y.G.Y., Zhou, W.X., Hsia, K.J. *et al.* (2005) Stamp collapse in soft lithography. *Langmuir*, **21** (17), 8058–8068.

[50] Wang, S., Xiao, J., Song, J. *et al.* (2010) Mechanics of curvilinear electronics. *Soft Matter*, **6** (22), 5757–5763.

[51] Hull, R. (1999) *Properties of Crystalline Silicon*, INSPEC, London.

[52] Wilder, E.A., Guo, S., Lin-Gibson, S. *et al.* (2006) Measuring the modulus of soft polymer networks via a buckling-based metrology. *Macromolecules*, **39** (12), 4138–4143.

[53] Newby, B.M.Z., Chaudhury, M.K., and Brown, H.R. (1995) Macroscopic evidence of the effect of interfacial slippage on adhesion. *Science*, **269** (5229), 1407–1409.

[54] Chaudhury, M.K. and Whitesides, G.M. (1991) Direct measurement of interfacial interactions between semispherical lenses and flat sheets of poly(dimethylsiloxane) and their chemical derivatives. *Langmuir*, **7** (5), 1013–1025.

14

Single-Molecule Pulling: Phenomenology and Interpretation

Ignacio Franco, Mark A. Ratner, and George C. Schatz
Northwestern University, USA

14.1 Introduction

In recent years, a number of techniques have been developed that allow us to access the properties of single molecules; for example, see Refs [1–8]. These windows into the single-molecule world are teaching us with unprecedented detail how molecules behave, including how they move, their mechanical behavior, their conductivity, reactivity, optical properties, and other properties. While measurements made on bulk matter are averages over a whole ensemble of molecules, single-molecule measurements highlight the contributions of the constituent parts to the ensemble. A key feature of the properties of single molecules is that fluctuations in the observables are comparable to the average values. This contrasts with macroscopic behavior, where such fluctuations are usually negligible; indeed, fluctuation-induced emergent phenomena can arise that, simply put, cannot be observed in the macroscale.

In this chapter we discuss basic aspects of single-molecule pulling experiments in which laser optical tweezers or atomic force microscopes (AFMs) are employed to exert mechanical forces on single molecules [2,9–12] and the resulting stress–strain response is used as a probe of the molecular potential energy surface. The focus will be on how to simulate single-molecule pulling experiments using molecular dynamics (MD), on what class of information can be extracted from the resulting force–extension isotherms, and on how to interpret the basic phenomena typically observed upon pulling.

Figure 14.1 shows a schematic of a single-molecule pulling experiment using AFMs (an equivalent setup exists for experiments that employ laser optical tweezers). In it, one end of a molecule is attached to a surface while the other end is attached to an AFM tip coupled to a cantilever. The distance L between the surface and the cantilever is controlled, while the molecular end-to-end distance $\xi(t)$ is allowed to fluctuate. The pulling is performed by varying $L(t)$ in some prescribed fashion. The force F exerted

Nano and Cell Mechanics: Fundamentals and Frontiers, First Edition. Edited by Horacio D. Espinosa and Gang Bao.
© 2013 John Wiley & Sons, Ltd. Published 2013 by John Wiley & Sons, Ltd.

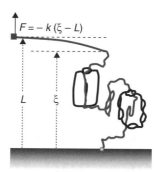

Figure 14.1 Schematic of a single-molecule pulling experiment using AFMs. One end of the molecular system is attached to a surface and the other end to an AFM tip attached to a cantilever. During the pulling, the distance L between the surface and the cantilever is controlled while the deflection of the cantilever from its equilibrium position measures the instantaneous applied force $F(t) = -k(\xi(t) - L)$, where k is the cantilever spring constant and $\xi(t)$ the fluctuating molecular end-to-end extension

during the process is determined by measuring the deflection of the cantilever from its equilibrium position $F = -k(\xi - L)$, where k is the cantilever stiffness. For a discussion of experimental aspects of single-molecule force spectroscopy, see Neuman and Nagy [13].

The experiments exert mechanical control over the molecular conformation while simultaneously making thermodynamic measurements of any mechanically induced unfolding events. The stress maxima observed during pulling have been linked to the breaking [11,14] of noncovalent interactions, thus providing insight into the secondary and tertiary structures of macromolecules. From the force–extension data it is possible to reconstruct the molecular free energy profile along the extension coordinate, quantifying in this way the thermodynamic changes undergone by the molecule during folding.

The structure of this chapter is as follows. Section 14.2 introduces the basic phenomenology observed during single-molecule pulling through a discussion of representative experiments. Section 14.3 establishes the statistical mechanical theory behind single-molecule pulling experiments and describes a method for reconstructing the molecular free energy profile along the extension coordinate from the force–extension data. Section 14.4, in turn, introduces direct and indirect methods that can be used to model single-molecule pulling experiments using MD. Section 14.5 focuses on the interpretation of single-molecule pulling experiments through an analysis of the basic structure of the molecular free energy profile. Section 14.6 summarizes the main topics discussed in this chapter.

14.2 Force–Extension Behavior of Single Molecules

In this section we consider a series of examples of single-molecule pulling experiments that illustrate the basic phenomenology that can be observed. Consider first the mechanically induced unfolding of an RNA hairpin [9]. The hairpin is attached to polystyrene beads via RNA–DNA hybrid handles and the pulling is performed in an optical

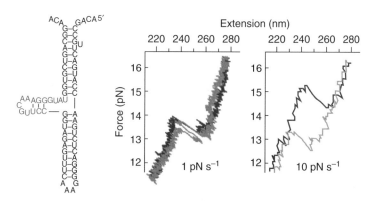

Figure 14.2 Force–extension behavior of an RNA hairpin obtained in an optical tweezer setup. The left panel shows the sequence and secondary structure of the main structural motif of the pulled RNA hairpin. The right panel plots detail the force–extension curves during stretching (black) and contraction (gray) when pulling under equilibrium (left) and nonequilibrium (right) conditions. Note that the x-axis is the molecular end-to-end distance ξ and not the distance L between the surface and the cantilever. Figure adapted from Liphardt *et al.* [9]

tweezers arrangement. Figure 14.2 shows the experimentally determined force–extension behavior and the details of the hairpin structure. For fast pulling speeds, hysteresis in the force–extension traces is evidenced by the fact that the traces observed upon pulling and subsequent contraction do not coincide. However, upon reduction of the pulling speed the traces obtained during extension and contraction do coincide, indicating that for all practical purposes the process occurs under reversible conditions. During pulling, the force initially increases monotonically, then at $\sim$14 pN it shows a drop, and then it increases again. In the process, the RNA hairpin undergoes a conformational transition from the hairpin structure to an extended coil. The drop in the force corresponds to the molecular unfolding event where the hydrogen bonding network holding the RNA hairpin together is broken. The local maximum in the force–extension isotherm is interpreted as the force required to break the secondary structure of the molecule.

The force–extension behavior of the RNA hairpin illustrates two basic phenomena that are commonly observed during single-molecule pulling. During pulling there is a region of *mechanical instability* where the average force decreases with the extension $\partial \langle F \rangle_L / \partial L < 0$. This region occurs when the pulled molecule undergoes a conformational transition from a stable folded structure to an extended system. Note that around the unfolding region the molecule hops between the folded and the unfolded states. That is, around the region of mechanical instability the molecule plus cantilever system exhibits *dynamical bistability* where the molecule unfolds and refolds, leading to blinking in the force measurement between a stronger force when the molecule is folded and a weaker force regime when extended. In Section 14.5 we discuss the origin of these two phenomena.

Figure 14.3 shows the force–extension characteristics of a related system, a 6838 base-pair long DNA hairpin, also obtained in an optical tweezers arrangement [15]. Even for this complex system it is possible to find experimentally accessible pulling speeds

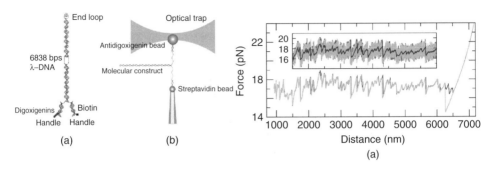

Figure 14.3   In this experiment the 6838 base-pair long DNA hairpin schematically shown in (a) is pulled in the optical tweezer arrangement in (b). The pulling is done by moving the optical trap relative to the pipette at low pulling speeds ($10\,\text{nm\,s}^{-1}$). (c) The force–extension profile during unfolding (in red) and refolding (in green). The inset shows the raw data obtained, while the smooth curves are time averaged. Except for some hysteresis in the last rip, the curves obtained during pulling and relaxing are almost identical. Figure adapted from Huguet *et al*. [15]

($10\,\text{nm\,s}^{-1}$) where reversible behavior is recovered in the mechanically induced melting process. Melting in this case refers to dehybridization of the base pairs, and we see that unlike Figure 14.2, there is an extended unfolding region where the average force is in the $15-19\,\text{pN}$ range while the length L increases. This corresponds to sequential dehybridization (believed to occur a few base pairs at a time) rather than a concerted breaking of all the base pairs at the same time. The inset in the figure shows the raw data obtained in the experiment, while the smooth curves shown are after further temporal averaging. The averaged data evidences the changes in the pulling force required to mechanically induce melting. It quantifies the strength of the type of hydrogen bonding encountered as the elongation proceeds.

The raw data evidence yet another fundamental aspect of this class of experiments: the substantial thermal fluctuations in the force measurements that are typically observed during pulling indicating that the system is not in the thermodynamic limit. This observation has the important consequence of leading to a nonequivalence between different statistical ensembles. In particular, it leads to a nonequivalence between the isometric ensemble (where L is controlled and F fluctuates) and the isotensional ensemble (where F is controlled and L fluctuates). Throughout, we will focus on the isometric ensemble, since this is the most common way to perform these experiments. For recent discussions of the nonequivalence between ensembles [16–18] for an interpretation of the pulling phenomenology in the isotensional ensemble.

As a third example, consider the reversible force–extension behavior of a synthetic polymer pulled in poor and good solvent conditions using an AFM [19]. Figure 14.4 shows the force–extension isotherm of an individual polystyrene (PS) chain pulled in toluene (good solvent) and water (poor solvent). In toluene (Figure 14.4a), the deformation of PS exhibits a worm-like chain [20] behavior with a persistence length of 0.25 ± 0.05 nm. In water (Figure 14.4b) the behavior is qualitatively different. In this case, the force initially rises linearly up to $\sim 13\,\text{pN}$, then exhibits a plateau force at $\sim 13\,\text{pN}$, followed by

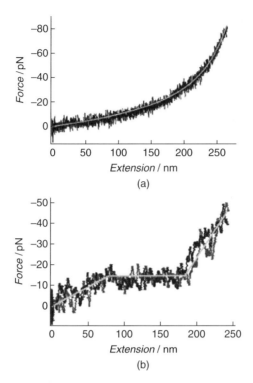

Figure 14.4   Reversible pulling of individual polystyrene chains in toluene (top panel) and water (bottom panel) obtained using an AFM setup. Figure adapted from Gunari *et al.* [19]

a third regime where the force rises again. This behavior is well predicted by a model [21] in which the initial linear restoring force is due to deformation of a collapsed spherical globule to an ellipsoid during which the polymer's surface energy increases. In this model, the plateau is associated with a conformational transition involving the coexistence of the globule and a stretched chain of polymer or polymer globules. For large stretching, where coexistence is no longer possible, the force extension behavior of a Gaussian chain [20] is recovered. Note that large-scale fluctuations in the force measurements are also evident in this system.

Consider last the forced unfolding of a recombinant polyprotein composed of eight repeats of the Ig27 domain of human titin obtained in an AFM setup [22]. Figure 14.5 shows the force–extension profile when pulling at a constant speed under nonequilibrium conditions. As the polyprotein is extended, the force exerted on the molecule increases until one of the Ig27 domains suddenly unfolds. Unfolding of a domain reduces the overall force, but with increasing extension another local subunit unfolds. This occurs until all the domains are unfolded and the protein–tip interaction is broken. This results in the sawtooth pattern shown in Figure 14.5, where each peak in the force–extension profile marks the unfolding of one subunit in the polyprotein and the last peak signals the detachment of the protein from the tip.

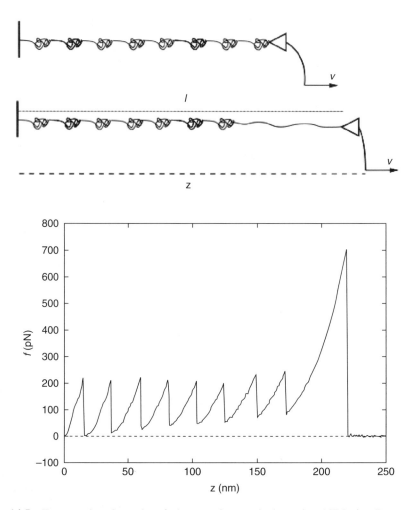

Figure 14.5 Top panel: schematic of the protein attached to the AFM tip. Bottom panel: force–extension behavior of the polyprotein composed of eight repeats of the Ig27 domain. Reprinted with permission from [22]. Copyright 2008 Institute of Physics

14.3 Single-Molecule Thermodynamics

In what follows we specify what we mean about thermodynamics in the context of single-molecule pulling experiments [12] and the class of thermodynamic information that is accessible from the force extension isotherms. The thermodynamic system consists of the molecule plus cantilever coupled to a thermal bath, typically the surrounding solvent. As such, the equilibrium state of the system is well described by the canonical ensemble with configurational partition function

$$Z(L) = \int d\mathbf{r} \, \exp[-\beta U_L(\mathbf{r})], \qquad (14.1)$$

where r denotes the coordinates of the N atoms of the molecule, $\beta = 1/k_B T$ is the inverse temperature, and

$$U_L(r) = U_0(r) + V_L[\xi(r)] \tag{14.2}$$

is the potential energy function of the molecule plus restraints. Here, $U_0(r)$ is the molecular potential energy plus potential restraints not varied during the experiment and $V_L[\xi(r)]$ is the potential due to the cantilever at extension L. To a good approximation, the quantity $V_L[\xi(r)]$ is given by a harmonic function

$$V_L[\xi(r)] = \frac{k}{2}[\xi(r) - L]^2 \tag{14.3}$$

of stiffness k. The instantaneous force exerted by the cantilever on the molecule is determined by measuring the deflection of the cantilever from its equilibrium position and it is given by

$$F = -\nabla V_L = \frac{\partial V_L}{\partial L} = -k[\xi(r) - L], \tag{14.4}$$

where the gradient ∇ is with respect to the $[\xi(r) - L]$ coordinate. In writing Equations (14.3) and (14.4), the vector nature of L, $\xi(r)$, and F has been obviated, since these quantities are typically collinear in the single-molecule pulling setup. For notational simplicity, in Equation (14.1) and throughout this chapter we ignore the constant multiplicative factor that makes partition functions dimensionless, since this is irrelevant for determining relative free energy changes.

14.3.1 Free Energy Profile of the Molecule Plus Cantilever

Because the state of the molecule plus cantilever under equilibrium conditions is well described by a canonical ensemble, the force exerted during the pulling yields the associated change in the Helmholtz free energy

$$\Delta A = A(L) - A(L_0) = -\frac{1}{\beta} \ln \frac{Z(L)}{Z(L_0)} \tag{14.5}$$

for the composite molecule plus cantilever system. To see this, note that in the canonical ensemble the average force at a given extension L is given by

$$\langle F \rangle_L = \frac{1}{Z(L)} \int dr \left(\frac{\partial V_L}{\partial L}\right) \exp[-\beta U_L(r)] = -\frac{1}{\beta} \frac{1}{Z(L)} \frac{\partial Z(L)}{\partial L} = \frac{\partial A}{\partial L}. \tag{14.6}$$

Therefore, if the pulling is done under reversible conditions, such that the state of the system is well described by a canonical ensemble at each step of the pulling, the change in A when pulling from L_0 to L is determined by the reversible work exerted in the process

$$\Delta A = \int_{L_0}^{L} \frac{\partial A}{\partial L'} dL' = \int_{L_0}^{L} \langle F \rangle_{L'} dL' = W_{\text{rev}}. \tag{14.7}$$

Remarkably, even when the pulling is performed under nonequilibrium conditions, it is still possible to determine the free energy changes during pulling by means of the Jarzynski nonequilibrium work fluctuation relation [23–25]

$$\exp(-\beta \Delta A) = \langle \exp(-\beta W) \rangle_{\text{noneq}} \tag{14.8}$$

where the average in this expression is over nonequilibrium realizations of a given pulling protocol starting from a system initially prepared in the canonical ensemble. Equation (14.8) relates the nonequilibrium work with the equilibrium free energy changes of the system and holds arbitrarily far away from equilibrium. Note that by using Jensen's inequality $(e^{\langle x \rangle} \leq \langle e^x \rangle)$, Equation (14.8) implies that the average nonequilibrium work is greater than or equal to the free energy change

$$\langle W \rangle_{noneq} \geq \Delta A, \tag{14.9}$$

which is the usual statement of the second law of thermodynamics.

The above analysis associates a free energy change to the pulling process although the system is not in the thermodynamic limit. This association relies on the assumption that the equilibrium state of the system is well described by a classical canonical ensemble. This feature has been experimentally tested: Liphardt *et al.* [26] experimentally demonstrated the validity of the Jarzynski equality (Equation (14.8)). The Jarzynski equality supposes that the initial equilibrium state of the system is given by a classical canonical partition function and, hence, the test of the Jarzynski equality also tests whether this is a good description of the equilibrium state of the system.

14.3.2 Extracting the Molecular Potential of Mean Force $\phi(\xi)$

The properties that are measured during the pulling are those of the molecule plus cantilever. Thus, direct integration of the force–extension isotherms yields the free energy profile of the composite system; see Equation (14.7). However, what one is really interested in are the properties of the molecule itself. Specifically, one is interested in the molecular Helmholtz free energy profile (also known as the potential of mean force (PMF) [27]) along the extension coordinate ξ. The PMF succinctly captures thermodynamic changes undergone by the molecule during folding and determines the basic phenomenology that is observed upon pulling [18,28].

A useful method to extract the molecular PMF $\phi(\xi)$ from the force versus extension measurements is the weighted histogram analysis method (WHAM) [29–33]. The WHAM analysis combines all the force–extension data collected during the pulling to properly remove the bias due to the cantilever and extract the PMF. In fact, one can view the single-molecule pulling process as an experimental realization of the WHAM methodology using harmonic potential restraints. We now describe how to extract the PMF from equilibrium force–extension data using the WHAM. A schematic of the process is shown in Figure 14.6.

The PMF is defined by [31]

$$\phi(\xi) = \phi(\xi^\star) - \frac{1}{\beta} \ln \left[\frac{p_0(\xi)}{p_0(\xi^\star)} \right], \tag{14.10}$$

where $\xi^\star$ and $\phi(\xi^\star)$ are arbitrary constants and

$$p_0(\xi) = \frac{\int d\mathbf{r}\, \delta[\xi - \xi(\mathbf{r})] \exp\left[-\beta U_0(\mathbf{r})\right]}{\int d\mathbf{r} \exp[-\beta U_0(\mathbf{r})]} = \frac{Z_0(\xi)}{Z_0} = \langle \delta[\xi - \xi(\mathbf{r})] \rangle. \tag{14.11}$$

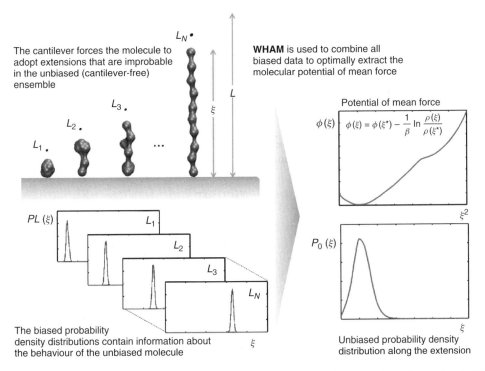

Figure 14.6 Scheme of the free energy reconstruction from single-molecule pulling. The PMF $\phi(\xi)$ along the extension coordinate ξ is determined by the probability density distribution $p_0(\xi)$ (Equation (14.10)). Sampling $p_0(\xi)$ directly is challenging, because there are regions along ξ that have vanishingly small probability of being observed. In a single-molecule pulling experiment, the molecule is forced to adopt those improbable extensions ξ by being subjected to the bias due to the cantilever. For each L visited ($L_1, L_2, \cdots, L_N$) during pulling, the experiments measure the probability density distribution $p_L(\xi)$ along ξ in the presence of the cantilever bias. The $p_L(\xi)$ contains information about the unbiased probability density distribution $p_0(\xi)$ (Equation (14.16)). The WHAM combines all sampled $p_L(\xi)$ to properly remove the cantilever bias and extract $p_0(\xi)$ with minimal variance in the estimate

Here,

$$Z_0 = \int d\mathbf{r} \exp[-\beta U_0(\mathbf{r})] \tag{14.12}$$

is the configurational partition function of the molecule and

$$Z_0(\xi) = \int d\mathbf{r}\, \delta[\xi - \xi(\mathbf{r})]\, \exp[-\beta U_0(\mathbf{r})]. \tag{14.13}$$

In the context of the pulling experiments, $p_0(\xi)$ is the probability density that the molecular end-to-end distance function $\xi(\mathbf{r})$ adopts the value ξ in the unbiased (cantilever-free) ensemble. The quantity $p_0(\xi)$ determines the PMF up to a constant and can be estimated from the force measurements, as we now describe.

At this point it is convenient to discretize the pulling process and suppose that M extensions $L_1, \ldots, L_i, \ldots, L_M$ are visited during pulling. The potential of the system plus cantilever at extension L_i is $U_i = U_0 + V_i$, where $V_i = \frac{k}{2}[\xi(r) - L_i]^2$ is the potential bias due to the cantilever. In the ith biased measurement, knowledge of the force F_i and length L_i gives the molecular end-to-end distance. The probability density of observing the value ξ at this given extension is given by

$$p_i(\xi) = \frac{\int d\mathbf{r}\, \delta[\xi - \xi(r)] \exp[-\beta U_i(r)]}{\int d\mathbf{r} \exp[-\beta U_i(r)]} = \frac{Z_i(\xi)}{Z_i} = \langle \delta[\xi - \xi(r)] \rangle_i, \qquad (14.14)$$

where $Z_i = \int d\mathbf{r} \exp[-\beta U_i(r)]$ is the configurational partition function for the system plus cantilever at the ith extension and

$$Z_i(\xi) = \int d\mathbf{r}\, \delta[\xi - \xi(r)] \exp[-\beta U_i(r)]. \qquad (14.15)$$

In principle, the unbiased probability density $p_0(\xi)$ can be reconstructed from each $p_i(\xi)$ since, by virtue of Equations (14.11) and (14.14), these two quantities are related by

$$p_0(\xi) = \exp[+\beta V_i(\xi)] \frac{Z_i}{Z_0} p_i(\xi), \qquad (14.16)$$

where we have exploited the fact that $Z_0(\xi) = \exp[+\beta V_i(\xi)]Z_i(\xi)$. In practice, this approach will not work because the range of ξ values where $p_0(\xi)$ and $p_i(\xi)$ differ significantly from zero need to overlap [32]. Nevertheless, the experiments do provide measurements at a wealth of values of L_i that can be combined to estimate $p_0(\xi)$:

$$p_0(\xi) = \sum_{i=1}^{M} w_i \exp[+\beta V_i(\xi)] \frac{Z_i}{Z_0} p_i(\xi), \qquad (14.17)$$

where the w_i are some normalized ($\sum_i w_i = 1$) set of weights. In the limit of perfect sampling any set of weights should yield the same $p_0(\xi)$. In practice, for finite sampling it is convenient to employ a set that minimizes the variance in the $p_0(\xi)$ estimate from the series of independent estimates of biased distributions. Such a set of weights is precisely provided by the WHAM prescription [29,30,32]

$$w_i = \frac{\exp[-\beta V_i(\xi)] \dfrac{Z_0}{Z_i g_i}}{\sum_{i=1}^{M} \exp[-\beta V_i(\xi)] \dfrac{Z_0}{Z_i g_i}}, \qquad (14.18)$$

where $g_i = 1 + 2\tau_i$ is the statistical inefficiency [34] and τ_i the integrated autocorrelation time [29,34]. The difficulty in estimating the τ_i for each measurement generally leads to further supposing that g_i are approximately constant and factor out of Equation (14.18), so that

$$p_0(\xi) = \frac{\sum_{i=1}^{M} p_i(\xi)}{\sum_{i=1}^{M} \exp[-\beta V_i(\xi)]Z_0/Z_i}. \qquad (14.19)$$

Neglecting g_i from Equation (14.19) does not imply that the resulting estimate of $p_0(\xi)$ is incorrect, but simply that the weights selected do not precisely minimize the variance in the estimate. Experience with this method indicates that, if the g_i's do not differ by more than an order of magnitude, their effect on $\phi(\xi)$ is small [30,35].

The computation of the PMF then proceeds as follows. Suppose that N_i force measurements are done for each extension i. Knowledge of the force F_i^j ($j = 1, \ldots, N_i$) and of L_i gives the molecular end-to-end distance ξ_i^j in each of these measurements. The numerator in Equation (14.19) is then estimated by constructing a histogram with all the available data:

$$\sum_{i=1}^{M} p_i(\xi) \approx \sum_{i=1}^{M} \sum_{j=1}^{N_i} \frac{C_i^j(\xi)}{N_i \Delta \xi}, \tag{14.20}$$

where $\Delta \xi$ is the bin size, and $C_i^j(\xi) = 1$ if $\xi_i^j \in [\xi - \Delta\xi/2, \xi + \Delta\xi/2)$ and zero otherwise. Estimating the denominator in Equation (14.19) requires knowledge of Z_i, the configurational partition functions of the system plus cantilever at all extensions $\{L_i\}$ considered. There are two ways to obtain these quantities. If reversible force–extension data are available, then the most direct one is to employ the Helmholtz free energies for the system plus cantilever obtained through the thermodynamic integration in Equation (14.7), as they determine Z_i up to a constant multiplicative factor. Alternatively, it is also possible to determine the ratio of the configurational partition functions between the ith biased system and its unbiased counterpart using $p_0(\xi)$:

$$\frac{Z_i}{Z_0} = \int d\xi \exp[-\beta V_i(\xi)] p_0(\xi). \tag{14.21}$$

Equations (14.19) and (14.21) can be solved iteratively. Starting from a guess for the Z_i/Z_0, $p_0(\xi)$ is estimated using Equation (14.19) and normalized. The resulting $p_0(\xi)$ is then used to obtain a new set of Z_i/Z_0 through Equation (14.21), and the process is repeated until self-consistency. This latter approach is the usual procedure to solve the WHAM equations. For computational implementations of the WHAM methodology [35–38].

14.3.3 Estimating Force–Extension Behavior from $\phi(\xi)$

The PMF is central to the interpretation of the phenomena observed upon pulling and in the estimation of the force–extension behavior when pulling with cantilevers of arbitrary stiffness [28]. Its utility relies on the fact that the configurational partition function of the molecule plus cantilever at extension L (Equation (14.1)) can be expressed as

$$Z(L) = \int d\xi \exp\{-\beta[\phi(\xi) + V_L(\xi)]\}, \tag{14.22}$$

where we have neglected the constant and L-independent multiplicative factor that arises in the transformation since it is irrelevant for the present purposes. The above relation implies that the extension process can be viewed as thermal motion along a one-dimensional effective potential determined by the PMF $\phi(\xi)$ and the bias due to the cantilever $V_L(\xi)$:

$$U_L(\xi) = \phi(\xi) + V_L(\xi). \tag{14.23}$$

Further, by means of the PMF it is possible to estimate the force versus extension characteristics for the composite system for any value of the force constant k. This is because the average force exerted on the system at extension L can be expressed as

$$\langle F \rangle_L = -\frac{1}{\beta} \frac{\partial \ln [Z(L)/Z_0]}{\partial L} = \frac{\int d\xi \, [\partial V_L(\xi)/\partial L] \exp\{-\beta[\phi(\xi) + V_L(\xi)]\}}{\int d\xi \exp\{-\beta[\phi(\xi) + V_L(\xi)]\}}, \quad (14.24)$$

where, for convenience, we have introduced Z_0 in the logarithm just to make the argument dimensionless. Since $\phi(\xi)$ is a property of the isolated molecule it is independent of k. Hence, Equation (14.24) can be employed to estimate the F–L isotherms for arbitrary k. Another quantity of interest is the probability density in the force measurements as a function of the extension L, given by

$$p_L(F) = \langle \delta[F - F(\xi)] \rangle_L = \frac{\int d\xi \, \delta[F - F(\xi)] \exp\{-\beta[\phi(\xi) + V_L(\xi)]\}}{\int d\xi \exp\{-\beta[\phi(\xi) + V_L(\xi)]\}}, \quad (14.25)$$

as this quantity highlights the fluctuations in the force measurements during pulling.

14.4 Modeling Single-Molecule Pulling Using Molecular Dynamics

14.4.1 Basic Computational Setup

Computationally, the forced unfolding of single molecules can be studied using "pulling" or "steered" MD simulations [39–41]. The general setup of this type of simulation is schematically described in Figure 14.7. The stretching computation begins by attaching

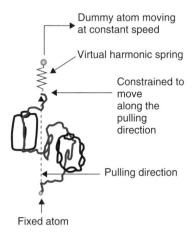

Dummy atom moving
at constant speed

Virtual harmonic spring

Constrained to
move
along the
pulling
direction

Pulling direction

Fixed atom

Figure 14.7 Setup of an MD pulling simulation. In it, one end of the molecule is fixed while the opposite end is attached to a dummy atom via a virtual harmonic spring. The position of the dummy atom is the simulation analogue of the cantilever position and is controlled throughout. The pulling is performed by moving the dummy atom away from the molecule at a constant speed. The fluctuating deflection of the harmonic spring measures the force exerted during the pulling and is the simulation analog of the cantilever deflection in the AFM pulling experiment

one end of the molecule to a rigid isotropic harmonic potential that mimics the molecular attachment to the surface. Simultaneously, the opposite molecular end is connected to a dummy atom via a virtual harmonic spring. The position of the dummy atom is the simulation analogue of the cantilever position L, and is controlled throughout. In turn, the varying deflection of the virtual harmonic spring measures the force exerted during the pulling. The stretching is caused by moving the dummy atom away from the molecule at a constant speed. The pulling direction is defined by the vector connecting the two terminal atoms of the complex. Since cantilever potentials are typically stiff in the direction perpendicular to the pulling, in the simulations the terminal atom that is being pulled is typically forced to move along the pulling direction by introducing appropriate additional restraining potentials.

In the simulations, the molecule plus any surrounding solvent are assumed to be weakly coupled to a heat bath and the effect of the heat bath is modeled through a thermostat. Thermostats that are useful in simulating the force spectroscopy include the Nosé–Hoover chain [32,42–44] and the Langevin thermostat [45]. In turn, velocity rescaling thermostats, like the Berendsen thermostat [46], can be problematic because they can lead to a spurious violation of the second law during the pulling because the ensemble generated by them does not satisfy the equipartition theorem [47]. The thermal dynamics of the molecule, cantilever, and any surrounding solvent are then followed in an NVT or NPT ensemble using MD. Freely available MD codes, such as NAMD [48] and GROMACS [49], have pulling routines implemented in them. We have, in addition, developed [28] a pulling routine for TINKER [50] and created a graphical interface for it called MOLpull [36]. MOLpull is accessible through the NanoHub (http://nanohub.org/) and can be used to perform the pulling and reconstruct the PMF along the extension coordinate of arbitrary molecules.

14.4.2 Modeling Strategies

One of the challenges in simulating pulling experiments using MD is to bridge the large disparity between the pulling speeds v that are employed experimentally and those that can be accessed computationally. While typical experiments often use $v \approx 1 - 10^{-3} \, \mu m \, s^{-1}$, current computational capabilities require pulling speeds that are several (approximately five to eight) orders of magnitude faster.

One possible strategy that can be used to overcome this difficulty is to strive for pulling speeds that are slow enough that reversible behavior is recovered. Under such conditions, the results become independent of v, and the simulations comparable to experimental findings. The way to determine if the pulling is done under quasi-static conditions is by making sure that the force–extension curves during extension and subsequent contraction essentially coincide, such that the net work exerted during a pulling–retraction cycle is negligible. The advantage of this approach is that there is a systematic way to test if quasi-static behavior is recovered by estimating the amount of work dissipated in a pulling–retraction cycle. Its limitation is that the pulling speeds that are required to obtain reversible behavior may be computationally unfeasible for pulling large macromolecules. In fact, the examples that exist in the literature where reversible behavior has been computationally recovered are for relatively small molecular systems [28,51,52] using pulling speeds between $10^{-4} - 10^{-7} \, nm \, ps^{-1}$. When pulling large macromolecules, present computational limitations require using pulling speeds of $\sim 10^{-3}\,\mathrm{nm\,ps^{-1}}$ [53], which is far too fast for recovering reversible behavior even in simple systems.

A second possible strategy is to pull under irreversible conditions. A single nonequilibrium trajectory can recover some qualitative structural features encountered during the pulling [41,53], but it is not sufficient to reconstruct the molecular PMF. In order to reconstruct the PMF it is necessary to pull many times and use a nonequilibrium work fluctuation theorem [23–25,54] to reconstruct the free energy profile [22,55–61]. The advantage of this strategy is that it can be used when equilibrium data is difficult to obtain; for example, during the forced unfolding of a complex protein like the one in Figure 14.5. Note, however, that this strategy requires adequately estimating the average of an exponential quantity (Equation (14.8)) that often has poor convergence properties [60,62]. For a computational implementation of some nonequilibrium methods to reconstruct the PMF, [38].

As a third *indirect* alternative, it is possible to reconstruct the molecular PMF without pulling continuously, but rather by sampling the fluctuating force at selected points along the extension coordinate. For each L the system is allowed to thermally equilibrate and the force is measured for a given amount of simulation time. The data obtained are used to reconstruct the PMF using WHAM as described in Section 14.3, and then the PMF is employed to estimate the force–extension behavior of the molecule plus cantilever. While the free energy reconstruction using continuous equilibrium force–extension data uses a data set with a large (essentially continuous) set of L values with very few data points for the force at each L, this strategy works in a different limit where one has few extensions L but extensive sampling of the force at each extension. The advantage of this procedure is that it can be used to reconstruct the PMF of molecules that are computationally challenging to pull under equilibrium or near-equilibrium conditions. The limitation is that there is no entirely satisfactory way to judge when the system has achieved a state of thermal equilibrium in the MD trajectory. This contrasts with the direct strategy involving explicit pulling, in which it is easy to judge when quasi-static behavior has been recovered by measuring the amount of dissipative work in extension–contraction cycles. As in the direct equilibrium pulling, the sampling for each L has to be sufficient to capture all relevant molecular events that are consistent with the extension.

Another important aspect that has to be taken into consideration when modeling single-molecule pulling is the cantilever stiffness employed. Experimentally accessible AFM cantilevers stiffnesses are in the 10^{-1}–$10^4\,\mathrm{pN\,\mathring{A}^{-1}}$ range [13]. While experiments tend to favor soft cantilever potentials because they provide high resolution in the force measurements, stiff cantilevers are preferable in the simulations. This is because, when employing stiffer cantilever potentials: (i) a shorter range of L values is required to explore all the molecular extension coordinates, thus reducing the simulation time; (ii) the possible values of ξ that are accessible for the molecule for a given L are further restricted, making it faster to sample all accessible conformational space. As a consequence of (ii), the amount of dissipative work in a pulling–retraction cycle decreases with increasing cantilever stiffness for fixed pulling speed [63]. In our experience with molecular pulling, a rule of thumb that we frequently use is that a cantilever stiffness of $\sim 1\,\mathrm{pN\,\mathring{A}^{-1}}$ can be regarded as soft and one with stiffness of $\sim 10^2\,\mathrm{pN\,\mathring{A}^{-1}}$ as stiff. These numbers, however, will clearly depend on the particular system that is being pulled.

14.4.3 Examples

Direct Modeling: Oligorotaxanes

As an example of an explicit simulation of a single-molecule pulling experiment, consider the forced unfolding of a [3] rotaxane [28,64,65]. Figure 14.8 shows the force–extension curves during the extension (blue) and subsequent contraction (red) for two different pulling speeds and a schematic of the structure of the molecule that is being pulled. The oligorotaxane[1] consists of two cyclobis(paraquat-p-phenylene) tetracationic cyclophanes (the blue boxes) threaded onto a linear chain (in red) composed of three naphthalene units linked by oligoethers with oligoether caps at each end. As the system is stretched, the oligorotaxane undergoes a conformational transition from a folded globular state to an extended coil. In the process, the force initially increases approximately linearly with L, then drops, and subsequently increases again. For the faster pulling speed, hysteresis in the force–extension cycle is evident, while for $v = 10^{-4}$ nm ps^{-1} the F–L curves obtained during extension and contraction essentially coincide and, for all practical purposes, the pulling occurs under equilibrium conditions. Note that thermal fluctuations in the force measurements are roughly comparable to the average values, indicating that the system is far from the thermodynamic limit.

The force–extension behavior predicted by the MD simulations for the oligorotaxane is qualitatively similar to that experimentally observed for the RNA hairpin (recall Figure 14.2). Specifically, there is a region of mechanical instability where the force drops with increasing extension ($\partial \langle F \rangle_L / \partial L < 0$). Further, if one fixes L around the region of mechanical instability ($L = 70.0$ Å in this case) and allows the system to evolve, the molecule undergoes conformational transitions between a folded globular state and an extended coil; see Figure 14.9. The right panel in Figure 14.9 shows the probability density

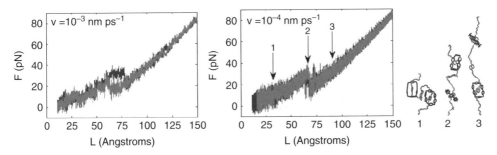

Figure 14.8 Simulated force–extension behavior of a [3] rotaxane immersed in a high dielectric medium at 300 K obtained using MD and the MM3 [66] force field. The plots show the force–extension curves during stretching (blue) and retraction (red) using pulling speeds v of 10^{-3} and 10^{-4} nm ps^{-1}. The cantilever employed has a soft spring constant of $k = 1.1$ pN Å^{-1}. Snapshots of structures (labels 1, 2 and 3) encountered during pulling are shown on the far right. Refer to Plate 13 for the colored version (in small)

---

[1] To be precise, we actually deal with pseudorotaxanes that lack the steric blocking units at the chain termini that force the cyclophanes to remain attached to the oligomeric chain. For simplicity, we will use the term rotaxane.

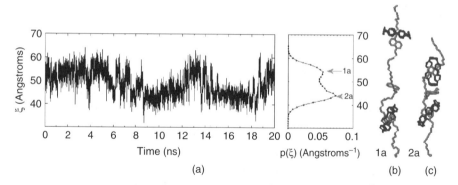

Figure 14.9   Dynamical bistability when L is fixed around the region of mechanical instability. The figure shows the time dependence and probability density distribution of the end-to-end molecular extension ξ when the [3] rotaxane plus cantilever is constrained to reside in the unstable region of Figure 14.8, with $L = 70.0$ Å. Typical structures encountered in this regime (labels 1a and 2a) are shown on the right

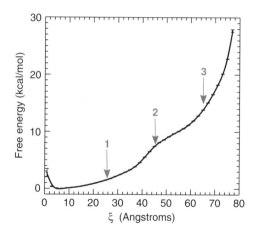

Figure 14.10   Molecular potential of mean force for the [3] rotaxane along the end-to-end distance ξ. The solid lines result from a spline interpolation of the available data points. Typical structures (labels 1–3) are shown in Figure 14.8

of the distribution of ξ values obtained from a 20 ns trajectory. The system exhibits a clear dynamical bistability along the ξ coordinate. As in the RNA case, the bistability along the end-to-end distance leads to blinking in the force measurements from a high-force to a low-force regime during the pulling.

The molecular PMF of the [3] rotaxane reconstructed from the force–extension data is shown in Figure 14.10. The thermodynamic native state of the oligorotaxane, i.e. the minimum in $\phi(\xi)$, corresponds to a globular folded structure with an average end-to-end distance of 7 Å. The PMF consists of a convex region (a region of positive curvature) for $\xi < 44$ Å when the molecule is folded, another convex region for $\xi > 52$ Å when the molecule is extended, and a region of concavity (where $\partial^2\phi/\partial\xi^2 < 0$) for $\xi = 44$–52 Å

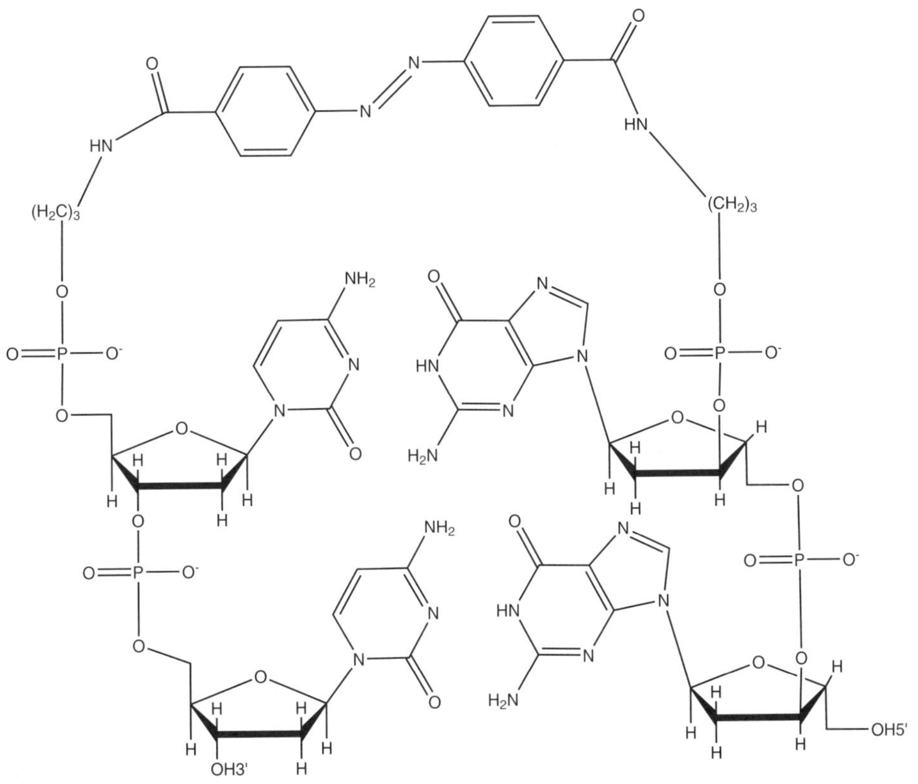

Figure 14.11 DNA dimer consisting of two guanine–cytosine base pairs connected by a *trans*-azobenzene linker

where the conformational transition occurs. The region of concavity along the PMF arises because the molecule is inherently bistable along some natural unfolding pathway. As discussed in Section 14.5, this characteristic structure of the PMF is responsible for many of the interesting features observed during pulling.

Indirect Modeling: DNA Dimer

As an example of indirect modeling of the force spectroscopy, we now consider the forced extension of the DNA dimer shown in Figure 14.11 along the O5′–O3′ coordinate in explicit water [67]. The dimer is composed of two guanine–cytosine base pairs capped by a *trans*-azobenzene linker. Computationally, it is described by the CHARMM27 force field and the total system is composed of the 178 DNA hairpin atoms, 4 sodium ions, and 1731 TIP3P water molecules in a $27.1 \times 34.7 \times 56.3$ $Å^3$ box in the NVT ensemble at 300 K (see [67] for details). Even for this relatively simple system it is challenging to computationally obtain equilibrium force–extension data by pulling directly because the pulling has to be slower than any characteristic time scale of the system. An indirect approach is thus more convenient.

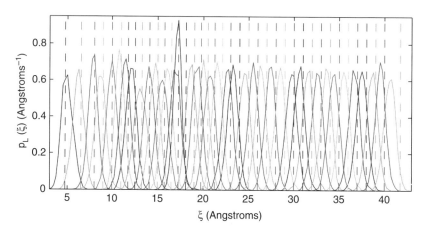

Figure 14.12 Probability density distribution of ξ when the cantilever bias is fixed at position L for the DNA dimer. Each curve corresponds to a different L. The dotted lines mark the L chosen in each case

The sampling along the extension coordinate required to reconstruct the PMF proceeded as follows. The distance L between the surface and the cantilever was fixed at several different values along the extension coordinate. The system was first allowed to equilibrate for 4 ns. Subsequently, the dynamics was followed for 8 ns and the end-to-end distance ξ recorded every 1 ps. The cantilever was taken to have a stiffness of $k_0 = 110$ pN Å^{-1}. In total, 155 extensions were simulated. Values of L ranged from 6.75 to 44.75 Å, while ξ ranged from 2.9 to 44.9 Å. Total analyzed simulation time was 1.24 μs.

Figure 14.12 shows some of the sampled $p_L(\xi)$, defined in Equation (14.14), that are used as input for the PMF reconstruction. The resulting Helmholtz free energy profile is shown in Figure 14.13. It consists of a convex region for $\xi < 17$ Å representing the folded state, a mostly concave region for $17 < \xi < 40$ Å where the molecule unfolds, followed by another region of convexity that corresponds to the unfolded state. The unfolding region exhibits some convex intervals that signal marginally stable intermediates encountered during the unfolding.

Using the PMF one can then estimate the force–extension characteristics of the system when pulling with cantilevers of arbitrary stiffness as described in Section 14.3.3. Figure 14.14 shows the predicted average and probability density in the force measurements when the DNA dimer is pulled using a cantilever stiffness of 1.1 pN Å^{-1}. Note the degree of detail of the pulling process revealed by the probability density.

14.5 Interpretation of Pulling Phenomenology

A striking feature of single-molecule pulling is that the elastic behaviors of vastly different molecules often have qualitatively similar features. In Secions 14.2 and 14.4 we discussed two of these features: (i) mechanically unstable regions where $\partial \langle F \rangle_L / \partial L < 0$ and (ii) regions of dynamical bistability where blinking in the force measurements for fixed L is observed. In this section we discuss the origin of both of these effects. Emphasis

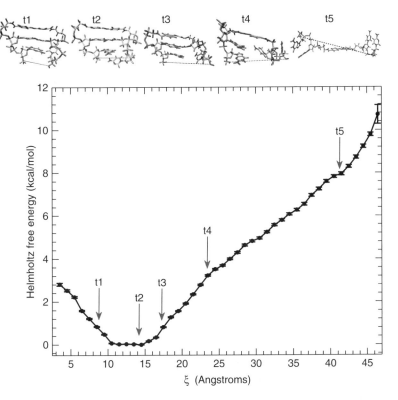

Figure 14.13 Potential of mean force along the end-to-end distance coordinate for the azobenzene-capped DNA hairpin. Snapshots of structures adopted during the unfolding are shown in the top panel. The error bars correspond to twice the standard deviation obtained from a bootstrapping analysis. Refer to Plate 14 for the colored version (in small)

will be placed on the basic requirements of the molecule and the pulling device for the emergence of the phenomena.

14.5.1 Basic Structure of the Molecular Potential of Mean Force

Recall the basic structure of the PMF for the oligorotaxanes (Figure 14.10): it consists of a convex region ($\partial^2 \phi(\xi)/\partial \xi^2 > 0$) where the molecule is folded, another region of convexity when the molecule is extended, and a region of concavity ($\partial^2 \phi(\xi)/\partial \xi^2 < 0$) where the molecule unfolds. This basic structure in the PMF is not unique to the rotaxanes, but is actually a common feature of the extension behavior of single molecules of sufficient structural complexity. Regions of convexity mark mechanically stable conformations, while regions of concavity mark molecular unfolding events. We had already encountered this very same basic structure when studying the elastic properties of the DNA dimer shown in Figure 14.13. Perhaps a more spectacular example of this structure is provided by the polyprotein composed of eight repeats of the Ig27 domain of human titin that we introduced in Figure 14.5. The PMF reconstructed from experimental force–extension

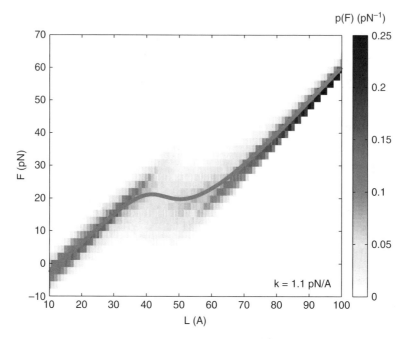

Figure 14.14 Force–extension profile of the DNA dimer when pulled with a cantilever of stiffness 1.1 pN Å^{-1} estimated from the PMF in Figure 14.13. The solid line represents the average force–extension profile and the color plot is the probability density distribution of the force measurements $p_L(F)$ (Equation (14.25))

data is shown in Figure 14.15. In this case also, mechanically stable conformations are described by regions of convexity in the PMF, while regions of concavity mark unfolding events. The reason why in this case there are eight regions of concavity along the extension pathway is that there are eight unfolding events, each marking the unfolding of one of the Ig27 domains in the polyprotein. Below, we discuss the implications of these regions of concavity along the PMF.

14.5.2 Mechanical Instability

A first consequence of the regions of concavity along the molecular PMF is that they lead to mechanically unstable regions in the $F–L$ isotherms. To see this, consider the configurational partition function of the molecule plus cantilever at extension L (Equation (14.22)) for large k. In this regime, most contributions to the integral will come from the region where $\xi = L$. Consequently, $\exp[-\beta\phi(\xi)]$ can be expanded around this point to give

$$\exp[-\beta\phi(\xi)] = \exp[-\beta\phi(L)]\left[1 - \beta A_1(L)(\xi - L) - \frac{\beta}{2}A_2(L)(\xi - L)^2 + \cdots\right]$$

$$(14.26)$$

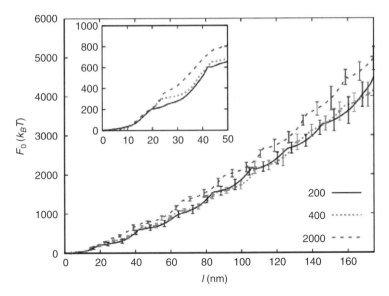

Figure 14.15 Molecular potential of mean force for a polyprotein composed of eight repeats of the Ig27 domain of human titin. The free energy profile was reconstructed using a nonequilibrium method and pulling at different pulling speeds of 200, 400 (dotted curve) and 2000 nm s^{-1} (dashed curve). The number of nonequilibrium unfolding trajectories used in the estimate was 66, 35, and 29, respectively. The force–extension profile of this molecule is shown in Figure 14.5. Reprinted with permission from [22]. Copyright 2008 Institute of Physics

where $A_1(L) = \partial\phi(L)/\partial L$ and

$$A_2(L) = \frac{\partial^2\phi(L)}{\partial L^2} - \beta\left[\frac{\partial\phi(L)}{\partial L}\right]^2.$$

Here, the notation is such that

$$\frac{\partial\phi(L)}{\partial L} = \left.\frac{\partial\phi(\xi)}{\partial\xi}\right|_{\xi=L}.$$

Introducing Equation (14) into Equation (14.22), integrating explicitly the different terms, and performing an expansion around $1/k = 0$ one obtains

$$Z(L) = \sqrt{\frac{2\pi}{k\beta}} \exp\left[-\beta\phi(L)\right]\left[1 - \frac{1}{2k}A_2(L) + \mathcal{O}(1/k^2)\right]. \qquad (14.27)$$

Using Equation (14.24), the derivative of the force can be obtained from this approximation to the partition function. To zeroth order in $1/k$ it is given by

$$\frac{\partial\langle F\rangle_L}{\partial L} = \frac{\partial^2\phi(L)}{\partial L^2} + \mathcal{O}(1/k). \qquad (14.28)$$

That is, an unstable region in the force versus extension where $\partial\langle F\rangle_L/\partial L < 0$ requires a region of concavity in the PMF.

Note, however, that since the properties that are measured during pulling are those of the molecule plus cantilever, the observed mechanical stability properties will depend on the cantilever stiffness employed. In fact, when employing very soft cantilevers, no mechanically unstable region can be observed independent of the molecule that is being pulled, provided that no bond breaking occurs during the pulling. To see this, consider the soft-spring approximation of the configurational partition function of the system plus cantilever (Equation (14.22)):

$$\frac{Z(L)}{Z_0} \approx \frac{\exp(-\beta k L^2/2)}{Z_0} \int d\xi \exp[-\beta\phi(\xi)]\exp(\beta k L\xi) = \exp(-\beta k L^2/2)\,\langle\exp(\beta k L\xi)\rangle$$

(14.29)

where the notation $\langle f \rangle$ stands for the unbiased (cantilever-free) average of f. In writing Equation (14.29) we have supposed that in the region of relevant ξ (in which the integrand is non-negligible) $k\xi^2/V_L(\xi) \ll 1$ and, hence, that the cantilever potential is well approximated by $V_L(\xi) \approx \frac{kL^2}{2} - kL\xi$. This approximation is valid provided that that no bond breaking is induced during pulling and permits the introduction of a cumulant expansion [68] in the configurational partition function. Specifically, the average $\langle\exp(\beta k L\xi)\rangle$ can be expressed as

$$\langle\exp(\beta k L\xi)\rangle = \sum_{n=0}^{\infty} \frac{(\beta k L)^n}{n!}\,\langle\xi^n\rangle = \exp\left[\sum_{n=1}^{\infty} \frac{(\beta k L)^n}{n!}\kappa_n(\xi)\right]$$

(14.30)

where $\kappa_n(\xi)$ is the nth-order cumulant. In view of Equations (14.24), (14.29) and (14.30) the slope of the F–L curves can be expressed as

$$\frac{\partial \langle F\rangle_L}{\partial L} = k - \beta k^2 \sum_{n=0}^{\infty} \frac{(\beta k L)^n}{n!}\kappa_{n+2}(\xi).$$

(14.31)

It then follows that, to lowest order in k,

$$\frac{\partial \langle F\rangle_L}{\partial L} \approx k > 0.$$

(14.32)

That is, no unstable region in the F–L curve can arise in the soft-spring limit, irrespective of the specific form of $\phi(\xi)$.

Figure 14.16 illustrates the dependence of the region of mechanical instability on the cantilever stiffness using the [3] rotaxane case discussed in Section 14.4.3 as an example. The figure shows the k-dependence of the local maximum F^+ and minimum F^- in the average force measurements that enclose the region of mechanical instability. The critical forces F^+ and F^- show a smooth and strong dependence on the cantilever spring constant. For large k, the system persistently shows a region of instability in the isotherms. However, as the cantilever spring is made softer, F^+ and F^- approach each other, and for small k the mechanical instability in the F–L isotherms is no longer present.

Another interesting aspect of single-molecule pulling is that there is a close relationship between the mechanical stability properties and the fluctuations in the force measurements. To see this, note that the average slope of the F–L curves can be expressed in terms of the fluctuations in the force:

$$\frac{\partial \langle F\rangle_L}{\partial L} = \left\langle\frac{\partial^2 V_L(\xi)}{\partial L^2}\right\rangle_L - \beta(\langle F^2\rangle_L - \langle F\rangle_L^2) = k - \beta(\langle F^2\rangle_L - \langle F\rangle_L^2)$$

(14.33)

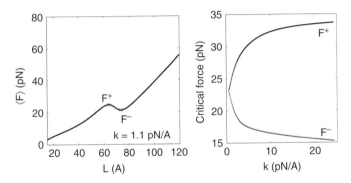

Figure 14.16 Dependence of the mechanical instability on the cantilever stiffness. Left panel: force–extension isotherm for a [3] rotaxane when pulled with a cantilever of stiffness $k = 1.1$ pN Å^{-1}. Here, F^+ and F^- correspond to the values of the force when the F–L curves exhibit a maximum and minimum, and enclose the region of mechanical instability. Right panel: dependence of F^+ and F^- on k. Note the decay of the region of mechanical instability for soft cantilevers

where we have employed Equation (14.24). The sign of $\partial \langle F \rangle_L / \partial L$ determines the mechanical stability during the extension and, hence, Equation (14.33) relates the thermal fluctuations in the force measurements with the stability properties of the F–L curves. Specifically, for the stable regions for which $\partial \langle F \rangle_L / \partial L > 0$ the force fluctuations satisfy

$$\langle F^2 \rangle_L - \langle F \rangle_L^2 < \frac{k}{\beta}. \tag{14.34a}$$

In turn, in the unstable regions the force fluctuations are larger than in the stable regions and satisfy the inequality

$$\langle F^2 \rangle_L - \langle F \rangle_L^2 > \frac{k}{\beta}. \tag{14.34b}$$

The fact that the fluctuations are larger around the region of mechanical instability is noticeable in some of the force–extension curves that we have discussed here (see Figures 14.8 and 14.14). Figure 14.17 illustrates these general observations in the specific case of the pulling of [3] rotaxane. As shown, for $k = k_0$ and $k = 2k_0$ (where $k_0 = 1.1$ pN Å^{-1}) there is a region where the force fluctuations become larger than k/β and, consequently, unstable behavior in F–L develops. For $k = 0.4k_0$ or less the fluctuations in the force are never large enough to satisfy Equation (14.34b) and no critical points in the average F–L curve develop, as can be confirmed in Figure 14.16.

14.5.3 Dynamical Bistability

The observation of force measurements that blink between a high-force and a low-force regime for selected extensions requires the composite molecule plus cantilever system to be bistable along the end-to-end distance for some L. That is, the effective potential $U_L(\xi)$ (Equation (14.23)) must have a double minimum. Since the molecular PMF is not usually bistable along ξ, the bistability must be introduced by the cantilever potential. To see how this bistability arises, consider the effective potential for the [3] rotaxane

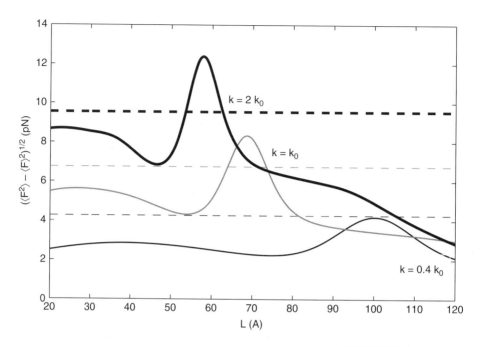

Figure 14.17 Standard deviation in the force measurements $\sigma_F = \sqrt{\langle F^2 \rangle_L - \langle F \rangle_L^2}$ as a function of L for three different cantilever spring constants k during the pulling of the [3] rotaxane. In each case, the dotted line indicates the value of $\sqrt{k/\beta}$, which sets the limit between the stable and unstable branches in the extension; see Equation (14.34). Here, $k_0 = 1.1$ pN Å^{-1}. Reprinted with permission from [28]. Copyright 2009 American Institute of Physics

for selected L shown in Figure 14.18. For L in the mechanically stable regions of the force–extension isotherms ($L = 40$ and $L = 90$ Å), the effective potential exhibits a single minimum along the ξ coordinate. However, when $L = 70$ Å, a bistability in the potential develops. At this extension the cantilever potential turns the region of concavity in $\phi(\xi)$ into a barrier between two minima. The secondary minimum is the cause for the blinking in the force measurements observed at this L (cf. Figure 14.9).

What are the minimum requirements for the emergence of bistability along ξ? A necessary condition is that the effective potential $U_L(\xi)$ is concave for some region along ξ; that is:

$$\frac{\partial^2 U_L(\xi)}{\partial \xi^2} = \frac{\partial^2 \phi(\xi)}{\partial \xi^2} + k < 0. \tag{14.35}$$

For Equation (14.35) to be satisfied it is required that both (i) the PMF of the isolated molecule has a region of concavity where $\partial^2 \phi(\xi)/\partial \xi^2 < 0$ and (ii) the cantilever employed is sufficiently soft such that

$$k < -\min\left[\frac{\partial^2 \phi(\xi)}{\partial \xi^2}\right] \tag{14.36}$$

for some ξ. Equation (14.36) imposes an upper bound on $k > 0$ for bistability to be observable. If k is very stiff then the inequality would be violated for all ξ and bistability

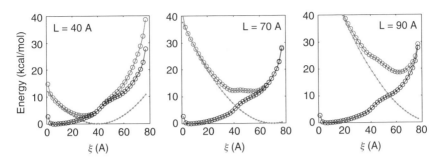

Figure 14.18 Effective potential $U_L(\xi) = \phi(\xi) + V_L(\xi)$ for the molecule plus cantilever for different values of the extension L. In the panels, the blue circles correspond to $U_L(\xi)$, the black circles to the PMF $\phi(\xi)$, and the solid line to the cantilever potential $V_L(\xi)$ with $k = 1.1$ pN Å^{-1}. Note the bistability in the effective potential for $L = 70$ Å

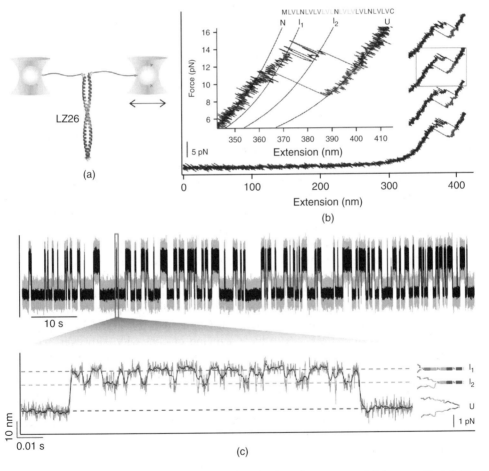

Figure 14.19 Bistability during the pulling of a leucine zipper measured in an optical lattice arrangement. (a) Schematic of the experimental setup; (b) Sample force–extension traces; (c) Force versus time record for fixed L in the region of dynamical bistability. Reprinted with permission from [70]. Copyright 2010 National Academy of Sciences USA

would not be manifest. Note, however, that there is no lower bound for k that prevents bistability along ξ.

Since the region of concavity in the PMF is a common feature of molecules that have stable folded conformations, this bistability for selected extensions is a common feature of the pulling. We have already encountered this phenomenon during the pulling of the oligorotaxane and the RNA hairpin (Figure 14.2). The DNA dimer (Figure 14.14) and the polyprotein in Figure 14.5 are also expected to show this behavior, since their PMF has regions of concavity along ξ. The bistability has also been predicted to be measurable when pulling π-stacked molecules [51,69]. For a discussion of the bistability in the isotensional ensemble, [18]. Figure 14.19 shows a striking recent experimental demonstration of the bistability obtained during the pulling of a leucine zipper [70].

14.6 Summary

The power of single-molecule pulling techniques is that they allow for mechanical control over the molecular conformation while simultaneously making thermodynamic measurements of any mechanically induced unfolding events. The force–extension data often exhibit mechanically unstable regions where the force decreases with increasing extension and bistable regions where, for fixed extension, blinks between a high-force and a low-force regime are observed.

From the force–extension isotherms it is possible to reconstruct the PMF along the extension coordinate using the WHAM. In fact, the pulling process can be seen as an experimental realization of the WHAM methodology. The PMF summarizes the changes in the free energy during folding and determines the elastic properties of the molecule. Quite remarkably, the PMFs of widely different molecules often share the same basic structure. It consists of regions of convexity that represent mechanically stable molecular conformations interspersed by regions of concavity that signal molecular unfolding events.

The starting point in the interpretation of single-molecule pulling experiments is Equation (14.22), which shows that the pulling process can be viewed as thermal motion along a one-dimensional potential $U_L(\xi) = \phi(\xi) + V_L(\xi)$ that is determined by the PMF $\phi(\xi)$ and the cantilever potential $V_L(\xi)$. Several important conclusions follow from this simple observation. The first and most obvious one is that the properties that are measured during pulling are those of the molecule plus cantilever, and hence that the basic phenomenon observed during pulling depends on the cantilever stiffness used. It also follows that in order for the mechanical instability and dynamical bistability to be observable, the molecular PMF has to have a region of concavity. However, while the dynamical bistability survives for a soft cantilever and decays for rigid cantilevers (see Equation (14.36)), the opposite is true for the mechanical instability (recall Equation (14.32)). The mechanical stability properties during pulling were also shown to be intimately related to the fluctuations in the force measurements (see Equation (14.34)).

One basic challenge in modeling the force spectroscopy using MD is to bridge the several orders of magnitude gap that exists between the pulling speeds used experimentally and those that are computationally feasible. We detailed two possible strategies that can be employed to overcome this difficulty: either strive for pulling speeds that are slow enough such that reversible behavior is recovered or use an indirect approach in which the PMF is first reconstructed and the force–extension behavior is then estimated from the PMF. Other strategies based on nonequilibrium pulling are also possible. From the simulation of the pulling one can gain insights into the conformations encountered during molecular unfolding and into structure–function relations responsible for the elastic behavior of single molecules.

Currently, the accurate reconstruction of the PMF of molecular systems using equilibrium sampling is a serious computational challenge except for relatively modest systems. The recent discovery of nonequilibrium work fluctuation relations has led to a surge of activity in seeking nonequilibrium methods to reconstruct the PMF that can be used in situations where proper equilibrium sampling is unfeasible. The full consequences of this avenue of research are still to be determined.

Acknowledgments

This work was supported by the Non-equilibrium Energy Research Center (NERC), which is an Energy Frontier Research Center funded by the US Department of Energy, Office of Science, Office of Basic Energy Sciences under Award Number DE-SC0000989. We thank Dr Martin McCullagh for useful discussions.

References

[1] Ashkin, A. (1997) Optical trapping and manipulation of neutral particles using lasers. *Proceedings of the National Academy of Sciences of the United States of America*, **94** (10), 4853–4860.

[2] Rief, M., Gautel, M., Oesterhelt, F. *et al.* (1997) Reversible unfolding of individual titin immunoglobulin domains by AFM. *Science*, **276** (5315), 1109–1112.

[3] Weiss, S. (1999) Fluorescence spectroscopy of single biomolecules. *Science*, **283** (5408), 1676–1683.

[4] Moerner, W.E. and Orrit, M. (1999) Illuminating single molecules in condensed matter. *Science*, **283** (5408), 1670–1676.

[5] Strick, T.R., Croquette, V., and Bensimon, D. (2000) Single-molecule analysis of DNA uncoiling by a type II topoisomerase. *Nature*, **404** (6780), 901–904.

[6] Nitzan, A. and Ratner, M.A. (2003) Electron transport in molecular wire junctions. *Science*, **300** (5624), 1384–1389.

[7] Ritort, F. (2006) Single-molecule experiments in biological physics: methods and applications. *Journal of Physics: Condensed Matter*, **18** (32), R531–R583.

[8] Camden, J.P., Dieringer, J.A., Wang, Y. *et al.* (2008) Probing the structure of single-molecule surface-enhanced Raman scattering hot spots. *Journal of the American Chemical Society*, **130** (38), 12616–12617.

[9] Liphardt, J., Onoa, B., Smith, S.B. *et al.* (2001) Reversible unfolding of single RNA molecules by mechanical force. *Science*, **292** (5517), 733–737.

[10] Oberhauser, A.F., Hansma, P.K., Carrion-Vazquez, M., and Fernandez, J.M. (2001) Stepwise unfolding of titin under force-clamp atomic force microscopy. *Proceedings of the National Academy of Sciences of the United States of America*, **98** (2), 468–472.

[11] Evans, E. (2001) Probing the relation between force-lifetime and chemistry in single molecular bonds. *Annual Review of Biophysics and Biomolecular Structure*, **30** (1), 105–128.

[12] Bustamante, C., Liphardt, J., and Ritort, F. (2005) The nonequilibrium thermodynamics of small systems. *Physics Today*, **7**, 43–48.

[13] Neuman, K.C. and Nagy, A. (2008) Single-molecule force spectroscopy: optical tweezers, magnetic tweezers and atomic force microscopy. *Nature Methods*, **5** (6), 491–505.

[14] Freund, L.B. (2009) Characterizing the resistance generated by a molecular bond as it is forcibly separated. *Proceedings of the National Academy of Sciences of the United States of America*, **106**, 8818–8823.

[15] Huguet, J.M., Bizarro, C.V., Forns, N. *et al*. (2010) Single-molecule derivation of salt dependent base-pair free energies in DNA. *Proceedings of the National Academy of Sciences of the United States of America*, **107** (35), 15431–15436.

[16] Kreuzer, H.J. and Payne, S.H. (2001) Stretching a macromolecule in an atomic force microscope: statistical mechanical analysis. *Physical Review E*, **63** (2), 021906.

[17] Süzen, M., Sega, M., and Holm, C. (2009) Ensemble inequivalence in single-molecule experiments. *Physical Review E*, **79** (5), 051118.

[18] Kirmizialtin, S., Huang, L., and Makarov, D.E. (2005) Topography of the free energy landscape probed via mechanical unfolding of proteins. *Journal of Chemical Physics*, **122** (23), 234915.

[19] Gunari, N., Balazs, A.C., and Walker, G.C. (2007) Force-induced globule–coil transition in single polystyrene chains in water. *Journal of the American Chemical Society*, **129** (33), 10046–10047.

[20] Rubinstein, M. and Colby, R.H. (2003) *Polymer Physics*, Oxford University Press, New York.

[21] Halperin, A. and Zhulina, E.B. (1991) On the deformation behavior of collapsed polymers. *Europhysics Letters*, **15** (4), 417–421.

[22] Imparato, A., Sbrana, F., and Vassalli, M. (2008) Reconstructing the free energy landscape of a polyprotein by single-molecule experiments. *Europhysics Letters*, **82** (5), 58006.

[23] Jarzynski, C. (1997) Nonequilibrium equality for free energy differences. *Physical Review Letters*, **78** (14), 2690–2693.

[24] Jarzynski, C. (2008) Nonequilibrium work relations: foundations and applications. *European Physical Journal B: Condensed Matter Physics*, **64**, 331–340.

[25] Jarzynski, C. (2011) Equalities and inequalities: irreversibility and the second law of thermodynamics at the nanoscale. *Annual Review of Condensed Matter Physics*, **2** (1), 329–351.

[26] Liphardt, J., Dumont, S., Smith, S.B. *et al*. (2002) Equilibrium information from nonequilibrium measurements in an experimental test of Jarzynski's equality. *Science*, **296** (5574), 1832–1835.

[27] Kirkwood, J.G. (1935) Statistical mechanics of fluid mixtures. *Journal of Chemical Physics*, **3** (5), 300–313.

[28] Franco, I., Schatz, G.C., and Ratner, M.A. (2009) Single-molecule pulling and the folding of donor–acceptor oligorotaxanes: phenomenology and interpretation. *Journal of Chemical Physics*, **131** (12), 124902.

[29] Ferrenberg, A.M. and Swendsen, R.H. (1989) Optimized Monte Carlo data analysis. *Physical Review Letters*, **63** (12), 1195–1198.

[30] Kumar, S., Bouzida, D., Swendsen, R.H. *et al*. (1992) The weighted histogram analysis method for free energy calculations on biomolecules: I. *The method. Jounal of Computational Chemistry*, **13** (8), 1011–1021.

[31] Roux, B. (1995) The calculation of the potential of mean force using computer simulations. *Computer Physics Communications*, **91**, 275–282.

[32] Frenkel, D. and Smit, B. (2002) *Understanding Molecular Simulation*, 2nd edn., Academic Press.

[33] Newman, M.E.J. and Barkena, G.T. (2007) *Monte Carlo Methods in Statistical Physics*, Oxford University Press, New York.

[34] Chodera, J.D., Swope, W.C., Pitera, J.W. *et al*. (2007) Use of the weighted histogram analysis method for the analysis of simulated and parallel tempering simulations. *Journal of Chemical Theory and Computation*, **3** (1), 26–41.

[35] Hub, J.S., de Groot, B.L., and van der Spoel, D. (2010) g_wham – a free weighted histogram analysis implementation including robust error and autocorrelation estimates. *Journal of Chemical Theory and Computation*, **6** (12), 3713–3720.

[36] Felberg, L., Franco, I., McCullagh, M. *et al*. (2010), MOLpull: a tool for molecular free energy reconstruction along a pulling coordinate, doi:10254/nanohub-r9583.2, http://nanohub.org/resources/9583?rev=87.

[37] Grossfield, A. (n.d.), The weighted histogram analysis method, http://membrane.urmc.rochester.edu/content/wham.

[38] Minh, D.D.L. (n.d.), FERBE free energy reconstruction from biased experiments, https://simtk.org/home/ferbe.

[39] Izrailev, S., Stepaniants, S., Isralewitz, B. *et al*. (1998) Steered molecular dynamics, in *Computational Molecular Dynamics: Challenges, Methods, Ideas* (eds. P. Deuflhard, J. Hermans, B. Leimkuhler, A.E. Marks, S. Reich, and R.D. Skeel), vol. 4 of *Lecture Notes in Computational Science and Engineering*, Springer-Verlag, pp. 39–65.

[40] Isralewitz, B., Gao, M., and Schulten, K. (2001) Steered molecular dynamics and mechanical functions of proteins. *Current Opinion in Structural Biology*, **11**, 224–230.

[41] Heymann, B. and Grubmüller, H. (1999) 'Chair–boat' transitions and side groups affect the stiffness of polysaccharides. *Chemical Physics Letters*, **305** (3–4), 202–208.

[42] Nosé, S. (1984) A molecular dynamics method for simulation in the canonical ensemble. *Molecular Physics*, **52**, 255–268.

[43] Hoover, W.G. (1985) Canonical dynamics: equilibrium phase-space distributions. *Physical Review A*, **31** (3), 1695–1697.

[44] Martyna, G.J., Klein, M.L., and Tuckerman, M. (1992) Nosé–Hoover chains: the canonical ensemble via continuous dynamics. *Journal of Chemical Physics*, **97** (4), 2635–2643.

[45] Thijssen, J.M. (2007) *Computational Physics*, 2nd edn., Cambridge University Press, New York.

[46] Berendsen, H.J.C., Postma, J.P.M., Vangunsteren, W.F. *et al*. (1984) Molecular-dynamics with coupling to an external bath. *Journal of Chemical Physics*, **81** (8), 3684–3690.

[47] Harvey, S.C., Tan, R.K.Z., and Cheatham, T.E. (1998) The flying ice cube: velocity rescaling in molecular dynamics leads to violation of energy equipartition. *Journal of Computational Chemistry*, **19** (7), 726–740.

[48] Phillips, J.C., Braun, R., Wang, W. *et al*. (2005) Scalable molecular dynamics with NAMD. *Journal of Computational Chemistry*, **26** (16), 1781–1802.

[49] Hess, B., Kutzner, C., van der Spoel, D., and Lindahl, E. (2008) GROMACS 4: algorithms for highly efficient, load-balanced, and scalable molecular simulation. *Journal of Chemical Theory and Computation*, **4** (3), 435–447.

[50] Ponder, J. (2004), TINKER: Software Tools for Molecular Design 4.2, http://dasher.wustl.edu/tinker/.

[51] Franco, I., George, C.B., Solomon, G.C. *et al*. (2011) Mechanically activated molecular switch through single-molecule pulling. *Journal of the American Chemical Society*, **133** (7), 2242–2249.

[52] Schäfer, L.V., Müller, E.M., Gaub, H.E., and Grubmüller, H. (2007) Elastic properties of photoswitchable azobenzene polymers from molecular dynamics simulations. *Angewandte Chemie*, **119** (13), 2282–2287.

[53] Lee, E.H., Hsin, J., von Castelmur, E. *et al*. (2010) Tertiary and secondary structure elasticity of a six-Ig titin chain. *Biophysical Journal*, **98** (6), 1085–1095.

[54] Crooks, G.E. (1999) Entropy production fluctuation theorem and the nonequilibrium work relation for free energy differences. *Physical Review E*, **60** (3), 2721–2726.

[55] Hummer, G. and Szabo, A. (2001) Free energy reconstruction from nonequilibrium single-molecule pulling experiments. *Proceedings of the National Academy of Sciences of the United States of America*, **98** (7), 3658–3661.

[56] Park, S. and Schulten, K. (2004) Calculating potentials of mean force from steered molecular dynamics simulations. *Journal of Chemical Physics*, **120** (13), 5946–5961.

[57] Hummer, G. and Szabo, A. (2005) Free energy surfaces from single-molecule force spectroscopy. *Accounts of Chemical Research*, **38**, 504–513.

[58] Harris, N.C., Song, Y., and Kiang, C.H. (2007) Experimental free energy surface reconstruction from single-molecule force spectroscopy using Jarzynski's equality. *Physical Review Letters*, **99**, 068101.

[59] Minh, D.D.L. and Adib, A.B. (2008) Optimized free energies from bidirectional single-molecule force spectroscopy. *Physical Review Letters*, **100** (18), 180602.

[60] Pohorille, A., Jarzynski, C., and Chipot, C. (2010) Good practices in free energy calculations. *Journal of Physical Chemistry B*, **114** (32), 10235–10253.

[61] Hummer, G. and Szabo, A. (2010) Free energy profiles from single-molecule pulling experiments. *Proceedings of the National Academy of Sciences of the United States of America*, **107** (50), 21441–21446.

[62] Jarzynski, C. (2006) Rare events and the convergence of exponentially averaged work values. *Physical Review E*, **73** (4), 046105.

[63] Marsili, S. and Procacci, P. (2010) Free energy reconstruction in bidirectional force spectroscopy experiments: the effect of the device stiffness. *Journal of Physical Chemistry B*, **114** (7), 2509–2516.

[64] Franco, I., Ratner, M.A., and Schatz, G.C. (2011) Coulombic interactions and crystal packing effects in the folding of donor–acceptor oligorotaxanes. *Journal of Physical Chemistry B*, **115** (11), 2477–2484.

[65] Basu, S., Coskun, A., Friedman, D.C. *et al*. (2011) Donor–acceptor oligorotaxanes made to order. *Chemistry: A European Journal*, **17** (7), 2107–2119.

[66] Allinger, N.L., Yuh, Y.H., and Lii, J.H. (1989) Molecular mechanics. The MM3 force-field for hydrocarbons. 1. *Journal of the American Chemical Society*, **111** (23), 8551–8566.

[67] McCullagh, M., Franco, I., Ratner, M.A., and Schatz, G.C. (2011) DNA-based optomechanical molecular motor. *Journal of the American Chemical Society*, **133** (10), 3452–3459.

[68] Kubo, R. (1962) Generalized cumulant expansion method. *Journal of the Physical Society of Japan*, **17** (7), 1100–1120.

[69] Kim, J.S., Jung, Y.J., Park, J.W. *et al*. (2009) Mechanically stretching folded nano-π-b;-stacks reveals pico-newton attractive forces. *Advanced Materials*, **21** (7), 786–789.

[70] Gebhardt, J.C.M., Bornschlögl, T., and Rief, M. (2010) Full distance-resolved folding energy landscape of one single protein molecule. *Proceedings of the National Academy of Sciences of the United States of America*, **107** (5), 2013–2018.

15

Modeling and Simulation of Hierarchical Protein Materials

Tristan Giesa[1,2], Graham Bratzel[1], and Markus J. Buehler[1]

[1]*Massachusetts Institute of Technology, USA*
[2]*RWTH Aachen University, Germany*

15.1 Introduction

From its origins in describing multibody motion and fluid dynamics using first-principles calculations [1,2], computational modeling of materials has sought to bridge complex analytical theory and experimental observation in order to explain phenomena and predict behavior. Advances in computer performance, networking, and parallelization have recently made possible simulations of length scales and particle numbers from the dark matter interactions of millions of galaxy clusters [3], to simulations with billions of atoms to describe the work hardening of copper wires [4,5], as well as simulation of instabilities in fluid mixing [6], dynamic fracture of amorphous silica glass [7], and even trillion-atom molecular dynamics (MD) simulations [8]. Simulations of proteins have experienced a similar advance in complexity [9], from early reports of two-dimensional Monte Carlo method folding predictions of small polypeptides [10] to fully three-dimensional atomistic MD simulations of a ribosome [11], the cellular component that reads RNA and synthesizes proteins. As computer performance continues to advance, computational modeling of synthetic and biological materials will further enable the top-down understanding of processes and the bottom-up design of advanced materials. Category theory is a mathematical abstraction technique that can be applied to utilize the lessons learned from one network (e.g., a hierarchical protein material) to a similarly structured network (e.g., linguistics) by describing the emergence of functionality from first principles, on the basis of fundamental interactions between building blocks [12,13]. This has major implications for our ability to design *de novo* materials for tailored material properties. The union of a

Nano and Cell Mechanics: Fundamentals and Frontiers, First Edition. Edited by Horacio D. Espinosa and Gang Bao.
© 2013 John Wiley & Sons, Ltd. Published 2013 by John Wiley & Sons, Ltd.

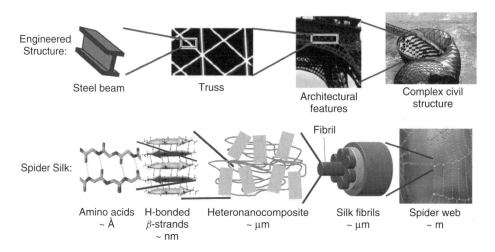

Engineered Structure:

Steel beam   Truss
                                                  Architectural   Complex civil
                                                  features   structure

                                                  Fibril

Spider Silk:

Amino acids H-bonded   Heteronanocomposite   Silk fibrils   Spider web
  ~ Å   β-strands   ~ μm   ~ μm   ~ m
             ~ nm

Figure 15.1   The compared hierarchical structures of engineered civil structures and protein materials such as spider silk. While civil engineering structures, for example, are designed from the bottom up by overcoming the limitations of structural beams, biological materials such as silk are typically investigated top down to understand how fibrillar networks may overcome fundamental strength limits at the nanoscale

structural and functional description of hierarchical biological materials on various length and time scales, realized within the framework of a generalized mathematical theory, enables the revelation of the source of superior material properties and the deduction of insights to the system under design from seemingly disparate fields.

While simulation tools exist to separately model atoms, proteins, oceans, and galaxies, a major hurdle in multiscale engineering is linking disparate length scales, as seen in Figure 15.1. For example, the limitations of a modern skyscraper are governed by the behavior of its parts at multiple length scales due to its hierarchical structure. The shape of structural beams is derived from steel's ductility, stiffness, and strength. These beams form the building blocks of the system and can be assembled in almost infinite variations. Vaults, cantilevers, and other architectural features based on trusses which comprise multiple structural beams define the final structure of the skyscraper. During an earthquake, the mechanical behavior of the building depends on the hierarchical arrangement of cantilevers, trusses, beams, and the microcrystalline structure of steel. Hierarchical structures of protein materials, such as spider silk, perform in a similar way. While all proteins are made of covalently and noncovalently bonded amino acids, their secondary and tertiary structures prescribe the behavior and assembly of the next hierarchical level [14]. In spider silk, polypeptides assemble into a cross-linked fibrillar network and are organized into a macroscopic orb web [15]. Like a skyscraper in an earthquake, the mechanical behavior of the orb web during prey impact depends on the fibrils, cross-links, and noncovalent hydrogen bonds at the respective levels of silk's hierarchical structure. While skyscrapers are designed from the bottom up with the knowledge of the possibilities and limitations of steel and structural beams taught by hundreds of years of architectural engineering experience, spider silk is currently largely understood top down by dissecting the macro-scopic structure. By establishing an understanding of how the building blocks of protein materials like spider silk assemble at each hierarchical level, it may soon become possible to engineer advanced biological and biomimetic fibers, composites, and other structural materials for applications that span from regenerative medicine to civil engineering.

15.2   Computational and Theoretical Tools

In this section, various computational and theoretical tools of multiscale engineering are reviewed to illustrate the linking of length and time scales to match the range accessible to experimental observations. First, quantum chemistry and molecular simulation are described as a method to elucidate atomistic and molecular mechanisms of protein assembly and deformation. Next, mesoscale methods and coarse-graining are reviewed as methods of modeling micro- and macro-scopic protein materials. Finally, mathematical approaches to biomateriomics – the study of structure–function correlations in biological materials – are presented using an analogy between the hierarchical structures of protein materials, language, and music to show how category theory, in particular ontology logs (ologs), provides a method for defining structural and functional properties within a system and for drawing connections to other fields or topics.

15.2.1   Molecular Simulation from Chemistry Upwards

The computational tools for multiscale simulation and engineering of protein materials cover the same length-scale range as experimental tools and provide critical insights into chemical and deformation mechanisms, as seen in Figure 15.2. Due to constraints on computational performance, each simulation tool carries with it a limited time scale. Capturing processes at longer length and time scales requires a sacrifice of fidelity for a more convenient user timeframe; for example, simulations lasting days rather than decades. Beginning from first principles, quantum mechanical methods such as Hartree–Fock theory solve many-electron wave functions, based on the Schrödinger equation, to derive bond energies and chemical interactions, but are limited to small molecules and describing processes lasting only femtoseconds [17]. Another quantum mechanical method, density functional theory (DFT), instead employs functionals based on the spatially dependent electron density to explore the electronic structure of atoms and molecules, such as the interactions of single peptides. To capture the formation and interactions of secondary structures of polypeptides (e.g., β-sheets and α-helices) and solvent-mediated processes on the order of nanoseconds, MD simulations use force fields trained by DFT results [9,18]. While reactive forces fields such as ReaxFF are capable of capturing covalent bond (re)formation, MD using nonreactive force fields such as CHARMM [19] can simulate noncovalent (re)formation (e.g., hydrogen bonding, electrostatics, and van der Waals forces) typical of secondary structure changes approaching microseconds.

Changes in secondary structure, such as the strain-induced transitions to β-strands [20], and the rupture of noncovalent bonds are especially important in solvated single-molecule studies, in which energetic and chemical pathways are tracked while whole or partial

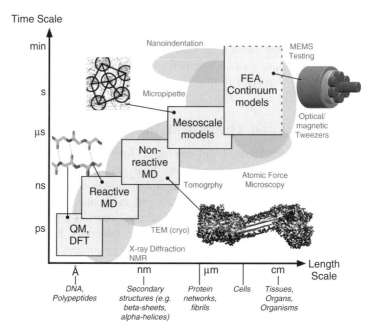

Figure 15.2   Approximate length and time-scale regimes of the tools for multiscale engineering. Computational tools predict and explain phenomena that are observed experimentally, but are limited to certain regimes due to constraints on computational performance. While mesoscale and continuum modeling cannot capture atomistic details, they are trained by atomistic results from DFT and MD simulations. Figure adapted from [16]

domains of proteins are unfolded. One method of simulated unfolding, steered molecule dynamics (SMD), complements experimental atomic force microscopy (AFM) by using a virtual spring moving at a constant velocity or maintaining a constant force to deform the protein [21]. This method often uncovers intermediate states during the unfolding process, as shown in Figure 15.3 for the case of fibronectin type III_1, a domain of the larger fibronectin dimer [22]. Mesoscale and continuum descriptions of protein materials at the level of fibrils and tissues, described more in the following section, cannot capture atomistic information but are trained by DFT and MD simulation results where the respective time and length scales overlap via scale linking.

15.2.2   Mesoscale Methods for Modeling Larger Length and Time Scales

Even at moderate time and length scales, atomistic methods quickly become computationally exhaustive. On the other hand, continuum analysis is not applicable, since small-scale matter cannot be regarded as infinitesimally divisible, the basic assumption of continuum theory [24]. Coarse-grained mesoscale models serve as a tool to bridge the dimensional gap, correlate mechanical properties at different scales, and encompass the basic features of the micromolecular behavior in regions of large biological macromolecules [25]. While

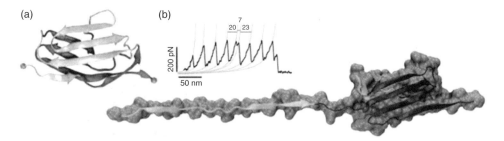

Figure 15.3 Single-molecule atomistic simulation of the unfolding events of fibronectin type III$_1$ (FN III$_1$) using SMD. (a) The structure of FN III$_1$ comprises a β-sandwich, shown here with a cartoon representation. With SMD, one terminus (shown by a sphere) is fixed while the other is pulled by a virtual spring. Adapted with permission from [21] Copyright 2007 American Association for the Advancement of Science. (b) The saw-tooth pattern of the loading curve shows many intermediate states during the unfolding process, where individual β-strands shear off from the molecule but remain attached by the continuous backbone. The solvent-accessible surface shows the extent of the side-chains that are hidden in the cartoon representation. Reprinted with permission from [23]. Copyright 2002 Elsevier

resembling the material's physical structure at larger scales, mesoscale models contain information about molecular mechanisms from lower scales by retaining atomistic degrees of freedom in regions of high spatial variation which cannot be modeled with methods provided by continuum mechanics; for example, near crack propagation and bifurcation, or dislocation [26,27]. Novel field-theoretic approaches based on the mean-field approximation deliver useful results in the calculation of mesoscopic polymer models [28]. Three basic methodologies are available: quasi-continuum methods, coarse-grained kinetic Monte Carlo simulations, and a coarse-grained MD approach [26]. For protein structures, bead–spring models for macromolecules (both single-bead and multi-bead models) successfully describe thermal fluctuation, aggregation, and unfolding in accordance to experimental results [24,29–32]. These large-scale coarse-grained models treat whole clusters of amino acids as beads with interactions often described by approximated harmonic bond and angle terms [33]. The MARTINI force-field calibrates coarse-grained building blocks against thermodynamic data and allows a broad range of applications (e.g., protein structures) without focusing on an accurate reproduction of structural details at a particular state [34,35]. Hence, there is no need to reparameterize the model for each case. Izvekov and Voth [36] and Bond *et al.* [37] applied the coarse-graining technique to the lipid bilayer of a cell membrane (Figure 15.4). Training the beads with atomisticly resolved information, such as hydrophobicity and/or charge, yielded sufficiently accurate results while reducing computational costs [36,37].

By extending chemical concepts to larger scales and funneling continuum mechanical concepts to smaller scales, a "handshaking" paradigm enables predictive modeling and material optimization over a wide range of length and time scales. Coarse-graining provides an accurate and reliable method for system-level analysis, for probing the structural and mechanical response, and for understanding the structure–property relationship of hierarchical materials while allowing for a more direct comparison between simulation and experimental techniques.

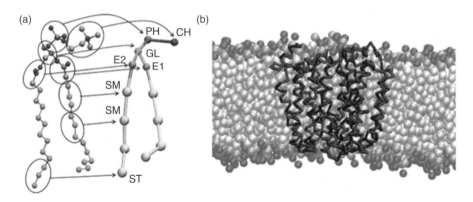

Figure 15.4 Coarse-graining of the lipid bilayer of a cell membrane. (a) The atomistic structure of the protein (here a DMPC phospholipid) is coarse-grained (CG) by representing the centers of mass of the amino acids by a heavy particle. These CG particles are also trained with other atomistic information, such as hydrophobicity and charge. Reprinted with permission from [34]. Copyright 2005 American Chemical Society. (b) A solvated CG model of a lipid bilayer is more computationally efficient than an atomistic model and yields sufficient results if trained correctly. Reprinted with permission from [35]. Copyright 2007 Elsevier

15.2.3 Mathematical Approaches to Biomateriomics

Analogy and Comparative Analysis of Linguistics, Music, and Protein Materials

Hierarchical structures result from the intricate assembly of basic building blocks that form the basis of the system's functionality. While classical material engineering often fails in rendering the combination of properties such as robustness and strength, multiscale systems exhibit a universality–diversity paradigm: the usage of a limited set of universal building blocks, often with inferior mechanical properties, in diverse structural arrangements at different hierarchical scales, can give rise to a surprising overall performance optimization [38]. This paradigm leads to highly adapted, robust, and multifunctional structures that are governed through self-regulatory processes and assemble at low energy cost [39].

Structural hierarchies are not solely found in biomaterials, but occur, often unperceived, in our daily lives. For example, language and music reveal similar abstracted structures and properties as hierarchical biomaterials [40,41]. The fundamental building blocks (for language, letters; and in music, basic waveforms) are universal and can form a seemingly infinite number of diverse assemblies at larger scales (for language, books and epic poems; and in music, symphonic arrangement and operas) which adapt to different situations and provide macroscale functionality, as illustrated in Figure 15.5. For example, only four basic oscillator forms constitute the basis for music as diverse as base drumming and countertenor singing. Concerning functionality, the diversification ranges from shopping mall music that creates a calm atmosphere to the frantic alarm siren of an ambulance. Similar patterns can be found in language. Talking to a baby is considerably different than talking to an adult, despite the reliance on the same limited set of building blocks [42].

The manipulation of single building blocks has fundamental influences on the overall system behavior. A single point-mutation in a DNA strand can make the difference

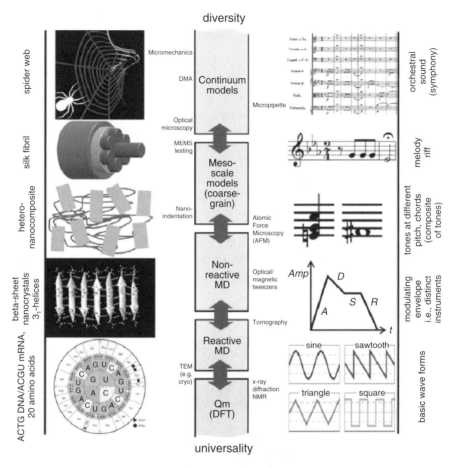

Figure 15.5 Multiscale hierarchical structures of protein materials, modeling, and experimental tools, and an analogy to music within the universality–diversity paradigm. In protein materials (left, spider silk), multifunctional materials are created via the formation of hierarchical structures. The synergistic interaction of structures and mechanisms at multiple scales provides the basis for enhanced functionality of biological materials despite the reliance on few distinct building blocks. Similarly, music (right) combines universal elements, such as basic waveforms or a set of available instruments, in hierarchical assemblies to provide macroscale functionality and eventually a particularly notable orchestral sound. Universality tends to dominate at smaller levels, whereas diversity is found predominantly at larger, functional levels. The integrated use of computational and experimental methods at multiple scales provides a powerful approach to elucidate the scientific concepts underlying the materiomics paradigm (center). Reprinted with permission from [14]. Copyright 2011 Elsevier

between healthy and diseased [43]. On the other hand, the localized failure of larger structural elements does not influence the total system behavior, in accordance with the robustness of hierarchical systems, owing to a high redundancy within the structure. Biological materials such as spider silk and diatom algae arrange in structural hierarchies and optimize their behavior in regard to the environmental requirements. Diatoms exhibit high structural stiffness combined with high robustness to protect against predators, and spider webs show high energy absorption and extensibility for catching prey while localizing web damage [44,45].

Category Theory and Ontology Logs

In order to make the advances in our understanding of multiscale mechanisms available for engineering and, for example, to understand the source of high strength and robustness, novel descriptive methods based on mathematical tools must be introduced. Mathematical theories have been successfully applied to many areas in science and to biological hierarchical systems. For example, graph theory has been used to describe the structure of biopolymers, disease spread, and neuronal activity [46–50]. Barabási and coworkers elucidated the role of individual molecules in various cellular processes with quantifiable tools of network theory and showed how the scale-freeness of these networks contribute to their topological, functional and dynamical robustness [51,52]. Figure 15.6 shows an example of the application of graph theory to the cellular network architecture and the

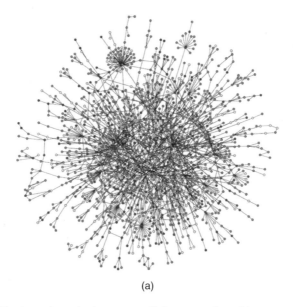

(a)

Figure 15.6 Application of graph theory to cellular network architecture and metabolism. (a) Map of yeast protein interactions in form of an inhomogeneous scale-free network [52]. Highly connected proteins (hub nodes) play a central role in mediating interactions among numerous, less connected proteins. Thus, the tolerance against random errors (mutations) and the fragility against the removal of the most connected nodes are increased. Reprinted with permission from [50] Copyright 2001 Nature Publishing Group. (b) Network characteristics of the metabolism for a simple pathway (catalyzed by Mg^{2+}-dependent enzymes). The links between nodes represent reactions that interconvert one substrate into another [51]. The three graph types depict different levels of system abstraction. The degree distribution is scale free and the clustering coefficient reveals a hierarchical architecture of the metabolism. Reprinted with permission from [49] Copyright 2004 Nature Publishing Group

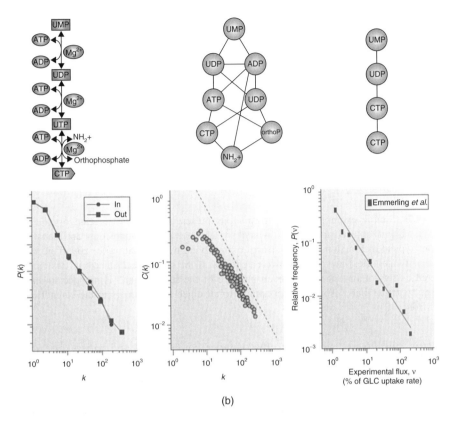

(b)

Figure 15.6 (*continued*)

metabolism. Both networks are scale free and ultra-small, which implies that their degree distributions follow a power law and their average path length is significantly shorter than $\log N$. Additionally, the clustering coefficient of hierarchical networks follows a power law, which indicates that sparsely connected nodes are part of highly clustered areas connected by hub nodes. Providing additional evidence for the concept of universality, Barabási and Albert also showed that the patterns present in biological networks reappear in a wide range of large networks, such as the World Wide Web and gene regulatory networks [53].

Current theoretic approaches to materials science focus either on structural aspects or on functional aspects. This single-faceted approach lacks the general description of how system elements behave and interact with each other in order to create functionality. Barabási and coworkers pointed out the need for enhanced data collection abilities and integrated studies. Category theory provides a means to overcome this limitation to conventional graphs and can be seen as an abstraction of graph theory. Category theory was originally introduced in the 1940s to rescind fundamental mathematical notions and describe structure while preserving the transformations of categories [54]. Since then it has been used to describe language, physics, philosophy, and other fields in a rigorous mathematical framework [55–58]. Categories are "algebras" that consist of objects and

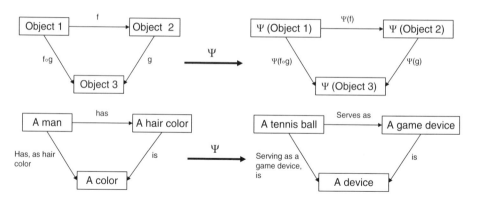

Figure 15.7 Simple examples of transformations preserving structure in category theory. Categories consist of objects and arrows which are closed under composition and satisfy certain conditions typical of functions. Ψ is a structure-preserving transformation (or covariant *functor*, or *morphism* of categories) between the two categories. If the categorical objects in this example are considered as sets of instances, then each instance of the set "A man" is mapped to an instance of the set "A tennis ball." This concept applies to all objects and arrows in the categories. Figure adapted from [12]

arrows (or *morphisms*) which are closed under composition and satisfy certain conditions typical of the composition of functions [59]. Transformations are thereby defined via *morphisms* between objects (or *functors*) among categories. In a linguistic version, category theory describes the essential features of a given subject and embeds them into a database framework. This concept is depicted for a simple example in Figure 15.7. Note how the structure of the category (i.e., the arrangement of objects and arrows) is preserved while the objects and the arrows themselves are subjected to a transformation. This means that if a certain property, such as the superior toughness of spider silk, can be described in a categorical framework, *functors* translate the components of the system into other systems, such as a wood- or concrete-based system, while the relations, and thus the functionality, within the category are maintained. The challenge for the scientist lies in the revelation and abstraction of the origin of the property. For protein materials this must be done by intensive materiomics studies that typically involve multiscale experiment and simulation.

Mathematicians have recently expanded category theory to so-called ontology logs (ologs) [12,60]. The linguistic categorical objects in ologs are sets, and the arrows represent unique functions between the objects. The composition of chains of arrows determines how several small-scale functional relationships form single large-scale relationships. For materials scientists, category theory and ologs in particular provide a useful methodology. Ologs offer a comprehensive toolset for the description of systems in all desired detail, the representation of knowledge gathered during research, and also for the revelation of the source and evolution of functionality at different scales. For example, the condition that geometric confinement of protein materials at the nanoscale leads to the rupture of clusters of three or four H-bonds in the β-sheet structures, and thus to an optimized shear strength, relates a functional property (the shear strength) to a structural condition (the geometric confinement) [61]. All these conditions have to be assembled and related first for deduction to other materials (such as concrete) or fields (such as social science) to be possible. The potential of an olog originates from the ultimate vicinity to computer implementation by well-established database tools (e.g., SQL) and ontology languages (e.g., OWL, KM), while providing the means to easily overview, extend, and manipulate data [62]. Yet, category theoretic approaches as reviewed above yield important advantages over plain databases [63]. In principle, databases also fall into the realm of category theory; but, in an olog, several fundamental category theoretic mathematical relations are found, such as *isomorphisms* (commutativity), *limits* (products, pullbacks, etc.), and *colimits* (coproducts, pushouts, etc.) [60].

Paths and definitions in the olog are supported and rendered more precisely with the help of the relations mentioned before. The condition for commutativity is that, in starting from the same instance in one set, two different paths lead to the same instance. Hence, labeling commutative paths adds further information to the olog and serves as a check by confining the intended meaning. Starting from the set "A mother" in Figure 15.8a, the path returns to the same instance of this set. In contrast, starting from the set "A child," the path does not necessarily lead to the same instance again. Therefore, the path might not be commuting. This concept can be abstracted to *functors*, where a *natural transformation* defines the commutativity of category transformations [64]. Other instances are illustrated in Figure 15.8b and c. Definitions in ologs are conducted by fiber products, a special type of *limit*. Fiber products precisely specify categorical objects in terms of other objects; that is, the objects are defined by *universal properties*. For example, every "beautiful object" can be defined as a flower with bright monochrome color (Figure 15.8b). Although not everyone may agree with this provocative definition (as the world views of authors differ), any olog represents a category itself and can be attached to and/or combined with other ologs via *morphisms* which are structure-preserving transformations. Of course, this compatibility is only possible if all ologs follow the same design rules of "good practice" [60]. For instance, the tendency to extravagate from important system-relevant facts to random generality, the need to "put in everything," should be avoided. Figure 15.8c shows an example for a pushout that allows the equation of terms.

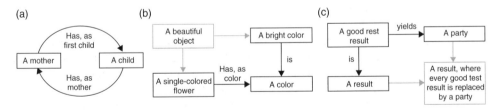

Figure 15.8 Examples for an isomorphism, a limit, and a colimit. (a) Starting from the set "A mother," the path returns to the same instance of this set and is therefore an *isomorphism*, whereas from the set "A child" the path does not necessarily lead to the same instance again. It might not be commuting. (b) Categorical objects can be defined in terms of other objects, that is, they are defined by *universal properties*. This kind of *limit* is called a fiber product. Here, all beautiful objects are understood as bright monochrome flowers. Note that the paths are also commuting. (c) Equivalencies of categorical objects are constructed via pushouts, a type of *colimit*. Here, every good test result is set equal to a party

With their rigorous foundation, ologs represent a unique way to store and share data, knowledge, and insights in structure and functionality among many disparate research groups. They offer the means to relate findings to previous results and thereby help to reveal the origin of the described system property and to connect them to other topics or fields. The principles from which the desired functionality arises have to be elucidated in order to define the hierarchical structure-function relationships or even synthesize biological materials. The insights of how inferior building blocks can create superior macroscopic functionality can be gained from the category theoretic analysis of protein materials by describing the emergence of functionality from first principles; for example, on the basis of fundamental interactions between building blocks. There may be no need for environmentally exhaustive superior building blocks (e.g., graphene, graphane, carbon nanotubes, and others) when simple building blocks with overall negative CO_2 balance, such as silica and water, can also create highly functional materials and structures [12]. This has a dramatic impact on the current general understanding of a material's design. The conditions for the design of hierarchically organized *de novo* materials with mutable properties, as strong as Kevlar™ and as extensible as rubber, made of renewable resources, become only comprehensible when they are systematically categorized and related to other features within and outside the material system.

15.3   Case Studies

In this section, case studies are presented to illustrate the multiscale study of protein materials. First, the folding and atomistic assembly of silk is discussed to show how hierarchical structures may overcome fundamental strength limits via noncovalent H-bond cooperativity. A second case study presents the derivation and mechanical analysis of a coarse-grained model of G-actin polymerization into F-actin filaments. A third case study presents an example of category theory as a mathematical abstraction technique to apply the lessons of one network (e.g., a hierarchical protein material) to a similarly structured network with seemingly disparate building blocks.

15.3.1   *Atomistic and Mesoscale Protein Folding and Deformation in Spider Silk*

Silk is a hierarchically structured protein fiber with a high tensile strength and great extensibility, making it one of the toughest materials known [65–67]. In contrast to synthetic polymers based on petrochemicals, silk is spun into strong and totally recyclable fibers at ambient temperatures, low pressures, and with water as the solvent. However, biomimetic reproductions of silk remain a challenge because of silk's characteristic microstructural features that can only be achieved by controlled self-assembly of protein polymers with molecular precision [68,69]. Unlike silkworms, some spiders can use different glands to create up to seven types of silk, from the strong dragline to the viscoeleastic capture silk and tough eggsack casing [67]. Dragline silk, containing a high fraction of densely H-bonded domains, is used to provide the structural frame for the web and has an elastic modulus of around 10 GPa [65]. Capture silk, on the other hand, is a viscid biofilament containing cross-linked polymer networks and has an elastic modulus that is comparable to that of other elastomers [70].

While rubber is extensible and Kevlar™ is stiff and strong, silks feature a combination of strength and toughness not typically found in synthetic materials. It is known that β-sheet crystals at the nanoscale play a key role in defining the mechanical properties of silk by providing stiff and orderly cross-linking domains embedded in a semi-amorphous matrix that consists predominantly of less orderly structures [71,72]. These β-sheet nanocrystals, bonded by means of assemblies of H-bonds, have dimensions of a few nanometers and constitute roughly 10–15% of the silk volume. When silk fibers are stretched, the β-sheet nanocrystals reinforce the partially extended and oriented macromolecular chains by forming interlocking regions that transfer the load between chains under lateral loading, similar to their function in other structural proteins [73]. The hierarchical network of spider draglines, contrasting synthetic elastomers like rubber, enables quick energy absorption and efficiently suppresses vibration during an impact [74]. Rubbers, composed of random polymer chains, display an elasticity regime that is primarily due to the change in conformational entropy of these chains. In contrast, the amorphous chains in silk filaments are extended and held in partial alignment with respect to the fiber axis in its natural state, resulting in remarkably different mechanical behaviors from rubbers [72,73].

At the base of the structural hierarchy, β-sheet nanocrystal size may be tuned though genetic modification and control of the self-assembly process. Computational atomistic structure predictions of the major ampullate spidroin 1 (MaSp1) protein aggregation in dragline silk, seen in Figure 15.9a, show that the size of the β-sheet nanocrystal may

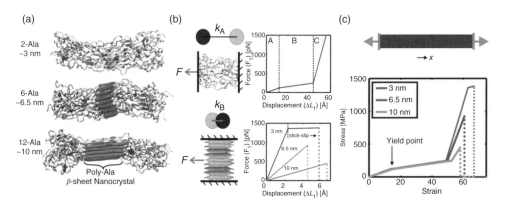

Figure 15.9 Atomistic structure predictions and training of a coarse-grain model of the MaSp1 cross-link in spider dragline silk. (a) Computational structure predictions of MaSp1 agglomeration show that the size of the β-sheet nanocrystal may be tuned by changing the poly-alanine length [75]. The structure predictions indicate a critical poly-alanine length of four to six alanine residues for the establishment of a well-defined nanocrystal. (b) To investigate the influence of nanocrystal size on the mechanics of the fibrillar network within a silk thread, the network is coarse-grained by forming two distinct particles: a strain-hardening semi-amorphous region (top) and a β-sheet nanocrystal (bottom). (c) A coarse-grained nodal network of these two bond types enables us to carry out larger scale simulations and explains the interplay of distinct protein phases, including information about the source of extensibility or flaw tolerance of silk fibrils. The result shown here reveals the dependence of the overall stress–strain response of silk fibrils depending on the size of constituting β-sheet nanocrystals. The result confirms the strong dependence of the overall stress–strain response of silk on the β-sheet nanocrystal size. Figures adapted from [71,75]

indeed be tuned by controlling the poly-alanine length [75], which in turn affects the stability and "quality" of the resulting nanostructures. These structure predictions used replica exchange MD to efficiently explore conformational space using parallel high-temperature replicas that exchange structure information with more accurate room-temperature replica simulations. A critical poly-alanine length of four to six residues was observed for the amyloidization of a defined and stable β-sheet nanocrystal and confirms the importance of size effects at the molecular scale. Indeed, a sampling of amino acid sequences of both orb-weaving and non-araneoid spiders shows that the MaSp1 and MaSp2 proteins that make up the strong and tough dragline silk typically have poly-alanine repeats six to eight residues in length, while the minor ampullate spidroin (MiSp) protein of viscoelastic capture silk has poly-alanine repeats of only two to four residues [76]. Combined with the structure predictions, these results provide concrete evidence that the size of the β-sheet nanocrystals has a fundamental effect on the mechanical properties of the macroscopic silk threads [71]. Further computational studies of spider silk proteins reveal that the nanoscale confinement of β-sheet nanocrystals in silks indeed has a fundamental role in achieving great stiffness, resilience, and fracture toughness at the molecular level of the structural hierarchy [15,71] H-bond cooperativity depends strongly on the size of the crystals and breaks down once β-sheet nanocrystals exceed a critical size, as visualized in Figure 15.9b. Larger β-sheet nanocrystals are softer and fail catastrophically at much lower forces owing to crack-like flaw formation. Furthermore, noncovalent H-bonds are able to reform during stick–slip deformation, allowing smaller crystals a self-healing ability until complete rupture occurs [77].

Advances in nanotechnology may allow for an optimization of silk and silk-like synthetic fibers for specific applications by modifying the governing network at critical length scales. By using multiscale simulations to understand and control how H-bonded silk threads can reach great strength and toughness and overcome strength limitations at the molecular scale, the behavior of a broader class of β-sheet-rich protein materials and similarly structured synthetic composite fibers can also be tuned and adapted.

15.3.2   Coarse-Grained Modeling of Actin Filaments

While the proteins in silk are suited for structural purposes outside of the organism, actin, in contrast, plays an important structural role in the mechanics of the eukaryotic cell cytoskeleton as well as muscle fibers. In particular, globular G-actin is the building block that rearranges into filamentous F-actin upon polymerization. To simulate F-actin filaments at the cellular scale, mesoscale modeling offers the means to understand and explain the origin of the properties of actin filaments with more accessible time and length scales than atomistic modeling and with more detail than continuum modeling could possibly provide. The mesoscale model described by Deriu *et al.* explored the formation of domains in G- and F-actin that determines the large-scale mechanical properties of this hierarchical biological material [78].

A bottom-up approach from MD simulation to normal-mode analysis with an elastic network model links molecular phenomena to macroscopic properties. Starting from the configuration of the fully water-solvated MD model of G-actin, a first level of coarse-graining was attained by replacing α carbon atoms C_α by point-mass nodes, as illustrated in Figure 15.10a. These nodes are then connected by springs using nonlinear interparticle

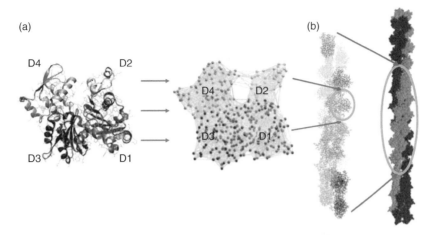

Figure 15.10   Coarse graining of F-actin filaments. (a) In a first level of coarse-graining, the α carbon atoms C_α in globular G-actin are represented as nodes that are interconnected by harmonic virtual springs to form an elastic network model. (b) In a second level of coarse-graining, four functional subdomains (D1–D4) are indentified and replaced by rigid blocks. G-actin rotates around and translates along the filament axis, forming the architectural spiral of F-actin and resembling the structure of a double helix. Reprinted with permission from [76]. Copyright 2011 Elsevier

potential functions. In a second level of coarse-graining, the functional subdomains are clustered into rigid blocks along a filament axis, shown in Figure 15.10b. These rigid blocks rotate around and translate along the filament axis, forming the architectural spiral of F-actin and resembling the structure of a double helix. Through this transition from atomistic to coarse-grained modeling, the system size, which correlates to the diagonalization time of the Hessian matrix for the normal-mode analysis, can be reduced by a factor of 100. The normal-mode analysis of the coarse-grained filament model then allows the estimation of the axial, bending, and torsional stiffnesses as well as the persistence length of F-actin with good agreement with experimental data.

By including atomistic structural information and training the interactions from molecular studies, mesoscale modeling provides a means to effectively overcome computational limits without losing insight into the protein dynamics. However, coarse-graining requires sufficient understanding of the initial structure and internal processes, such as restructuring, bond rupture and formation, and so on, in order to quantify parameters. In the absence of molecular studies, methods of mathematical abstraction such as category theory may further aid in the estimation of coarse-graining parameters.

15.3.3   Category Theoretical Abstraction of a Protein Material and Analogy to an Office Network

The final case study presented here focuses on category theory, used here to showcase the analysis of complex hierarchical biological structures like proteins based on the fundamental interaction of its building blocks. Once the origin of functionality in one system is understood – for example, by a combined computational–experimental analysis using the tools reviewed above – the building blocks for another material or system can be identified. In such a transition from one set of building blocks to another, functionality is preserved but assigned to a new structural carrier. The following example developed by Spivak *et al.* of the *isomorphism* of an α-helix/amyloid structure illustrates the possibility of a completely new approach to materials design [12]. An *isomorphism*, a type of *functor*, is defined as the one-to-one correspondence of both the objects and their *morphisms* of two categories; for example, a biological and a synthetic system [60].

Figure 15.11 shows the abstraction process used for the protein material. First, the desired system property (failure extension after system is subjected to axial load) is generically described for a protein material, and the three fundamental building blocks (brick, glue, and lifeline) are defined (Figure 15.11a). For protein materials, bricks refer to the polypeptide chain, glue to the H-bond clusters, and the lifeline can be understood as a hidden length of the polypeptide chain after bond rupture typical for α-helix structures. Category theory (and ologs in particular) aims to abstract of the parameters that govern properties and conditions and to reveal universal patterns by the comparison of characteristics within the system by defining properties in terms of more basic and self-sufficient concepts. Applying the mathematical tools of category theory, a linguistic version of an olog can be constructed. In the olog in Figure 15.11b, "ductile" and "brittle" are defined in terms of the relative failure extension of the building blocks compared with the whole system and "one-dimensional" is defined as an assembly of alternating building blocks in a chain graph. In addition to the graphical representation, a complete olog consists of a table of commutative paths and fiber products.

One distinctive feature of the example presented by Spivak *et al.* [12] is that the *morphisms* for this biological system are simplified yet still contain basic conditions which determine the system's behavior concerning the desired properties. Its relations are based on molecular simulations from previous studies [77,79,80]. Bricks, glue, and lifeline serving as building blocks may not reflect an ultimate physical reality but are sufficiently abstract to allow the translation to other fields such as other biological or synthetic materials or even an office hierarchy (Figure 15.11c). An *isomorphism* relates all the objects and the arrows from the polymer network to a social network. The chain of polypeptides (bricks) is transformed into a group of persons, the H-bonds (glue) refer to the wireless communication between the persons, and the lifeline can be understood as doors permitting a direct communication in case the glue fails. Within this framework, all fiber product definitions in the protein olog (e.g., for "ductile," "brittle," and "one-dimensional") are still valid. For instance, the "design principle" that bricks and glue have to alternate in a linear chain ("one-dimensional") in order to form a stable structure is transferred from the protein to the social network and "axial tension" is now understood as communication noise. Although the names and interpretations of objects and arrows are different in the case of the protein versus the social network, it is possible to construct a system where the fundamental properties, their relations, and definitions are similar. In particular, the way how functional properties emerge from the interplay of simple building blocks is identical in both networks. Spivak *et al.* [12] explained that the general presentation of such relationships in networks is what is missing in current theories, and is where ologs present a powerful paradigm for *de novo* design of biologically inspired systems that span multiple hierarchical levels. This is because ologs achieve a rigorous

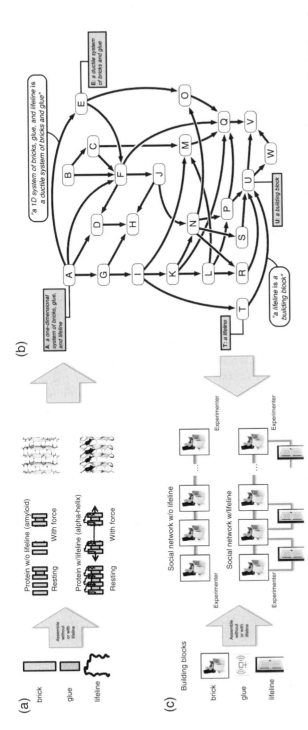

Figure 15.11 Abstraction of the mechanical properties of protein materials under axial extension and analogy to a social network. (a) Overview of fundamental building blocks of the representative protein materials. The protein materials considered here are composed of a linear arrangement of three elements: "bricks," "glue," and in some cases "lifeline." For proteins, these may represent the polypeptide backbone, H-bond clusters, and an unfolding hidden length, respectively. (b) Olog capturing the mechanical behavior of the system in (a). The boxes contain an intended set of instances and the arrows are *morphisms* (here unique functions). (c) Analogy to a social network that is modeled with precisely the same olog. The one-to-one correspondence is called *isomorphism* of the two categories protein and social network. The chain of polypeptides (bricks) is transformed into a group of persons, the H-bonds (glue) refer to the wireless communication between the persons, and the lifeline can be understood as doors permitting a direct communication in case the glue fails. Within this constructed framework, all fiber product definitions in the protein olog (e.g., for "ductile," "brittle," and "one-dimensional") are still valid. Although the names and interpretations of objects and arrows are different, the fundamental properties and how they relate to each other are similar. Figure adapted from [12]

description of the synergistic interactions of structures and mechanisms at multiple scales which provides the basis for enhanced functionality despite the reliance on few distinct building blocks.

15.4 Discussion and Conclusion

Computational modeling of materials provides us with a powerful tool to form a bridge between analytical theory and experimental observation in order to explain phenomena and to predict material behavior. Emerging developments in computational tools suitable for multiscale simulation and engineering of protein materials now enable the coverage of the same length-scale range as experimental tools (e.g., from nanometer to meters and from femtoseconds to hours) and provide many additional insights to chemical and deformation mechanisms. As computer performance continues to advance, computational modeling of synthetic and biological materials will further enable our understanding of processes across all relevant scales. To date, hierarchical protein materials like spider silk are being understood top down by dissecting the macroscopic structure via imaging, experimental analysis, and computational simulations. In addition, ologs represent a unique way to store and share data, knowledge, and insights in structure and functionality among many disparate research groups. Ologs offer a means to relate findings to previous results and thereby help to reveal the origin of the described system property and to connect seemingly unrelated topics and fields. By understanding how the building blocks of protein materials like spider silk assemble at each hierarchical level, and by relating the features of other hierarchical materials and organizations through the use of ologs and category theory, multiscale engineering and advances in nanotechnology may soon allow the bottom-up design and optimization of advanced biological, biocompatible, and biomimetic fibers, composites, and other structural materials by controlling the governing networks at critical length scales.

Acknowledgments

This work was supported by ONR, ARO-MURI, and the German Academic Foundation (Studienstiftung des deutschen Volkes). Additional support from NSF is acknowledged.

References

[1] Alder, B.J. and Wainwright, T.E. (1959) Studies in molecular dynamics. I. General method. *The Journal of Chemical Physics*, **31** (2), 459–466.

[2] Bernal, J.D. (1964) The Bakerian Lecture, 1962. The structure of liquids. *Proceedings of the Royal Society of London, Series A: Mathematical and Physical Sciences*, **280** (1382), 299–322.

[3] Springel, V., Yoshida, N., and White, S.D.M. (2001) GADGET: a code for collisionless and gasdynamical cosmological simulations. *New Astronomy*, **6** (2), 79–117.

[4] Abraham, F.F., Walkup, R., Gao, H. *et al.* (2002) Simulating materials failure by using up to one billion atoms and the world's fastest computer: work-hardening. *Proceedings of the National Academy of Sciences of the United States of America*, **99** (9), 5783–5787.

[5] Buehler, M.J., Hartmaier, A., Gao, H. *et al.* (2004) Atomic plasticity: description and analysis of a one-billion atom simulation of ductile materials failure. *Computer Methods in Applied Mechanics and Engineering*, **193** (48–51), 5257–5282.

[6] Kadau, K., Rosenblatt, C., Barber, J.L. *et al.* (2007) The importance of fluctuations in fluid mixing. *Proceedings of the National Academy of Sciences of the United States of America*, **104** (19), 7741–7745.

[7] Nomura, K.-i., Chen, Y.-C., Wang, W. *et al.* (2009) Interaction and coalescence of nanovoids and dynamic fracture in silica glass: multimillion-to-billion atom molecular dynamics simulations. *Journal of Physics D: Applied Physics*, **42** (21), 214011.

[8] Germann, T.C. and Kadau, K. (2008) Trillion-atom molecular dynamics becomes a reality. *International Journal of Modern Physics C*, **19** (9), 1315–1319.

[9] Park, S., Yang, X., and Saven, J.G. (2004) Advances in computational protein design. *Current Opinion in Structural Biology*, **14** (4), 487–494.

[10] Hao, M.-H. and Scheraga, H.A. (1994) Monte Carlo simulation of a first-order transition for protein folding. *The Journal of Physical Chemistry*, **98** (18), 4940–4948.

[11] Ratje, A.H., Loerke, J., Mikolajka, A. *et al.* (2010) Head swivel on the ribosome facilitates translocation by means of intra-subunit tRNA hybrid sites. *Nature*, **468** (7324), 713–716.

[12] Spivak, D.I., Giesa, T., Wood, E., and Buehler, M.J. (2011) Category theoretic analysis of hierarchical protein materials and social networks. *PLoS ONE*, **6** (9), e23911.

[13] Chandrasekaran, B. (1988) Generic tasks as building blocks for knowledge-based systems: the diagnosis and routine design examples. *The Knowledge Engineering Review*, **3**, 183–210.

[14] Buehler, M.J. (2011) Multiscale aspects of mechanical properties of biological materials. *Journal of the Mechanical Behavior of Biomedical Materials*, **4** (2), 125–127.

[15] Keten, S. and Buehler, M.J. (2010) Nanostructure and molecular mechanics of spider dragline silk protein assemblies. *Journal of the Royal Society Interface*, **7** (53), 1709–1721.

[16] Buehler, M.J. and Yung, Y.C. (2009) Deformation and failure of protein materials in physiologically extreme conditions and disease. *Nature Materials*, **8** (3), 175–188.

[17] Murphy, R.B., Philipp, D.M., and Friesner, R.A. (2000) A mixed quantum mechanics/molecular mechanics (QM/MM) method for large-scale modeling of chemistry in protein environments. *Journal of Computational Chemistry*, **21** (16), 1442–1457.

[18] Klepeis, J.L., Lindorff-Larsen, K., Dror, R.O. *et al.* (2009) Long-timescale molecular dynamics simulations of protein structure and function. *Current Opinion in Structural Biology*, **19** (2), 120–127.

[19] Brooks, B.R., Brooks III,, C.L., MacKerell Jr.,, A.D. *et al.* (2009) CHARMM: the biomolecular simulation program. *Journal of Computational Chemistry*, **30** (10), 1545–1614.

[20] Qin, Z. and Buehler, M.J. (2010) Molecular dynamics simulation of the α-helix to β-sheet transition in coiled protein filaments: evidence for a critical filament length scale. *Physical Review Letters*, **104** (19), 198304.

[21] Carrion-Vazquez, M., Oberhauser, A.F., Fisher, T.E. *et al.* (2000) Mechanical design of proteins studied by single-molecule force spectroscopy and protein engineering. *Progress in Biophysics and Molecular Biology*, **74** (1–2), 63–91.

[22] Sotomayor, M. and Schulten, K. (2007) Single-molecule experiments *in vitro* and *in silico*. *Science*, **316** (5828), 1144–1148.

[23] Oberhauser, A.F., Badilla-Fernandez, C., Carrion-Vazquez, M., and Fernandez, J.M. (2002) The mechanical hierarchies of fibronectin observed with single-molecule AFM. *Journal of Molecular Biology*, **319** (2), 433–447.

[24] Tozzini, V. (2005) Coarse-grained models for proteins. *Current Opinion in Structural Biology*, **15** (2), 144–150.

[25] Ghoniem, N.M. and Cho, K. (2002) The emerging role of multiscale modeling in nano- and micro-mechanics of materials. *Computer Modeling in Engineering & Sciences*, **3** (2), 147–173.

[26] Vvedensky, D.D. (2004) Multiscale modelling of nanostructures. *Journal of Physics: Condensed Matter*, **16** (50), 1537–1576.

[27] Ghoniem, N.M., Busso, E.B., Kioussis, N., and Huang, H. (2003) Multiscale modelling of nanomechanics and micromechanics: an overview. *Philosophical Magazine*, **83** (31), 3475–3528.

[28] Baeurle, S. (2009) Multiscale modeling of polymer materials using field-theoretic methodologies: a survey about recent developments. *Journal of Mathematical Chemistry*, **46** (2), 363–426.

[29] Dietz, H. and Rief, M. (2008) Elastic bond network model for protein unfolding mechanics. *Physical Review Letters*, **100** (9), 098101.

[30] West, D.K., Brockwell, D.J., Olmsted, P.D. *et al.* (2006) Mechanical resistance of proteins explained using simple molecular models. *Biophysical Journal*, **90** (1), 287–297.

[31] Nguyen, H.D. and Hall, C.K. (2004) Molecular dynamics simulations of spontaneous fibril formation by random-coil peptides. *Proceedings of the National Academy of Sciences of the United States of America*, **101** (46), 16180–16185.

[32] De Pablo, J.J. (2011) Coarse-grained simulations of macromolecules: from DNA to nanocomposites. *Annual Review of Physical Chemistry*, **62** (1), 555–574.

[33] Launey, M.E., Buehler, M.J., and Ritchie, R.O. (2010) On the mechanistic origins of toughness in bone. *Annual Review of Materials Research*, 40, 25–53

[34] Marrink, S.J., Risselada, H.J., Yefimov, S. *et al*. (2007) The MARTINI force field: coarse grained model for biomolecular simulations. *The Journal of Physical Chemistry B*, **111** (27), 7812–7824.

[35] Monticelli, L., Kandasamy, S.K., Periole, X. *et al*. (2008) The MARTINI coarse-grained force field: extension to proteins. *Journal of Chemical Theory and Computation*, **4** (5), 819–834.

[36] Izvekov, S. and Voth, G.A. (2005) A multiscale coarse-graining method for biomolecular systems. *The Journal of Physical Chemistry B*, **109** (7), 2469–2473.

[37] Bond, P.J., Holyoake, J., Ivetac, A. *et al*. (2007) Coarse-grained molecular dynamics simulations of membrane proteins and peptides. *Journal of Structural Biology*, **157** (3), 593–605.

[38] Buehler, M.J. (2010) Tu(r)ning weakness to strength. *Nano Today*, **5** (5), 379–383.

[39] Ackbarow, T. and Buehler, M.J. (2008) Hierarchical coexistence of universality and diversity controls robustness and multi-functionality in protein materials. *Journal of Computational and Theoretical Nanoscience*, 5, 1193–1204.

[40] Cranford, S. and Buehler, M.J. (2010) Materiomics: biological protein materials, from nano to macro. *Nanotechnology, Science and Applications*, **3**, 127–148.

[41] Buehler, M.J. (2010) Computational and theoretical materiomics: properties of biological and *de novo* bioinspired materials. *Journal of Computational and Theoretical Nanoscience*, **7** (7), 1203–1209.

[42] Kuhl, P.K., Andruski, J.E., Chistovich, I.A. *et al*. (1997) Cross-language analysis of phonetic units in language addressed to infants. *Science*, **277** (5326), 684–686.

[43] Buehler, M.J. and Yung, Y.C. (2009) Deformation and failure of protein materials in physiologically extreme conditions and disease. *Nature Materials*, **8** (3), 175–188.

[44] Garcia, A., Sen, D., and Buehler, M. (2010) Hierarchical silica nanostructures inspired by diatom algae yield superior deformability, toughness, and strength. *Metallurgical and Materials Transactions A*, **42**, 1–9.

[45] Cranford, S.W., Tarakanova, A., Pugno, N., and Buehler, M.J. (2012) Nonlinear constitutive behaviour of spider silk minimizes damage and begets web robustness from the molecules up. *Nature*, **482**, 72–76.

[46] Mason, O. and Verwoerd, M. (2007) Graph theory and networks in biology. *IET Systems Biology*, **1** (2), 89–119.

[47] Verdasca, J., Telo da Gamaa, M.M., Nunes, A. *et al*. (2005) Recurrent epidemics in small world networks. *Journal of Theoretical Biology*, **233** (4), 553–561.

[48] Pastor-Satorras, R. and Vespignani, A. (2001) Epidemic spreading in scale-free networks. *Physical Review Letters*, **86** (14), 3200–3203.

[49] Rodriguez, E., George, N., Lachaux, J.P., *et al*. (1999) Perception's shadow: long-distance synchronization of human brain activity. *Nature*, **397** (6718), 430–433.

[50] Jeong, H., Tombor, B., Albert, R. *et al*. (2000) The large-scale organization of metabolic networks. *Nature*, **407** (6804), 651–654.

[51] Barabasi, A.-L. and Oltvai, Z.N. (2004) Network biology: understanding the cell's functional organization. *Nature Reviews Genetics*, **5** (2), 101–113.

[52] Jeong, H., Mason, S.P., Barabási, A.-L., and Oltvai, Z.N. (2001) Lethality and centrality in protein networks. *Nature*, **411** (6833), 41–42.

[53] Barabasi, A.-L. and Albert, R. (1999) Emergence of scaling in random networks. *Science*, **286** (5439), 509–512.

[54] Eilenberg, S. and Maclane, S. (1945) General theory of natural equivalences. *Transactions of the American Mathematical Society*, **58** (Sep), 231–294.

[55] Ellis, N.C. and Larsen-Freeman, D. (eds.) (2009) *Language as a Complex Adaptive System*, Wiley–Blackwell, Chichester.

[56] Croft, W. (2010) Pragmatic functions, semantic classes, and lexical categories. *Linguistics*, **48** (3), 787–796.

[57] Croft, W. (2003) *Typology and Universals*, 2nd edn., Cambridge University Press, Cambridge.

[58] Sica, G. (2006) What is Category Theory? *Advanced Studies in Mathematics and Logic*, vol. **3**, Polimetrica, Monza.

[59] Awodey, S. (2010) *Category Theory*, 2nd edn., Oxford Logic Guides, vol. **52**, Oxford University Press.

[60] Spivak, D.I. and Kent, R.E. (2012) Ologs: a categorical framework for knowledge representation. *PLoS ONE*, **7** (1), e24274.

[61] Keten, S. and Buehler, M.J. (2008) Geometric confinement governs the rupture strength of H-bond assemblies at a critical length scale. *Nano Letters*, **8** (2), 743–748.

[62] Staab, S. and Studer, R. (2009) *Handbook on Ontologies*, 2nd edn., Springer, Berlin.

[63] Spivak, D.I. (2009) Simplical databases, http://arxiv.org/abs/0904.2012v1.

[64] Mac Lane, S. (1998) *Categories for the Working Mathematician*, 2nd edn., Graduate Texts in Mathematics 5, Springer, New York.

[65] Denny, M. (1976) The physical properties of spider's silk and their role in design of orb-webs. *Journal of Experimental Biology*, **65** (2), 483–506.

[66] Hayashi, C.Y. and Lewis, R.V. (1998) Evidence from flagelliform silk cDNA for the structural basis of elasticity and modular nature of spider silks. *Journal of Molecular Biology*, **275** (5), 773–784.

[67] Hu, X., Vasanthavada, K., Kohler, K. *et al*. (2006) Molecular mechanisms of spider silk. *Cellular and Molecular Life Sciences*, **63** (17), 1986–1999.

[68] Huemmerich, D., Scheibel, T., Vollrath, F. *et al*. (2004) Novel assembly properties of recombinant spider dragline silk proteins. *Current Biology*, **14** (22), 2070–2074.

[69] Foo, C.W.P., Bini, E., Henseman, J. *et al*. (2006) Solution behavior of synthetic silk peptides and modified recombinant silk proteins. *Applied Physics A: Materials Science & Processing*, **82** (2), 193–203.

[70] Gosline, J.M., Guerette, P.A., Ortlepp, C.S., and Savage, K.N. (1999) The mechanical design of spider silks: from fibroin sequence to mechanical function. *Journal of Experimental Biology*, **202** (23), 3295–3303.

[71] Nova, A., Keten, S., Pugno, N.M. *et al*. (2010) Molecular and nanostructural mechanisms of deformation, strength and toughness of spider silk fibrils. *Nano Letters*, **10** (7), 2626–2634.

[72] Grubb, D.T. and Jelinski, L.W. (1997) Fiber morphology of spider silk: the effects of tensile deformation. *Macromolecules*, **30** (10), 2860–2867.

[73] Gosline, J., Lillie, M., Carrington, E. *et al*. (2002) Elastic proteins: biological roles and mechanical properties. *Philosophical Transactions of the Royal Society of London Series B: Biological Sciences*, **357** (1418), 121–132.

[74] Du, N., Yang, Z., Liu, X.Y. *et al*. (2011) Structural origin of the strain-hardening of spider silk. *Advanced Functional Materials*, **21** (4), 772–778.

[75] Bratzel, G.H. and Buehler, M. (2012) Sequence–structure correlations and size effects in silk nanostructure: poly-Ala repeat of *N. clavipes* MaSp1 is naturally optimized. *Journal of the Mechanical Behavior of Biomedical Materials*, **7**, 30–40.

[76] Gatesy, J., Hayashi, C., Motriuk, D. *et al*. (2001) Extreme diversity, conservation, and convergence of spider silk fibroin sequences. *Science*, **291** (5513), 2603–2605.

[77] Keten, S., Xu, Z., Ihle, B., and Buehler, M.J. (2010) Nanoconfinement controls stiffness, strength and mechanical toughness of beta-sheet crystals in silk. *Nature Materials*, **9** (4), 359–367.

[78] Deriu, M.A., Bidone, T.C., Mastrangelo, F. *et al*. (2011) Biomechanics of actin filaments: a computational multi-level study. *Journal of Biomechanics*, **44** (4), 630–636.

[79] Ackbarow, T., Chen, X., Keten, S., and Buehler, M.J. (2007) Hierarchies, multiple energy barriers, and robustness govern the fracture mechanics of α-helical and β-sheet protein domains. *Proceedings of the National Academy of Sciences of the United States of America*, **104** (42), 16410–16415.

[80] Paparcone, R. and Buehler, M.J. (2011) Failure of $A\beta(1–40)$ amyloid fibrils under tensile loading. *Biomaterials*, **32** (13), 3367–3374.

16

Geometric Models of Protein Secondary-Structure Formation

Hendrik Hansen-Goos[1] and Seth Lichter[2]
[1]*Yale University, USA*
[2]*Northwestern University, USA*

16.1 Introduction

This chapter reviews recent models which determine protein equilibrium configurations by minimizing solvation free energy using novel geometric formulations. In distinction to numerical simulations, the analytical methods reviewed here use a simplified set of global geometric parameters (such as volume and surface area) to represent the protein. These methods are exciting. They give new insight into the forces which drive protein secondary structure formation. Other reviews more widely cover reduced-order methods of protein folding or discuss solvation effects more generally [1–8].

The review is organized as follows. Protein folding has traditionally been discussed in terms of the hydrophobic effect proportional to the area of protein interface exposed to the solvent, as discussed in Section 16.2. In this way, interfacial area has become recognized as an important geometric measure. The work to create an inert space within a solvent is proportional to volume, introducing a second geometric measure, as further discussed in Section 16.2. Section 16.3 briefly scans computational and alternative coarse-grained methods. The tube model is introduced in Section 16.4. Here, we first see that a bendy impenetrable tube with homogeneous properties (that is, without accounting for the differences of one amino acid from another) generates α helical and β strand conformations. These conformations appear for a tube subject to a constraint; for example, by being forced to fit into a small space or by having sections of the tube attract one another. In Section 16.4.3, these constraints are replaced by a more realistic model of the solvent, by placing the tube-like protein in a thermal bath of small, hard spheres and then adding additional parameters to more accurately model the protein and solvent (Section 16.4.4). For these more realistic cases, the radius of the protein is not at its impenetrable surface,

Nano and Cell Mechanics: Fundamentals and Frontiers, First Edition. Edited by Horacio D. Espinosa and Gang Bao.
© 2013 John Wiley & Sons, Ltd. Published 2013 by John Wiley & Sons, Ltd.

but at a radius enlarged by the solvent radius (as shown in Figure 16.5). This additional sheath of volume surrounding the protein is the space which dictates the translational entropic contribution to the solvent free energy. Though the tube model has also been modified such that its properties vary along the length of the tube [9], these are not discussed here. Rather, we focus on the remarkable feature of the tube model that it generates secondary structure conformations for homogeneous properties.

Equilibrium protein conformations are identified by their having a minimum of free energy. Hadwiger's theorem, introduced in Section 16.5, states that the free energy of a protein can be expressed as the sum of four terms, each of which is conveniently the product of a term which is dependent only on protein geometry (including volume and surface area) times a coefficient which is dependent only on solvent properties (and the protein–solvent interaction potential). Thus, the volume and area dependencies, mentioned above, and which are well known from observation of real proteins in solution, can be formally derived using Hadwiger's theorem. In Section 16.5.2, Hadwiger's theorem is combined with the tube model to show that solvent free energy can be accurately calculated, and to show how protein stability depends on solvent and protein properties. The major result of this approach is summarized in Figure 16.7, which shows how helix and β-sheet formation depends on solvent properties and the properties of a homogeneous tube-like protein. The review closes with a summary of the findings from the tube model plus Hadwiger's theorem, and speculation on what lies ahead (Section 16.6).

In explaining the figures, which come from a variety of research groups, we use the nomenclature in the figure, which may differ from that used in another figure. Please pay attention. The solvent radius is denoted R_s, r, or ε; the helix radius is R_h or simply R, which is also defined as $R = R_p + R_s$, the radius of the parallel surface of a spherical solute with radius R_p; the surface tension is γ in Figure 16.1 and σ elsewhere.

16.2 Hydrophobic Effect

We begin with a description of the hydrophobic effect, to show how geometric measures (in this case surface area and volume) enter as important parameters.

The hydrophobic effect is the low affinity of solutes for water, relative to water–water affinity, and the consequent tendency for solutes to pack together to decrease their exposure to water [10]. The hydrophobic effect considers one particular substance: water. The singling out of water for special treatment is warranted owing to its preeminence as a laboratory and *in vivo* solvent. Early investigations of the hydrophobic effect used hydrocarbons, in which it was observed that their free energy of transfer, from pure hydrocarbon to water, was proportional to their area [11]. Consequently, one way for a solute to reduce its free energy is to adopt a conformation with low surface area.

It is instructive to consider the free-energy cost ΔG of creating a spherical (vapor-filled) cavity of radius R in water. Given that under ambient condition water is close to its triple point, the volume term $(4\pi/3)R^3 \Delta p$, where Δp is the cavity transmural pressure difference, can be neglected. Hence, for R sufficiently large such that a continuum description is valid, we must expect $\Delta G = 4\pi R^2 \gamma$, where γ is the liquid–vapor surface tension. This is indeed what is found in detailed macroscopic calculations [13]; see Figure 16.1. However, in the limit where the cavity is very small ($R \lesssim 1$ nm) the energetic cost is

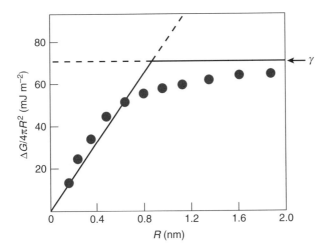

Figure 16.1 The solvation free energy divided by area for a spherical vapor-filled cavity approaches a constant (equal to the liquid–vapor surface tension γ) as cavity size grows larger than approximately 1 nm. Solvation energy at small scales is approximately proportional to cavity volume. Different proportions of these two scalings (area and volume) will make up the first two terms in Hadwiger's expression for solvation energy in Section 16.5. However, in Section 16.5 we drop the assumption of the cavity being filled with vapor, which makes $p4\pi R^3/3$ the leading behavior of ΔG for $R \to \infty$. Reprinted with permission from [12]. Copyright 2005 Nature Publishing Group

significantly lower than $4\pi R^2\gamma$. In fact, calculations reveal that $\Delta G \propto R^3$ as $R \to 0$. How can this behavior be explained? For a small cavity, a continuum description is no longer appropriate. Even the very concept of a vapor-filled cavity becomes problematic. Instead, the cavity should now be defined as blocking a certain volume proportional to R^3 from the configuration space of the solvent. Considering, in addition, that a very small solute does not cause hydrogen bonds in the solvent to break, we can understand that ΔG vanishes as R^3 for small cavities [12].

It is tempting to visualize hydrophobicity of nonpolar groups (such as hydrocarbon chains) as being a consequence of energy loss due to hydrogen bonds between water molecules being disrupted in the vicinity of the surface of the solute. Contrary to the case of an ionic or strongly polar solute, this loss would not be compensated by the formation of bonds between water molecules and the solute. This picture is too simple. Measurements of the change in enthalpy and entropy upon transferring hydrocarbons from organic solvents to water (see Tanford [14]) show that water is even energetically favorable (enthalpy change ΔH is negative) compared with organic solvents. It is rather the entropy which decreases ($\Delta S < 0$) as hydrocarbons are transferred to water, causing the difference in solvation free energy $\Delta H - T\Delta S$ to be positive. It can be concluded that it is much more the reduction in entropy related to the *distortion* of hydrogen bonds in the vicinity of a nonpolar solute, rather than the energetic loss of breaking hydrogen bonds which lies at the origin of the hydrophobic effect. In this sense the hydrophobic effect has been termed entropic, as early as in 1945 by Frank and Evans [15]. However, the strength of this entropic effect is proportional to the surface area consistent with Figure 16.1.

Recent molecular dynamics simulations show that the picture of hydrophobicity as an entropic effect is no longer valid in the case of *concave* solutes that impose hydrophobic confinement on the solvent [16]. For a model system of a ligand binding to a cavity it is found to be the favorable *enthalpy* change related to releasing the water molecules from the hydrophobic cavity to the bulk which drives the hydrophobic association between receptor and ligand [16].

The importance of the hydrophobic effect for protein folding was first promoted by Kauzmann in 1959 [10]. In recent work, Levy *et al.* discuss that the fraction of the solvation free energy of proteins that is due to solvent–solute van der Waals interactions is approximately given by $\Delta G_{vdW} = \gamma A + b$, where A is the surface area, γ is an effective surface tension, and b is the (hypothetical) free energy of a solute of zero volume [17]. This scaling confirms the picture conveyed in Figure 16.1.

There is an important difference between the scenario where the change in solvation free energy is measured upon transferring hydrocarbons from an organic solvent to water (see the experimental results compiled in Tanford [14]) and protein solvation, where the change in solvation free energy upon conformational transitions of the protein, *without* changing the solvent, is of interest. This can be illustrated by considering the volume term pV_{exc} of solvation free energy, where p is the solvent pressure and V_{exc} is the volume excluded to the solvent by the presence of the solute, to be defined in Section 16.4.3. The transfer of a hydrocarbon from an organic solvent to water leaves this term mostly unchanged. This follows because: (i) the different solvents are at the same pressure (usually 1 atm) and, hence, p does not change; and (ii) the hydrocarbon does not undergo significant change in conformation and, hence, V_{exc} remains (approximately) the same. Only solvent size effects contribute to a possible change in V_{exc}, as will be introduced in Section 16.4.3 and discussed more extensively in Section 16.5. The problem of protein folding, however, relies on comparing various protein conformations in a given solvent with the aim of identifying the native structure. As we have discussed above, there is a surface contribution of mostly entropic origin to the solvation free energy which varies as a function of the surface area of different protein conformations. Furthermore, there is the possibility that V_{exc} changes as the protein conformation is modified. The corresponding contribution to the free energy results from the *translational* entropy of the solvent (as opposed to the entropy related to the distortion of hydrogen bonds, which is at the origin of the surface contribution to ΔG). It has been shown by Harano and Kinoshita that, for sufficiently large peptides or proteins (of the order of 50 residues or more), the gain in translational solvent entropy is in fact enough to compensate for the loss in conformational entropy which a biomolecule incurs upon folding into a compact state [18]; see Figure 16.4. The magnitude of the effect depends critically on the size of the solvent molecule. For solvents with a molecular size significantly larger than that of water, translational entropy is no longer able to compete with conformational entropy. A related phenomenon is the depletion interaction that has been measured in colloidal suspensions [19] and which is also thought to play a role for the folding of proteins in the crowded cellular environment [20]. However, in the latter scenario the entropy which creates the effect stems from macromolecular crowding rather than from the water itself, as in the approach taken by Harano and Kinoshita [18].

In summary, the hydrophobic effect can be understood as a combination of two different entropic effects. First, hydrogen bonds in water next to the nonpolar solute (of surface area *A*) have to be distorted, which results in a loss of solvent entropy without an energetic compensation through the formation of solute–solvent hydrogens bonds [14]. Second, the presence of the solute reduces the translational entropy of the solvent by excluding solvent from a volume V_{exc}. Depending on the relative importance of the two effects, the first one scaling with *A* and the second with V_{exc}, a protein will seek to minimize a linear combination of *A* and V_{exc}, as far as solvent contributions to the protein free energy are concerned. These two contributions, and two additional terms, will be treated in detail in Section 16.5.

16.2.1 Variable Hydrogen-Bond Strength

In determining the contribution to the hydrophobic effect from hydrogen bonds, it is frequently assumed that the strength of hydrogen bonds, before and after folding, is unchanged. With this assumption, losing a hydrogen bond with water while gaining a hydrogen bond in the interior yields no net increase in protein stability. The assumption of constancy of hydrogen-bond strength has come under scrutiny. In molecular dynamics simulations, atomic force fields are ascribed fixed parameters. For example, the often-used Lennard–Jones force field is parameterized by a fixed energy and length scale. New proposals depart from this view, and propose that hydrogen-bond strength cannot be based on fixed parameters, but rather on parameters which depend upon the presence of neighboring molecules. Experimental evidence indicates that hydrogen bonds in apolar or hydrophobic environments can be up to $2k_BT$ stronger than in the aqueous environment [21]. In theoretical treatments, neighboring molecules affect hydrogen-bond strength through the dielectric constant: apolar environments reduce the shielding of the hydrogen bond's electrostatic charge–charge interaction, resulting in a stronger hydrogen bond [22].

In an assessment of hundreds of proteins in the Protein Data Bank, it has been found that, compared with hydrophilic residues, hydrophobic residues are more sensitive to the presence of neighboring residues [23]. That is, the relative decrease of solvent-accessible area due to neighboring residues is greater for hydrophobic residues than fot hydrophilic residues. This suggests that hydrogen bonds made by hydrophobic residues within the folded protein are more influenced by the hydrophobic environment than those from hydrophilic residues.

Protein folding is known to be a cooperative process in which folding of one part of the protein facilitates hydrogen bonding of other parts by bringing them into closer proximity. The coupling between neighboring residues and hydrogen-bond strength provides a new type of cooperativity: as neighboring residues fold, they strengthen the original hydrogen bond at fixed distance. By being able to more strongly pull the protein together, the strengthening of the bond aids further folding [21].

Microenvironment may affect other aspects of the hydrogen bond, such as the degree to which the proton donor and proton acceptor are deprotonated and protonated, respectively. These and other effects of microenvironment are just beginning to be investigated [24].

16.3 Prior Numerical and Coarse-Grained Models

In 1958, on the elucidation of the structure of myoglobin, the complexity of protein conformation was made apparent [25]. The basic double-helix structure of DNA, determined a few years earlier, was much simpler. Additionally, while DNA was constructed of four nucleotides, proteins utilize 20 amino acid monomers, each with unique structure and properties. As computational techniques developed and as computer storage increased, the complexity of protein structure encouraged models of protein folding which incorporated detailed accounting of atomic interactions. The great strides made in computational techniques are discussed in recent reviews [26–28].

Another methodological line of attack has been to simplify the folding problem, purposely omitting burdensome details, leaving a streamlined problem which retains essential interactions. One particularly well-developed and useful model reduced the properties of the 20 different amino acid residues along a protein's primary structure down to two types of monomers: "H" for a hydrophobic type and "P" for a polar type, each characterized by a single energy [29]. The HP-modeled proteins would fold or not, resisted by the chain's conformational entropy depending on the relative number of H versus P monomers and the overall chain length. For this approach, and other coarse-grained models, see the reviews [3, 5, 6, 8]. We mention the HP model here for two reasons. First, to note that the simplification to hydrophobic and hydrophilic monomers follows an insightful line of thought arising from observations of the behavior of amphiphiles, molecules which are significantly apolar at one end and polar at the other. These molecules aggregate into structures in which the apolar ends congregate together, leaving the polar ends directed toward the solvent. Second, the HP model serves as a foil to highlight the geometric models reviewed here, which have uniform properties along their length, and so break with this long tradition of retaining even a binary remnant of amino acid specificity.

16.4   Geometry-Based Modeling: The Tube Model

16.4.1   Motivation

The tens of thousands of different protein structures can be decomposed into a sequence of α helices, β strands, loops and several other types of helices and other structures. The total number of these basic conformational elements (the set of secondary structures) is small, $O(10)$. As secondary structures occur ubiquitously, it is argued that their folding might be insensitive to many of the details ensconced in the primary sequence. Most simply, secondary structure folding might be described by a protein denuded of its amino acid specificity and represented via only a few geometric measures. The tube-like protein model is such a model, which reduces the protein to a tube of constant thickness. The tube model can be further motivated by analogy with atomic crystals and through observations of the distribution of hydrophobic and hydrophilic residues at the solvent-exposed surface, as we now discuss.

Motivation: Crystals

Atomic crystals are particular arrangements of atoms that can be infinitely repeated periodically throughout space to generate the entire crystal. We note three characteristics of crystal structure. First, the prevalence of crystals as equilibrium structures suggests that they are low energy. Second, their regularity further suggests that diffusion of atoms into their crystalline positions occurs without encountering a rate-limiting step; that is, without a bottleneck with a time constant greater than the observation time. For example, rate-limited aggregation produces tree-like branching structures very different from the regularity of crystalline arrays. Third, the same set of crystal structures is available independent of the precise spatial dependenceof attraction between atoms. So, for example, the cubic crystal structure can form from ionic interactions, as for the salt crystal NaCl, as well as for covalently bonded atoms. So, crystal structures suggest a process of creation (i) directed towards a low free-energy arrangement with (ii) little frustration, and (iii) independent of the form of atomic attraction. Do these characteristics hold, as well, for protein folding?

There are only two infinitely repetitive protein structures: the α helix and β sheets [11]. These secondary structures provide an optimal arrangement of hydrogen bonds, having the lowest energy as provided by hydrogen bonding [30]. Second, Ramachandran plots show that avoiding steric clashes between pairs of amino acids favors α helices and β sheets [31]. So, α helices and β sheets in proteins may be likened to atomic crystalline structures: we may hypothesize that proteins will tend towards these types of structures with little frustration due to steric clashes, and that this tendency is generic, independent of the particular form of attraction. In the case of proteins, this implies that secondary structure formation is at most weakly dependent on primary sequence.

Motivation: Buried Residues and Area Change

The driving force of protein folding is often attributed to the presence of hydrophobic residues which are preferentially "buried" in the native state interior, inaccessible to the solvent. However, there is roughly a 50/50 split in the number of hydrophobic and hydrophilic residues on the accessible surface [11].

Also, the change in accessible area on the formation of secondary structures is greater than that during the aggregation of secondary structures to form tertiary structure [32]. The nonparticular choice of residues to be buried plus the large change in surface area during secondary structure formation suggest that a generic force acting to minimize surface area, combined with a tendency to minimize the excluded volume in order to increase solvent translational entropy, might be effective in secondary structure formation.

16.4.2   Impenetrable Tube Models

It has been found that compaction by itself is insufficient for secondary structure formation [33, 34]. By this is meant that a protein modeled, say, as a string of spheres will compact when subject to a simple intra-sphere attractive force into conformations nearly devoid of recognizable secondary structure [33]. On compaction, the tube geometry overcomes this deficiency.

The protein is modeled as a tube of constant thickness Δ centered on a curve along the protein backbone [35–37]. This simple thickening of the one-dimensional space curve confers two significant constraints on possible configurations. First, tight bends which result in a cusp or crease in the tube are not allowed. This condition establishes a minimum radius of curvature which can be interpreted as a characteristic angle (defined along the curve which runs through the center of the tube). This angle plays the role of the angles in the Ramachandran plot. That is, the tube thickness is a proxy for local steric effects. Second, the thickness establishes nonlocal steric effects: sections of the tube lying obliquely to one another and sections lying parallel can approach only to within contact of the tube surfaces. In brief, two sections of tube cannot occupy the same spatial volume, and so tube thickness enforces the necessity that the protein be self-avoiding. The tube model administers these local and nonlocal properties homogeneously along the protein. Variations along the tube arising from amino acid specificity are not included.

For an infinitely long tube model, compaction is administered by constraining each interval (of some chosen length) along the protein to have a radius of gyration less than an assigned value. It is found that the tube with maximum Δ is a helix with $P/R \approx 2.5$, where P is pitch and R is radius (Figure 16.2). This value is close to that of helices occurring in real proteins [35].

For tubes of finite length, compaction can be established by constraining the tube to lie within a box whose length is less than the arc length along the tube centerline (which corresponds to the protein backbone). The backbone is then subject to simulated annealing, in which the curve is allowed to twist and writhe. Configurations are accepted or rejected, where thicker tubes are accepted with a higher probability. As the temperature of the Monte Carlo simulation is slowly decreased, a stable distribution of conformations emerges. Curved tubes in the shape of helices and saddles are obtained [35].

Homogeneous attraction of one section of the tube for another is an approximate means to model solvation forces. The tube model was augmented by assigning an attractive

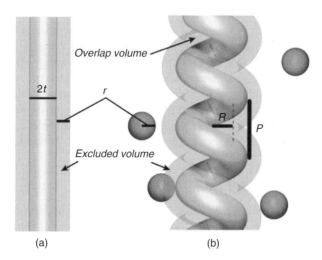

(a)   (b)

Figure 16.2   Consider a protein backbone as a tube of radius t. (a) Solvent particles of radius r can approach the centerline of the protein only to within $t + r$. Thus, there is a sheath of volume from which solvent is excluded. In the fully extended protein, the excluded volume is $\pi[(t + r)^2 - t^2]$ per unit length. (b) For a curved protein, the sheath of adjacent sections can overlap and so the excluded volume will decrease. Consequently, the surrounding solvent volume is increased, resulting in a gain in solvent translational entropy. As solvent size decreases, the pitch-to-radius ratio of the helix matches that seen in naturally occurring α helices $P/R \approx 2.5$. Reprinted with permission from [38]. Copyright 2005 American Association for the Advancement of Science

energy between proximal tube sections separated by greater than 2Δ, but still sufficiently close, say $< R_1$, to other parts of the chain. Separations within this range are denoted as protein–protein *contact*. Note that this model is still homogeneous; the properties are uniform along the entire length of the tube. As opposed to the previous model, here the thickness of the tube Δ is an assigned parameter. It is found that if the tube thickness is chosen too small, $\Delta/R_1 \ll 1$, then equilibrium configurations have many contacts between different parts of the chain, but these conformations show little secondary structure. If $\Delta/R_1 > 1$, then the large tube thickness prevents any significant number of contacts. The range $\Delta/R_1 \approx 1$ is denoted the marginally compact regime. Secondary structure appears within this regime, where the prevalent conformations are helices, saddles, and planar hairpins [35].

In a separate set of studies, tubes of $N = 8, \ldots, 13$ stiff monomers interact through a Lennard–Jones potential, constrained such that the global radius of curvature was greater than a certain value ρ [39–41]. It is found that helical structures are favored for ρ small enough; and for $\rho \approx 0.686$, an 8-mer nearly reproduces the geometry of the naturally occurring α helix. Larger values of ρ are flatter, and in that sense have a similarity to β hairpins. As ρ increases further, rings and rods are encountered. A phase diagram can be generated in (ρ, T), where T is temperature. Boundaries between the helical, hairpin, ring, and rod phases are sharply defined ridges in specific heat. It is suggested that the thickness ρ portrays the effect of side chains, and so indicates how they might help determine secondary structure preferences.

The tube model has been applied to a chain of links modeling 24 (or 48 [42]) C_2 atoms with a bending stiffness along the protein backbone and pairwise interactions [9, 43]. The pairwise interactions model intra-residue hydrogen bonding. In keeping with the point of view of the tube model, there is no variation of either the bending stiffness or the pairwise interactions with location along the tube, as there would be if the differences among the 20 different amino acids and their interactions were to be accounted for. The pairwise interactions, though, are taken to vary with the separation distance between the interacting pairs, and there is an energetic reinforcement of consecutive hydrogen bonds to account for cooperativity; see Refs. [9, 43].

Compaction and protein–protein attraction have been used as a proxy for solvation forces, and a summary of much of this work can be found in Banavar and Maritan [37]. As will be described below, compaction and attraction can be replaced by a more realistic model of the translational-entropy component of solvation forces [44].

16.4.3 Including Finite-Sized Particles Surrounding the Protein

Solvation effects of finite-sized solvent particles, providing a thermal bath surrounding large solute particles, were explored by Asakura and Oosawa [45]. They showed that the solvent will produce an attractive potential between spherical particles, proportional to the number density of the solvent [45, 46]. In their work, solvent molecules were spheres of radius R_s. Owing to their finite size, the center of the solvent molecules can be only as close as R_s to the surface of the large particles. The volume surrounding the particle surface outward to a thickness of R_s is called the excluded volume. When widely separated, then the total excluded volume in a solution of many such solute particles is simply equal to the excluded volume around one particle times the number of solute particles. But, when the large particles are within $2R_s$, then their regions of excluded volume overlap; see Figure 16.5. So, as particles come into close proximity, the excluded volume decreases, the volume available for solvent motion increases, and hence the entropy of the solvent (which in the limit of strong dilution is proportional to volume) also increases. Larger entropy implies that the free energy will decrease for smaller excluded volume. The free energy decrease with separation leads to an attractive force which tends to drive the solute particles together.

Initial work considered only simple geometries such as spheres. Furthermore, the original work of Asakura and Oosawa treated a continuum solvent containing finite-sized particles. Snir and Kamien used a tube model of a protein immersed in a bath of finite-sized solvent particles [38]. Consider a completely straight tube with a annular volume surrounding it out to a radius of $\Delta + \varepsilon$, into which the centers of solvent molecules cannot penetrate (Figure 16.2). When the tube curves, the annular region from one portion of the tube can overlap that of another portion [38]. Tube curvature results in overlapping volumes; that is, of excluded volumes. This enlarges the volume available to the solvent particles and, hence, their entropy is increased. Therefore, a helical tube conformation has larger translational solvent entropy than a stretched-out tube (Figure 16.2). Additionally, for small enough solvent molecules, the geometry of the helix was once again found to reproduce the value found in the α helices of naturally occurring proteins, $P/R \approx 2.5$.

For finite-length tubes driven to compact conformations through minimization of either excluded volume or surface area, it is found that helices, saddles, and planar β hairpins are formed [44]. The choice of conformation depends upon ε/Δ; that is, on the relative size of the solvent molecule and on the peptide length. Larger values of ε/Δ favor helical conformations, while small sizes, even those for $\varepsilon/\Delta \rightarrow 0$, favor planar β sheet conformations. In between these two ranges of values is a crossover regime of values in which both types of conformations, helices and β sheets, appear (Figure 16.3).

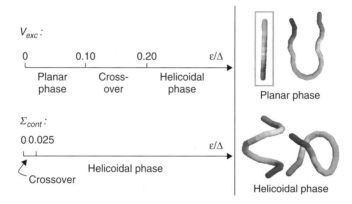

Figure 16.3 The low-energy conformations of a tube of thickness Δ and length 24Δ, in a bath of solvent hard spheres of radius ε. Minimizing the excluded volume V_{exc} yields planar β hairpin structures for small-sized solvent, β helical structures for larger sized solvent, and both types in the intermediate crossover region. Alternatively, minimizing the surface area Σ_{cont} does not yield a β hairpin regime. Reprinted with permission from [44]. Copyright 2008 American Physical Society

16.4.4 Models Using Real Protein Structure

The results discussed above do not include configurational entropy of the protein itself. Competition between the loss of protein configurational entropy and solvent translational entropy was considered by Harano and Kinoshita [18]. Excluded volume and protein configurational entropy were measured using a molecular model of several real proteins, in which the atoms of the residues were represented by Lennard–Jones spheres. The magnitude of the Asakura–Oosawa entropic force depends on the ratio of solvent size to excluded volume – where, for the same excluded volume, smaller solvent size yields a larger force [18]. As water is one of the smallest of all solvents, its small size confers upon it the ability to generate large translational entropy forces. It was found that to effectively compete against the loss of protein configurational entropy requires sufficiently large peptide or protein molecules (Figure 16.4). The magnitude of the excluded volume per length of protein was found to correlate with its thermal stability [48].

The stability of four proteins, including a thermophilic protein, was determined in water solvent, modeled as hard spheres with dipole and quadrupole moments [49]. Stability was determined as gain of solvent entropy minus loss of protein configurational entropy per

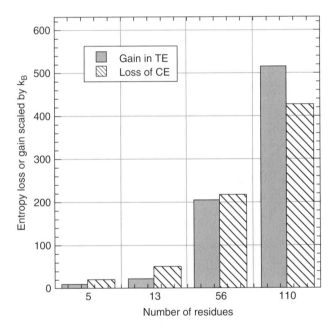

Figure 16.4 For Met-enkephalin, a 13-residue fragment of ribonuclease A, protein G, and barnase, the number of residues are 5, 13, 56, and 110, respectively. The peptide or protein molecule is modeled with only its conformational entropy (CE), excluding all other possibly stabilizing interactions such as intra-residue hydrogen bonds. As the size of the solute increases, the gain in solvent translational entropy (TE) exceeds the loss of peptide or protein CE. The results thus argue for the importance of translational entropy alone in stabilizing proteins and in folding. Reprinted with permission from [47]. Copyright 2005 Elsevier

residue from the denatured to native states. Larger values conferred greater stability, and the values were found to increase in the same order as the thermal denaturation temperature. Differences in stability arose from extent of compaction in the native state, and so it was argued that solvent entropy was the dominant contribution to protein stability.

The tube model has not been without its critics. The folding of two small (approximately 60-residue) α-helical proteins using an intra-residue interaction potential was compared versus folding using that same potential minus hydrogen bonding [50]. The finding that folding without hydrogen bonding did not yield α-helical structure was used to warn against the tube model's claim that minimizing excluded volume alone would result in α-helix secondary structure formation. (See the rejoinder [51].) The interaction potential did not include the effects of the solvent, so these results might be interpreted, not as a caveat against the tube model, but rather as an indication of the critical role of solvent in secondary structure formation.

16.5 Morphometric Approach to Solvation Effects

16.5.1 Hadwiger's Theorem

Hadwiger's theorem from integral geometry states that a given mapping φ from the set of three-dimensional bodies B_i to the real numbers can, under certain conditions, be expressed as a linear combination of only four integral measures, the so-called Minkowski measures, of the bodies B_i [52]. This abstract definition can be made concrete by thinking of the bodies B_i as different protein conformations; that is, folds among which we seek to identify the native state. This requires obtaining the solvation free energy $F_{sol}(B_i)$ for each possible conformation or, speaking in terms of Hadwiger's theorem, for each given body B_i. Obviously, $F_{sol}(B_i)$ is a mapping to the real numbers, which makes it a candidate for applying Hadwiger's theorem. Let us examine the conditions imposed on the mapping by the theorem. First, φ must be invariant to rotations and translations of the body. This condition is satisfied, as the orientation of the protein in the solvent does not affect F_{sol}, assuming that the solvent is isotopic. Second, φ on a sequence of approximate shapes as they better approximate a given convex shape is required to better approximate its value on the given shape. This continuity condition is satisfied for F_{sol} except for academic cases such as solvents that are close packed or in the vicinity of a surface phase transition [1, 53]. Finally, φ must be additive; that is, $\varphi(B_i \cup B_j) = \varphi(B_i) + \varphi(B_j) - \varphi(B_i \cap B_j)$. This condition does not strictly hold for F_{sol} [53, 54]. Consider two spheres (or globular proteins) at large separation. In this case, the solvation free energy of each solute sphere contributes the same as that of an isolated particle, and additivity holds. However, as the two spheres are at a distance that is comparable to the correlation length of the solvent, they experience a solvent-mediated interaction. Additivity implies that the interaction between the solutes must be contained in the overlap term $\varphi(B_i \cap B_j)$. This seems to make it impossible to realize any kind of interaction that goes beyond a mere contact force within the framework of Hadwiger's theorem.

The possible breakdown of additivity at small scales appears to be a serious impediment to applying Hadwiger's theorem to the solvation free energy. However, let us have a closer look at the four Minkowski measures of B_i. These are the volume V, the surface area A, the integral of mean curvature $C \equiv \int_A [(1/R_1 + 1/R_2)/2] dA$, and the integral of Gaussian curvature $X \equiv \int_A [1/(R_1 R_2)]dA$, where R_1 and R_2 are the principal radii of curvature. The question that first arises is how a given protein conformation should be mapped onto a geometric body B_i for which the measures can be calculated. After all, proteins do not have a sharply defined surface as expected of a mathematical body such as a sphere, whose surface can be precisely located as all points equidistant from the center. This is the type of surface that we encounter from hard macroscopic objects. We anticipate that a protein molecule might be fuzzier, as its spatial extent is somehow defined by the cloud of electrons surrounding it. But, the repulsive potential between atoms is steep, even on a molecular scale, which makes the definition of a protein surface meaningful [11, 57]. Proteins, though, do achieve a "fuzzy" boundary, due to the finite size of the surrounding solvent. Consider a globular protein which, for simplicity, is a sphere with radius R_p in a solvent of molecules with a radius R_s. The relevant body B_i for which to obtain the Minkowski measures is then a sphere of radius $R = R_p + R_s$; that is, not the physical surface of the protein, but rather a parallel surface at distance R_s, the so-called solvent-accessible surface [55], has to be considered. This is the surface "seen" by the solvent molecules. Consider now a second protein of the same radius R_p at distance d from the first, where d is measured from center to center. As shown shaded in Figure 16.5, the solvent-accessible surfaces of the proteins overlap for $2R_p \leq d < 2R$. Within this range,

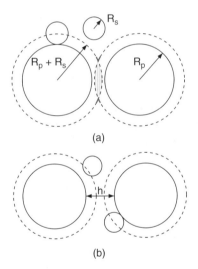

(a)

(b)

Figure 16.5 Two spherical solutes with radii R_p (for instance, globular proteins) in a solvent of spherical particles with radius R_s. The solvent-accessible surface [55] (dashed lines) is defined as the surface at distance R_s from the physical surface of the solutes (solid line). (a) If the distance h between the physical surfaces of the solutes is below $2R_s$, the solvent-accessible surfaces overlap, as shown shaded. The Minkowski measures which are used in the geometrical approach are calculated for the body defined by the solvent-accessible surface. The resulting dumbbell shape has a concave groove contributing negatively to C (indicated by the upward-and downward-facing "v" at the top and bottom of the shaded region). For $h > 2R_s$, as in (b), no overlap occurs and the Minkowski measures are calculated for two isolated spheres. Hence, they contain no information about the distance between the solutes. Figure modified from [56], copyright (2009) by *EPL*

the use of the solvent-accessible surface instead of the physical protein surface yields an F_{sol} that is additive while being more than just the sum of the F_{sol} for the isolated proteins. As a consequence, for protein surface-to-surface separations that are smaller than the solvent particle diameter, a nonvanishing solvation force is compatible with additivity.

After these remarks concerning the requirements of Hadwiger's theorem, we are ready to introduce the form of F_{sol} as it follows if we assume the theorem to be applicable (at least as a reasonable approximation). We then have [53, 54]

$$F_{sol} = pV + \sigma A + \kappa C + \bar{\kappa} X \qquad (16.1)$$

where the coefficients p and σ can be readily identified as the solvent pressure and the surface tension of the planar interface. The coefficients κ and $\bar{\kappa}$ which are conjugated to the integrals of curvature can be interpreted as curvature corrections to the surface tension. This can be illustrated for a spherical solute of radius R for which the last three terms of F_{sol} in Equation (16.1) can be written as $(\sigma + \kappa/R + \bar{\kappa}/R^2)A$. Hence, in the spherical geometry, κ is related to the Tolman length δ [58] through $\kappa = -2\delta$ and $\bar{\kappa}$ determines the magnitude of higher order corrections. There is strong numerical evidence showing that the morphometric form for F_{sol} is extremely accurate for solutes that are convex [54].

The importance of Equation (16.1) lies in the fact that the solvent coefficients p, σ, κ, and $\bar{\kappa}$ (sometimes also referred to in the literature as thermodynamic coefficients) depend only on the solvent properties (such as the solvent–solvent interactions, the temperature, and the chemical potential) and on the solvent–solute interaction potential, while they are *independent* of the shape of the solute. This separation of computational tasks – solvent properties on one side in the coefficients p, σ, κ, and $\bar{\kappa}$ and geometric measures for a given protein configuration on the other in V, A, C, and X – has the potential of reducing CPU time for ground-state calculations dramatically.

In a first step, the approach based on Hadwiger's theorem (which is also referred to as the morphometric approach) requires the determination of the four solvent coefficients. This can be achieved in various ways which involve established methods of treating the given solvent, such as computer simulations [59], density functional theory [60], or analytical methods such as scaled particle theory [61, 62]. Given the independence of the solvent coefficients from solute shape, it is computationally expedient to calculate the coefficients on a simple solute; for instance, spheres of different size. In principle, it is sufficient to calculate F_{sol} for four different spherical solutes in order to obtain the solvent coefficients by fitting Equation (16.1) to the data.

16.5.2 Applications

Solvent-Mediated Interactions

How well does the linear decomposition work in the previous example of two interacting globular proteins? Obviously, in the overlap regime ($2R_p \leq d < 2R$), along the line of intersection the solvent accessible surface is concave (i.e., one of the principal curvatures is negative); see Figure 16.5. The concavity increases as $d \to 2R$. A typical result for both the solvation free energy and the solvation force of solutes in a hard-sphere solvent as obtained by Oettel *et al.* [56] is shown in Figure 16.6. The morphometric approach

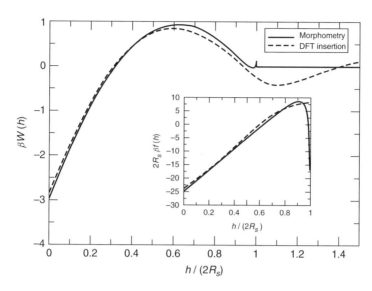

Figure 16.6 The solvation free energy $W(h) = F_{sol}(h) - F_{sol}(\infty)$ of two large spheres with radii $R_p = 5R_s$ (which can be thought of as representing globular proteins) in a hard-sphere solvent with particle radius R_s is plotted as a function of the surface-to-surface distance $h = d - 2R_p$ of the large spheres [56]. The dashed line shows the result from a density functional theory based approach. The result exhibits the oscillatory nature of W, which reflects the correlation shells of the solvent. The solid line shows the result from the morphometric approach of Equation (16.1). As long as the solvent-accessible surfaces of the two solutes overlap sufficiently, Equation (16.1) yields W very accurately. However, the approach breaks down as the overlap vanishes, resulting in a failure of the morphometric approach to account for any solvent-mediated interaction when $h > 2R_s$. The inset shows the corresponding solvation force, where negative (positive) values of force f are attractive (repulsive). Figure modified from [56]. Copyright (2009) by EPL

matches the result from a density functional theory calculation surprisingly well, even close to the breakup point $d = 2R$ (which corresponds to $h = 2R_s$ in the figure). However, as anticipated, the approach fails to yield the oscillations that occur in F_{sol} for $d > 2R$, where it "approximates" F_{sol} of the interacting proteins by the sum of the constant contributions from two isolated proteins. It should be emphasized that confinement of the solvent fluid on a length scale comparable to the correlation length (for which the morphometric approach shows its limitations, as it cannot yield the then-occurring characteristic oscillations of F_{sol}) is equivalent to the solvent-accessible surface exhibiting concave domains. As soon as one of the principal radii of curvature is negative and on the order of the correlation length, the fluid has to be considered confined. On the other hand, even a small convex solute particle, with radius equal to that of the solvent particles, requires as many as six solvent particles to be fully 'surrounded.' For a correlation length that is approximately the size of the solvent particles, this is largely sufficient for correlations to decay so that no oscillatory behavior of F_{sol} will be observed as the solute size is increased.

The theory developed by Asakura and Oosawa in the 1950s is related to keeping only the first term of Equation (16.1), with the further assumption that p is computed in the dilute-solvent approximation [45, 46]. As in atomic orbital theory, Equation (16.1) takes into account that translational entropy generates a pressure force p which acts through an excluded volume V. However, Equation (16.1) also includes that the protein–solvent interface with the protein defines an area A acted upon by a surface tension σ. These two effects are modified by the distribution of curvature such that, for example, even if the hydrophobic effect dominates the first term of Equation (16.1), then the minimum solvation free energy may not be at the minimum area, but rather at some greater area with negative contributions from the curvature terms C and X. However, note that $X/2\pi$ equals the Euler characteristic of the solute; that is, $X = 4\pi$ for all solutes that are topologically equivalent to a sphere. Hence, the term $\bar{\kappa} X$ impacts the solvation free energy only upon topological transitions (for instance, from a spherical to a toroidal shape). The integral of Gaussian curvature X, which can be calculated as $4\pi(1 - N_h + N_c)$, where N_h is the number of holes and N_c is the number of cavities in the solute, is expected to be important for the characterization of fluids in porous media [63], while it is insignificant for a protein in compact conformation.

Proteins in the Tube Model

The morphometric approach for F_{sol} has been used to calculate native states of idealized proteins [53] represented in the tube model [36]. As a function of the properties of a purely entropic solvent, protein configurations with the lowest free energy are shown in Figure 16.7. These configurations can be interpreted as native states. For a given

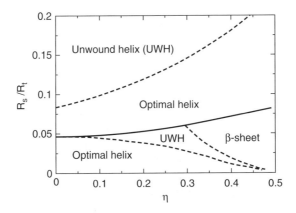

Figure 16.7   Protein conformations in the tube model (tube radius R_t) that minimize the solvation free energy F_{sol} as a function of the properties of a hard-sphere solvent (packing fraction η, particle radius R_s) [53]. In the dilute limit ($\eta \to 0$), a destabilization (unwound helix) of a tightly packed helical structure (optimal helix) upon increasing R_s/R_t is observed (cf. [38]). For small R_s/R_t but large η, planar structures, indentifiable as β-sheets, minimize F_{sol}. These conformations correspond to taking the helix radius R_h to infinity while keeping the pitch minimal. Solid (dashed) lines correspond to transitions that are discontinuous (continuous) in R_h. Figure modified from [53], copyright (2007) by The American Physical Society

solvent, the minimum of F_{sol} is obtained by scanning over all possible helical protein configurations which are defined by helix pitch and radius. As the helix radius is allowed to go to infinity, this search comprises planar arrangements of parallel tubes at distances defined by the pitch. The diagram extends previous work by Snir and Kamien [38]. They determined that a tightly packed helical conformation with a pitch-to-radius ratio close to that found in naturally occurring proteins, which is stable for small solvent particles (see Figure 16.2), is destabilized upon increasing the size of the solvent particles relative to the radius of the protein tube. This transition from an optimal helix to an unwound helix is indicated in Figure 16.7 for dilute solvents (i.e., $\eta \rightarrow 0$). By extending Snir and Kamien's work to solvents with a finite density η, a multitude of conformational transitions is revealed. Perhaps most interestingly, the work shows that stable β sheet-type structures can be realized in the model for certain solvent properties.

Proteins in the Fused-Sphere Model

Finally, we want to discuss the usefulness of Equation (16.1) for calculations involving actual proteins. The power of the morphometric approach in this regard has been illustrated quite impressively by Roth *et al.* [57], who calculated F_{sol} for 600 different structures of protein G in a hard-sphere solvent using a three-dimensional integral-equation approach which is able to resolve the microscopic structure on the scale of the solvent particle size under the effect of the different protein geometries. On the other hand, they obtained F_{sol} from Equation (16.1) by using the solvent-accessible surface of the atoms in the protein which are represented through spheres with suitable Lennard–Jones diameters. The solvent radius $R_s = 1.4$ Å and density $\rho_s (2R_s)^3 = 0.7$ are set to mimic the values of water. The resulting solvent-accessible surface of the protein gives rise to a fused-sphere structure for which the geometric measures required in Equation (16.1) have to be computed. The deviation of the morphometric result for F_{sol} from the three-dimensional integral-equation approach is always less than 0.7%, and for most structures is significantly smaller than this. However, the computational cost of the morphometric approach is more than four orders of magnitude smaller than the full three-dimensional calculation.

How can these results be used in order to properly identify the native state of a protein from a given set of configurations? We have to recall that it is not the solvation free energy alone which determines the free energy of a protein in a solvent. There can be a significant amount of energy stored in intramolecular hydrogen bonds of the protein. This is compensated to a degree by the dehydration penalty, due to a donor or an acceptor engaged in an intramolecular bond that cannot form a hydrogen bond with the water molecules of the solvent. However, given that donors and acceptors can be potentially buried in the protein and, hence, not be engaged in a bond at all, this cancelation is incomplete. A reliable free energy can be constructed for a given protein conformation as $F = -TS_{sol} + \xi$, where S_{sol} is the solvent entropy and ξ denotes the dehydration penalty [47]. The solvent entropy S_{sol} under the isochoric condition is obtained as $S_{sol} = -(\partial F_{sol}/\partial T)_V$, where F_{sol} is calculated efficiently from Equation (16.1) along the lines of Roth *et al.* [57]. The solvent coefficients are determined using a realistic computer model for water, which

includes the effect of the dipole and quadrupole moments as well as the tetrahedral symmetry. Molecular polarizability is included on the mean-field level. The protein–water
interaction is set to be very simple: the protein is modeled as a set of fused hard-spheres.
This is acceptable, as under the isochoric condition S_{sol} is mostly determined by the
excluded volume effect and not significantly influenced by the protein–solvent interaction potential [47]. The term $-TS_{sol}$ in F reflects the hydrophobic effect. A compact
protein configuration causes the solvent entropy S_{sol} to be large, which helps reduce F.
The dehydration penalty ξ is computed by imposing a penalty of $7k_BT$ for each donor
or acceptor that is buried in the protein and not engaged in an intramolecular hydrogen
bond. The free energy F is computed for seven proteins with native states that display
a large variation of α-helix and/or β-sheet structures. More than 600 decoy structures
are considered for each of the proteins. The free energy F is able to identify the native
state for all seven proteins (see Figure 16.8). Interestingly, there are many structures with
either a lower dehydration penalty or a larger solvent entropy than native state. However,
simply speaking of a large solvent entropy implies a compact protein, which increases
the chances for dehydration penalty. Increasing S_{sol}, and thereby lowering the first term
of F, tends also to lead to an increase in ξ, which works against lowering F. Only in the
native state are the two contributions optimally tuned, and the free energy F constructed
in Harano et al. [47] is capable of capturing this interplay.

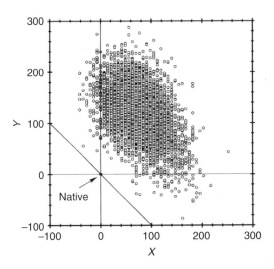

Figure 16.8   The free energy $F = -TS_{sol} + \xi$ has been calculated by Harano et al. [47] for seven
different proteins in aqueous solution. The hydration entropy S_{sol} is obtained using Hadwiger's
theorem for F_{sol} (Equation (16.1)) and ξ is the dehydration penalty for proton acceptors or donors
that are buried and not engaged in a hydrogen bond. The results for the native state and about
600 decoy structures for each protein are plotted in terms of $k_BTY = -TS_{sol} - (-TS_{sol})_{native}$ and
$k_BTX = \xi - \xi_{native}$, where $T = 298$ K. Various decoy structures have a larger hydration entropy
($Y < 0$) or a smaller dehydration penalty ($X < 0$) than the native state. However, the free energy
F is minimized by the native state as none of the other structures lies below the line shown at
$X + Y = 0$. Reprinted with permission from [47]. Copyright 2007 Elsevier

16.6 Discussion, Conclusions, Future Work

16.6.1 Results

The hydrophobic effect singles out water for its especially low affinity for hydrocarbons. The results reviewed here recapitulate the special character of water as a solvent. Water is special in its small size as a solvent, with almost all other solvents being larger. As the magnitude of the folding force due to translational entropy is proportional to the excluded volume multiplied by the solvent number density, water's size helps generate folding forces which can successfully compete with the loss of protein configurational entropy [18].

While maximizing translational entropy of the solvent introduces an effect which is proportional to excluded volume, thereby driving the solute to maximize overlap of excluded volume, the hydrophobic effect also has a component that can be expressed in terms of minimizing surface area. Water has a high dielectric constant, so there is a large difference in hydrogen-bond strength in aqueous versus apolar environments. Consequently, as the microenvironment of a hydrogen bond changes, its strength can change greatly. In this way, neighboring apolar residues in the folded state reduce the electrostatic shielding, resulting in hydrogen bonds within the apolar core of proteins which are stronger than those that were present in the unfolded state with water. So, even if the number of bonds in the folded and unfolded state were to remain constant, hydrogen bonding can stabilize the folded state through its increased strength. The strengthening of hydrogen bonds due to the microenvironment is a new type of cooperativity, in which (rather than a bond assisting other neighboring bonds) neighbors assist the original hydrogen bond, strengthening its folding force.

The tube model treats proteins as homogeneous tubes of constant thickness. In possessing a thickness Δ, the tube model incorporates a simplified Ramachandran diagram. That is, there are a range of bends which are possible and those that are sterically disallowed. Steric clashes in the tube model arise from radii of curvature which are so small that they would crease the tube. Self-avoidance is included by not allowing tube centers to be closer than 2Δ.

The tube model has been combined with a solvent model which decomposes the solvent free energy of a bath of hard spheres into four contributions according to Hadwiger's theorem [54]; see Equation (16.1). The first two contributions are proportional to the tube excluded volume and surface area, respectively, and correspond to contributions of the solvent translational entropy and the interfacial free energy due to the hydrophobic effect. The third and fourth contributions account for energy required to deform the planar solvent–protein interface.

The tube model together with the Hadwiger-theorem-based morphometric approach to treat the solvent can thus be used to find the conformations of a homopolymer composed of a common average monomer that arise from minimizing the solvent free energy. It is found that conformations similar to α helices and β sheets are the favored conformations, depending upon the size of the solvent, the volume fraction of solvent, and the length of the tube [53].

The tube model in a hard-sphere solvent generates secondary-structure-like conformations. This suggests that protein secondary structure may arise independent of the specificity of the different amino acid residues of which a protein is composed. This type of nonspecific mechanism differs from the usual hydrophobic effect, which differentiates hydrophilic from hydrophobic residues and posits that folding relies on the differential burying of hydrophobic residues.

16.6.2   Discussion and Speculations

A mechanism such as this – which forms secondary structure without regard to the detailed primary structure information – leads to several speculative consequences. The independence of secondary structure folding on the primary sequence leaves the information encoded in the specificity of particular primary sequence to be used for other purposes. These purposes might include organizing tertiary conformation and protein function. For example, recall that α helices and β sheets are the lowest energy structures in terms of hydrogen bonding. This suggests that they would be the most structurally rigid. If the entire protein were so constructed, then it would be static, remaining fixed in its low-energy conformation. But, by linking these rigid structures together, the protein could then form itself into a mechanism, joining the rigid parts by less rigid linkages. These higher energy, more flexible linkages would permit the protein to act as a mechanism. So, the nonspecific solvation effect – sterically steering toward the lowest energy structures – might lead to the accumulation of rigid parts, which are then assembled into a flexible tertiary structure as specified by details of the primary sequences of amino acids.

With regard to the time history of protein folding, the results suggest that secondary structure formation proceeds independently of tertiary structure formation, in the sense that the mechanisms of each are specified by different sets of parameters. This helps explain the robustness of structure in the face of primary-structure variation, namely that a protein with many amino acid substitutions still folds into the native-state configuration. And, these results further suggest that proteins use generically folded structures to make folding reliable and fast: the amino-acid residue primary structure needs only orchestrate pre-folded secondary structures rather than the full number of degrees of freedom of the chain of amino acids [37].

How accurate is the use of a hard-sphere solvent? For small-sized solutes, namely less than about 1 nm, the encompassing hydrogen-bonding network is not greatly disrupted [12]. Hence, models which neglect hydrogen bonding might still achieve accurate approximations, as the missing effect may not contribute much to the free-energy change. However, protein interfaces with solvent and protein–protein interfaces between secondary structures can exceed the 1 nm size. On these larger scales, bulk hydrogen bonding is disrupted by the presence of the protein. In these cases as well, van der Waals attractive interactions become important [64].

In addition to hydrogen bonds both within the protein and between the protein and water, which have been properly taken into account by a morphometry-based free-energy model for proteins [47], there are additional effects relating to other aspects of the protein–water interaction. These include changes in the intra-water bonding and in the conformational entropy of the solvent due to the presence of the protein. While Harano et al. [47] reduce the protein–solvent interaction to the formation of hydrogen bonds between the protein and the water, the morphometric approach is in principle capable of accounting for more complex interactions. One route, which could be explored in the future, is to let the solvent coefficients in Equation (16.1) vary over the surface of the protein. In this way, specific interactions of the different chemical groups with the solvents could be modeled. In order to compute the solvent coefficients, a task that is usually performed in a simple geometry such as provided by spherical solutes, it would be crucial to make use of a realistic potential through which the protein acts on the solvent. While the application of the morphometric approach is straightforward once this potential is defined, the bulk of work lies in determining a suitable potential to create the manifold effects that a protein has on the solvent.

One important new contribution might arise from the boundaries between patches on the protein with different solvent–solute interaction (hydrophilic versus hydrophobic, say). It could be expedient to define a line tension associated with the length of the boundary. Determining the value of this new coefficient from a density functional or simulation treatment of the solvent, however, would no longer be possible from a simple, spherically symmetric geometry.

A consequence of the morphometric approach in Equation (16.1) being based on the solvent-accessible surface is the failure to model long-ranged solvent-induced interactions between two solutes, which we discussed in Section 16.5. Curing this shortcoming within the framework provided by integral geometry, however, is an ambitious program. A first step toward capturing the oscillations in the solvation free energy due to solvent correlations for short-ranged solvent interactions is achieved in the work by König *et al.*, who introduced a curvature expansion of the density profile around a convex solute [65]. Different from the morphometric form (Equation (16.1)) of the solvation free energy (a thermodynamic "integrated" quantity), the expansion of the density profile (a microscopic "local" quantity) can only be achieved properly if higher order powers of the mean and Gaussian curvature are employed. Therefore, strictly speaking, the approach is no longer morphometric. Based on the insertion method, the density profiles can be used to calculate the solvation free energy of two arbitrarily shaped solutes even for intermediate and long separations [66]. It should be noted that the limitations of the morphometric approach to account for long-ranged solute–solute interactions are only of minor relevance when Equation (16.1) is applied to protein solvation. Proteins in compact conformation mostly exhibit convex surfaces, which are excellently treated by morphometry, with occasional concave domains on the solvent-accessible surface inducing narrow confinement of the solvent. Owing to solvent correlations, confinement gives rise to oscillations of the solvation free energy as the length scale of confinement is varied. Morphometry is unable to yield these oscillations. However, it has been shown for a hard-sphere fluid in a cylindrical pore that Equation (16.1) provides a smooth curve that averages over the oscillations that occur as the pore radius is decreased [67]. This property appears to be the key to the reliability of the morphometric approach in the calculation of solvation free energies for large proteins. Considering that a given protein conformation gives rise to multiple confining geometries of different widths, it seems reasonable to assume that local under- and over-estimation of the solvation free energy cancel out as the global integral over the whole protein is taken. As a result, highly efficient morphometric calculations are in excellent agreement with time-consuming full three-dimensional calculations which involve the microscopic structure of the solvent [57].

The tube model treats proteins as uniform cylinders, omitting any of the branching structures, which can be quite extensive, such as the side chains of lysine, arginine, phenylalanine, tryptophan, and other residues. The general result from the tube model is that the backbone taking on α helix and β sheet conformation leads to a decrease in free energy. However, including the solvation of side chains modifies this result, as the side chains may pack very efficiently when folded [68]. There are several lines along which the effect of side chains could be accounted for in possible refinements of the tube model or the fused-spheres model. In the latter model, this has been partly achieved by imposing an energetic penalty for proton donors and receptors within the protein that are unmatched and, hence, not engaged in a hydrogen bond [47]. There remains the factor of conformational entropy change incurred by the side chains as they pack densely in a compactly folded protein to be accounted for in the fused-sphere model. As far as the tube model is concerned, a modulated tube (that is, with a radius varying along the backbone) could help to account for side chains. Reducing the radius of the tube locally (within certain bounds dictated by the nature of the given amino acid at a particular location) would have to be penalized by an entropy reduction. Hydrogen bonds are related to using solvent coefficients in Equation (16.1) that vary over the surface of the tube, thereby mimicking solvent–solute hydrogen bonds in certain locations. However, hydrogen bonds within the protein would have to be accounted for separately, as the morphometric approach to the solvation free energy does not take internal contributions of the solute into account. So far, the only internal energetic contribution that has been studied in the tube model is a biomolecule's tendency to resist bending. A classical approach implementing a stiffness is provided by the wormlike chain model, which imposes an energetic cost of bending proportional to the square of the curvature of the centerline of a chainlike molecule. The model has been successfully applied to force–extension curves measured on DNA [69] and in the context of helix formation in the tube model [38].

While all these extensions of the tube model are certainly important and necessary if one aims at understanding the folding of a particular naturally occurring protein, we would like to emphasize that the origin of the tube model, and its success, is based on the idea to create a minimal model that is able to yield secondary motifs as they occur in proteins. Using a tube to model mimics two important properties of proteins: (i) the protein backbone and the side chains require a certain volume to be accommodated and (ii) the protein backbone cannot be bent too strongly, which is reflected in the fact that the radius of curvature of the tube in the model cannot be smaller than the radius of the tube itself in order to prevent self-intersection. The second important constituent of the model is compaction, either directly by confining the tube "protein" in a box [35] or indirectly by maximizing the translational entropy of a hard-sphere solvent [38, 44, 53]. The success of this minimal model in generating helices, β-sheets, and other principal secondary motifs that occur in folded proteins underlines the significance of simple principles, such as close-packing and maximum entropy, for the understanding even of complex biological systems. Refining the tube model is likely to increase its ability to predict native states of actual proteins, but these folds will still be essentially determined by the same basic and fundamental principles that are revealed by the original tube model of protein folding.

Here, we have reviewed modeling which reduces the description of protein conformations to only a few integral measures over a flexible tube of constant diameter in a solvent of hard spheres. The successes of this approach are encouraging. It would be interesting to apply the model to protein–protein and protein–drug interactions, and to the design and self-assembly of molecular-scale devices.

Acknowledgments

HHG is grateful for support from the Helmholtz Association through the research alliance "Planetary Evolution and Life." HHG would like to thank John Wettlaufer for support and stimulating discussions. HHG has enjoyed discussions and collaborations with Roland Roth, Klaus Mecke, Martin Oettel, and Siegfried Dietrich which resulted in some of the work reported in this contribution. SL gratefully acknowledges the Lillian Sidney Foundation, and the collegial support of Ashlie Martini, Andreas Matouschek, and Igal Szleifer.

References

[1] Lum, K., Chandler, D., and Weeks, J.D. (1999) Hydrophobicity at small and large length scales. *Journal of Physical Chemistry B*, **103**, 4570–4577.

[2] England, J.L. and Haran, G. (2011) Role of solvation effects in protein denaturation: from thermodynamics to single molecules and back. *Annual Review of Physical Chemistry*, **62**, 257–277.

[3] Baker, D. (2000) A surprising simplicity to protein folding. *Nature*, **405**, 39–42.

[4] Muñoz, V. (2001) What can we learn about protein folding from Ising-like models? *Current Opinion in Structural Biology*, **11**, 212–216.

[5] Pande, V.S., Grosberg, A.Y., and Tanaka, T. (2000) Heteropolymer freezing and design: towards physical models of protein folding. *Reviews of Modern Physics*, **72**, 259–314.

[6] Kolinski, A. and Skolnick, J. (2004) Reduced models of proteins and their applications. *Polymer*, **45**, 511–524.

[7] Onuchic, J.N. and Wolynes, P.G. (2004) Theory of protein folding. *Current Opinion in Structural Biology*, **14**, 70–75.

[8] Dill, K.A., Ozkan, S.B., Shell, M.S., and Weikl, T.R. (2008) The protein folding problem. *Annual Review of Biophysics*, **37**, 289–316.

[9] Hoang, T.X., Trovato, A., Seno, F. *et al.* (2004) Geometry and symmetry presculpt the free-energy landscape of proteins. *Proceedings of the National Academy of Sciences of the United States of America*, **101**, 7960–7964.

[10] Kauzmann, W. (1959) Some factors in the interpretation of protein denaturation. *Advances in Protein Chemistry*, **14**, 1–63.

[11] Richards, F.M. (1977) Areas, volumes, packing, and protein structure. *Annual Review of Biophysics and Bioengineering*, **6**, 151–176.

[12] Chandler, D. (2005) Interfaces and the driving force of hydrophobic assembly. *Nature*, **437**, 640–647.

[13] Huang, D.M., Geissler, P.L., and Chandler, D. (2001) Scaling of hydrophobic solvation free energies. *Journal of Physical Chemistry B*, **105**, 6704–6709.

[14] Tanford, C. (1980) *The Hydrophobic Effect: Formation of Micelles and Biological Membranes*, 2nd edn., John Wiley & Sons, Inc., New York.

[15] Frank, H.S. and Evans, M.W. (1945) Free volume and entropy in condensed systems: III. Entropy in binary liquid mixtures; partial molal entropy in dilute solutions; structure and thermodynamics in aqueous electrolytes. *Journal of Chemical Physics*, **13**, 507–532.

[16] Setny, P., Baron, R., and McCammon, J.A. (2010) How can hydrophobic association be enthalpy driven? *Journal of Chemical Theory and Computation*, **6**, 2866–2871.

[17] Levy, R.M., Zhang, L.Y., Gallicchio, E., and Felts, A.K. (2003) On the nonpolar hydration free energy of proteins: surface area and continuum solvent models for the solute–solvent interaction energy. *Journal of the American Chemical Society*, **125**, 9523–9530.

[18] Harano, Y. and Kinoshita, M. (2005) Translational-entropy gain of solvent upon protein folding. *Biophysical Journal*, **89**, 2701–2710.

[19] Yodh, A.G., Lin, K.H., Crocker, J.C. *et al.* (2001) Entropically driven self-assembly and interaction in suspension. *Philosophical Transactions of the Royal Society of London A*, **359**, 921–937.

[20] Marenduzzo, D., Finan, K., and Cook, P.R. (2006) The depletion attraction: an underappreciated force driving cellular organization. *Journal of Cell Biology*, **175**, 681–686.

[21] Gao, J., Bosco, D.A., Powers, E.T., and Kelly, J.W. (2009) Localized thermodynamic coupling between hydrogen bonding and microenvironment polarity substantially stabilizes proteins. *Nature Structural and Molecular Biology*, **7**, 684–691.

[22] Fernández, A., Zhang, X., and Chen, J. (2008) Folding and wrapping soluble proteins: exploring the molecular basis of cooperativity and aggregation. *Progress in Molecular Biology Translational Science*, **83**, 53–87.

[23] Moret, M.A. and Zebende, G.F. (2007) Amino acid hdyrophobicity and accessible surface area. *Physical Review E*, **75**, 011920.

[24] Donnini, S., Tegeler, F., Groenhof, G., and Grubmüller, H. (2011) Constant pH molecular dynamics in explicit solvent with λ-dynamics. *Journal of Chemical Theory and Computation*, **7**, 1962–1978.

[25] Kendrew, J.C., Dintzis, B.H.M., Parrish, R.G. *et al.* (1958) A three-dimensional model of the myoglobin molecular obtained by X-ray analysis. *Nature*, **181**, 662–666.

[26] Hsin, J., Strümpfer, J., Lee, E.H. *et al.* (2011) Molecular origin of the hierarchical elasticity of titin: simulation, experiment, and theory. *Annual Review of Biophysics*, **40**, 187–203.

[27] Schlick, T. (2010) *Molecular Modeling and Simulation: An Interdisciplinary Guide*, 2nd edn., Springer.

[28] Kamerlin, S.C.L. and Warshel, A. (2011) Multiscale modeling of biological functions. *Physical Chemistry Chemical Physics*, **13**, 10401–10411.

[29] Dill, K.A. (1985) Theory for the folding and stability of globular proteins. *Biochemistry*, **24**, 1501–1509.

[30] Pauling, L., Corey, R.B., and Branson, H.R. (1951) The structure of proteins: two hydrogen-bonded helical configurations of the polypeptide chain. *Proceedings of the National Academy of Sciences of the United States of America*, **37**, 205–211.

[31] Ramachandran, G.N. and Sasisekharan, V. (1968) Conformation of polypeptides and proteins. *Advances in Protein Chemistry*, **28**, 283–437.

[32] Chothia, C. (1976) The nature of the accessible and buried surfaces in proteins. *Journal of Molecular Biology*, **105**, 1–14.

[33] Socci, N.D., Bialek, W.S., and Onuchic, J.N. (1994) Properties and origins of protein secondary structure. *Physical Review E*, **49**, 3440–3443.

[34] Maritan, A., Micheletti, C., and Banvar, J.R. (2000) Role of secondary motifs in fast folding polymers: a dynamical variational principle. *Physical Review Letters*, **84**, 3009–3012.

[35] Maritan, A., Micheletti, C., Trovato, A., and Banavar, J.R. (2000) Optimal shapes of compact strings. *Nature*, **406**, 287–290.

[36] Banavar, J.R. and Maritan, A. (2003) Colloquium: Geometrical approach to protein folding: a tube picture. *Reviews of Modern Physics*, **75**, 23–34.

[37] Banavar, J.R. and Maritan, A. (2007) Physics of proteins. *Annual Review of Biophysics and Biomolecular Structure*, **36**, 261–280.

[38] Snir, Y. and Kamien, R.D. (2005) Entropically driven helix formation. *Science*, **307**, 1067–1068.

[39] Vogel, T., Neuhaus, T., Bachmann, M., and Janke, W. (2009) Thickness-dependent secondary structure formation of tubelike polymers. *Europhysics Letters*, **85**, 10003.

[40] Vogel, T., Neuhaus, T., Bachmann, M., and Janke, W. (2009) Thermodynamics of tubelike flexible polymers. *Physical Review E*, **80**, 011802.

[41] Vogel, T., Neuhaus, T., Bachmann, M., and Janke, W. (2009) Ground-state properties of tubelike flexible polymers. *European Physical Journal E*, **30**, 7–18.

[42] Hoang, T.X., Trovato, A., Seno, F. *et al.* (2006) Marginal compactness of protein native structures. *Journal of Physics: Condensed Matter*, **18**, S297–S306.

[43] Banavar, J.R., Hoang, T.X., Maritan, A. *et al.* (2004) Unified perspective on proteins: a physics approach. *Physical Review E*, **70**, 041905.

[44] Poletto, C., Giacometti, A., Trovato, A. *et al.* (2008) Emergence of secondary motifs in tubelike polymers in a solvent. *Physical Review E*, **77**, 061804.

[45] Asakura, S. and Oosawa, F. (1954) On the interaction between two bodies immersed in a solution of macromolecules. *Journal of Chemical Physics*, **22**, 1255–1256.

[46] Asakura, S. and Ooswa, F. (1958) Interaction between particles suspended in solutions of macromolecules. *Journal of Polymer Science*, **33**, 183–192.

[47] Harano, Y., Roth, R., Sugita, Y. *et al.* (2007) Physical basis for characterizing native structures of proteins. *Chemical Physics Letters*, **437**, 112–116.

[48] Harano, Y. and Kinoshita, M. (2004) Large gain in translational entropy of water is a major driving force in protein folding. *Chemical Physics Letters*, **399**, 342–348.

[49] Amano, K.I., Yoshidome, T., Harano, Y. *et al.* (2009) Theoretical analysis on thermal stability of a protein focused on the water entropy. *Chemical Physics Letters*, **474**, 190–194.

[50] Hubner, I.A. and Shakhnovich, E.I. (2005) Geometric and physical considerations for realistic protein models. *Physical Review E*, **72**, 022901.

[51] Banavar, J.R., Cieplak, M., Flammini, A. *et al.* (2006) Geometry of proteins: hydrogen bonding, sterics and marginally compact tubes. *Physical Review E*, **73**, 031921.

[52] Hadwiger, H. (1957) *Vorlesungen über Inhalt, Oberfläche und Isoperimetrie*, Springer, Berlin.

[53] Hansen-Goos, H., Roth, R., Mecke, K., and Dietrich, S. (2007) Solvation of proteins: linking thermodynamics to geometry. *Physical Review Letters*, **99**, 128101.

[54] König, P.M., Roth, R., and Mecke, K.R. (2004) Morphological thermodynamics of fluids: shape dependence of free energies. *Physical Review Letters*, **93**, 160601.

[55] Lee, B. and Richards, F.M. (1971) The interpretation of protein structures: estimation of static accessibility. *Journal of Molecular Biology*, **55**, 379–400.

[56] Oettel, M., Hansen-Goos, H., Bryk, P., and Roth, R. (2009) Depletion interaction of two spheres – full density functional theory vs. morphometric results. *Europhysics Letters*, **85**, 36003.

[57] Roth, R., Harano, Y., and Kinoshita, M. (2006) Morphometric approach to the solvation free energy of complex molecules. *Physical Review Letters*, **97**, 078101.

[58] Blokhuis, E.M. and Kuipers, J. (2006) Thermodynamic expressions for the Tolman length. *Journal of Chemical Physics*, **124**, 074701.

[59] Henderson, J.R. and van Swol, F. (1984) On the interface between a fluid and a planar wall – theory and simulations of a hard-sphere fluid at a hard-wall. *Molecular Physics*, **51**, 991–1010.

[60] Evans, R. (1979) Nature of the liquid–vapor interface and other topics in the statisical-mechanics of nonuniform, classical fluids. *Advances in Physics*, **28**, 143–200.

[61] Reiss, H. (1977) Scaled particle theory of hard sphere fluids to 1976, in *Statistical Mechanics and Statistical Methods in Theory and Application: A Tribute to Elliott W. Montroll* (ed. U. Landman), Plenum Press, New York, pp. 99–140.

[62] Hansen-Goos, H. and Roth, R. (2006) Density functional theory for hard-sphere mixtures: the White Bear version Mark II. *Journal of Physics: Condensed Matter*, **18**, 8413–8425.

[63] Mecke, K. and Arns, C.H. (2005) Fluids in porous media: a morphometric approach. *Journal of Physics: Condensed Matter*, **17**, S503–S534.

[64] Ashbaugh, H.S. and Paulaitis, M.E. (2001) Effect of solute size and solute–water attractive interactions on hydration water structure around hydrophobic solutes. *Journal of the American Chemical Society*, **123**, 10721–10728.

[65] König, P.M., Bryk, P., Mecke, K., and Roth, R. (2005) Curvature expansion of density profiles. *Europhysics Letters*, **69**, 832–838.

[66] König, P.M., Roth, R., and Dietrich, S. (2006) Depletion forces between nonspherical objects. *Physical Review E*, **74**, 041404.

[67] Hansen-Goos, H. (2008) Entropic forces on bio-molecules, Ph.D. thesis, University of Stuttgart.

[68] Yasuda, S., Yoshidome, T., Oshima, H. *et al.* (2010) Effects of side-chain packing on the formation of secondary structures in protein folding. *Journal of Chemical Physics*, **132**, 065105.

[69] Marko, J.F. and Siggia, E.D. (1995) Stretching DNA. *Macromolecules*, **28**, 8759–8770.

17

Multiscale Modeling for the Vascular Transport of Nanoparticles

Shaolie S. Hossain[1], Adrian M. Kopacz[2], Yongjie Zhang[3],
Sei-Young Lee[4], Tae-Rin Lee[2], Mauro Ferrari[1], Thomas J.R. Hughes[5],
Wing Kam Liu[2], and Paolo Decuzzi[1]

[1] *The Methodist Hospital Research Institute, USA*
[2] *Northwestern University, USA*
[3] *Carnegie Mellon University, USA*
[4] *Samsung-Global Production Technology Center, Korea*
[5] *The University of Texas at Austin, USA*

17.1 Introduction

Several research groups are developing nanoparticles (NPs) for a variety of biomedical applications, including the early detection of diseases, imaging, conventional molecular therapy and thermal ablation therapy, and follow-up on the therapeutic intervention [1–5]. NPs are man-made materials, sufficiently small to be safely administered systemically and, transported by the blood flow, can reach virtually any site along the circulatory system [6].

Differently from single-molecule-based therapeutic and contrast agents, NPs are *engineerable* and *multifunctional*. NPs can be fabricated in a variety of sizes, shapes, and surface properties (engineerable) to finely tailor their transport within the circulatory system and their interaction with the target cells, with the objective of improving biodistribution, drug bioavailability, and toxicity [7,8]. Also, NPs can be loaded simultaneously (multifunctional) with molecular agents providing the unique opportunity of codelivering multiple therapeutic (polipharmacy) and imaging (multimodal) agents at the same site.

Nano and Cell Mechanics: Fundamentals and Frontiers, First Edition. Edited by Horacio D. Espinosa and Gang Bao.
© 2013 John Wiley & Sons, Ltd. Published 2013 by John Wiley & Sons, Ltd.

These extraordinary advantages have fostered the design and development of a plethora of NPs over the last 20 years exhibiting differences in (i) *sizes*, ranging from a few tens (as for dendrimers, gold, and iron oxide NPs) to hundreds of nanometers (as for polymeric and lipid-based particles) up to a few micrometers; (ii) *surface* functionalizations, offering a variety of molecular coatings and electrostatic charges; and more recently, it has been even reported differences in (iii) *shapes*, from classical spherical beads to discoidal, hemispherical, cylindrical, and conical particles [9–11].

In cancer, the prominent delivery strategy exploits the well-known leakiness of the tumor vessels, which have openings (*fenestrations*) smaller than about 300 nm [12,13]. Therefore, NPs are mostly designed to exhibit long circulation time (to increase the number of passages within the tumor vasculature) and to be smaller than about 300 nm (to passively cross the fenestrations and be retained within the tumor parenchyma) [14]. This strategy is generally termed the *enhanced permeability and retention* (EPR) effect. As a consequence, most of the NPs used in cancer imaging and treatment are small spheres (∼50–200 nm) coated with special polymer (polyethylene glycol) chains, extending their permanence in the vascular compartment. Despite the success of a few NP-based formulations, the EPR delivery strategy suffers two major limitations: first, the fenestration size is time and tumor dependent, and is not homogeneous across the tumor mass; second, fenestrations are specific for neoplastic lesions and are absent in several other pathologies, including cardiovascular and hemorrhagic diseases [15]. On the other hand, new *in-vivo* screening technologies, such as, for instance, those based on phage display libraries [16], are providing direct evidence on the extraordinary biological diversity offered by the endothelial cells lining the vascular walls. It is becoming clear that disease-specific stimuli can induce the expression of molecules (receptors) on the membrane of endothelial cells. Typical examples are the specific expression of $\alpha_v\beta_3$ receptors (integrins) on the tumor endothelial cells and the overexpression of ICAM-1 and E-selectin molecules on atherosclerotic plaques [17]. These receptors can be recognized by counter-molecules (ligands) decorating the NP surface and employed effectively as vascular docking sites for blood-borne NPs. This approach, known as *vascular targeting*, is more general than the EPR delivery strategy, in that it is applicable to several pathologies, including indeed cancer [7].

With such a complex scenario, where a plethora of NP types can be fabricated and administered systemically using different delivery strategies, mathematical modeling and computational methods can provide formidable tools to rationally design disease- and patient-specific NPs. The objective of this chapter is to briefly describe some of the most recently developed predictive tools for analyzing the vascular transport, wall adhesion, and accumulation of circulating NPs: Section 17.2 focuses on the application of isogeometric analysis (IA) to the patient-specific macrocirculation and Section 17.3 discusses the use of semi-analytical models and the immersed finite-element method (IFEM) for analyzing the NP behavior in the microcirculation.

17.2 Modeling the Dynamics of NPs in the Macrocirculation

Blood flow and transport problems in the macrocirculation are influenced by the actual vascular geometry. Predictive tools can only be accurate if accounting for the authentic three-dimensional (3D) vascular structure. This poses serious challenges in combining computational modeling and geometrical design. In an effort to break down the barrier between design and analysis, IA was proposed by Hughes *et al.* [18] as an alternate approach to the classical finite-element analysis (FEA). The IA framework uses geometric primitives developed in computer-aided design and computational geometry, which allows one to represent geometrical objects with a higher degree of precision and smoothness (higher order basis functions, order C^3) than the piecewise polynomial (continuous basis functions, order C^0) commonly used in FEA. NURBS (nonuniform rational B-splines) are predominantly used in IA as basis functions [19], and several structural, fluid and solid mechanical, fluid–structure interaction, and biomedical problems have been solved [20–28] following this same approach. The ability of NURBS to accurately represent complex geometries (i.e., arterial vascular system) renders fluid and structural computations more physiologically realistic [21]. Compared with FEA, the IA approach has significant benefits, both in terms of solution accuracy and implementational convenience. First, in the simple case of a flow in a straight circular pipe subjected to a constant pressure gradient, a pointwise exact solution to the incompressible Navier–Stokes equations can be obtained with C^2 NURBS, which calls for the overall accuracy of the approach [23]. Second, the parametric definition of the NURBS mesh facilitates boundary-layer refinement near the arterial wall, which is crucial for the overall accuracy and for precisely estimating wall quantities (wall shear rate, permeability, adhesion, etc.). In contrast, unstructured meshes for FEA lead to much less accurate solutions for a comparable number of degrees of freedom requiring the use of complex and computationally demanding adaptive boundary-layer meshing [29].

Below, IA is applied to the analysis of blood flow and transport problems in coronary arteries, within the context of NP delivery for the localized treatment of atherosclerotic plaques.

17.2.1   The 3D Reconstruction of the Patient-Specific Vasculature

A chain of specific procedures (*vascular modeling pipeline*) developed by Zhang *et al.* [23] was used to create hexahedral solid NURBS for patient-specific geometric modeling of the left coronary artery (LCA) of a healthy, over-55 volunteer, obtained from 64-slice CT angiography. Figure 17.1 highlights the main steps of this modeling procedure. This hexahedral solid NURBS model for the lumen (see Figure 17.1d) generated by Zhang *et al.* [23] was further modified to include an artery wall of uniform thickness surrounding the lumen using isogeometric modeling techniques [18]. There are four patches along the length of the LCA with each patch again split into two concentric patches at approximately one-third depth into the artery wall thickness. The inner patch represents the *intima*, while the outer patch represents the *media*. To simulate the diseased condition, an atherosclerotic plaque characterized by a large lipid core and a thin fibrous cap was placed in the circumflex branch. Plaque features and dimensions vary over a wide range of values depending on the status of the disease and the individual patient. In this case, the plaque was chosen to be larger than average, approximately 40 mm in length with a peak wall thickness (excluding the *adventitia*) of approximately 1.1 mm. The model can be found in Figure 17.2.

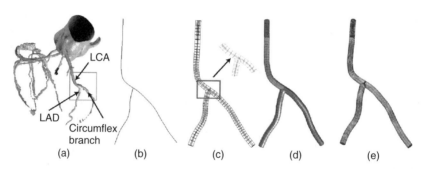

Figure 17.1   The *vascular modeling pipeline*: (a) isocontour of LCA; (b) arterial path; (c) sweeping along the arterial path – templated circle translated and rotated to each cross-section; (d) solid NURBS mesh; (e) NURBS wire-mesh

17.2.2   Modeling the Vascular Flow and Wall Adhesion of NPs

In local catheter-based drug delivery, a solution containing NPs is released into the arterial lumen. Carried by the blood flow and then aided by their special surface properties (ligand molecules and electrostatic charge), some of the NPs adhere to the arterial wall. Figure 17.3 shows the details of the simulation setup.

A Navier–Stokes solver coupled to the scalar advection–diffusion equation was used to determine the time evolution of NP concentration in the lumen. For spatial discretization, the IA employing quadratic NURBS was used. Blood was modeled as a Newtonian incompressible fluid driven by a time-dependent pulsatile inflow condition (Figure 17.3a). As a first approximation, the arterial wall was assumed rigid for simplicity [26]. A residual-based multiscale method [30] was used to solve the coupled equations with the generalized-α method [31] adopted for time advancing. The highly advective nature of the blood flow necessitated implementation of stabilization such as the YZβ discontinuity capturing method [24]. Simulations were run for six cardiac cycles with a time step of 0.05 s. The wall deposition data thus obtained were then normalized against the highest NP concentration to get the spatial distribution of NPs at the lumen–wall interface (Figure 17.3b). Quite expectedly, the NPs accumulate near the bifurcation area, which is a recirculation zone. Such flow separation has also been reported to enhance platelet deposition [32].

17.2.3   Modeling NP Transport across the Arterial Wall and Drug Release

Following their tissue uptake, the NPs get transported through the wall mainly via diffusion and advection while releasing the encapsulated drug. The released drug then propagates through the wall and provides a therapeutic effect to the surrounding region (Figure 17.4). While the radial release of NPs from the catheter into the bloodstream takes place over

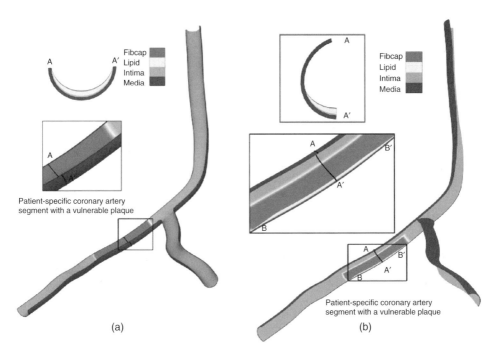

Figure 17.2   (a) A cut-away view of the diseased patient-specific left coronary artery segment with an idealized atherosclerotic plaque placed in the circumflex branch of the LCA. A cross-section is taken along A–A′ midway through the plaque highlighting the thickening of the intima and lipid accumulation. (b) Another cut-away view of the same diseased patient-specific LCA segment showing the composition of both the diseased and healthy portion of the transverse cross-section taken at A–A′

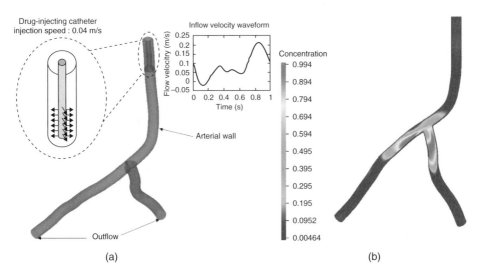

Figure 17.3   (a) A schematic describing the problem setup for the simulation of localized catheter-based NP delivery. Reprinted with permission from [26]. Copyright 2008 Springer. (b) The NP surface concentration on the arterial wall, at $t = 0$, which acts as the lumen-side boundary condition for the transport problem across the arterial wall

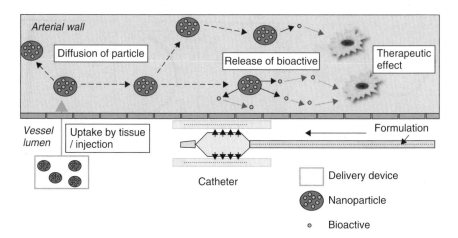

Figure 17.4 Schematic for the NP transport and adhesion at the lumen side, and NP/drug transport within the arterial wall. Created by Dariush Davalian (Abbott Vascular Inc.) and Jinping Wan. Reproduced with permission

a few cardiac cycles, the NP transport in the arterial wall and drug release from the NP can continue for days to weeks. Because of such a significant difference in time-scales, the transport problem in the arterial wall is decoupled from the luminal transport problem. On the lumen side, appropriate boundary conditions are introduced that take into account findings from blood flow calculations (i.e., the concentration of NPs adhering to the arterial wall).

A coupled NP (species I) and drug (species II) transport model was developed within the arterial wall [28]. The governing equations can be expressed in cylindrical coordinates as follows:

$$\frac{\partial C_I^*}{\partial t^*} = \frac{1}{r^*}\frac{\partial}{\partial r^*}\left(D_{r,I}^* r^* \frac{\partial C_I^*}{\partial r^*}\right) + \frac{1}{r^2}\frac{\partial}{\partial \theta}\left(D_{\theta,I}^* \frac{\partial C_I^*}{\partial \theta}\right) + \frac{\partial}{\partial z^*}\left(D_{z,I}^* \frac{\partial C_I^*}{\partial z^*}\right)$$

$$-\mathrm{Pe}\left[\frac{1}{r^*}\frac{\partial(r^* V_r^* C_I^*)}{\partial r^*} + \frac{1}{r^*}\frac{\partial(V_\theta^* C_I^*)}{\partial q} + \frac{\partial(V_z^* C_I^*)}{\partial z^*}\right] - \mathrm{Da_I} C_I^* \qquad (17.1)$$

$$\frac{\partial C_{II}^*}{\partial t^*} = \frac{1}{r^*}\frac{\partial}{\partial r^*}\left(D_{r,II}^* r^* \frac{\partial C_{II}^*}{\partial r^*}\right) + \frac{1}{r^2}\frac{\partial}{\partial \theta}\left(D_{\theta,II}^* \frac{\partial C_{II}^*}{\partial \theta}\right) + \frac{\partial}{\partial z^*}\left(D_{z,II}^* \frac{\partial C_{II}^*}{\partial z^*}\right)$$

$$-\mathrm{Pe}\left[\frac{1}{r^*}\frac{\partial(r^* V_r^* C_{II}^*)}{\partial r^*} + \frac{1}{r^*}\frac{\partial(V_\theta^* C_{II}^*)}{\partial q} + \frac{\partial(V_z^* C_{II}^*)}{\partial z^*}\right] - \mathrm{Da_{II}} C_{II}^* + C_I^* f^* \qquad (17.2)$$

where the Peclet number is

$$\mathrm{Pe} = \frac{V_{r,0} b}{D_{r,II}^0} \qquad (17.3)$$

and the Damkohler numbers are

$$\mathrm{Da_I} = \frac{\sigma_I b^2}{D^0_{r,II}} \quad \text{and} \quad \mathrm{Da_{II}} = \frac{\sigma_{II} b^2}{D^0_{r,II}} \tag{17.4}$$

To derive these dimensionless equations, the following definitions of nondimensional space and time were adopted:

$$r^* = \frac{r}{b} \quad \text{and} \quad t^* = \frac{t b^2}{D^0_{r,II}} \tag{17.5}$$

Here, b denotes the thickness of the arterial wall. The NP concentration C_I was nondimensionalized by the initial concentration of NPs in the formulation $C_{I,0}$, while that for the drug C_{II} was nondimensionalized by the initial concentration of free drug in the formulation $C_{II,0}$. Omitting subscripts I and II for simplicity, D_r, D_θ, and D_z are the diffusivities of the respective species in the radial, circumferential, and axial directions, respectively, such that $D_{i,j} = D^0_{i,j} \varphi_i(\mathbf{r}, t)$, where $i = r, \theta, z$ and $j = I, II$. Similarly, V_r, V_θ, and V_z are the corresponding advective velocities in the artery wall, such that $V_i = V_{i,0} \varphi_i(\mathbf{r}, t)$, where $i = r, \theta, z$. Here, D^0_i and $V_{i,0}$ are reference values of the diffusivity and the advective velocity, respectively, and $\varphi_i(\mathbf{r}, t)$ is a space- and time-varying function accounting for the spatial and temporal variation of these quantities. In Equations (17.1) and (17.2), the first three terms on the right-hand sides represent diffusion of scalars in the r, θ, and z directions, respectively. These are followed by three terms describing transport by advection in the three coordinate directions. The seventh term is the reaction term and denotes metabolic decay in the tissue. The two transport equations are coupled through the source term $C_I^* f^*$, where f^* denotes drug release rate from the NPs.

Drug release from the NPs was assumed to occur solely by diffusion. A biphasic diffusion model [33], originally devised for predicting drug release from a drug eluting stent coating [34], was adopted and extended to polymeric NPs. This biphasic model is based on the assumption that transport through the dispersed drug phase within the NPs takes place via two modes: the fast mode and the slow mode. The fast mode is the release of drug from a highly percolated structure of drug phase within the polymer, and the slow mode is the release of drug from a nonpercolated polymer-encapsulated phase of the drug. The mathematical expression of the biphasic drug-release-rate model for the polydisperse case is

$$f^* = \frac{\mathrm{d}M/M_0}{\mathrm{d}t^*} = f \frac{C_{I,0}}{C_{II,0}} \frac{b^2}{D^0_{r,II}}$$

$$= \frac{6}{D^0_{r,II}/b^2} \left[\sum_i \alpha_i f_1 \frac{D_1}{R_i^2} \sum_n e^{-n^2 \pi^2 t D_1 / R_i^2} + \sum_i \alpha_i (1 - f_1) \frac{D_2}{R_i^2} \sum_n e^{-n^2 \pi^2 t D_2 / R_i^2} \right] \tag{17.6}$$

where M/M_0 is the percentage release of drug from all the NPs at time t with M_0 denoting the total weight of drug encapsulated in all the NPs, N_0 (total number of NPs) at $t = 0$. There are three major design parameters, effective diffusivity of slow drug D_1, effective diffusivity of fast drug D_2, and fraction of drug in slow phase f_1, along with R, the size of

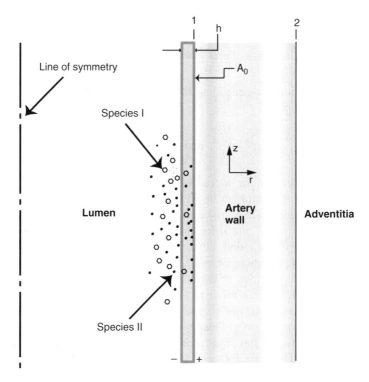

Figure 17.5 Schematic diagram of the cross-section of an artery to explain the derivation of boundary conditions. A very thin volume of thickness h and area A_0 is considered to perform an unsteady-state mass balance of NPs at the lumen–wall interface to determine the loss of NPs to blood flow as it appears in Equation (17.8). See Hossain *et al.* [28] for a more detailed derivation

the ith NP in the mixture and α_i, the fraction by weight of the ith NP in the formulation. By manipulating these parameters during the NP synthesis and loading with drug molecules, the desired release-rate profile can be engineered. Indeed, the computational model can easily implement different release profiles.

To determine appropriate boundary conditions for the lumen side, the following factors were considered (see Figure 17.5):

1. the loss of particles and free drug at the wall due to their exposure to blood flow that tends to wash them back into the bloodstream;
2. the propensity of the NP to stick to the wall, depending on the surface reaction rate (sticking coefficient) for the particular species; and
3. the NP solubility in the tissue compared with the blood through partition coefficients for each species.

The number of NPs that will stick to the artery wall tissue will depend on the competing influences of sticking and unsticking due to the special surface characteristics. This can be modeled by the following unsteady first-order rate equation:

$$\frac{\partial C_I^{*+}}{\partial t^*} = K_I' C_I^{*-} - K_{II}'' C_I^{*+} \tag{17.7}$$

Here, C_I^{*-} is the loss of NPs to blood flow, expressed as

$$C_I^{*-} = \exp\left[\left(-\frac{k_{m,I} b^2}{D_{r,I}^0} \frac{\text{Factor}}{h}\right) t^*\right] \tag{17.8}$$

with Factor $= (A_0 + A_0')/A_0$. Equation (17.8) was obtained by performing a mass balance at the lumen-side boundary volume. Here, Factor ≥ 1, since it scales flux area from the sides A_0' of the hollow cylinder, in addition to the circumferential area A_0 (see Hossain et al. [28] for a detailed derivation). Solving Equation (17.7) for C_I^{*+} we get

$$C_I^{*+} = \frac{K_I'}{K_I'' - K_I^+}[\exp(-K_I^+ t^*) - \exp(K_I'')] \tag{17.9}$$

where

$$K_I^+ = \frac{k_{m,I} b^2}{D_{II}^0} \frac{\text{Factor}}{h}$$

such that the NP tissue concentration at the lumen-side boundary becomes

$$C_I^*|_1 = C_I^{*+} \tag{17.10}$$

In the above equations, K_I' and K_I'' are forward (sticking) and backward (unsticking) reaction rates for the NPs and the drug, respectively, and $k_{m,I}$ is the mass transport coefficient for the NPs. Similarly, the loss of drug to blood flow at the lumen side is modeled as

$$C_{II}^{*+} = C_{II}^{*-} = \exp\left[\left(\frac{k_{m,II} b^2}{D_{r,II}^0} \frac{\text{Factor}}{h}\right) t^*\right] \tag{17.11}$$

After partitioning into the tissue, the drug tissue concentration at the lumen-side boundary becomes,

$$C_{II}^*|_1 = \bar{K}_{II} C_{II}^{*+} \tag{17.12}$$

where $\bar{K}_{II}$ is the drug tissue partition coefficient.

On the adventitia side, a Robin-type boundary condition was implemented:

$$-D_{r,I}^* \left.\frac{\partial C_I^*}{\partial r^*}\right|_2 = \frac{k_{m,I} b}{D_{r,II}^0}\left(C_I^*|_2 - C_{I,\infty}^*\right) \tag{17.13}$$

$$-D_{r,II}^* \left|\frac{\partial C_{II}^*}{\partial r^*}\right|_2 = \frac{k_{m,II} b}{D_{r,II}^0}\left(C_{II}^*|_2 - C_{II,\infty}^*\right) \tag{17.14}$$

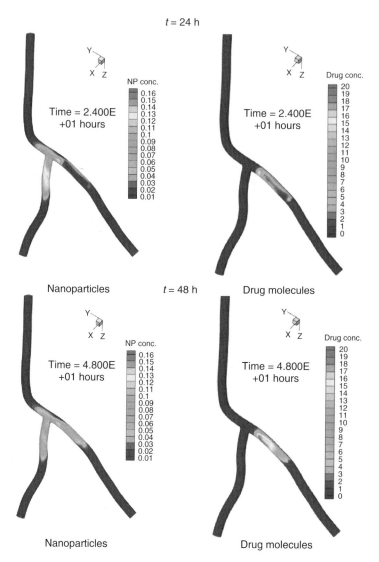

Figure 17.6   NP and drug distribution in terms of concentration (normalized) at one-third depth from the lumen side of the diseased patient-specific coronary artery wall at times $t = 24$ h (top) and $t = 48$ h. Notice that the concentration scales used for the NPs and drug molecules are different

The species transport across the internal elastic lamina (IEL) between the two homogeneous parts, the intima and the media, was simulated by implementing Kedem and Katchalsky's model for solute transport across a tissue barrier [35] as follows:

$$J_s = J_v(1 - \sigma_i)\frac{C_1 + C_2}{2} + P(C_1 - C_2) \qquad (17.15)$$

where J_s is the solute flux per unit area, J_v is the filtration rate or flow of solute, σ_i is the reflection or retardation factor, C_1 and C_2 are the solute concentrations in solvent on the intima and media sides (normalized), respectively, P is the diffusive permeability of the membrane to the solute and $k_{m,\text{II}}$ is the mass transport coefficient for the drug. $C_{\text{I},\infty}$ and $C_{\text{II},\infty}$ denote the concentration of NPs and the drug in the adventitia, respectively.

Applying the same general solution strategy adopted for the blood flow calculations, the simulations were run for 7 days with a time step of 2 min using appropriate parameters (see Hossain [36] for the parameters chosen and the rationale behind their selection). Figures 17.6 and 17.7 show the resulting time evolution of NP and drug distribution, side

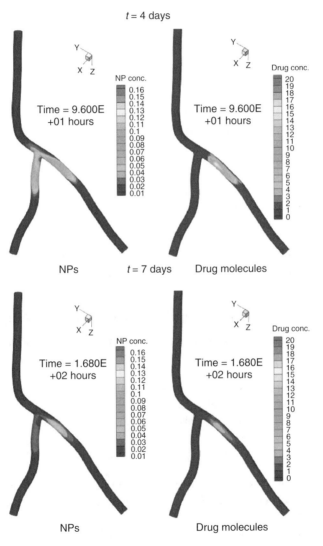

Figure 17.7   NP and drug distribution in terms of concentration (normalized) at one-third depth from the lumen side of the diseased patient-specific coronary artery wall at times $t = 4$ days (top) and $t = 7$ days. Notice that the concentration scales used for the NPs and drug molecules are different

by side. The 3D nature of transport is quite apparent here. The NP luminal wall deposition pattern and its intensity have a significant bearing on the overall drug distribution. Another important observation is that arterial and plaque heterogeneity modulate drug transport. Results indicate that both the IEL and the diseased part pose barriers to transport. At $t = 24$ h, we see that radial NP penetration in the diseased area is lagging behind that in the neighboring healthy parts. Consequently, there appears a darker region with a significantly lower concentration surrounded by regions of higher intensity, so much so that one can essentially make out the location and size of the vulnerable plaque. On the drug side, we see a similar slower radial penetration through the diseased part. However, despite the lack of NP availability in this diseased part, we see a hint of drug accumulating near the atherosclerotic plaque. This is largely because of the faster planar diffusion (compared with its radial counterpart) of drug released from the NPs in the neighboring healthy regions, demonstrating the anisotropic nature of drug transport. By day 2 there is evidence of considerable radial penetration of the NPs through the lipid core, leading to a higher NP concentration at this plane of the circumflex branch. As a result, drug concentration, too, reaches a maximum before diminishing to about 1/20 of this highest intensity by the end of day 7 as it depletes through the surrounding outer intima with a two orders of magnitude faster diffusivity. Important insights can be gained from the simulations that are unattainable in an experimental setting. For instance, we observed that the lipid core attracts and recruits the hydrophobic drug from its neighboring healthy branches and retains it over a long period of time, which is encouraging from a therapeutic point of view as it may lead to sustained reduction in plaque progression. Furthermore, this indicates that NPs that are off-target may still contribute significantly to the overall drug tissue concentration in the target region. Observations such as this make patient-specific geometry an essential ingredient in simulating realistic transport forces and flow features that are crucial in drug delivery system design.

17.3 Modeling the NP Dynamics in the Microcirculation

Blood is a complex fluid composed of cells suspended in plasma, an aqueous solution rich in proteins and molecules. The cellular component of blood comprises:

1. circulating leukocytes, or white blood cells (WBCs), which are the first level of body surveillance against external pathogens;
2. erythrocytes, or red blood cells (RBCs), which are responsible for the delivery of oxygen; and
3. platelets, which are mostly involved in the complex process of homeostasis.

RBCs are by far the most abundant, with 4–6 million cells per microliter of human blood and, on average, constitute 40–45% of the total blood volume in the macrocirculation and about 30–35% of total blood volume in the microcirculation. WBCs and platelets are present in less than 11 000 and 400 000 cells per microliter of human blood, respectively.

The size, shape, and deformability of these cells are also different. RBCs have a biconcave shape with a characteristic size ranging between 7 and 10 μm, and are designed to circulate for a few months (100–120 days), crossing through the smallest capillaries

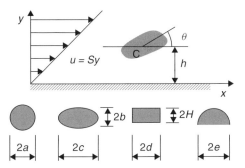

Figure 17.8 Schematic representation of an arbitrarily shaped particle in a linear laminar flow. The flow is in the x direction with a wall shear rate S, the separation distance of the particle centroid from the wall is h in the y direction normal to the flow, and θ is the particle orientation with respect to x. Reprinted with permission from [43]. Copyright 2009 Institute of Physics

in the lungs (4–5 μm) and the spleen. WBCs have in the flow a rounded shape and are larger than RBCs, with a characteristic diameter of 15–20 μm. On the other hand, platelets exhibit a quasi-elliptical shape and are smaller than RBCs, with a characteristic size of 2–4 μm.

RBCs are deformable and tend to accumulate in the core of the blood capillaries, leaving a so-called *cell-free layer*, in the proximity of the vessel walls [37–39]. Differently, WBCs and platelets are pushed laterally by the fast-moving RBCs and tend to interact more often with the walls of the blood vessels, sensing for abnormalities. In particular, the lateral drifting of WBCs is a well-characterized phenomenon in physiology, called *margination* [40].

NPs should be designed to preferentially drift towards the vessel walls (marginate as leukocytes), rather than staying in the core of the blood vessels (as RBCs). Indeed, a marginating NP can more efficiently sense the vessel walls for biological and biophysical diversities, such as the level of expression of receptors molecules and the presence of fenestrations.

17.3.1 Semi-analytical Models for the NP Transport

The margination propensity of nonspherical NPs has been characterized by Decuzzi and coworkers using a fairly simple and effective approach [41,42]. The geometry of the problem is presented in Figure 17.8, where an arbitrary-shaped particle is immersed within a linear laminar flow in a region bounded by a rigid flat plane (at $y = 0$). Although the approach is general, four different particle shapes were considered: a spherical bead with a radius a; an ellipsoidal particle with minor and major axes semi-lengths b and c, respectively, and aspect ratio $\gamma = b/c$; a discoidal particle with radius $2d$, height $2H$, and aspect ratio $\gamma = H/d$; and a hemispherical particle with radius e (Figure 17.8).

The governing equations are:

1. The Navier–Stokes equations for the fluid domain, including mass conservation

$$\nabla \cdot \mathbf{u} = 0 \tag{17.16}$$

and momentum conservation

$$\rho_f \left[\frac{\partial \mathbf{u}}{\partial t} + (\mathbf{u} \cdot \nabla) \mathbf{u} \right] = -\nabla p + \mu \nabla^2 \mathbf{u} \tag{17.17}$$

where $\mathbf{u}$ is the fluid velocity vector, ρ_f is the fluid density, μ is the fluid dynamic viscosity, and p is the dynamic pressure in the fluid.

2. Newton's laws for the particle, giving

$$M \frac{d\mathbf{V}}{dt} = \mathbf{F} \tag{17.18}$$

$$\mathbf{I} \frac{d\mathbf{\Omega}}{dt} = \mathbf{T} \tag{17.19}$$

where $\mathbf{V}$ is the particle velocity vector, $\mathbf{F}$ is the sum of the hydrodynamic force exerted by the fluid on the particle surface and the buoyancy force; M is the particle mass, $\mathbf{I}$ is the moment of inertia tensor, $\mathbf{\Omega}$ is the angular velocity of the particle, and $\mathbf{T}$ is the hydrodynamic moment vector.

The above system of equations is solved for the boundary conditions

$\mathbf{u} = \mathbf{0}$ on the wall $(y = 0)$

$\mathbf{u} = Sy\mathbf{e}_x$ away from the particle in the flow direction $\mathbf{e}_x = x/|x|$

$\mathbf{u} = \mathbf{V} + \mathbf{\Omega} \times (\mathbf{x} - \mathbf{X})$ on the particle surface ∂S (17.20)

where $\mathbf{X}$ is the position vector of the particle centroid and $\mathbf{x}$ is the position vector of a point over the particle surface.

For particles moving in a laminar flow with small Reynolds numbers $(Re_p < 1)$, the particle motion can be decoupled from the fluid flow so that equations $(18–19)$ become [42]:

$$\frac{4\pi}{3} St_a \frac{d}{dt} \begin{bmatrix} U_x \\ U_y \\ P\,\Omega \end{bmatrix} = \mathbf{D} + \mathbf{G} - \mathbf{R} \begin{bmatrix} U_x \\ U_y \\ \Omega \end{bmatrix} \tag{17.21}$$

where St_a is the Stokes number $(St_a = \rho_p a 2S/\mu)$ for a reference spherical bead of radius a and density ρ_p; U_x and U_y and Ω are the translational and angular velocities of the particles. The quantities $\mathbf{D}$ and $\mathbf{R}$ are the normalized drag vector and resistance matrix and $\mathbf{G}$ is the normalized buoyancy force, whose modulus is given as $G = (\rho_p - \rho_f) g a/S\mu$. In Equation (17.21), the coefficient P depends only on the shape of the particle, as listed in Table 17.1.

The drag vector $\mathbf{D}$ and the resistance matrix $\mathbf{R}$ are computed as a function of the particle size, shape, and separation distance from the wall using a commercial package code, such as Fluent V.6.3 [42]. The geometry of the particle and the mesh for the flow field are generated by Gambit V.2.4. A tet/hybrid mesh is used to discretize the geometry of the problem. The full 3D Navier–Stokes and mass continuity equations are solved using a pressure-based solver with the SIMPLEC algorithm and the QUICK scheme is used for space and time discretizations. Symmetric boundary conditions are imposed on

Table 17.1 The parameter P and rotational inertia L for the four NP shapes considered, for a fixed particle volume ($L_s = 4ma/5$: rotational inertia for a spherical particle)

	γ	L/L_s	P
Sphere	–	1	2/5
Ellipse	$\gamma_e = \dfrac{b}{c}$	$\dfrac{1 + \gamma_e^2}{2\gamma_e^{4/3}}$	$\dfrac{1 + \gamma_e^2}{5\gamma_e^{4/3}}$
Disc	$\gamma_d = \dfrac{H}{d}$	$\dfrac{5(3 + 4\gamma_d^2)}{24(2\gamma_d)^{2/3}}$	$\dfrac{3 + 4\gamma_d^2}{12(2\gamma_d)^{2/3}}$
Hemisphere	–	0.408	0.1163

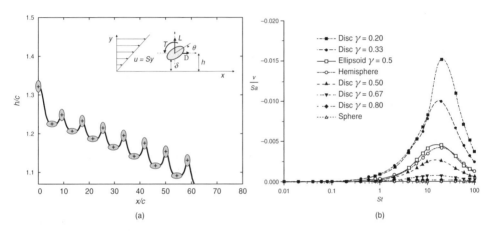

(a) (b)

Figure 17.9 (a) The dynamics of an ellipsoidal particle (aspect ratio 0.5) drifting towards the vessel wall (bottom); Reprinted with permission from [41]. Copyright 2009 Elsevier. (b) Normalized drift velocity $v/(Sa)$ of differently shaped NPs in the absence of gravitational force ($\mathbf{G} = 0$); Reprinted with permission from [42]. Copyright 2009 Institute of Physics

the longitudinal middle plane of the particle. The shear flow is generated by imposing a linear velocity boundary condition through a built-in user-defined function. Once **D** and **R** are known, the particle trajectory can be derived by integrating Equation (17.21) with proper initial conditions. In the present case, a Runge–Kutta scheme has been employed in MATLAB with a self-adjusted time interval that ensures numerical convergence.

Following this approach, it has been shown that nonspherical particles under the concurrent action of inertial and hydrodynamic forces can drift laterally (*hydrodynamic margination*) across the streamlines towards the rigid flat wall. A sample result of an ellipsoidal particle drifting towards the wall is shown in Figure 17.9a. Also, a lateral drift velocity has been calculated as a function of the particle Stokes number St_a and for different shapes of the particles, in the case of zero buoyancy force ($G = 0$) (Figure 17.9b). This shows that the drifting velocity is maximized for St_a ranging between 1 and 100, and for discoidal shapes with the lowest aspect ratio (thin disks).

The trajectories of discoidal particles with different aspect ratios ($\gamma_d = 0.20$, 0.33, 0.50, 0.67, and 0.80) as a function of the Stokes number St_a and buoyancy force $\mathbf{G}$ are presented in Figure 17.10. It is confirmed that the drift velocity increases as the Stokes number grows from 1 to 10, for a fixed $\mathbf{G}$. As the buoyancy force $\mathbf{G}$ grows, the margination speed increases for horizontal capillaries with $\mathbf{G}$ pointing to the wall, as well as for vertical capillaries. The largest drift velocity is achieved in descending capillaries, where $\mathbf{G}$ is aligned with the flow. The discoidal particles with lowest aspect ratio have the highest margination speed, even in the presence of a gravitational force. Very interestingly, the discoidal particles with ($\gamma_d = 0.20$ and 0.33; Figure 17.10b) do marginate towards the proximal wall even for a gravitational force pointing away from the wall.

17.3.2 An IFEM for NP and Cell Transport

In this section, the IFEM is briefly described for solving the fluid–structure interaction problem describing the dynamics of NPs and cells in a capillary flow.

The IFEM [43–49] was developed by integrating the concept of immersed boundary [50–52], finite element [53–55], and mesh-free methods [56,57]. Following the IFEM framework, the immersed bodies (NPs and cells) are modeled as a flexible and highly deformable structure using a Lagrangian mesh, whereas the fluid (plasma) is solved in a fixed Eulerian mesh. The details of the derivation can be found in Zhang *et al.* [43], and a brief explanation of the procedure is provided in the following. The cell-to-cell and cell-to-NP interactions are modeled using a Morse potential, and the Kramer reaction rate is employed for describing the ligand–receptor binding at the interface [58,59].

Within the IFEM framework, the dynamics of immersed bodies is coupled with the surrounding incompressible viscous fluid, but avoiding expensive remeshing or mesh adaptivity of the fluid domain.

Consider an incompressible three-dimensional deformable structure Ω^s completely immersed in an incompressible fluid domain Ω^f (Figure 17.11). Both, the fluid and the solid occupy the domain Ω, but do not intersect, where both $\Omega^f \cup \Omega^s = \Omega$ and $\Omega^f \cap \Omega^s = 0$ hold true. With these assumptions, an Eulerian fluid mesh is adopted that spans the entire domain Ω and a Lagrangian solid mesh is constructed on top of the Eulerian fluid mesh. Following Newtonian physics, the inertial force of a particle is balanced with the derivative of the Cauchy stress σ and the external force $\mathbf{f}^{ext}$ exerted on the continuum and the following equation holds true:

$$\rho \frac{dv_i}{dt} = \sigma_{ij,j} + f_i^{ext} \tag{17.22}$$

Noting that the solid density ρ^s is different from the fluid density ρ^f – that is, $\rho = \rho^s$ in Ω^s and $\rho = \rho^f$ in Ω^f – the inertial forces can be divided into two components:

$$\rho \frac{dv_i}{dt} = \rho^f \frac{dv_i}{dt} \qquad \text{in } \Omega/\Omega^s \tag{17.23}$$

$$\rho \frac{dv_i}{dt} = \rho^f \frac{dv_i}{dt} + (\rho^s - \rho^f) \frac{dv_i}{dt} \qquad \text{in } \Omega^s \tag{17.24}$$

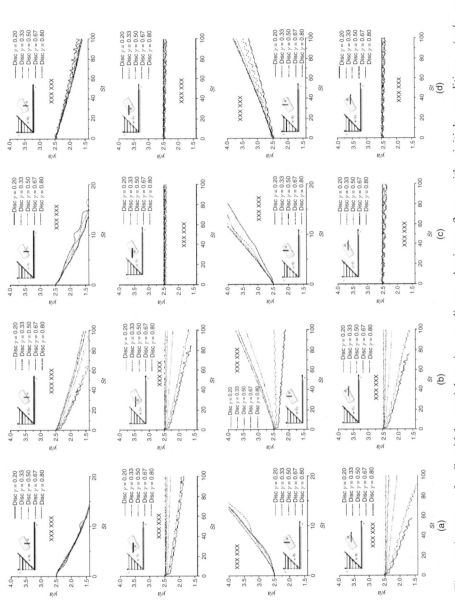

Figure 17.10 The trajectory of discoidal particles in a linear laminar flow with initial conditions $(x_0/a, u_0/(Sa), y_0/a, v_0/a, \theta_0, \Omega_0/S) = (0, 0, 2.5, 0, \pi/2, 0)$ in the presence of a gravitational force ($\mathbf{G} \neq 0$): (a) $St_a = 10$, $\mathbf{G} = 1$; (b) $St_a = 10$, $\mathbf{G} = 0.1$; (c) $St_a = 1$, $\mathbf{G} = 1$; (d) $St_a = 1$, $\mathbf{G} = 0.1$. The inset shows the different orientation of $\mathbf{G}$ with respect to the flow direction; from Lee et al. Reprinted with permission from [43]. Copyright 2009 Institute of Physics

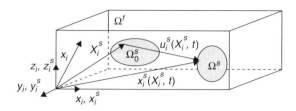

Figure 17.11 The Eulerian coordinates in the computation fluid domain Ω^f are described with the time-invariant position vector x_i. The solid positions in the initial configuration Ω^s_0 and the current configuration Ω^s are represented by X^s_i and $x^s_i(X^s_i, t)$ respectively

Following the same concept, both the external forces f^{ext}_i can be decomposed as

$$f^{ext}_i = 0 \qquad\qquad \text{in } \Omega/\Omega^s$$
$$f^{ext}_i = (\rho^s - \rho^f)g_i \qquad \text{in } \Omega^s \qquad\qquad (17.25)$$

Analogously, the Cauchy stress is decomposed as

$$\sigma_{ij,j} = \sigma^f_{ij,j} \qquad\qquad\qquad \text{in } \Omega/\Omega^s$$
$$\sigma_{ij,j} = \sigma^f_{ij,j} + \sigma^s_{ij,j} - \sigma^f_{ij,j} \qquad \text{in } \Omega^s \qquad\qquad (17.26)$$

It is important to note that the fluid stress in the solid domain, in general, is much smaller than the corresponding solid stress. The fluid–structure interaction (FSI) force within the Ω^s is then defined as

$$f^{FSI,s}_i = -(\rho^s - \rho^f)\frac{dv_i}{dt} + \sigma^s_{ij,j} + \sigma^f_{ij,j} + (\rho^s - \rho^f)g_i, \qquad \mathbf{x} \in \Omega^s \qquad (17.27)$$

The fluid–structure interaction force is calculated with the Lagrangian description, where a Dirac delta function δ is used to distribute the interaction force from the solid domain onto the computational fluid domain as

$$f^{FSI}_i(\mathbf{x}, t) = \int_{\Omega^s} f^{FSI,s}_i(\mathbf{X}^s, t)\delta[\mathbf{x} - \mathbf{x}^s(\mathbf{X}^s, t)]d\Omega \qquad\qquad (17.28)$$

Hence, the governing equation for the fluid domain can be derived by combining the fluid terms and the interaction force as:

$$\rho^f\frac{dv_i}{dt} = \sigma^f_{ij,j} + f^{FSI}_i, \qquad \mathbf{x} \in \Omega \qquad\qquad (17.29)$$

Since we consider the entire domain Ω to be incompressible, we only need to apply the incompressibility constraint ($v_{i,i} = 0$) once on the entire domain Ω. To define the Lagrangian description for the solid and Eulerian description for the fluid, different velocity field variables v^s_i and v_i are introduced to represent the motions of the solid in the domain Ω^s and the fluid within the entire domain Ω. The coupling of both velocity fields is accomplished with the Dirac delta function provided as

$$v^s_i(\mathbf{X}^s, t) = \int_{\Omega} v_i(\mathbf{x}, t)\delta[\mathbf{x} - \mathbf{x}^s(\mathbf{X}^s, t)]d\Omega \qquad\qquad (17.30)$$

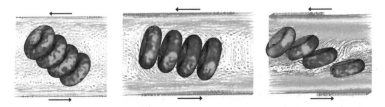

Figure 17.12 The shear of a four-RBC cluster at low (left), intermediate (middle), and high (right) shear rates S. The vectors represent the fluid velocity field. Refer to Plate 15 for the colored figure

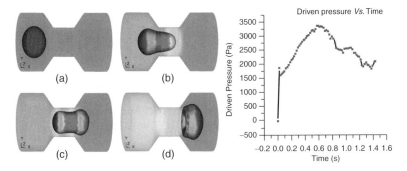

Figure 17.13 A 3D simulation of a single cell squeezing through a capillary vessel (left). The history of the driven pressure during the squeezing process (right). The cell diameter is 1.2 times the diameter of the constriction. Refer to Plate 16 for the colored figure

The coupling between the fluid and solid domains is enforced via the Dirac delta functions [44,45]. The nonlinear system of equations is then solved using the standard Petrov–Galerkin method and the Newton–Raphson [46] solution technique.

The IFEM framework has been applied to several problems in biology and biomedical sciences, and in particular for the analysis of blood flow and RBC transport. In a set of numerical experiments, clusters of RBCs were exposed at different values of the wall shear rate (low, intermediate, and high) and different behaviors were observed in agreement with experimental data (Figure 17.12). At low shear rates, the RBC clusters rotated as a bulk; at intermediate shear rates, the clusters were partially disrupted, with cells aligning normally to the shear direction; at high shear rates, the cell clusters were completely disintegrated, with individual cells orienting parallel to the flow direction. With the same technique, the large deformation of a cell passing through a vascular constriction was modeled and the hydrodynamic pressure drop along the conduit was also predicted (Figure 17.13). Four snapshots illustrating the large elongation and squeezing of the deformable cell are shown in Figure 17.13. The increase in pressure required to squeeze the cell through the capillary constriction was responsible for the increase in the apparent blood viscosity (Figure 17.13, right), in agreement with the Fahraeus–Lindqvist effect.

In addition, IFEM can be used to model the transport of NPs, with different sizes, shapes, and surface properties, in whole blood. A typical geometrical representation is sketched in Figure 17.14, which presents RBCs, occupying 30% of the capillary volume,

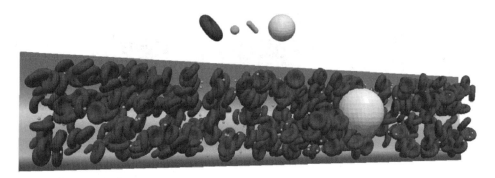

Figure 17.14 Model of RBCs with a diameter of 7–8 μm (top, left), NPs of various shapes and sizes ranging between 10 and 1000 nm (top, middle), and a WBC with a diameter of 20 μm (top, right). Fluid domain (40 × 40 × 200 μm³) with RBCs (red), leukocytes (white), and NPs (green). Refer to Plate 17 for the colored figure

a white blood cell (white sphere) and spherical and cylindrical NPs (green) distributed along the conduit.

17.4 Conclusions

Mathematical modeling and computational analysis are formidable tools for predicting and optimizing the vascular behavior of circulating agents, such as molecular agents, NPs, and cells. IA, presented in Section 17.2, can accurately and effectively model blood flow and transport problems in patient-specific vascular networks. The IFEM, presented in Section 17.3, can predict the dynamics of NPs in complex multiphase flows, such as blood, accounting for intercellular and RBC–NP interactions. Eventually, semi-analytical models can analyze the motion of individual particles under simple flow conditions.

Integrating IA with particle-based methods, such as IFEM, and semi-analytical models will allow us to develop truly multiscale and multi-physics computational tools for optimizing the NP behavior in the vascular compartment. The proper combination of size, shape, and surface properties (the 3S problem) for an NP could be identified *in silico* as a function of the disease type and patient-specific features. Sophisticated mathematical and computational methods could help in predicting the efficacy of systemically injected NPs and suggest proper therapeutic regimens.

Acknowledgments

SSH and TJRH acknowledge the support from Abbott Vascular Inc. through Grant No. UTA05-663, the Texas Advanced Research Program of the Texas Higher Education Coordinating Board through Grant No. 003658-0025-2006 and Portuguese CoLab through Grant No. 04A. PD and MF acknowledge the partial support of the Telemedicine and Advanced Technology Research Center (TATRC) through the pre-centre grant W81XWH-09-2-0139, the National Cancer Institute through the grants U54CA143837 and U54CA151668, and the Department of Defense through the grant W81XWH-09-1-0212. WKL, AMK, and TRL acknowledge the support from NSF CMMI 0856492 and NSF CMMI 0856333.

References

[1] Torchilin, V.P. (2005) Recent advances with liposomes as pharmaceutical carriers. *Nature Reviews. Drug Discovery*, **4**, 145–160.

[2] Ferrari, M. (2005) Cancer nanotechnology: opportunities and challenges. *Nature Reviews Cancer*, **5**, 161–171.

[3] Gabizon, A. and Papahadjopoulos, D. (1988) Liposome formulations with prolonged circulation time in blood and enhanced uptake by tumors. *Proceedings of the National Academy of Sciences of the United States of America*, **85**, 6949–6953.

[4] Allen, T.M. and Cullis, P.R. (2004) Drug delivery systems: entering the mainstream. *Science*, **19**, 1818–1822.

[5] Peer, D., Karp, J.M., Hong, S. *et al.* (2007) Nanocarriers as an emerging platform for cancer therapy. *Nature Nanotechnology*, **2**, 751–760.

[6] Jain, R.K. (1999) Transport of molecules, particles, and cells in solid tumors. *Annual Review of Biomedical Engineering*, **1**, 241–263.

[7] Decuzzi, P., Pasqualini, R., Arap, W., and Ferrari, M. (2009) Intravascular delivery of particulate systems: does geometry really matter? *Pharmaceutical Research*, **26** (1), 235–243.

[8] Godin, B., Driessen, W.H., Proneth, B. *et al.* (2010) An integrated approach for the rational design of nanovectors for biomedical imaging and therapy. *Advances in Genetics*, **69**, 31–64.

[9] Tasciotti, E., Liu, X., Bhavane, R. *et al.* (2008) Mesoporous silicon particles as a multistage delivery system for imaging and therapeutic applications. *Nature Nanotechnology*, **3**, 151–157.

[10] Gratton, S.E., Pohlhaus, P.D., Lee, J. *et al.* (2007) Nanofabricated particles for engineered drug therapies: a preliminary biodistribution study of PRINT nanoparticles. *Journal of Controlled Release*, **121** (1–2), 10–18.

[11] Champion, J.A., Katare, Y.K., and Mitragotri, S. (2007) Particle shape: a new design parameter for micro- and nanoscale drug delivery carriers. *Journal of Controlled Release*, **121** (1–2), 3–9.

[12] Matsumura, Y. and Maeda, H. (1986) A new concept for macromolecular therapeutics in cancer chemotherapy: mechanism of tumoritropic accumulation of proteins and the antitumor agent smancs. *Cancer Research*, **46**, 6387–6392.

[13] Maeda, H., Wu, J., Sawa, T. *et al.* (2000) Tumor vascular permeability and the EPR effect in macromolecular therapeutics: a review. *Journal of Controlled Release*, **65**, 271–284.

[14] Yuan, F., Leunig, M., Huang, S.K. *et al.* (1994) Microvascular permeability and interstitial penetration of sterically stabilized (stealth) liposomes in a human tumor xenograft. *Cancer Research*, **54**, 3352–3356.

[15] Hobbs, S.K., Monsky, W.L., Yuan, F. *et al.* (1998) Regulation of transport pathways in tumor vessels: role of tumor type and microenvironment. *Proceedings of the National Academy of Sciences of the United States of America*, **95**, 4607–4612.

[16] Pasqualini, R., Arap, W., and McDonald, D.M. (2002) Probing the structural and molecular diversity of tumor vasculature. *Trends in Molecular Medicine*, **8**, 563–571.

[17] Pasqualini, R., Koivunen, E., and Ruoslahti, E. (1997) αv integrins as receptors for tumor targeting by circulating ligands. *Nature Biotechnology*, **15** (6), 542–546.

[18] Hughes, T.J.R., Cottrell, J.A., and Bazilevs, Y. (2005) Isogeometric analysis: CAD, finite elements, NURBS, exact geometry and mesh refinement. *Computer Methods in Applied Mechanics and Engineering*, **194** (39–41), 4135–4195.

[19] Cottrell, J.A., Hughes, T.J.R., and Bazilevs, Y. (2009) *Isogeometric Analysis: Toward Integration of CAD and FEA*, 1st edn., John Wiley & Sons Ltd, Chichester.

[20] Bazilevs, Y., da Veiga, L.B., Cottrell, J.A. *et al.* (2006) Isogeometric analysis: approximation, stability and error estimates for h-refined meshes. *Mathematical Models & Methods in Applied Sciences*, **16** (7), 1031–1090.

[21] Bazilevs, Y., Calo, V.M., Zhang, Y., and Hughes, T.J.R. (2006) Isogeometric fluid–structure interaction analysis with applications to arterial blood flow. *Computational Mechanics*, **38** (4), 310–322.

[22] Cottrell, J.A., Reali, A., Bazilevs, Y., and Hughes, T.J.R. (2006) Isogeometric analysis of structural vibrations. *Computer Methods in Applied Mechanics and Engineering*, **8** (15), 5257–5296.

[23] Zhang, Y., Bazilevs, Y., Goswami, S. *et al.* (2007) Patient-specific vascular NURBS modeling for isogeometric analysis of blood flow. *Computer Methods in Applied Mechanics and Engineering*, **196** (29–30), 2943–2959.

[24] Bazilevs, Y., Calo, V.M., Tezduyar, T.E., and Hughes, T.J.R. (2007) YZβ discontinuity capturing for advection-dominated processes with application to arterial drug delivery. *International Journal for Numerical Methods in Fluids*, **54** (6–8), 593–608.

[25] Gomez, H., Calo, V.M., Bazilevs, Y., and Hughes, T.J.R. (2008) Isogeometric analysis of the Cahn–Hilliard phase-field model. *Computer Methods in Applied Mechanics and Engineering*, **197** (49–50), 4333–4352.

[26] Calo, V.M., Brasher, N., Bazilevs, Y., and Hughes, T.J.R. (2008) Multiphysics model for blood flow and drug transport with application to patient-specific coronary artery flow. *Computational Mechanics*, **43** (1), 161–177.

[27] Bazilevs, Y., Gohean, J.R., Moser, R.D. *et al.* (2009) Patient-specific isogeometric fluid–structure interaction analysis of thoracic aortic blood flow due to implantation of the Jarvik (2000) left ventricular assist device. *Computer Methods in Applied Mechanics and Engineering*, **198**, 3534–3550.

[28] Hossain, S.S., Hossainy, S.F.A., Bazilevs, Y. *et al.* (2012) Mathematical modeling of coupled drug and drug-encapsulated nanoparticle transport in patient-specific coronary artery walls. *Computational Mechanics*, **49**, 213–242, doi: 10.1007/s00466-011-0633-2.

[29] Sahni, O., Muller, J., Jansen, K.E. *et al.* (2006) Efficient anisotropic adaptive discretization of the cardiovascular system. *Computer Methods in Applied Mechanics and Engineering*, **195**, 5634–5655.

[30] Bazilevs, Y., Calo, V.M., Cottrell, J.A. *et al.* (2007) Variational multiscale residual-based turbulence modeling for large eddy simulation of incompressible flows. *Computer Methods in Applied Mechanics and Engineering*, **197** (1–4), 173–201.

[31] Chung, J. and Hulbert, G.M. (1993) A time integration algorithm for structural dynamics with improved numerical dissipation: the generalized method. *Journal of Applied Mechanics*, **60**, 371–375.

[32] Zarins, C.K., Giddens, D.P., Bharadvaj, B.K. *et al.* (1983) Carotid bifurcation atherosclerosis. Quantitative correlation of plaque localization with flow velocity profiles and wall shear stress. *Circulation Research*, **53** (4), 502–514.

[33] Hossainy, S.F.A., Prabhu, S., Hossain, S.S. *et al.* (eds.) (2008) *Mathematical Modeling of Bi-phasic Mixed Particle Drug Release from Nanoparticles. Proceedings of the 8th World Biomaterials Congress*, Amsterdam.

[34] Hossainy, S.F.A. and Prabhu, S. (2008) A mathematical model for predicting drug release from a biodurable drug-eluting stent coating. *Journal of Biomedical Materials Research Part A*, **87** (2), 487–493.

[35] Saltzman, W.M. (2001) *Drug Delivery: Engineering Principles for Drug Therapy*, Oxford University Press, New York, NY.

[36] Hossain, S.S. (2009) Mathematical modeling of coupled drug and drug-encapsulated nanoparticle transport in patient-specific coronary artery walls. Dissertation, University of Texas at Austin, Austin, TX.

[37] Ong, P.K., Jain, S., and Kim, S. (2011) Temporal variations of the cell-free layer width may enhance NO bioavailability in small arterioles: effects of erythrocyte aggregation. *Microvascular Research*, **81** (3), 303–312.

[38] Kim, S., Kong, R.L., Popel, A.S. *et al.* (2006) A computer-based method for determination of the cell-free layer width in microcirculation. *Microcirculation*, **13** (3), 199–207.

[39] Tateishi, N., Suzuki, Y., Soutani, M., and Maeda, N. (1994) Flow dynamics of erythrocytes in microvessels of isolated rabbit mesentery: cell-free layer and flow resistance. *Journal of Biomechanics*, **27** (9), 1119–1125.

[40] Goldsmith, H.L. and Spain, S. (1984) Margination of leukocytes in blood flow through small tubes. *Microvascular Research*, **27** (2), 204–222.

[41] Lee, S.Y., Ferrari, M., and Decuzzi, P. (2009) Design of bio-mimetic particles with enhanced vascular interaction. *Journal of Biomechanics*, **42** (12), 1885–1890.

[42] Lee, S.Y., Ferrari, M., and Decuzzi, P. (2009) Shaping nano-/micro-particles for enhanced vascular interaction in laminar flows. *Nanotechnology*, **20** (49), 495101.

[43] Zhang, L., Gerstenberger, A., Wang, X., and Liu, W.K. (2004) Immersed finite element method. *Computer Methods in Applied Mechanics and Engineering*, **193** (21–22), 2051–2067.

[44] Liu, W.K., Jun, S., and Zhang, Y.F. (1995) Reproducing kernel particle methods. *International Journal for Numerical Methods in Fluids*, **20**, 1081–1106.

[45] Liu, W.K., Jun, S., and Zhang, Y.F. (1996) Generalized multiple scale reproducing kernel particle methods. *Computer Methods in Applied Mechanics and Engineering*, **139**, 91–158.

[46] Tezduyar, T.E. and Sathe, S. (2003) Stabilization parameters in SUPG and PSPG formulations. *Journal of Computational and Applied Mechanics*, **4** (1), 71–88.

[47] Saad, Y. and Schultz, M.H. (1986) GMRES: a generalized minimal residual algorithm for solving non-symmetric linear systems. *SIAM Journal on Scientific and Statistical Computing*, **7** (3), 856–869.

[48] Zhang, L.T., Wagner, G.J., and Liu, W.K. (2002) A parallelized meshfree method with boundary enrichment for large-scale CFD. *Journal of Computational Physics*, **176**, 483–506.

[49] Liu, W.K., Kim, D.W., and Tang, S. (2007) Mathematical foundations of the immersed finite element method. *Computational Mechanics*, **39**, 211–222.

[50] Peskin, C.S. and McQueen, D.M. (1989) A three-dimensional computational method for blood flow in the heart. Immersed elastic fibers in a viscous incompressible fluid. *Journal of Computational Physics*, **81** (2), 372–405.

[51] Peskin, C.S. and McQueen, D.M. (1993) Computational biofluid dynamics. *Contemporary Mathematics*, **141**, 161–186.

[52] Wang, X. and Liu, W.K. (2004) Extended immersed boundary method using FEM and RKPM. *Computer Methods in Applied Mechanics and Engineering*, **193** (12–14), 1305–1321.

[53] Hughes, T.J.R. (1987) *The Finite Element Method*, Prentice Hall.

[54] Liu, W.K., Chang, H., Chen, J., and Belytschko, T.B. (1988) Arbitrary Lagrangian–Eulerian Petrov–Galerkin finite elements for nonlinear continua. *Computer Methods in Applied Mechanics and Engineering*, **68**, 259–310.

[55] Belytschko, T., Liu, W.K., and Moran, B. (2000) *Nonlinear Finite Elements for Continua and Structures*, John Wiley and Sons.

[56] Li, S. and Liu, W.K. (2002) Meshfree and particle methods and their applications. *Applied Mechanics Review*, **55**, 1–34.

[57] Li, S. and Liu, W.K. (2004) *Meshfree Particle Method*, Springer.

[58] Liu, W.K., Liu, Y., Farrell, D. *et al.* (2006) Immersed finite element method and its applications to biological systems. *Computer Methods in Applied Mechanics and Engineering*, **195**, 1722–1749.

[59] Kopacz, A.M., Liu, W.K., and Liu, S.Q. (2008) Simulation and prediction of endothelial cell adhesion modulated by molecular engineering. *Computer Methods in Applied Mechanics and Engineering*, **197** (25–28), 2340–2352.

Index

ab initio MD, 133
actin
    function of, 37
    in protrusion creation, 10
actin filaments
    coarse-grained modeling of, 402–3,
        403*f*
    described, 257, 257*f*
actin polymerization
    tension increasing, 50, 52, 52*f*
activation entropy
    and Meyer–Neldel compensation rule,
        322–4
adherens junctions, 259
adhesion(s). *see* two-dimensional
        adhesion; *specific types, e.g.,* focal
        adhesions
adjacent graphic layers
    cross-linking
        in hierarchical carbon-based
            materials, 110–13, 110*f*,
            111*f*
    weak shear interactions between
        in hierarchical carbon-based
            materials, 106–9
AFM-based methods
    in nanoindentation experiments on
        CNT composites, 114–15, 116*f*
AFM testing. *see* atomic force
        microscopy (AFM) testing
Ag
    TB and stacking fault energies of, 130,
        130*t*
"all-or-none" event, 39

alloy(s)
    twinned, 130*t*, 131–2
anistropic rigidity
    substrates with, 303
antibody(ies)
    function of, 4
Aplysia neurites
    reducing mechanical tension in,
        49–50, 51*f*
apoptosis, 3
    survival signaling and, 13
area change
    tube model for, 417
aspiration
    micropipette, 268–70, 268*f*, 281*t*
assay(s)
    bulk, 262–7. *see also* bulk assays
    cell mechanics. *see* cell mechanics
            assays
    microfluidic pore and deformation,
        278–9, 283*t*, 284*f*
    substrate stretching, 277, 283*t*
atomic force microscopy (AFM) testing,
        96, 98, 270–2, 270*f*, 282*t*
    force sensitivity of, 101–2, 101*f*
    in measuring ultraflow forces, 101–2,
        101*f*
    in MWNTs, 106–9
atomistic and mesoscale protein folding
        and deformation
    in spider silk, 400–2, 401*f*
atomistic reaction pathway sampling,
        313–38. *see also* nudged elastic
        band (NEB) method

Nano and Cell Mechanics: Fundamentals and Frontiers, First Edition. Edited by Horacio D. Espinosa and Gang Bao.
© 2013 John Wiley & Sons, Ltd. Published 2013 by John Wiley & Sons, Ltd.

atomistic simulation
    extending time scale in, 314–15
ATP hydrolysis, 20
Au
    TB and stacking fault energies of, 130,
        130*t*
axon(s), 36*f*, 37
    force actuators in, 53, 54*f*
    force thresholds for, 53, 53*f*
    large force relaxation of, 44
    low relaxation of, 44
    mechanical behavior of
        modeling of, 52–7, 53*f*, 54*f*, 57*f*
    in vivo
        force relaxation and generation in,
            57, 57*f*
        stiffening of, 44, 45*f*
axonal branches
    elimination of unnecessary
        tension in, 46–7
axonal growth, 40, 40*f*

BBS protein, 29
beacon(s)
    molecular. *see* molecular beacons
Beacon Designer, 243
bead-embedded gels
    traction force microscopy via, 275,
        283*t*
bending
    nanoindenter-based
        TEM-STM and, 201–4
binding proteins
    functions of, 14, 14*t*
binodal line, 87
biological measurements
    mechanical measurements combined
        with, 261–2
biological phenomena, 1–61
biomateriomics
    mathematical approaches to, 394–400,
        395*f*–9*f*
bistability
    dynamical, 361
        molecular potential of mean force
            and, 381–4, 383*f*

blood
    cellular components of, 448–9
body-centered cubic structures, 180–3,
    181*f*
bridging modeling with experiment,
    327–8, 328*f*, 328*t*
bulk assays, 262–7, 263*f*, 265*f*, 266*f*,
    281*t*–3*t*
    ektacytometry of, 266–7, 266*f*, 281*t*
    microfiltration of, 262–4, 263*f*, 281*t*
    parallel-plate flow chambers of, 267,
        281*t*
    rheometry of, 264–6, 265*f*, 281*t*
bulk metals
    nanotwinned
        deformation mechanisms in,
            147–53, 148*t*, 149*f*, 150*f*,
            150*t*, 152*f*, 153*f*

cadherin(s)
    function of, 4
    in survival signaling, 13
calcium-binding regulators
    of myosins and kinesins, 23–6, 25*f*
calmodulin
    in binding of myosins and kinesins,
        23–5
cancer
    cell mechanics in, 280, 284
carbon-based materials
    assemblage into high-performance
        hierarchical structures, 96
    hierarchical
        multiscale experimental mechanics
            of, 95–128. *see also*
            hierarchical carbon-based
            materials
carbon nanotube(s) (CNTs), 95–6
    AFM methods of measuring ultraflow
        forces of, 101–2, 101*f*
    composites and fibers with high
        volume fractions based on,
        115–20, 117*f*, 119*f*
    within larger networks
        local mechanical responses of, 114
    mechanical properties of, 105–6

micromechanical testing methods of
  collective and local behavior of,
  103–6, 105*f*
multi-walled
  in *in situ* TEM testing, 213–15,
    214*f*
in situ SEM/AFM methods in
  investigating shear interactions,
  102–3, 102*f*
in situ TEM methods of revealing
  atomic-level mechanics of,
  98–101, 100*f*
tension tests of, 104–5, 105*f*
carbon nanotube (CNT) yarns
  low ductility of, 117–18, 117*f*
Cassie–Baxter angle
  deviation from
    reason for, 81–2, 81*f*, 82*f*
  experimental data compared with, 80,
    80*f*
Cassie–Baxter formula, 79
Cassie–Baxter state, 65–6, 66*f*
  in hydrophobic to superhydrophobic
    behavior, 71–3, 72*f*
  designing surface roughness for, 73
  to Wenzel state, 73–7, 74*f*, 84
    apparent contact angle of drop in,
      77–9
    predicting transition from, 73–7,
      74*f*
      correlation between energy- and
        geometry-based criteria in,
        75–6
      energy formulation in, 75
      geometric considerations in, 74
    pressure sources for transition, 76–7
category theoretical abstraction of protein
    materials
  analogy to office network, 403–6,
    405*f*
category theory
  ontology logs and, 396–400, 398*f*,
    399*f*
cationic transfection agents
  in molecular beacon delivery, 245

CDKs. *see* cyclin-dependent kinases
    (CDKs)
cell(s)
  basic functions of, 3
  cytoskeleton of, 256–9, 258*f*
  diseased
    probing elastic properties of, 305,
      306*f*
  forces applied by, 259–60
  functions of, 293–4
  levels of attachment for, 3
  magnetic micropillars for local force
    application on, 304
  responses to force and environment,
    260–1
cell adhesion, 3
  probing of
    cell mechanics studies of, 306–7
    two-dimensional, 7–9, 8*f*
cell differentiation
  cell mechanics studies of, 304
  micropatterned substrates in, 298–9
cell differentiation signaling, 13–14, 14*t*
  integrins in, 13
cell division
  micropatterned substrates in, 299–300,
    299*f*
cell environment, 3
cell mechanics
  cancer and, 280, 284
  described, 293
  in disease pathophysiology, 279–84
  malaria and, 280, 284
  overview of, 255–62
  studies of
    microfabricated technologies for,
      293–309. *see also*
      microfabricated technologies,
      for cell mechanics studies
  techniques used for, 262–7, 263*f*,
    265*f*, 266*f*, 281*t*–3*t*
cell mechanics assays
  bulk assays, 262–7, 263*f*, 265*f*, 266*f*,
    281*t*–3*t*. *see also* bulk assays
  existing high-throughput, 274–9, 276*f*,
    282*t*–3*t*

cell mechanics assays (*continued*)

    magnetic twisting cytometry, 277–8, 283*t*

    microfluidic pore and deformation assays, 278–9, 283*t*, 284*f*

    optical stretchers, 274–5, 282*t*

    substrate stretching assays, 277, 283*t*

    traction force microscopy via bead-embedded gels, 275, 283*t*

    traction force microscopy via micropost arrays, 275–7, 276*f*, 283*t*

    single-cell techniques, 268–74, 268*f*, 270*f*, 273*f*, 281*t*–2*t*. *see also* single-cell techniques, in measuring mechanical properties of cells

    towards high-throughput

       for research and clinical applications, 255–92

cell migration

    micropatterned substrates in, 300–1

    in three-dimensional micropatterns, 303–4

cell-penetrating peptides (CPPs)

    described, 231*t*

    in molecular beacon delivery, 246–7, 247*f*

cell-receptor interactions, 3–18

    apoptosis and survival signaling, 13

    cell differentiation signaling, 13–14, 14*t*

    integrins' mechanics, 4–7, 5*f*, 6*f*

    introduction, 3–4

    three-dimensional adhesion, 11

    three-dimensional motility, 12–13

    two-dimensional adhesion, 7–9, 8*f*

    two-dimensional motility, 9–11, 9*f*

cell sensing

    overview of, 256–9, 258*f*

cell sorting

    deformability and size-based approach to

cell mechanics studies of, 305–6, 306*f*

cell substrate traction forces

    characterization of

       cell mechanics studies of, 301–2, 301*f*

cell–matrix adhesion

    cytoplasmic proteins and, 7

cellular neuromechanics, 35–61. *see also* tension, in neural growth

    unresolved issues in, 58

CENP-E, 23

CHARMM, 391

chemical permeabilization

    in molecular beacon delivery, 246

chemical reactions

    stress-mediated, 326–7, 327*f*

chemical synapses

    composition of, 38

circuit mesh

    maximum strain in, 351–3, 354*f*

classical MD (CMD), 133

CMD. *see* classical MD (CMD)

CNT/graphene composites

    local mechanical properties of, 113–15, 116*f*

CNT-based fibers and composites

    high volume fraction, 115–20, 117*f*, 119*f*

    tension tests of, 118–19

CNTs. *see* carbon nanotube(s) (CNTs)

coarse-grained model(s)

    prior, 415–16

coarse-grained modeling

    of actin filaments, 402–3, 403*f*

codon, 231*t*

coherent twin boundary (CTB)

    structure of, 134–5, 134*f*, 135*f*

columnar simulation geometry, 133

compression

    nanoindenter-based

       TEM-STM and, 201–4

    pilar

       nanoindentation and

          in mechanical and

             electromechanical testing of

nanotubes and
nanowires, 202
computational and theoretical tools
in modeling and simulation of
hierarchical protein materials,
391–400, 392*f*–9*f*
mathematical approaches to
biomateriomics, 394–400,
395*f*–9*f*
mesoscale methods for modeling
larger length and time scales,
392–3, 394*f*
molecular simulation from
chemistry upwards, 391–2,
392*f*, 393*f*
contact-line approach
vs. energy minimization approach,
78–9
copper
nanotwinned
fracture of, 155–6, 155*f*
Couette flow, 266
CPPs. *see* cell-penetrating peptides
(CPPs)
crack tip dislocation emission, 324–5,
325*f*, 326*f*
crystal(s)
cubic, 178–83, 181*f. see also* cubic
crystals
non-cubic single, 183–4
tube model for, 416–17
crystalline defects
evolution of
atomistic approach to, 313–18
crystalline nanospecimens
in situ HRTEM testing of, 215–16
CTB. *see* coherent twin boundary (CTB)
Cu
TB and stacking fault energies of,
130–1, 130*t*
CuAl
TB and stacking fault energies of,
130*t*, 131–2
cubic crystals, 178–83, 181*f*
body-centered cubic structures, 180–3,
181*f*

face-centered cubic metals, 178–80
statistical evaluation of nano-scale
cubic single-crystal deformation,
183
curvilinear electronics
described, 339–42, 340*f*, 341*f*
mechanics of, 339–57
buckling of interconnect bridges,
347–50, 349*f*–54*f*
deformation of elastomeric transfer
elements during wrapping
processes, 342–7, 343*f*, 344*f*,
346*f*
maximum strain in circuit mesh,
351–3, 354*f*
cyclin-dependent kinases (CDKs), 13
cytometry
magnetic twisting, 277–8, 283*t*
cytoplasmic proteins
cell–matrix adhesion and, 7
integrins and
function of, 9–10
cytoskeletal motors
generalized mechanism of, 19–21, 21*f*
cytosol, 3

deformability index (DI), 267
deformation experiments
uniaxial, 175–8. *see also* uniaxial
deformation experiments
deformation mechanisms
in nanotwinned metals, 145–56. *see
also* nanotwinned metals,
deformation mechanisms in
dendrite(s)
of neuron, 36–7, 36*f*
density-function–based tight binding
(DFTB) method, 107
DI. *see* deformability index (DI)
Dictyostelium myosin-II
phospho-regulatory mechanism for, 27
differentiation, 3
diffraction
in TEM testing of one-dimensional
nanostructures, 216–17, 217*f*

directional solidification
  in size-dependent strength testing in
      single-crystalline metallic
      nanostructures, 172–3
disclosure nucleation
  strain-rate dependence of, 329–30,
      329*f*
dislocation(s)
  TBs and
      interaction between, 146–7, 146*f*
DNA (deoxyribonucleic acid), 230*t*
DNA dimer
  indirect modeling of, 375–6, 375*f*–7*f*
double-wall nanotube(s) (DWNTs)
  cross-linking adjacent graphic layers,
      110, 110*f*, 112
  weak shear interactions between
      adjacent graphitic layers, 108
double-wall nanotube (DWNT) yarns
  stress–strain curve of, 118–20, 119*f*
drop
  static equilibrium state of, 78
drop deposited on surface in ambient air
  apparent contact angle of
      in hydrophobic to superhydrophobic
          behavior, 77–9
  classical configuration of, 66*f*, 67–8
drop on pillar-type roughness geometry
  schematic of, 68–9, 68*f*
dual-FRET molecular beacons
  for RNA detection, 232*f*, 236–7, 237*f*
DualBeam (FEI Nova Nanolab 200)
      system, 177
ductility
  of nanotwinned metals, 140, 141*f*
durotaxis, 276
DWNTs. *see* double-wall nanotubes
      (DWNTs)
dynamical bistability, 361
  molecular potential of mean force and,
      381–4, 383*f*
dynein, 20

EDS. *see* electron dispersive spectroscopy
      (EDS)

LS. *see* electron-energy-loss
      spectroscopy (EELS)D

ektacytometry
  of bulk assays, 266–7, 266*f*, 281*t*
elastin(s)
  function of, 4
elastomeric transfer elements
  deformation of
      during wrapping processes, 342–7,
          343*f*, 344*f*, 346*f*
  deformed shape of, 344–7, 346*f*
  stretched
      strain distribution in, 342–4, 343*f*,
          344*f*
electrical conductivity
  of nanotwinned metals, 145
electrical signaling
  neuronal, 39
electromigration resistance
  of nanotwinned metals, 145
electron dispersive spectroscopy (EDS)
  x-ray, 217–18
electron-energy-loss spectroscopy
      (EELS), 217–18
electronics
  curvilinear, 339–57. *see also*
      curvilinear electronics
electroplating
  templated
      in size-dependent strength testing in
          single-crystalline metallic
          nanostructures, 173, 173*f*
electroporation
  in molecular beacon delivery, 245–6
energy minimization approach
  vs. contact-line approach, 78–9
enhanced permeability and retention
      (EPR) effect, 438
entropy
  activation
      and Meyer–Neldel compensation
          rule, 322–4
environment
  cell responses to, 3, 260–1
enzyme(s)

switch-I-containing, 22
EPR effect. *see* enhanced permeability
    and retention (EPR) effect
EST. *see* expressed sequence tag (EST)
etching
    in size-dependent strength testing in
        single-crystalline metallic
        nanostructures, 172–3
experimentation, 191–309
expressed sequence tag (EST)
    described, 231*t*
    in measuring gene expression level
        within cell population, 228
extracellular matrix, 4
    in integrin-based cell motility, 11
    in stiffness, 13, 14*t*

face-centered cubic metals, 178–80
FAK, 10, 13
fatigue properties
    of nanotwinned metals, 143–5, 144*f*
FIB-based systems
    in size-dependent strength testing in
        single-crystalline metallic
        nanostructures, 176–7
FIB method. *see* focused ion beam (FIB)
    method
fibrillar adhesions, 8*f*, 9
fibroblast(s)
    motility of, 11
filament(s)
    actin, 257, 257*f*
    intermediate, 257, 257*f*
finite-element method
    in numerical modeling of nanotwinned
        metals, 132–3
FISH. *see* fluorescence *in situ*
    hybridization (FISH)
fish epithelial keratocytes
    motility of, 11
fluorescence *in situ* hybridization (FISH),
    230*t*
fluorescence resonance energy transfer
    (FRET), 230*t*
fluorescence resonance energy transfer
    (FRET) probes

linear, 232–3, 232*f*
fluorescent linear probes
    for RNA detection, 229, 231, 232*f*
fluorescent protein-based probes
    for RNA detection, 232*f*, 237–8
fluorophore(s), 230*t*
    of molecular beacons, 241–2, 243*f*
focal adhesions, 8–9, 8*f*, 258–9
focal complexes, 8
focused ion beam (FIB) method
    in size-dependent strength testing in
        single-crystalline metallic
        nanostructures, 163, 168, 170–2,
        171*f*
    advantages of, 171–2
force(s)
    cell responses to, 260–1
    cell substrate traction
        cell mechanics studies of, 301–2,
            301*f*
    by cells, 259–60
    intercellular adhesion, 302–3, 302*f*
    in neurons
        in vitro measurements of, 41–3, 42*f*
    van der Waals, 106
force–extension behavior
    of single molecules, 360–3, 361*f*–4*f*
fracture
    Griffith theory of, 330–2, 331*f*
    of nanotwinned copper, 155–6, 155*f*
    size and loading effects on, 330–2,
        331*f*
fracture toughness
    of nanotwinned metals, 143–5, 144*t*
free-end NEB method, 317–19, 318*f*
FRET. *see* fluorescence resonance energy
    transfer (FRET)
FRET probes
    linear, 232–3, 232*f*
fused-sphere model
    proteins in, 427–8, 428*f*

G proteins
    activation of
        controller of, 21–3, 24*f*
        described, 22

G proteins (*continued*)
  heads of
    motifs in, 21–2
  switch I in, 23
  "switch" regions of, 21–2
GDI. *see* GDP dissociation inhibitor
  (GDI)
GDP dissociation inhibitor (GDI), 22
gels
  bead-embedded
    traction force microscopy via, 275,
    283*t*
gene expression
  live-cell imaging of
    engineering nano-probes for,
    227–53
geometry
  columnar simulation, 133
GFP. *see* green fluorescent protein (GFP)
Gibbs free energy of system, 69
GoLoco proteins, 22–3
graph theory, 396–7, 396*f*–7*f*
graphene, 95–6
graphitic materials
  experimentally determined interfacial
    properties of, 120–1, 121*t*
green fluorescent protein (GFP), 231*t*
Griffith theory of fracture, 330–2, 331*f*
growth-cone guidance
  tension in, 46, 47*f*
"growth dashpot," 53, 53*f*

H-bonds, 96
Hadwiger's theorem, 422–4, 423*f*
Hall–Petch relation, 138, 147–50, 150*f*,
  150*t*, 153*f*
Hartree–Fock theory, 391
HCP metals. *see* hexagonal close-packed
  (HIP) metals
Helmholtz free-energy profile, 365–6,
  369
Hertzian mechanics equation, 271
hexagonal close-packed (HCP) metals,
  183–4
hierarchical carbon-based materials
  CNTs, 95–7

collective and local behavior
  micromechanical testing methods,
    103–6, 105*f*
cross-linking adjacent graphitic layers,
  110–13, 110*f*, 111*f*
described, 106
in development of mechanically
  superior and multifunctional
  materials, 120–2, 121*t*
graphene, 95–7
high volume fraction CNT fibers and
  composites, 115–20, 117*f*, 119*f*
investigating shear interactions
  in situ SEM/AFM methods, 102–3,
    102*f*
local mechanical properties of
  CNT/graphene-based materials,
  113–15, 116*f*
measuring ultralow forces
  AFM methods, 101–2, 101*f*
modeling of
  requirements for, 121–2
multiscale experimental mechanics of,
  95–128
  introduction to, 95–7, 97*f*
  tools in, 97–106, 99*f*–102*f*, 105*f*
revealing atomic-level mechanics
  in situ TEM methods, 98–101,
    100*f*
weak shear interactions between
  adjacent graphitic layers, 106–9
hierarchical metals
  nanotwinned, 129–62. *see also*
    nanotwinned hierarchical metals
hierarchical protein materials
  modeling and simulation of, 389–409
    atomistic and mesoscale protein
     folding and deformation in
     spider silk, 400–2, 401*f*
    case studies, 400–6, 401*f*, 403*f*,
     405*f*
    category theoretical abstraction of
     protein material and analogy
     to office network, 403–6,
     405*f*

coarse-grained modeling of actin
    filaments, 402–3, 403*f*
computational and theoretical tools
    in, 391–400, 392*f*–9*f*. *see
    also* computational and
    theoretical tools, in modeling
    and simulation of hierarchical
    protein materials
introduction, 389–91, 390*f*
hierarchical structures
    in tendons, 96, 97*f*
HRTEM testing
    of crystalline nanospecimens, 215–16
    of one-dimensional nanostructures,
        212–16, 214*f*
    nanotubes, 213–15, 214*f*
human umbilical vein endothelial cells
    (HUVEC)
    image of, 256, 257*f*
HUVEC. *see* human umbilical vein
    endothelial cells (HUVEC)
hydrogen-bond strength
    variable, 415
hydrophilic
    to superhydrophobic, 84–6, 85*f*
hydrophilic contact, 65–6, 66*f*
hydrophobic
    defined, 65
    to superhydrophobic, 67–84. *see also*
        thermodynamic analysis,
        hydrophobic to superhydrophobic
hydrophobic effect, 412–15, 413*f*
hysteresis
    modeling of, 79–84, 80*f*–2*f*
        Cassie–Baxter angle compared with
            experimental data in, 80, 80*f*
        described, 82–4
        deviation from Cassie–Baxter angle
            in, 81–2, 81*f*, 82*f*

IFEM
    for NP and cell transport, 452, 454–6,
        454*f*, 455*f*, 456*f*
IFT particles. *see* intraflagellar transport
    (IFT) particles
immunoglobulin(s)

function of, 4
in situ hybridization (ISH) methods, 229
in situ MEMS–based testing, 96
in situ sample preparation
    for TEM of one-dimensional
        nanostructures, 211–12
in situ SEM/AFM methods
    in investigating shear interactions,
        102–3, 102*f*
in situ specimen modification
    in *in situ* TEM testing of
        one-dimensional nanostructures,
        218–19, 219*f*
in situ systems
    in size-dependent strength testing in
        single-crystalline metallic
        nanostructures, 176–8
in situ TEM tensile testing, 111, 111*f*
in situ TEM testing, 193–226
inside-out adhesion signals, 8
instability
    mechanical, 361
        molecular potential of mean force
            and, 378–81, 381*f*, 382*f*
integrin(s)
    activated, 4–6, 5*f*
    in cell differentiation signaling, 13
    configurations of, 4–6, 5*f*
    cytoplasmic proteins and
        function of, 9–10
    mechanosensitive properties of, 7
    deactivated, 4–5, 5*f*
    described, 4, 258
    in differentiation of cell lines, 13
    force-sensitive signals independent of
        dissociation produced by, 6–7,
        6*f*
    forces effects on, 9
    function of, 4, 10
    integrin–ligand bond of, 6
    matrix binding by, 4
    mechanics of, 4–7, 5*f*, 6*f*
    motility with, 10
    outside-in activation, 8
    strain-sensing mechanism of, 7
    subunits of, 4

integrin-based cell motility
  extracellular matrix in, 11
intercellular adhesion forces
  cell mechanics studies of, 302–3, 302*f*
interconnect bridges
  buckling of, 347–50, 349*f* –54*f*
intermediate filaments, 257, 257*f*
intraflagellar transport (IFT) particles, 29
ion(s)
  in membrane potential, 38–9
ISH methods. *see in situ* hybridization
    (ISH) methods

Jarzynski equality, 366

KCBP. *see* kinesin calmodulin–binding
    protein (KCBP)
Kelvin effect, 87
keratocyte(s)
  fish epithelial
    motility of, 11
kinesin(s)
  C-terminal, 25
  described, 19
  function of, 19
  kinesin-1, 22–3, 26
  kinesin-5, 23, 24*f*
    function of, 27
    regulation of
      Wee1 and M-Cdk in, 27–8, 28*f*
  regulatory mechanisms of, 19–33
    calcium-binding, 24*f*, 25–6, 25*f*
    cooperative action of, 28–9
    phospho-regulation, 26–8, 27*f*
kinesin calmodulin–binding protein
    (KCBP), 25–6

lamellipodium, 10
lamellum, 10
lathe milling
  FIB-based sample preparation in, 171,
    172*f*
linear FRET probes
  for RNA detection, 232–3, 232*f*
linguistics
  protein materials and

analogy and comparative analysis
    of, 394–6, 395*f*
liquid–vapor interface
  plot of availability as function of
    position of, 90, 90*f*
  at top of cylindrical cavity, 89, 89*f*
liquid–vapor phase transition
  schematic diagram for, 87–8, 87*f*
lithography
  soft
    for cell mechanics studies, 295–7,
      296*f*
live-cell imaging
  of gene expression
    engineering nano-probes for,
      227–53. *see also* probe(s);
      specific types and molecular
      beacons
  of mRNAs, 227–8, 228*f*
  of RNA, 227
living cells
  imaging of. *see* live-cell imaging
lotus effect, 66
lyophobic
  defined, 65

M-Cdk
  in kinesin-5 regulation, 27–8, 28*f*
macrocirculation
  NPs in
    modeling dynamics of, 438–48. *see*
      also nanoparticles (NPs),
      multiscale modeling for
      vascular transport of, in
      macrocirculation
magnetic micropillars for local force
    application on cells
  cell mechanics studies of, 304
magnetic twisting cytometry, 277–8, 283*t*
malaria
  cell mechanics in, 280, 284
MAPK pathway. *see* mitogen-activated
    protein kinase (MAPK) pathway
MARTINI force-field, 393
material(s)
  ageing of

nanomechanics in problems of,
334–6, 335*f*
MD. *see* molecular dynamics (MD)
mean force
molecular potential of. *see* molecular
potentional of mean force
mechanical instability, 361
molecular potential of mean force and,
378–81, 381*f*, 382*f*
mechanical measurements
biological measurements combined
with, 261–2
mechanical tension
in growth, guidance, and function of
neurons in, 35
mechanotransduction, 259
MEMS-based testing
in mechanical and electromechanical
testing of nanotubes and
nanowires, 204–6, 205*f*
MEMS–based testing. *see*
micro-electrical–mechanical
systems (MEMS)–based testing
MEP. *see* minimum-energy path (MEP)
mesoscale methods
for modeling larger length and time
scales, 392–3, 394*f*
messenger RNAs (mRNAs)
imaging of
in living cells, 227–8, 228*f*
metal(s). *see specific metals*
Meyer–Neldel compensation rule
activation entropy and, 322–4
mfold, 243
microelectrical–mechanical systems
(MEMS)–based testing, 100–1,
100*f*
cross-linking adjacent graphic layers,
110–11
in MWNTs, 106–9
*in situ,*96
in situ SEM/AFM methods in
investigating shear interactions,
102–3, 102*f*
microcirculation
NPs in

modeling dynamics of, 448–56. *see
also* nanoparticles (NPs),
multiscale modeling for
vascular transport of, in
microcirculation
microfabricated technologies
for cell mechanics studies, 293–309
applications, 298–307. *see also
specific applications*
introduction to, 293–4
microfluidic devices, 304–7, 306*f*
micropatterned substrates, 298–301,
299*f*. *see also* micropatterned
substrates, cell mechanics
studies of
microphotopatterning, 297, 297*f*
micropillared substrates, 301–4,
301*f*, 302*f*. *see also*
micropillared substrates, cell
mechanics studies of
photolithography, 294–5, 295*f*,
296*f*
soft lithography, 295–7, 296*f*
microfiltration
of bulk assays, 262–4, 263*f*, 281*t*
microfluidic devices
cell mechanics studies of, 304–7, 306*f*
cell adhesion probing, 306–7
cell sorting–related, 305–6, 306*f*
probing elastic properties of
diseased cells, 305, 306*f*
microfluidic pore and deformation assays,
278–9, 283*t*, 284*f*
microinjection
in molecular beacon delivery, 245
micromechanical testing methods
in collective and local behavior,
103–6, 105*f*
micropattern(s)
three-dimensional
cell migration in, 303–4
micropatterned substrates
cell mechanics studies of, 298–301,
299*f*
in cell differentiation, 298–9
in cell division, 299–300, 299*f*

micropatterned substrates (*continued*)
  in cell migration, 300–1
  in polarization, 299–300, 299*f*
microphotopatterning
  for cell mechanics studies, 297, 297*f*
micropillared substrates
  cell mechanics studies of, 301–4,
      301*f*, 302*f*
    anistropic rigidity–related, 303
    cell differentiation, 304
    cell migration in, 303–4
    cell substrate traction forces
        characterization, 301–2, 301*f*
    intercellular adhesion forces,
        302–3, 302*f*
    magnetic micropillars for local force
        application on cells, 304
micropipette aspiration, 268–70, 268*f*,
    281*t*
microplate stretcher, 272, 282*t*
micropost arrays
  traction force microscopy via, 275–7,
      276*f*, 283*t*
microstructure
  evolution at long times, 332–6, 333*f*,
      335*f*
microtubule(s)
  described, 257–8, 257*f*
  function of, 37, 37*f*
minimum-energy path (MEP), 314
Miro (mitochondrial Rho GTPase), 26
mitogen-activated protein kinase (MAPK)
    pathway, 13
MLCK. *see* myosin light chain kinase
    (MLCK)
molecular beacons
  challenges related to, 248–9
  delivery of, 244–7, 247*f*
    cationic transfection agents in, 245
    chemical permeabilization in, 246
    CPPs in, 246–7, 247*f*
    electroporation in, 245–6
    microinjection in, 245
  dual-FRET
    for RNA detection, 232*f*, 236–7,
        237*f*

fluorophores of, 241–2, 243*f*
  future directions in, 248–9
  quenchers of, 241–2, 243*f*
  for RNA detection, 232*f*, 234–6, 235*f*
  signal-to-background ratio of, 241–2,
      243*f*
  specificity of, 239–40, 240*f*, 241*f*
  structure–function relations of,
      239–40, 241*f*
  target accessibility of, 242–4, 244*f*
molecular dynamics (MD)
  ab initio, 133
  classical, 133
  single-molecule pulling modeling
      using, 370–6, 370*f*, 373*f*–7*f*
    basic computational setup, 370–1,
        370*f*
    examples, 373–6, 373*f*–7*f*
    strategies, 371–2
  steered, 392
molecular dynamics (MD) simulations,
    132–4
  in numerical modeling of nanotwinned
      metals, 132–4
molecular motors
  as force actuators in axons, 53, 54*f*
molecular potential of mean force
  basic structure of, 377–8, 377*f*–9*f*
  dynamical bistability related to,
      381–4, 383*f*
  extracting, 366–9, 367*f*
  mechanical instability related to,
      378–81, 381*f*, 382*f*
molecular probes
  for RNA detection, 229, 231–8
    dual-FRET molecular beacons,
        232*f*, 236–7, 237*f*
    fluorescent linear probes, 229, 231,
        232*f*
    fluorescent protein-based probes,
        232*f*, 237–8
    linear FRET probes, 232–3, 232*f*
    molecular beacons, 232*f*, 234–6,
        235*f*
    QUAL probes, 233–4
molecule(s)

single. *see* single molecule(s)
motility
 integrin-based cell
  extracellular matrix in, 11
 three-dimensional, 12–13
 two-dimensional, 9–11, 9*f*
motor proteins
 activation of
  controller of, 21–3, 24*f*
 heads of
  motifs in, 21–2
 "switch" regions of, 21–2
mRNAs. *see* messenger RNAs (mRNAs)
multi-walled CNTs (MWCNTs)
 in *in situ* TEM testing, 213–15, 214*f*
music
 protein materials and
  analogy and comparative analysis
   of, 394–6, 395*f*
MWCNTs. *see* multi-walled CNTs
 (MWCNTs)
myosin(s)
 described, 19
 function of, 19
 myosin-II, 23, 24*f*
  Dictyostelium
   phospho-regulatory mechanism
    for, 27
  function of, 27
 myosin-V
  calcium-dependent regulatory
   mechanisms for, 24–5
  regulatory mechanisms of, 19–33
  calcium-binding, 23–5, 24*f*
  cooperative action of, 28–9
  phospho-regulation, 26–8, 27*f*
myosin light chain kinase (MLCK), 27

nanoprobes
 engineering of
  for live-cell imaging of gene
   expression, 227–53. *see also*
   probe(s); *specific types and*
   molecular beacons
nanoscale cubic single-crystal deformation
 statistical evaluation of, 183

nanoimprinting
 in size-dependent strength testing in
  single-crystalline metallic
  nanostructures, 173–4, 174*f*
nanoindentation
 AFM-based methods in, 114–15, 116*f*
 of CNT-based materials, 105–6
 pillar compression and
  in mechanical and
   electromechanical testing of
   nanotubes and nanowires, 202
nanoindenter-based compression, tension,
 and bending
 TEM-STM and, 201–4
nanoindenter-based systems
 in size-dependent strength testing in
  single-crystalline metallic
  nanostructures, 176
nanomachine(s)
 examples, 19
nanomanipulation
 for TEM of one-dimensional
  nanostructures, 209–11, 210*f*
nanomechanics
 case studies, 324–32
  bridging modeling with experiment,
   327–8, 328*f*, 328*t*
  crack tip dislocation emission,
   324–5, 325*f*, 326*f*
  size and loading effects on fracture,
   330–2, 331*f*
  stress-mediated chemical reactions,
   326–7, 327*f*
  temperature and strain-rate
   dependence of disclosure
   nucleation, 329–30, 329*f*
 in materials ageing–related problems,
  334–6, 335*f*
 reaction pathway sampling in, 314
 recurring themes in, 313
nanoparticles (NPs)
 described, 437–8
 multiscale modeling for vascular
  transport of, 437–59
  in macrocirculation, 438–48

nanoparticles (NPs) (*continued*)
  across arterial wall and drug
      release, 440, 442–8, 442*f*,
      444*f*, 446*f*, 447*f*
    described, 438–9
    three-dimensional reconstruction
      of patient-specific
      vasculature, 439, 440*f*,
      441*f*
    vascular flow and wall adhesion,
      440, 441*f*
  in microcirculation, 448–56
    described, 448–9
    IFEM for, 452, 454–6, 454*f*–6*f*
    semi-analytical models, 449–52,
      449*f*, 451*f*, 453*f*
nanopillar(s)
  nanotwinned
  deformation mechanisms in, 154–5
nanospecimen(s)
  crystalline
    in situ HRTEM testing of, 215–16
nanostructure(s). *see also specific types*
  mechanical and electromechanical
      characterization of, 196–7, 197*f*
  one-dimensional
    characterization of mechanical and
      electromechanical properties
      of, 194
    described, 193
    HRTEM testing of, 212–16, 214*f*
    outlook on, 220–1
    relevance of mechanical and
      electomechanical testing for,
      194–5
    in situ TEM testing of, 193–226.
      see also nanotube(s);
      nanowire(s)
      analytical techniques, 217–18
      capabilities of, 212–19, 214*f*,
        217*f*, 219*f*
      diffraction, 216–17, 217*f*
      sample preparation for, 208–12,
        210*f*
      in situ specimen modification in,
        218–19, 219*f*

single-crystalline metallic
    size-dependent strength in, 163–90.
      see also singlecrystalline
      metallic nanostructures,
      size-dependent strength in
nanotube(s). *see also* nanostructure(s),
    one-dimensional
  carbon. *see* carbon nanotube(s) (CNTs)
  described, 193
  double-wall. *see* double-wall
      nanotube(s) (DWNTs)
  mechanical and electromechanical
      testing methods for, 200–8,
      201*f*, 205*f*, 207*f*, 208*t*
    MEMS-based testing in, 204–6,
      205*f*
    resonance, 200, 201*f*
    TEM-STM in, 201–4
  in situ HRTEM testing of, 213–15,
      214*f*
  in situ TEM testing of
    capabilities of, 212–19, 214*f*, 217*f*,
      219*f*
    experimental methods, 197–212
    introduction to, 193–4
    sample preparation for, 208–12,
      210*f*
  uses of, 193
nanotwinned bulk metals
  deformation mechanisms in, 147–53,
      148*t*, 149*f*, 150*f*, 150*t*, 152*f*,
      153*f*
nanotwinned copper
  fracture of, 155–6, 155*f*
nanotwinned hierarchical metals. *see also*
      nanotwinned metals
  characteristics of, 129
  described, 129
  mechanics of, 129–62
    studies of, 129
  numerical modeling of, 132–4
    MD simulations, 132–4
  overview of, 129–34, 130*t*
nanotwinned materials
  described, 130–2, 130*t*

metals. *see* nanotwinned hierarchical
    metals; nanotwinned metals
microstructures of, 135–7, 136*f*
  characterization of, 134–45
  CTB structures, 134–5, 134*f*, 135*f*
  nanotwinned metals, 135–7, 136*f*
  nanotwinned nanowires, 136–7
  semiconductor nanowires, 130*t*, 131
  TBs in, 129–33, 130*t*
  twinned alloys, 130*t*, 131–2
nanotwinned metals. *see also*
    nanotwinned hierarchical metals
  deformation mechanisms in, 145–56
    interaction between dislocations and
      TBs, 146–7, 146*f*
    nanotwinned bulk metals, 147–53,
      148*t*, 149*f*, 150*f*, 150*t*, 152*f*,
      153*f*
    nanowires and nanopillars, 154–5
  described, 130–1, 130*t*
  fracture of nanotwinned copper,
    155–6, 155*f*
  mechanical and physical properties of,
    137–45
    ductility, 140, 141*f*
    electrical conductivity, 145
    electromigration resistance, 145
    fatigue properties, 143–5, 144*f*
    fracture toughness, 143–5, 144*t*
    strain hardening, 140–2, 141*f*
    strain-rate sensitivity, 142–3, 142*t*,
      143*f*
    yield strength, 137–8, 137*f*, 139*f*
  microstructures of, 135–7, 136*f*
  numerical modeling of
    finite-element method, 132–3
  strengthening and softening
    mechanisms in, 147–55
nanotwinned nanowires
  deformation mechanisms in, 154–5
  microstructures of, 136–7
nanotwinned semiconductor nanowires
  described, 130*t*, 131
nanowire(s). *see also* nanostructure(s),
    one-dimensional
  described, 193

growth of
  in size-dependent strength testing in
    single-crystalline metallic
    nanostructures, 175
mechanical and electromechanical
  testing methods for, 200–8,
  201*f*, 205*f*, 207*f*, 208*t*
  MEMS-based testing in, 204–6,
    205*f*
  resonance, 200, 201*f*
  TEM-STM in, 201–4
nanotwinned
  deformation mechanisms in, 154–5
  microstructures of, 136–7
nanotwinned semiconductor, 130*t*, 131
in situ TEM testing of
  capabilities of, 212–19, 214*f*, 217*f*,
    219*f*
  experimental methods, 197–212
  introduction to, 193–4
  sample preparation for, 208–12,
    210*f*
  uses of, 193
NEB. *see* nudged elastic band (NEB)
neurite(s)
  Aplysia
    reducing mechanical tension in,
      49–50, 51*f*
  force evolution of network in, 53–4
  initiation and axonal specifications of,
    45–6, 46*f*
neurofilament(s)
  function of, 37
neuromechanics
  cellular, 35–61. *see also* tension, in
    neural growth
neuron(s)
  described, 36–8, 36*f*, 37*f*
  force in
    in vitro measurements of, 41–3, 42*f*
  function of
    processes in, 38–40
  growth of, 40–1, 40*f*
  parts of, 36–7, 36*f*
  tension in

neuron(s) (*continued*)
  function-related, 48–52, 49*f*–52*f*.
    see also tension, in neuron
    function
  structural development–related,
    45–8, 46*f*, 47*f*
  in vitro measurements of, 41–3, 42*f*
  in vivo measurements of, 43–5, 44*f*
  types of, 36
  in vivo
    studies on mechanical behavior of,
    54
neuronal development
  mechanical cues in, 47–8
neuronal growth
  described, 40–1, 40*f*
  tension in, 41–8. *see also* tension, in
    neural growth
neuronal transport, 38–9
neurotransmission
  process of
    function in, 38–9
    tension increasing, 48, 49*f*
Ni
  TB and stacking fault energies of, 130*t*
non-cubic single crystals, 183–4
Northern blotting
  in measuring gene expression level
    within cell population, 228
Northern hybridization
  in measuring gene expression level
    within cell population, 228
NPs. *see* nanoparticles (NPs)
nudged elastic band (NEB), 314
nudged elastic band (NEB) method
  described, 315–17, 316*f*
  free-end, 317–19, 318*f*
  for stress-driven problems, 315–24
numerical models
  prior, 415–16

ODN probes. *see* oligonucleotide (ODN)
  probes
oleophobic, 65
oligonucleotide (ODN), 230*t*
oligonucleotide (ODN) probes, 229

oligorotaxanes
  direct modeling of, 373–5, 373*f*, 374*f*
omniphobic
  defined, 65
ontology logs
  category theory and, 396–400, 398*f*,
    399*f*
optical stretchers, 274–5, 282*t*
optical tweezers, 273–4, 273*f*, 282*t*
outside-in signaling, 3

parallel-plate flow chambers
  of bulk assays, 267, 281*t*
PCR. *see* polymerase chain reaction
    (PCR)
Pd
  TB and stacking fault energies of, 130*t*
PDMS. *see* polydimethylsiloxane (PDMS)
peptide(s)
  cell-penetrating
    described, 231*t*
    in molecular beacon delivery,
      246–7, 247*f*
permeabilization
  chemical
    in molecular beacon delivery, 246
phospho-regulation
  of myosins and kinesins, 26–8, 27*f*
phosphoinositide 3-kinase (PI3K)
    pathway, 13
phosphorylation
  described, 26
photolithography
  for cell mechanics studies, 294–5,
    295*f*, 296*f*
PI3K pathway. *see* phosphoinositide
    3-kinase (PI3K) pathway
pilar compression
  nanoindentation and
    in mechanical and
      electromechanical testing of
      nanotubes and nanowires, 202
plasticity
  synaptic, 39–40
Poisson effect, 118
polarization

micropatterned substrates in, 299–300, 299*f*
polydimethylsiloxane (PDMS), 295–6, 296*f*
polymerase chain reaction (PCR)
  described, 230*t*
  in measuring gene expression level within cell population, 228
polymorphism(s)
  single nucleotide, 229
potential of mean force
  molecular. *see* molecular potential of mean force
probe(s)
  design, imaging, and biological issues related to, 239–44
  fluorescent linear, 229, 231, 232*f*
  fluorescent protein-based, 232*f*, 237–8
  linear FRET, 232–3, 232*f*
  molecular
    for RNA detection, 229, 231–8. *see also specific probes and molecular probes, for RNA detection*
  nano-
    for live-cell imaging of gene expression, 227–53
  ODN, 229
  QUAL, 233–4
  in tube model, 426–7, 426*f*
profilin
  action of, 10
protein(s). *see also* motor proteins; *specific types, e.g.,* binding proteins
  finite-sized particles surrounding tube model including, 419–20, 420*f*
  in fused-sphere model, 427–8, 428*f*
  structure of
    models using, 421–2, 421*f*
protein materials
  category theoretical abstraction of
    analogy to office network, 403–6, 405*f*
  hierarchical. *see* hierarchical protein materials
  linguistics and

analogy and comparative analysis of, 394–6, 395*f*
music and
  analogy and comparative analysis of, 394–6, 395*f*
protein secondary-structure formation
  geometric models of, 411–35
    discussion, 430–2
    introduction, 411–12
    morphometric approach to solvation effects, 422–8, 423*f*, 425*f*, 426*f*, 428*f*
    prior numerical and coarse-grained models, 415–16
    results, 429–30
    speculations, 430–2
    tube model, 416–22, 418*f*, 420*f*, 421*f*. *see also* tube model

QIAKPIRP sequence in the tail, 22–3
quasi-three-dimensional (Q3D) simulation
  geometry, 133
quenched auto-ligation (QUAL) probes
  for RNA detection, 233–4
quencher(s)
  described, 231*t*
  of molecular beacons, 241–2, 243*f*

random sample preparation
  for TEM of one-dimensional nanostructures, 208–9
RDA. *see* representational difference analysis (RDA)
reaction pathway sampling
  in nanomechanics, 314
regulatory light chain (RLC) domain
  activation of, 26–7
representational difference analysis (RDA), 231*t*
residues
  buried
    tube model for, 417
resonance
  in mechanical and electromechanical testing of nanotubes and nanowires, 200, 201*f*

rheometry (*continued*)
rheometry
  of bulk assays, 264–6, 265*f*, 281*t*
Rho kinase (ROCK)
  stiffness for, 13, 14*t*
ribonucleoprotein (RNP), 230*t*
rigidity
  anistropic
    substrates with, 303
RNA (ribonucleic acid)
  described, 230*t*
  detection of
    molecular probes for, 229, 231–8.
      see also specific probes and
      molecular probes, for RNA
      detection
  in living cells
    imaging of, 227
RNP. *see* ribonucleoprotein (RNP)
ROCK. *see* Rho kinase (ROCK)
roughness-induced superhydrophobicity,
    65–94. *see also*
    superhydrophobicity,
    roughness-induced

SAED. *see* selected-area electron
    diffraction (SAED)
SAGE. *see* serial analysis of gene
    expression (SAGE)
selected-area electron diffraction (SAED),
    100–1
selectin(s)
  function of, 4
SEM-based systems
  in size-dependent strength testing in
    single-crystalline metallic
    nanostructures, 176–7
SEMentor, 177
semiconductor nanowires
  nanotwinned, 130*t*, 131
sensitivity
  strain-rate
    of nanotwinned metals, 142–3,
      142*f*, 143*f*
serial analysis of gene expression (SAGE)
  described, 231*t*

in measuring gene expression level
    within cell population, 228
signal-to-background ratio
  of molecular beacons, 241–2, 243*f*
silk
  spider
    atomistic and mesoscale protein
      folding and deformation in,
      400–2, 401*f*
singlecell techniques
  in measuring mechanical properties of
    cells, 268–74, 268*f*, 270*f*, 273*f*,
    281*t*–2*t*
    AFM, 270–2, 270*f*, 282*t*
    micropipette aspiration, 268–70,
      268*f*, 281*t*
    microplate stretcher, 272, 282*t*
    optical tweezers, 273–4, 273*f*, 282*t*
singlecrystalline metallic nanostructures
  size-dependent strength in, 163–90
    background material, 164–9, 166*f*,
      169*f*
    cubic crystals, 178–83, 181*f*
    directional solidification and etching
      in, 172–3
    discussion of, 178–84, 181*f*
    experimental foundation, 164–7,
      166*f*
    FIB method in, 163, 168, 170–2,
      171*f*
    HCP metals, 183–4
    introduction, 163–4, 164*f*
    models, 167–9, 169*f*
    nanoimprinting in, 173–4, 174*f*
    nanowire growth in, 175
    non-cubic single crystals, 183–4
    outlook on, 184–5
    sample fabrication, 170
    templated electroplating in, 173,
      173*f*
    tetragonal metals, 184
    uniaxial deformation experiments,
      175–8. *see also* uniaxial
      deformation experiments
    VLS growth in, 174, 175*f*
single molecule(s)

force–extension behavior of, 360–3, 361*f*–4*f*

thermodynamics of, 364–70, 367*f*

estimating force–extension behavior from Φ, 369–70

extracting molecular potential of mean force, 366–9, 367*f*

free-energy profile of molecule plus cantilever, 365–6

singlemolecule experiment(s)

described, 359–60

schematic of, 359–60, 360*f*

singlemolecule pulling, 359–88

described, 359–60, 360*f*

modeling of

MD in, 370–6, 370*f*, 373*f*–7*f*. *see also* molecular dynamics (MD), single-molecule pulling modeling using

phenomenology of, 376–84, 377*f*–9*f*, 381*f*–3*f*

single nucleotide polymorphisms (SNPs), 229, 230*t*

SLO. *see* streptolysin O (SLO)

small nuclear RNA (snRNA), 231*t*

SMD. *see* steered molecular dynamics (SMD)

smooth-muscle myosin motors

activation of, 26–7

SNPs. *see* single nucleotide polymorphisms (SNPs)

snRNA. *see* small nuclear RNA (snRNA)

soft lithography

for cell mechanics studies, 295–7, 296*f*

softening mechanisms

in nanotwinned metals, 147–55

solvation effects

morphometric approach, 422–8, 423*f*, 425*f*, 426*f*, 428*f*

applications, 424–8, 425*f*, 426*f*, 428*f*

Hadwiger's theorem, 422–4, 423*f*

proteins in fused-sphere model, 427–8, 428*f*

proteins in tube model, 426–7, 426*f*

solvent-mediated interactions, 424–6, 425*f*

soma

of neuron, 36–7, 36*f*

"source exhaustion hardening," 167

"source truncation," 167

specimen modification

in situ

in *in situ* TEM testing of one-dimensional nanostructures, 218–19, 219*f*

spectroscopy

electron dispersive, 217–18

electron-energy-loss, 217–18

spider silk

atomistic and mesoscale protein folding and deformation in, 400–2, 401*f*

Src family proteins, 13

H. *see* suppression subtractive hybridization (SSH)D

steered molecular dynamics (SMD), 392

stiffness

extracellular matrix in, 13, 14*t*

strain hardening

of nanotwinned metals, 140–2, 141*f*

strain-rate sensitivity

of nanotwinned metals, 142–3, 142*t*, 143*f*

strength

size-dependent

in single-crystalline metallic nanostructures, 163–90. *see also* single-crystalline metallic nanostructures, size-dependent strength in

strengthening mechanisms

in nanotwinned metals, 147–55

streptolysin O (SLO), 231*t*
stress-dependent activation energy and
    activation volume, 320–2, 320*f*
stress-driven problems
    NEB method for, 315–24
stress-mediated chemical reactions,
    326–7, 327*f*
stress sensor, 7
stretcher(s)
    microplate, 272, 282*t*
    optical, 274–5, 282*t*
structural development
    tension in, 45–8, 46*f*, 47*f*
substrate stretching assays, 277, 283*t*
superhydrophobic
    from hydrophilic to, 84–6, 85*f*
    from hydrophobic to, 67–84. *see also*
        thermodynamic analysis,
        hydrophobic to superhydrophobic
superhydrophobicity
    roughness-induced, 65–94. *see also*
        specific components
    applications, 90–1
    background of, 65–7, 66*f*
    future challenges facing, 90–1
    from hydrophilic to
        superhydrophobic, 84–6, 85*f*
    from hydrophobic to
        superhydrophobic, 67–84. *see
        also* thermodynamic analysis,
        hydrophobic to
        superhydrophobic
    vapor stabilization, 86–90, 87*f*,
        89*f*, 90*f*
suppression subtractive hybridization
    (SSH), 231*t*
survival-pathway signals
    types of, 13
survival signaling
    apoptosis and, 13
switch I
    function of, 22
    in G proteins, 23
    inhibitors of, 23
switch-I-containing enzymes, 22
"switch" regions, 21–2

SWNTs
    weak shear interactions between
        adjacent graphitic layers, 108
synapse(s)
    chemical
        composition of, 38
    malfunctioning
        diseases related to, 38
    of neuron, 36*f*, 37–8
    "remembering" by, 39
synaptic plasticity, 39–40

TBAZ model. *see* twin boundaries
    (TBs)–affected zone (TBAZ) model
TBs. *see* twin boundaries (TBs)
TEM. *see* transmission electron
    microscopy (TEM)
TEM-AFM
    in mechanical and electromechanical
        testing of nanotubes and
        nanowires, 203–4
TEM-based nanomechanical instruments
    in size-dependent strength testing in
        single-crystalline metallic
        nanostructures, 177–8
TEM-STM
    in mechanical and electromechanical
        testing of nanotubes and
        nanowires, 203
    nanoindenter-based compression,
        tension, and bending and
    in mechanical and
        electromechanical testing of
        nanotubes and nanowires,
        201–4
temperature
    strain-rate dependence of dislocation
        nucleation and, 329–30, 329*f*
templated electroplating
    in size-dependent strength testing in
        single-crystalline metallic
        nanostructures, 173, 173*f*
tendon(s)
    hierarchical structure in, 96, 97*f*
tensegrity model, 259
tension

nanoindenter-based
    TEM-STM and, 201–4
in neural growth, 41–8
    described, 41
    in structural development, 45–8,
        46f, 47f
    in vitro measurements of, 41–3, 42f
    in vivo measurements of, 43–5, 44f
in neuron(s)
    in vitro measurements of, 41–3, 42f
    in vivo measurements of, 43–5, 44f
in neuron function, 48–52, 49f–52f
    affecting vesicle dynamics, 48–52,
        50f–52f
    increasing neurotransmission, 48,
        49f
tension tests
    of CNT composites, 118–19
tetragonal metals, 184
thermodynamic(s)
    single-molecule, 364–70, 367f
thermodynamic analysis
    hydrophilic to superhydrophobic,
        84–6, 85f
    hydrophobic to superhydrophobic,
        67–84
        Cassie–Baxter state, 71–3, 72f
        modeling hysteresis in, 79–84,
            80f–2f
        problem formulation, 68–71, 68f
        vapor stabilization, 86–90, 87f, 89f,
            90f
three-dimensional adhesion, 11
three-dimensional micropatterns
    cell migration in, 303–4
three-dimensional motility, 12–13
TiAl
    TB and stacking fault energies of,
        130t, 131
top-down approach
    in FIB method, 170, 171f
traction force microscopy
    via bead-embedded gels, 275, 283t
    via micropost arrays, 275–7, 276f,
        283t
transition-state theory, 315

transmembrane proteins, 3
transmission electron microscopy (TEM),
    98
    basic principles of, 198–9, 198f
    of CNTs, 98–101, 100f
    in situ
        of nanowires and nanotubes,
            193–226. *see also*
            nanotube(s); nanowire(s)
        in revealing atomic-level mechanics,
            98–101, 100f
    specimen holders in, 199–200, 199f
TSP trajectories
    sampling, 333–4, 333f
tube model, 416–22, 418f, 420f, 421f
    impenetrable, 417–19, 418f
    including finite-sized particles
        surrounding protein, 419–20,
        420f
    motivation for, 416–17
    proteins in, 426–7, 426f
    using real protein structure, 421–2,
        421f
tweezer(s)
    optical, 273–4, 273f, 282t
twin boundaries (TBs)
    dislocations and
        interaction between, 146–7, 146f
    in nanotwinned materials, 129–33,
        130t
twin boundaries (TBs)–affected zone
    (TBAZ) model, 132
twinned alloys, 130t, 131–2
two-dimensional adhesion
    integrin-associated, 7–9, 8f
    types of, 8–9, 8f
two-dimensional motility, 9–11, 9f
    factors affecting, 10–11

uniaxial deformation experiments
    in size-dependent strength testing
        in single-crystalline metallic
            nanostructures, 175–8
            FIB-based systems, 176–7
            nanoindenter-based systems, 176
            SEM-based systems, 176–7

uniaxial deformation experiments
  (*continued*)
    in situ systems, 176–8
    TEM-based nanomechanical
      instruments, 177–8

van der Waals (vdW) forces, 106
vapor stabilization, 86–90, 87*f*, 89*f*, 90*f*
vapour–liquid–solid (VLS) growth
  in size-dependent strength testing in
    single-crystalline metallic
    nanostructures, 174, 175*f*
vapour–liquid–solid (VLS) method, 131
vesicle(s)
  tension effects on, 48–52, 50*f*–52*f*
VLS. *see* vapour–liquid–solid (VLS)
"weakest link theory," 167

Wee1
  in kinesin-5 regulation, 27–8, 28*f*

weighted histogram analysis method
  (WHAM), 366–8
Wenzel formula, 79
Wenzel state, 65–7, 66*f*
  from Cassie–Baxter state to, 73–7,
    74f, 84. *see also* Cassie–Baxter
    state, to Wenzel state
  in hydrophobic to superhydrophobic
    behavior, 72–3, 72*f*
WHAM. *see* weighted histogram analysis
  method (WHAM)

x-ray EDS, 217–18

yield strength
  of nanotwinned metals, 137–8, 137*f*,
    139*f*